Springer Texts in Statistics

Advisors:
Stephen Fienberg   Ingram Olkin

Springer Texts in Statistics

Alfred	Elements of Statistics for the Life and Social Sciences
Berger	An Introduction to Probability and Stochastic Processes
Blom	Probability and Statistics: Theory and Applications
Chow and Teicher	Probability Theory: Independence, Interchangeability, Martingales, Second Edition
Christensen	Plane Answers to Complex Questions: The Theory of Linear Models
Christensen	Linear Models for Multivariate, Time Series, and Spatial Data
Christensen	Log-Linear Models
Creighton	A First Course in Probability Models and Statistical Inference
du Toit, Steyn and Stumpf	Graphical Exploratory Data Analysis
Finkelstein and Levin	Statistics for Lawyers
Jobson	Applied Multivariate Data Analysis, Volume I: Regression and Experimental Design
Jobson	Applied Multivariate Data Analysis, Volume II: Categorical and Multivariate Methods
Kalbfleisch	Probability and Statistical Inference: Volume I: Probability, Second Edition
Kalbfleisch	Probability and Statistical Inference: Volume 2: Statistical Inference, Second Edition
Karr	Probability
Keyfitz	Applied Mathematical Demography, Second Edition
Kiefer	Introduction to Statistical Inference
Kokoska and Nevison	Statistical Tables and Formulae
Lindman	Analysis of Variance in Experimental Design

Continued at end of book

J H C C R E I G H T O N

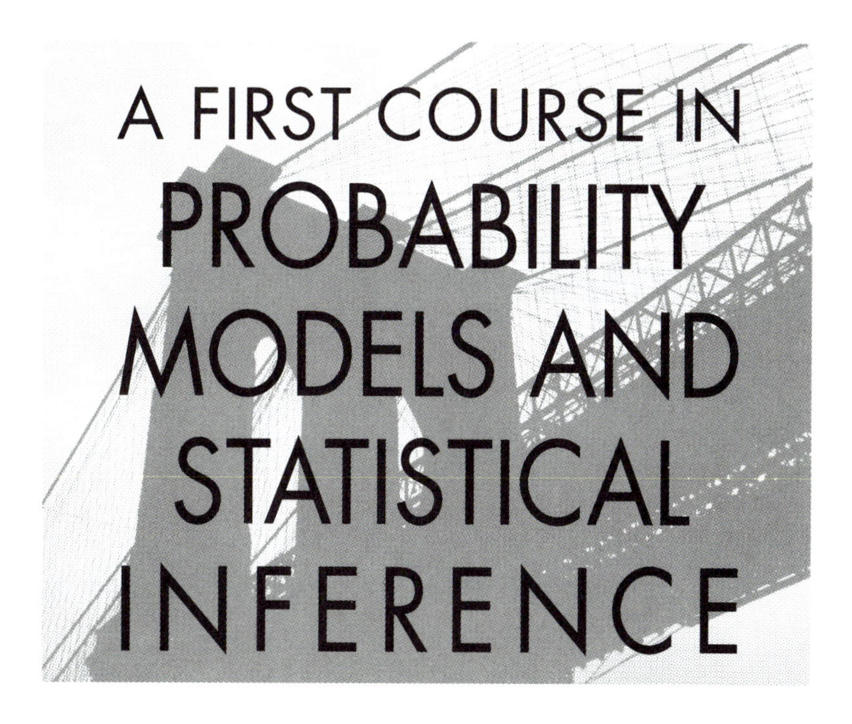

Springer-Verlag
New York Berlin Heidelberg London Paris
Tokyo Hong Kong Barcelona Budapest

J H C Creighton
San Francisco State University
1600 Holloway Avenue
San Francisco, CA 94132
USA

Mathematical Subject Classification (1991): 60-01, 62-01

Library of Congress Cataloging-in-Publication Data
Creighton, James H.C.
An first course in probability models and statistical inference / Creighton, James H.C. -- 1st ed.
p. cm. -- (Springer texts in statistics)
Includes bibliographical references and index.
ISBN 0-387-94114-2
1. Probabilities. 2. Mathematical statistics. I. title. II. Series.
QA273.C847 1994
519.5--dc20 93-25369

Printed on acid-free paper.

Production managed by Karen Phillips; manufacturing supervised by Jacqui Ashri.
Typeset by Digital Graphics Inc., East Lansing, MI.
Printed and bound by R.R. Donnelley & Sons, Harrisonburg, VA.
Printed in the United States of America.

9 8 7 6 5 4 3 2 1

ISBN 0-387-94114-2 Springer-Verlag New York Berlin Heidelberg
ISBN 3-540-94114-2 Springer-Verlag Berlin Heidelberg New York

आदिगुरवे नमः

Katherine Mayo Cowan
1883–1975

Student's Introduction

Welcome to new territory: A course in probability models and statistical inference. The concept of probability is not new to you of course. You've encountered it since childhood in games of chance—card games, for example, or games with dice or coins. And you know about the "90% chance of rain" from weather reports. But once you get beyond simple expressions of probability into more subtle analysis, it's new territory. And very foreign territory it is.

You must have encountered reports of statistical results in voter surveys, opinion polls, and other such studies, but how are conclusions from those studies obtained? How can you interview just a few voters the day before an election and still determine fairly closely how HUNDREDS of THOUSANDS of voters will vote? That's statistics. You'll find it very interesting during this first course to see how a properly designed statistical study can achieve so much knowledge from such drastically incomplete information. It really is possible—statistics works! But HOW does it work? By the end of this course you'll have understood that and much more. Welcome to the enchanted forest.

So now, let's think about the structure of this text. It's designed to engage you actively in an exploration of ideas and concepts, an exploration that leads to understanding. Once you begin to understand what's going on, statistics becomes interesting. And once you find it interesting, you don't mind working at it. It does require work. Statistics is hard; there's no getting around that. But it's interesting. Once it becomes interesting, the "hard" part is not too onerous.

The text is divided into three parts. The first part presents some basic information and definitions and leads you immediately to a set of exercises called "Try Your Hand." The second and third parts of the text are solutions to these exercises: "Solutions Level I" and "Solutions Level II." Leaf through the text and you'll see that the solutions comprise over half the book. They're much more than just "the right answer." They provide complete discussions for the issues raised in the exercises.

Over the first week or so you'll develop your own method for working with the Try Your Hand exercises. Different students learn in different ways, after all, and so you must find what works best for you. Still, a few words of orientation will help. First, it WON'T work to just read the problems and then read the solutions. As you read you'll be saying, "Yes, yes, yes. That sounds right. Yes, and that's right too. Yes, yes, yes. . . . " But afterward, when your instructor mentions some item from that list (on a quiz?), you'll swear you've never seen anything like it! In one eye and out the other.

Here's something else that won't work: Sometimes students are diligent in an unproductive way. One form of wrong diligence is to think you must master each step before going on to the next step. That's possible alright—if you have 30 or 40 hours a week to devote to this course. But it's not an efficient way to learn. The human mind is not a linear machine. Efficient learning is always grasping for the "total picture," grabbing something here and something there, leaving the details to be filled in later. Think how a small child learns. How she learns to talk, for example. Children are master learners, very efficient learners. They don't proceed in a logical step-by-step manner. They grab for everything at once. Of course, there does come a time when you're ready for careful detail and when you'll go over each step with a fine-tooth comb. But that's the last step in the learning process. In the beginning, trying for that kind of detail is counterproductive.

All of this tells you something about how to use this book. When you begin a new section, read over the text material quickly just to get an idea of what's there. Aim for the next set of exercises. Read the first question in the exercise set, reflect for a moment (30 seconds?) on what that question wants from you. Try to grasp the meaning of the question. Often you'll go back into the text for some detail. Realize that you may or may not be able to answer the question. Don't worry about that. Learning happens in the ATTEMPT. Whether or not you can answer the question right then is secondary. When you've given the question as much effort as seems appropriate, turn to the Level I solution. There you'll find help of some form—a hint, a clarification, the beginning of a solution, further information, and so on. Now, process this Level I information. Give a bit more time attempting to formulate a complete answer. You'll succeed if you've understood everything up to that point. But . . . if you're studying properly you WON'T have understood everything up to that point! That means even with the help of Level I, you may or may not be able to give a complete answer to the question. Try. Then turn to Level II for a complete solution.

So you see how the solutions are structured. Level I is help in some form, Level II provides a complete solution. Don't omit Level I! Level II will be meaningless without Level I if for example, Level I helps you, by providing a start on the solution, as sometimes it does. Or suppose Level I provides more information. If you skip Level I, you'll not have that information. Occasionally (not often) the Try Your Hand exercises ask a question that, as asked, CAN'T be answered. Ideally, you would respond by saying, "This question can't be answered." And you would explain: "It requires more information. We need to know this, this, and this." Or maybe you would say, "It's ambiguous, it could mean A or could mean B." As a practical matter, you'll struggle with questions like this, suffering a vague sense that something's wrong. And maybe—just maybe—after you've encountered a few such challenges, you'll have gained enough clarity to suspect something's missing or that there's an ambiguity. Great! You're learning. In any case, when you turn to the Level I solution, you'll get the clarification you need. But notice, it's in your struggle with the question that you learn, even though you don't "succeed" on your own in giving an answer or even in understanding the question!

Remember the principle: Grasp for everything at once. If you find you just can't understand some problem, GO ON TO THE NEXT PROBLEM! You can come back to this one later.

One more bit of advice: Find someone to study with! Better yet, form a study group with two or three other students. You'll be surprised how much you learn from each other. And it takes less time because the very point YOU get stuck on will be the thing that seems clear to someone else. Then, just around the next bend, what some other student finds totally mysterious seems clear to you. And both of you gain clarity trying to explain to the other person. Studying should be a social enterprise!

So success in answering questions is not your goal. The goal is to develop your understanding. And that happens as a result of your STRUGGLE to understand. That's why the exercises are called "Try Your Hand" instead of something like "Do It." Maybe we should have called the whole book "Try Your Hand," that's what it's all about.

Finally, a word about the technical terms: You will notice that technical terms are given in boldface italics in the very sentence where they first appear. Be sure you learn the definition for each such term PRECISELY, otherwise your thinking will necessarily be vague and confused everytime you encounter that term. Well, that's it—that's all the advice I can give. Good luck and have fun!

Instructor's Introduction

The concepts encountered in a first course in statistics are subtle, involving quite sophisticated logic. Understanding the statistical techniques presented in that course, understanding their application and the real-world meaning of the conclusions depends very critically on understanding those concepts. This text is my attempt to structure an exploration for the student leading to that kind of understanding.

The fundamental idea underlying the entire structure is the concept of a probability distribution as a model for real-world situations, a concept that is not in any sense elementary. Probability itself is hardly elementary. Still, even children grasp (naively!) the idea of probability as expressed in phrases such as "a fair coin" or, in card games, "the chance of drawing a spade." That being true, games of chance seem a good place to begin. That's Chapter 1. And on the first page of Chapter 1 comes the idea of a probability model. Thus the student has time—the entire course—to assimilate this concept with all its subtlety and later mathematical elaboration.

Before discussing chapter content in detail, a few general observations will be helpful. This "first course" covers the standard topics of introductory precalculus statistics, but with a very nonstandard presentation. The text itself is brief, leading the student quickly to "Try Your Hand" exercises in which the student actively explores concepts and techniques. Often, no more than a page or two passes before the student sees the next set of exercises. Much of the exposition usually given discursively in the text itself is presented here through these Try Your Hand exercises. This approach is possible only because the exercises have complete solutions with full discussions. The solutions are given in two levels. Level I gives hints, clarifications, further information, partial solutions, and so on. Having completed Level I, the student turns to Level II to find a full solution.[1] A number of problems with real data have been included to give some idea of the variety of applications of statistics and some feel for the unexpected questions which arise in specific situations.

By judicious use of the Try Your Hand exercises, the instructor can focus the course according to the students' needs and abilities. Some exercises might be omitted altogether; others might be presented by the instructor in class. For example, probability formulas for the discrete distributions of Chapter 3 are derived in the exercises. The student is led by the hand through the derivation with the help of Level I. An instructor who chooses to omit formula derivations can simply omit those exercises. The student will be totally unencumbered by the derivations. She won't even see them because they are not included in the main body of the text itself.

The presentation given in this text is more sophisticated with respect to the underlying logic of statistics than most introductory books. In Problem 6.2.21, for example, the student sees that two very different p-values could arise from the same objective data, depending on how that data is "modeled." The ramifications of that are discussed in Level II. In Problem 5.5.12, two different prediction intervals for the same problem—one parametric, the other nonparametric—are compared. The discussion of statistical testing in Chapter 6 is more thorough than in any book I know of at this level. The presentation of regression in Chapter 7 avoids the usual list of unmotivated assumptions. Instead, I give a natural characterization which reveals simple linear regression as the next logical increment in complexity beyond previous chapters. For further detail on all of this, please see the individual chapter discussions given below.

Finally, I give much in the way of informative heuristics, such as the "elementary errors" interpretation for the normal distribution introduced in Chapter 4. That criterion, which is just an intuitive formulation of the Central Limit Theorem, is used systematically throughout the course to explain many details otherwise left obscure. To give one example, the criterion explains why only in the small sample case for inferences about means we must assume we are sampling from a normal distribution (see Problems 5.3.2 and 5.3.3).

[1] The instructor should remind the students (often) not to neglect the Level I solutions. Level II will seldom be complete in itself.

This text should be appropriate for students requiring an elementary introduction to statistics, assuming little mathematical sophistication, who anticipate more than a passive involvement with statistics and who will be learning more sophisticated statistical techniques in later courses. It should be appropriate for engineering, economics, computer science, psychology, sociology, and education majors. It would probably be appropriate for students in fields such as geography, ecology, and so on. It certainly would be an ideal freshman introduction to statistics for mathematics majors with the expectation of later follow-up in a thorough Mathematical Statistics course. In my experience, it is difficult to teach a meaningful Mathematical Statistics course to students with no prior exposure to statistics and, consequently, with no intuitive orientation.

Chapter 1

The discussion above suggests that the probabililty distribution of a random variable is a central idea. Certainly it is. It's a sophisticated concept at the heart of virtually every technique of statistics. Even for exploratory and nonparametric techniques a probability distribution often lurks in the background (at least) as a standard of comparison. And, of course, the prior probability distribution is fundamental to Bayesian statistics.

To ease the student's initial exposure to this abstract concept, I introduce probability distributions at the very beginning through simple examples of real-world situations, namely, games of chance with coins, dice, cards, and so on. This term "real world" is used throughout the text in contrast to "abstract theory" (not in contrast to "artificial" or "contrived"). Random variables and linear functions of random variables are used in this chapter to model these simple games of chance. For example, if you receive two dollars for each dot on the uppermost face of a die and pay six dollars to play, your gain/loss random variable is $G = 2X - 6$, where X models the die. Thus, linear functions of random variables are also explored in some depth with the simple examples of Chapter 1. Most of this chapter relies on the student's intuitive idea of probability together with some simple ad hoc rules. After "reminding" them of what they already know, a more rigorous development of probability is presented in Section 1.4.

Betting games with dice, loaded in various ways, are effective for investigating the mean and variance of a random variable and for understanding the variance as a measure of risk, predictability, accuracy, and so on. The significance of the variance is much more readily appreciated by students in the dynamic context of random variables than in the static context of observed data, another reason for introducing random variables at the very beginning of the course. The die—in general, the random mechanism for the game—need not be fair. Loaded dice offer a variety of interesting situations and an opportunity to understand interesting concepts. For example, we can ask the student to load a given die differently so the mean stays the same, within a specified degree of accuracy, but the standard deviation is smaller or larger by a specified amount. Or again, we can ask which is more predictable, a die with a given loading or a fair die? Which game would you prefer to play, the one with the fair die or the one with the loaded die? The answer depends not only on how the die is loaded but also on one's motivation for playing. Students readily appreciate the relevance of this kind of analysis to more realistic situations such as portfolio analysis or variability in a manufactured product, or variety within a genotype, or any of many other situations of possible interest to the student.

Random variables are presented as providing a "bridge" from the real-world situation with all its complexity to the relatively simple world of theory (see the picture of this bridge on page 7). This metaphor is much more than a clever hook—it is a significant pedagogical device. I have the picture of this bridge on the blackboard every day for the first half of the course and am constantly surprised, and surprised again, how often misconceptions can be resolved by reference to this picture. Many errors arise from confusing the outcomes with the values of the random variable. I simply point to the picture and students catch their error immediately (well, almost)! For example, isn't a "constant random variable" a contradiction in terms? After all, "it" is predictable! Look at the picture: "It" (the value) is predictable, but "it" (the outcome) is not. Or, if we have a pair of fair dice, isn't the number of dots on the top faces uniformly distributed? Since the dice are fair, "they" are equally likely. Yes, "they" (the outcomes) are equally likely, but "they" (the values) are not. The bridge metaphor is particularly helpful in resolving the confusion of values with outcomes because it places them symmetrically on opposite sides of the River Enigma.

Chapter 2

This chapter covers topics of descriptive statistics, introducing frequency and relative frequency distributions, their histograms, and related ideas. In this chapter, we introduce random sampling: first sampling from a probability distribution, then sampling from numeric and dichotomous populations. After Chapter 1, sampling from a distribution seems quite natural to students. It's conceptually simpler than sampling from a population. The key question for sampling from a distribution is simply the independence for repetitions of the underlying random experiment. For example, if we're interested in monitoring "fill" for cups from a soft drink vending machine, 10 cups taken in succession will be a simple random sample from the distribution of "fill" provided only that the amount of drink dispensed is independent from one cup to the next. Students are readily able to suggest how that might or might not be true depending on the circumstances.

Most, if not all, introductory statistics texts seem to avoid sampling from distributions—a wise choice if the students don't understand what a distribution is—and, consequently, many examples throughout such a text force very artificial interpretations of sample data by reference to some hypothetical nonexistent population where certainly the data was NOT selected through any sampling plan. Did we use a random number table to select 10 cups from a population of "all possible cups?" In exactly what warehouse are "all possible cups" to be found? If that's not the procedure, what justifies calling those 10 cups a random sample? Interpreting such examples as "sampling from a population" not only does not help, it's a serious obstacle to clarity. You dare not ask the student if the assumption of randomness would, under the circumstances, seem justified. The definition before them is so artificial that any practical discussion of its relevance is impossible. An intelligent student can only conclude that one blindly assumes whatever one wants in order to make the theory work. I prefer the idea that one makes assumptions only where those assumptions seem reasonable and where they can later be verified.

Chapter 3

This chapter presents nine "models," nine classes of discrete probability distributions of varying degrees of concreteness, interrelated in various ways. I strongly urge that none of these models be omitted. Understanding a sophisticated, abstract concept—here, probability distributions—requires more than one or two examples. The goal of this chapter, at least from the point of view of the statistical material to be presented later, is to develop the student's skill in recognizing an appropriate model for a real-world problem. This requires experience with a number of different models. The instructor can mitigate the difficulty of this chapter without compromising its principal thrust by omitting some or all of the probability formulas, focusing instead on model recognition. Skill at model recognition can be developed and tested without probability questions per se, by simply restricting to questions about the mean and variance, questions such as, "How many cups should we get from this drink machine before the machine malfunctions?" (the mean of a geometric random variable).

The statistical topics of this course—random sampling, sampling distributions, estimation, statistical testing, and the regression model—cannot be understood if the underlying theoretical models are not understood in their roles as models, models for the sampling process or for the more complex situation of regression. In my experience, Chapter 1 and the nine models of this chapter will indeed bring the students to the requisite understanding of probability distributions as models for real-world situations.

The discussion from Chapter 1 is continued at a more sophisticated level with the nine models of this chapter. Constant and uniformly distributed random variables form two very simple classes, already familiar from Chapter 1. In particular, Chapter 1 has already shown how constant random variables arise very naturally through combinations of other random variables. For example: $X + Y = 7$, where X and Y are, respectively, the number of dots on the top and hidden faces of a six-sided die. These two simple examples—constant and uniformly distributed random variables—help us to establish what we mean by a "class" of random variables and set the pattern for the rest of this chapter.

The classes of this chapter are interrelated in various interesting ways. There are two sampling distributions: Sampling with or without replacement from a dichotomous population form one group. The binomial, geometric, and negative binomial form another group (with the geometric a special case of the negative binomial). The binomial has the previous "sampling with replacement" model as a special case. The Poisson model is the most abstract of the models in this chapter, having been derived abstractly through a purely mathematical process from the binomial. It becomes a model for real-world situations only after the fact and for that reason has a less concrete feel about it. The student is alerted to this "abstract" versus "concrete" consideration. That idea is picked up again in Chapter 4 where we introduce continuous distributions. The understanding that some models are more abstract than others is helpful in understanding the normal and chi-squared distributions which are indeed quite abstract.

An important challenge in this chapter (and again in Chapter 6) comes in the set of mixed review problems at the end. In these problems, the student is on her own to identify a correct model for a given problem. A few problems can be correctly modeled in more than one way. This review is very important and should not be omitted. A real difficulty for students, which will show up in these review problems, is their single-minded focus on the abstract part of the model. Students complain about abstraction, but, in fact, they love it—it's easier. The abstraction is precise and clear; equations and formulas can be learned. The real world, by unhappy contrast, is messy, ambiguous, and confusing. However, to identify an appropriate model for a real-world problem we have to look where the problem is—in the real world. That requires focusing on the real-world description of the random experiment, the real-world component of the model, and matching that description with what's going on in the problem. The skill to do this is developed through these review problems at the end of the chapter.

Chapter 4

This chapter extends the presentation of the previous chapter to continuous distributions. First are the uniform and exponential distributions; then the most abstract model so far encountered, the normal distribution, modeling random error or, by extension, any situation where the difference in two values "looks like" random error (see the criterion for normality on page 148. Finally, we see the chi-squared distribution, the most abstract of all among the distributions of this text.

Chapter 5

This chapter presents sampling distributions, the Central Limit Theorem, and interval estimates as a unified topic. We do three types of interval estimates: confidence intervals, prediction intervals, and tolerance intervals.

For understanding sampling distributions, variability from one sample to the next is not the really difficult concept. People with no knowledge at all of statistics see this variability very clearly. That's why they're so ready to criticize statistical surveys, complaining that "it's all just based on a sample!" This is the thinking students come to us with. We must show them that

They're right if they think one sample alone can't tell them anything.

BUT . . . They're wrong if they think sampling is useless or statistics a sham or if they think only very large samples are legitimate. And they're especially wrong if they think a very large sample carries any information by itself.

BECAUSE . . . The missing ingredient which makes sense of one sample is the *entire context of that sample*. That "entire context" is the sampling distribution, a sophisticated theoretical construct not easily understood.

For example, a sample mean by itself tells you nothing. On the other hand, a sample mean seen as just one of the many possible values of a normal distribution centered on the unknown true mean, with most of the probability concentrated there and with a standard deviation intimately related to the standard deviation of the original distribution, TELLS YOU A LOT!

When asked, "What assumption must you make about the data here?" (that it's a random sample) students will respond, "That it's more or less typical of the population." Well, if it's typical, you don't require statistics! But, in fact, you'll never know whether it's typical or not and YOU HAVE NO THEORY FOR THAT. There is, however, a theory for random sampling and that theory controls the error which could arise from a possibly atypical sample. Control of error is the theme, probability distributions the tool.

There may be an objection to the presentation in this chapter which treats only interval estimates and does not allow for point estimates. But point estimates are only appropriate when the estimator is in some sense "best" for the problem at hand. That more advanced discussion involves everything in the discussion above and more. For this reason, I present only interval estimates with the understanding that the point estimate is incomplete to the point of being meaningless if no further investigation is carried out.

There may also be an objection to interpretations of confidence intervals which begin "There's a 95% probability that. . . . " The usual argument tells the student to replace the wrong expression by another one where the offending term "probability" is replaced by the undefined term "confidence." This just replaces error by ignorance, hardly an improvement! No wonder highly intelligent people say they never could understand statistics. I prefer to use a natural probability expression, but acknowledge openly that it's *ambiguous*. One reading is wrong (with the 95% probability referring to the parameter), the other correct (it refers to the interval). The student is held responsible for understanding the two readings, understanding why the one is wrong and the other correct.

This approach is consistent with my exposition throughout the text, where I hold the student responsible for certain standard ambiguities or misstatements. For example, the question "What are the chances for a female on this committee?" almost certainly is asking for $P(X \geq 1)$, although technically it asks for $P(X = 1)$. Or I leave out the phrase "on average" in situations where it's clearly implied. Or, again, I ask for a count where a proportion is all that's possible. This significantly challenges the student's clarity of thought because she is on her own to discriminate among possible meanings. "How many," for example, might mean "how many on average" or "what proportion." There's no getting around it, she has to understand the context! In this way, the student of this text grows accustomed to dealing with ambiguity and the resolution of ambiguity as a part of the natural intellectual environment. All of this is possible through judicious use of the Level I answers to the Try Your Hand exercises.

Chapter 6

This chapter presents tests of statistical hypotheses. This vexed topic is presented with in-depth discussion of the possible misinterpretations, misuses, and limitations of the technique as well as a careful discussion of the correct interpretation of p-values, conclusions, and errors. The first section of the chapter gives an overview of two testing procedures, the "test of significance" (p-values) and the "hypothesis test," and introduces some terminology and some comparisons (the details of which are deferred until later sections). The second section presents tests of significance, including chi-squared tests. The third section of the chapter presents the "hypothesis test" as providing a decision procedure for a monitoring process as, for example, in quality control.

So the "test of significance" is presented first, with its formal answer (a p-value) and its not-so-straightforward real-world interpretation. The logic of statistical testing is quite subtle, involving considerable controversy. A significant amount of confusion has been introduced into the topic by not distinguishing between the "test of significance" and the "hypothesis test" proper, with its error probabilities, power considerations, and so on. Consequently, I have separated these two approaches to statistical testing.

The student's understanding is enhanced by first clearly understanding p-values as measuring consistency between the data and the hypothesis. The p-value calculation is relatively easy. Nevertheless, the real-world interpretation is not so straightforward. For instance, although small p-values are usually what we look for, with not small p-values being inconclusive, just the reverse may hold for the practical interpretaion. For cases of "discriminatory selection," for example, where the hypothesis to be challenged is "random choice," a NOT small p-value is quite conclusive—it's impossible to maintain an accusation of discrimination if the choice is consistent with having been random—whereas a small p-value may be relatively conclusive or relatively inconclusive, depending on the context. See the discussion beginning on page 245.

There are a number of advantages to presenting tests of significance as a separate topic. Certain issues concerning statistical testing are more clearly presented with reference to p-value calculations, unburdened by the heavy-handed and irrelevant machinery of null and alternative hypotheses, error probabilities, rejection regions, decision rules, power, and so on. For example, the test of significance already highlights the distinction between practical and statistical significance. It also reveals the asymmetry inherent in statistical testing, the asymmetry seen in the difference between "small p-value" (the hypothesis is challenged) and "not small p-value" (the data is inconclusive). Further, the test of significance, seen as a "probabilistic argument by contradiction," clarifies the underlying logic of statistical tests in general. Again, chi-squared tests are more appropriately introduced as tests of significance than as hypothesis tests because there is often no meaningful "alternative" hypothesis. Finally, by presenting tests of significance as a separate technique, instructors who prefer to spend less time on statistical testing can omit the complexity of "hypothesis tests" altogether and confine their discussion to this much simpler case.

Hypothesis tests are first studied without control of type II error. Our point of view is that when you "fail to reject H_0," you take no action based on the test itself since in that case you have exercised no control over the possible error. This asymmetry is exactly parallel, of course, to the asymmetry of "small" versus "not small" *p*-values. The important new idea here as compared with tests of significance is the "control of error" for type I error. Aside from the pedagogical advantages, leaving discussion of type II error until later is justified by the impossibility in some testing situations of finding a model for the alternative hypothesis. In other words, there are, indeed, situations where control of type II error is not practical.

This chapter gives an in-depth discussion of the role of hypothesis tests. For example, we see how the logic of hypothesis tests compares with classical inductive inference, leading to the distinction between null hypotheses which are sometimes true and sometimes false (in monitoring situations) and null hypotheses which are either always true or always false (the classic situation of inductive inference, where "accumulation of evidence" is the motivation for repetitions of the experiment).

There may be an objection to so much emphasis on the logic of testing. It has even been suggested that statistical tests should be omitted altogether because the same conclusions can be obtained from a confidence interval.[2] But statistical testing is too pervasive in statistical practice to justify omitting it. The student is not well served in being left ignorant of terms like "*p*-value," "null hypothesis," and so on. Given that we're going to teach the topic at all, surely we must teach it clearly so that common confusions and misunderstandings do not arise.

The distinction between tests of significance and hypothesis tests is certainly not artificial. Failing to make that distinction leads to a number of points of confusion. To name only one: Are you allowed to look at the data before setting up the test? For a test of significance where the question is "Does this data seem to challenge our hypothesis?" there is no "setting up" of the test. The data is part of the original question.

[2] As a matter of technical fact, this is not true in the case of proportions since the standard error will differ. If the hypothesis to be tested is false that difference could be significant.

How can you avoid looking at it in advance? For an hypothesis test, on the other hand, which is properly a monitoring procedure, the data will change from one run of the test to the next, so, OF COURSE, you have to set the test up without reference to the data. Most textbooks tell the student not to look at the data and then, in every example and problem, give the data in the problem statement. How can the student avoid looking at it? To add insult to injury, the solution—which here means simply choosing a direction for the test—will always be correct if you base it on the data. Never does the student see data which would be in the "wrong" tail of a properly determined test.

For other points of confusion which arise when the distinction between tests of significance and hypothesis tests is not made, see the text of Chapter 6. None of this says the distinction remains necessary for someone who clearly understands the entire logic of tests. But we should distinguish between what is logically correct and what is pedagogically clear.

At the end of this chapter, as at the end of Chapter 3, there is a critically important set of mixed review problems, the most challenging set of problems in the text. It should not be omitted. I give about a week of class time to these problems. It is through this set of problems that students assimilate the statistical techniques of Chapters 5 and 6.

Chapter 7

This chapter is a brief introduction to simple linear regression. In Chapter 4, anticipating the present chapter, the normal distribution is described along these lines: Suppose you have a fixed systematic "effect" for which any variation is due solely to something that "looks like" random error. Then you should expect a normal distribution. Take, for example, diameters of machine parts where the systematic "effect" is the manufacturing process itself which attempts to meet specifications (diameter 3.2 mm). The variability in diameters is purely random unless there's something wrong in the process. So diameters are $D = \mu + \epsilon$. Here $\mu = 3.2$ is the systematic part with ϵ being "like random error" so that $\epsilon = N(0, \sigma^2)$.

When we come to regression, we make ONE SIMPLE STEP FORWARD IN COMPLEXITY for the model. The "effect" which determines the mean is no longer fixed, but variable. But not variable in just any way at all; that's much too complex. Instead, the mean is determined through a linear function of the effect, a linear function being the simplest nontrivial function. So we obtain a model with a variable "effect," X, which is as simple as possible and which "affects" (not necessarily causally!) only the mean of Y, not affecting Y in any other way. The usual long, intimidating and unmotivated list of assumptions for the regression model follows quite naturally from this characterization [see problem 7.1.4(c)].

Of course, one may object, X can certainly affect more than just the mean of Y. For example, it can affect the percentiles. But that's not a *different* effect. If you tell me the effect on the mean of Y, I can determine the effect on the percentiles. This is parallel to what we say about the parameters of a model. For example, σ and σ^2 are not two different parameters; give me one I can calculate the other. For the hypergeometric model, p is not a fourth parameter because I can calculate it from two of the other three: $p = R/N$.

Here again, as in many other instances—confidence intervals in particular—I allow possibly ambiguous statements when it's convenient and natural, making a point of the ambiguity and its proper resolution. Understanding comes in being clear about the ambiguity. So, in the regression model, X affects only the mean of Y, but not in an absolute sense, rather in the sense that any other effect from X can be calculated from the effect on the mean.

With some hesitation, I will describe how I currently use this text in my classroom. I hesitate simply because I would not want to prescribe a "right" way to use the book. Indeed, I hope various instructors will find various effective ways of using the text.

At the beginning of each class, I make an assignment for the next class. The students are expected to read that material and process the problems on their own with no preliminary in-class discussion. At the beginning of the next class, there is a brief, very routine quiz on the assignment. These quizzes serve many useful purposes, not the least of which is to encourage students to actually do the assignment. Needless to say, when the students have already thought about the material, the class discussion can deal with issues in much greater depth and subtlety. For sections which meet three days a week in 50 minute classes, there are approximately 20 such quizzes per semester of which I drop the lowest three or four. These quizzes count 20% of the course grade. The rest of the course grade is determined by two 100-minute tests (each given over two successive class periods) and the final examination.

The grading policy for the quizzes is very lenient because the material is quite new at the time of the quiz. The quizzes provide an opportunity for the students to catch errors or misunderstandings without incurring a serious penalty. Thus I count off only for gross errors or completely wrong approaches which would indicate that the student did not really do the assignment. I often write "OKT" beside a mistake, meaning, "No penalty THIS time, but be alerted: this is an error!"

I allow questions before the quiz. If the question is relevant to the quiz, I am obliged to answer. If the question is not specifically relevant to the quiz, I may postpone discussion of that question until later. Attention in the class is never quite so clear and focused as during that question period before the quiz! In practice, the quiz is sometimes at the very beginning of the class, sometimes fifteen or twenty minutes into the period, sometimes not until the very end. Rarely—in the interests of time—I may omit a quiz (unannounced in advance) to spend the entire class going carefully over some topic or conducting a review.

There is always initial resistance from the students to this approach. It seems to be quite a novelty that they should be expected to read material and assimilate it on their own. But usually after a week or so they begin to accept the responsibility to work on their own and begin as well to appreciate the value of developing their skill for independent study. The structure of this text, with its complete solutions to the problems, facilitates this approach.

My experience shows that students coming through this course develop valuable skills for independent study and for critical, analytical thinking. In fact, for many students those are possibly the most valuable results of the course.

Acknowledgments

It is a pleasure to acknowledge the help of many people in the writing of this book. First of all, I would like to thank my colleague Al Schainblatt for his continuing encouragement. More than anyone else, he has been close to this project from the beginning, sharing his ideas and insights and providing very necessary critique. I would also like to thank Gerry Manning who encouraged me in this project at the outset. And I thank Jamie Eng for cheerfully responding to odd-hour phone calls and knocks on her office door with questions on arcane points. Not at all arcane have been her excellent problem ideas which I have used. Both Gerry Manning and Jamie Eng have been very cooperative as department chairs in arranging class schedules to accommodate the writing and classroom testing of this text. I would like to thank two other colleagues, Paul Schmidbauer and Rick Wing, who offered helpful and detailed suggestions on chapters which they read. I would like to thank Edward Tufte and Tim Redmond for permission to reproduce some very instructive material. My friend and colleague Bharat Sarath of New York University has been a valuable source of information and discussion. Finally, I would also like to thank David S. Moore for his help.

Of course, I must thank the many unnamed students who have been my audience over a period of six years and for whom I really wrote the book. Many other people have read parts of the manuscript providing feedback or have helped in other ways. Among them I would like to thank David Axel, Dana Dixon, James Lipka, Kenneth McCormick, and Carol Whitfield.

I would like to thank Martin Gilchrist, my editor at Springer-Verlag, and the book production staff. All authors should have editors so cheerful, patient, and helpful. Finally, I would like to thank the series editors, Stephen Fienberg and Ingram Olkin. At the very beginning, they recognized some value in my proposal before there was any text to substantiate it. But I thank them with a special sense of gratitude for their patience and encouragement at a particularly difficult time when I thought this book would never see the light of day.

I would like to acknowledge the reversal—often—of traditional gender roles of English pronouns throughout this book. If the reader finds the effect jarring, as I sometimes have, prehaps she will also find it amusing (or even instructive) to reflect on exactly why such "gender bending" seems so disturbing. I certainly have. Finally, I would like to acknowledge tireless support and encouragement throughout six years of tedium and toil from my loving helpmate, if only I had one.

Contents

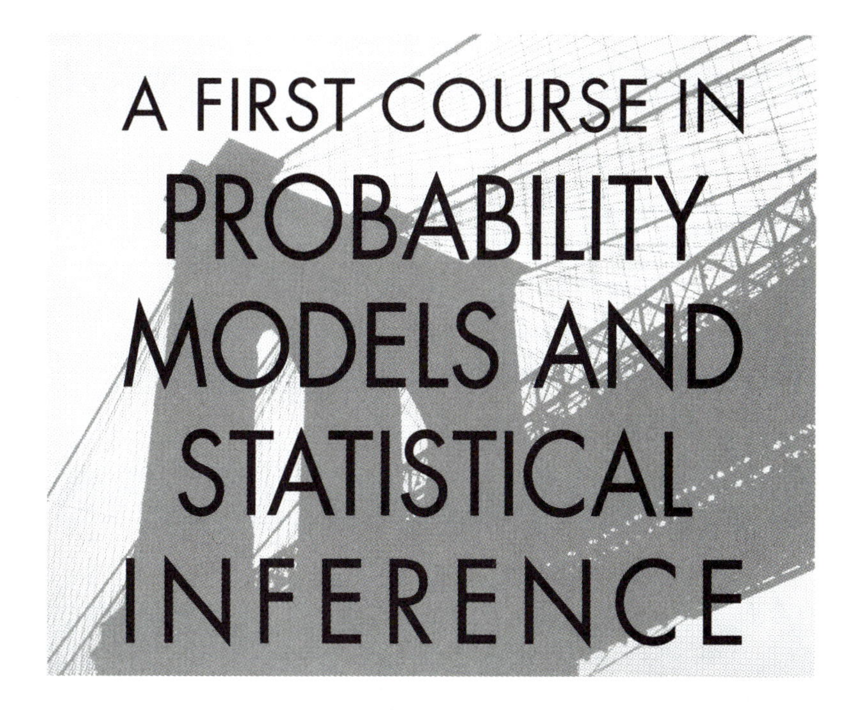
A FIRST COURSE IN
PROBABILITY
MODELS AND
STATISTICAL
INFERENCE

Chapter 1

Introduction to Probability Models of the Real World

1.1 Probability Distributions of Random Variables

Probability Models

There are situations so complex—you might even think chaotic—that any analysis would seem impossible. But often such situations yield to the techniques of statistics. With proper data and with proper analysis of that data, you may be able to say a lot. As this course proceeds, it will be interesting for you to see exactly how statistics is able to deal with these seemingly impossible situations. But it won't be at all what you expect. The answers provided by statistics are never exact; there's always error. Statistics does not eliminate error; rather, it controls it. In short, statistics allows you to be precise about imprecision! It's the *precise control of statistical error* that's at the heart of every statistical technique.

Error in the statistical sense arises from uncertainty in the play of chance, and so we'll begin our study with some simple chance mechanisms—coin tosses, rolls of dice, and so on—and with abstract models for those mechanisms. A roulette wheel is another such chance mechanism. Again, there are games which depend on the draw of one or more cards from a deck of 52 cards. If the cards are well shuffled, the draw is, indeed, a chance mechanism.

All such games of chance illustrate simply and very concretely some of the most basic concepts in statistics. Sometime in the period before 1600 in Europe, people who loved games of chance began to discover and formulate the mathematical laws for those games. These origins of statistics before 1600 are somewhat obscure, but from the time of Fermat and Pascal in the mid-seventeenth century, there's a continuous and well-documented development of the mathematical theory for games of chance. Let's see how that theory goes. We'll begin with

the fair die:

X	$P(X)$
1	1/6
2	1/6
3	?
4	?
5	?
6	?

and

the fair coin:

X	$P(X)$
H	?
T	?

Obviously, what we're giving here is not the physical object itself, not a die nor a coin, but rather an abstract mathematical model of that object—a table, in this case. A table like the ones above is an abstract "probability model" for such a chance mechanism. For the moment, we are asking you to guess how to interpret the models. In the tables, $P(X)$ is the probability that the variable X takes on the indicated value. As you can see, the tables are not complete. We've left it to you to complete the tables by filling in the missing probabilities. Think about a fair die. Or a fair coin. What do these tables suggest to you in real-world terms? Try to complete the tables before reading further (the answer is in Problem 1.1.1 below).

If the die is fair, each of the six faces comes up equally often on average. That's what the "one-sixth" in the table for the fair die means. That "one-sixth" is the ***theoretical relative frequency*** with which we would expect to roll each of the various faces of the die. It's "theoretical" because it's based on the theory that the die is fair. If the die is not fair, some one (or more) of the faces comes up too often—more than one-sixth of the time on average. Of course, if one face comes up too often, then some other one (or more) of the faces must come up *less* often.

The term "theoretical relative frequency" is just another name for probability. The ***probability*** of an event is, by definition, the relative frequency with which we expect, theoretically, to observe an occurrence of that event. There are other possible definitions of probability, but unless otherwise stated, we'll always intend the theoretical relative frequency definition. In Section 1.4 you'll find a more detailed presentation of probability. For now, your intuitive idea of probability, derived from your understanding of games and other real-life situations involving the idea of chance, will suffice.

But now, before we go any further, why don't you just . . .

Try Your Hand

1.1.1 (**a**) In the tables above for the fair die and the fair coin, give a verbal description of the variable X.

(**b**) Complete the tables by giving the probabilities for a fair die and a fair coin.

(**c**) What does the word "probability" mean? Explain it in terms of the probability to draw the ace of spades when you take the top card from a well-shuffled deck of 52 playing cards.

1.1.2 Later we'll develop precise rules for working with probabilities, but can you guess the probabilities of getting

(a) a number less than three on one roll of a fair die?

(b) a pair of sixes on a roll of two fair dice?

(c) a pair of heads on a toss of two fair coins?

1.1.3 Can you guess a rule that distinguishes a situation which requires adding probabilities from one which requires multiplying them?

Random Variables and Their Random Experiments

Rolling dice, tossing coins, drawing cards from a deck are all examples of random experiments. It's not easy to give a precise definition of the term "random experiment" because the word "random" leads into a deep (and fascinating) philosophical quagmire. But we can easily get along without a precise definition. It's enough to have a heuristic definition which will not lead us astray. Here it is: a ***random experiment*** is something you do that is repeatable, with clearly specified outcomes which cannot be predicted in advance. Although we're only giving a heuristic definition of the term random experiment, it's necessary to pay careful attention to the details.

What you must verify to show that you have a random experiment:

- the doing
- the repeatability
- the clearly specified outcomes
- the unpredictability

If the "doing" is not repeatable, if it represents an entirely unique occurrence, statistics can provide no help at all. Now, because the phrase "something you do" is hopelessly vague, it's necessary to pin the "doing" down more precisely. This is accomplished by specifying clearly the outcomes. This much gives the definition of a *scientific* experiment: something you do which is repeatable with clearly specified outcomes. By insisting that the outcomes cannot be predicted in advance, we capture the idea of randomness. This is not a very adequate definition of randomness from a philosophical point of view of course, but you get the idea!

For example, for the die, the "doing" is to "roll the die." Clearly, that's repeatable. Suppose we specify TWO possible outcomes: either the die lands on the table or it lands somewhere else—the floor, for example. That's probably not the random experiment you had in mind. But now you see why it's necessary to be clear about the outcomes. For this experiment, the outcome may or may not be predictable, depending on exactly how you roll the die. Suppose "rolling the die" means dropping it from a height of about two feet above the table top. In that case, we do indeed have a random experiment because the outcome would be unpredictable: You can't say in advance whether it will land on the table or the floor. Notice how we've gone through each of the four items listed above to verify that "rolling a die"—in the exact sense we've specified—is a random experiment. This is what you must do if asked to "verify that such and such is a random experiment."

Of course, the usual random experiment with a die assumes the die will remain on the table top. If not, you abort that attempt and do it again. For that experiment, an outcome is "the die lying on the table top in some position." Notice that we are very physical in our description of the outcomes. You should avoid any reference to numbers in describing outcomes. By following this rule, you'll have a much easier time in understanding random experiments and their random variables (which we are about to define). So, if you were inclined to describe the outcomes as "the numbers from one to six," think again, THAT'S NOT IT!

A ***random variable*** is a rule which associates a number to each of the possible outcomes of some random experiment. For a roll of the die, we can define the random variable—let's call it X—which assigns to an outcome the number of dots on the uppermost face of the die. Here's a description of this random variable:

an outcome: the die resting in some position on the table top;

association of a number to outcome: look at the top face of the die, count the number of dots on that face;

possible values: the positive integers one through six.

So you see, if you identified the outcomes of the experiment as the numbers from one to six, you confused the outcomes with the values of the random variable. The outcomes of the experiment are NOT the same thing as the values of the random variable.

An outcome for a random experiment is the physical, real-world situation which results from performing the experiment once. Here, after one roll of the die is completed, the real-world result, the outcome, is the die sitting in a certain position on the table. For that outcome (the "sitting die"), X counts the number of dots on the uppermost face. That count, that NUMBER, is the value of the random variable for that outcome. This is why we say you should avoid giving numbers as outcomes. If you never give numbers as outcomes, you won't make the mistake of confusing the outcomes with the values of the random variable.

The outcomes of the random experiment live in the real world; the values of the random variable live in the theoretical world of numbers. Again and again we'll see this contrast between the real world and theory. It's the interplay between the real world and theory that makes statistics such a powerful tool. Throughout this course, our point of view is that questions and answers live in the real world whereas, statistical tools live in the world of theory. What makes statistics interesting is to see how those theoretical tools can be made to yield real-world answers to real-world questions.

So random variables provide abstract mathematical models, probability models, for real-world situations. And they're very useful models, as you'll see. In fact, random variables are at the center of everything we do in this course. They provide a *bridge* between real-world situations (the random experiment) and mathematics (the numeric values of the random variable). For problems in the real world involving uncertainty, an appropriately defined random variable may provide a bridge to the powerful analytic tools of mathematics.

Here's a picture of a random variable:

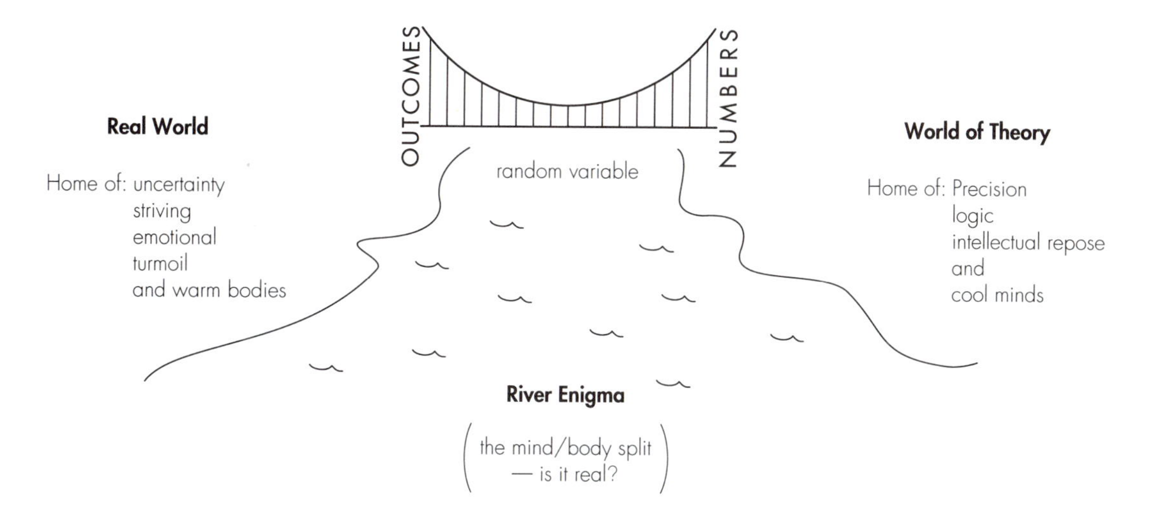

Now let's think about tossing a coin. Note that in the earlier table for the toss of a fair coin, X is NOT a random variable because it's not numeric-valued. There, X assigns the letters H and T to the outcomes. Letters are not subject to the laws of mathematics. We've insisted that random variables should be *numeric*-valued because we want to draw on the powerful analytic possibilities of mathematics. Still, it's easy to invent a random variable for the toss of a coin. If we're particularly interested in "heads" when we toss the coin, we might let $X = 1$ if a head comes up, and otherwise let $X = 0$. Note that X has a simple verbal description: It's the number of heads on one toss of the coin.

Just to see how this X might be useful, suppose you toss the coin ten times during some game and need to keep a record of the results of these ten tosses. In terms of X, we can say that there are ΣX heads. Here we're showing you a simple notation which is very convenient in statistical discussions. It's not hard: ΣX simply means sum the values of X. You can read it "sigma X," or just "sum X." For example, if you toss the sequence

$$\mathrm{H, H, T, H, T, H, H, H, T, H}$$

X would be

$$1, 1, 0, 1, 0, 1, 1, 1, 0, 1.$$

Because these zeros and ones add to seven, ΣX is seven, telling us that there are seven heads.

Note how so simple a random variable as X, taking only the values zero and one, is a very convenient abstraction. Using it and the summation notation, we can write ΣX, a mathematical expression which represents the number of heads tossed, no matter what that number might be.

Before doing the exercises, let's look at one final definition: The ***probability distribution of a random variable*** is a presentation of the possible values of the random variable together with the corresponding probabilities. The presentation may be in the form of a table, as we gave earlier for the fair die or it may be in a graphical form, as we'll see later. In Chapters 4 and 5, we'll have probability distributions presented through a set of equations. Note that in a probability distribution, the sum of all the probabilities must be exactly one; that is,

$$\sum \mathrm{P}(X) = 1.$$

You actually already know this. Think about tossing a coin. Suppose someone tells you a particular coin comes up heads a third of the time and comes up tails a third of the time. What happens the *other* third of the time? Does the coin land on its edge a third of the time? A strange coin indeed! On the other hand, suppose you had been told simply that the coin comes up heads a third of the time. Wouldn't you have immediately concluded that it comes up tails two-thirds of the time? You should have. Analytically,

$$\begin{aligned} \text{P(heads)} + \text{P(tails)} &= 1, \\ 1/3 + \text{P(tails)} &= 1, \end{aligned}$$

and so P(tails)$= 1 - 1/3 = 2/3$. Now please . . .

Try Your Hand

1.1.4 **(a)** Compare and contrast the terms "scientific experiment" and "random experiment."

(b) What in the definition of the term "random experiment" captures the idea of randomness?

1.1.5 Let X be the random variable which counts the number of dots on the uppermost face for one roll of a die:

(a) Give the probability distribution for X assuming the die yields two dots on the uppermost face 50% of the time on average, with all other faces equally likely.

(b) Give the probability distribution for X assuming the die yields two dots on the uppermost face 40% of the time and five dots 20% of the time while the other faces are all equally likely.

(c) Note that we have two different random variables in parts (a) and (b) although they have the same verbal description and, for convenience, we're using the same symbol, X. Now, look at the abstract definition of the technical term "random variable" and pinpoint exactly what part of the definition differs for these two random variables. You'll get some help if you think in terms of the picture of a random variable as a bridge.

1.1.6 Suppose you have a deck of playing cards which is *missing two hearts*. So the deck contains only 50 cards. Give the probability distribution for the random variable Y defined by the rule below for the draw of one card from this deck after thorough shuffling.

$$\begin{aligned} &\text{spades} && \longrightarrow 1 \\ &\text{clubs} && \longrightarrow 2 \\ &\text{hearts} && \longrightarrow 3 \\ &\text{diamonds} && \longrightarrow 4 \end{aligned}$$

1.1.7 Consider the six random experiments described in Problem 1.1.2 (a)–(c), Problem 1.1.5(a) and (b), and Problem 1.1.6.

(a) Specify the "doing" for the random experiments and then state the rule which defines the corresponding random variable.

(b) Now specify clearly the outcomes for each of the six random experiments.

(c) For each random experiment, explain how the conditions in the definition of the term "random experiment" are verified.

1.1.8 (a) Why is the table below not the probability distribution of a random variable?

X	$P(X)$
H	0.5
T	0.5

(b) Define a new X by letting $X = 1$ if a head comes up when you toss the coin, and otherwise let $X = 0$. Is this new X a random variable?

(c) Give a probability distribution for X in part (b).

1.1.9 We need to develop some precise techniques for questions like this, but based on your experience with games of chance, can you guess the answers to these simple questions?

(a) If you roll a fair die repeatedly (n times, let's say), how many dots altogether would you expect to roll on average?

(b) Suppose you toss a fair coin many times, how many heads would you expect on average?

(c) Suppose you toss a fair coin and that you'll be given two dollars if you toss a head and three dollars if you toss a tail. How much money would you expect to take in on average?

(d) Now suppose you toss a coin which comes up heads 90% of the time. If you're given two dollars for a head and three dollars for a tail, how much money would you expect to take in on average?

(e) Verify that your "take" in part (c) is a random variable.

(f) Intuitively speaking, why should you expect to take in less on average for the loaded coin of part (d) than for the fair coin?

1.1.10 Here is a problem to help you understand the definition of the term "random variable." Look at the result of one toss of a fair coin; if heads comes uppermost, call it H, otherwise call it T. Let X be the following rule which associates a number to each of the outcomes:

$$\begin{aligned} \mathrm{H} &\longrightarrow 1 \\ \mathrm{T} &\longrightarrow 1 \end{aligned}$$

Is X a random variable? This looks like a simple question, but be sure you address the question as it is put: Is X a *random* variable? Can you predict in advance that X will take the value "1"?

1.1.11 Suppose you are betting with a die which comes up two, 40% of the time and five, 20% of the time while the other faces are all equally likely. Suppose further that you'll be paid one dollar for each dot you roll and that you pay four dollars to play the game. Make a probability distribution for your gain/loss on one roll. Be sure to verify that your gain/loss is really a random variable so that it makes sense to give a probability distribution.

[With Problem 1.1.11, we introduce a new convention. The problem itself has no parts, but the Level I solution does. In Level I, parts (a), (b), (c), and (d) are introduced. The first three parts are intermediate steps; part (d) is the answer to the question.]

1.1.12 In the very beginning of the chapter we referred to ". . . some simple chance mechanisms—coin tosses, rolls of dice, and so on—and abstract models for those mechanisms."

(a) What's the technical term for such a chance mechanism?

(b) What's the technical term for the abstract models of those mechanisms?

1.2 Parameters to Characterize a Probability Distribution

The Expected Value or Mean of a Random Variable

In Problem 1.1.9, we asked about the number of dots to be expected on average for one roll of a fair die, or the number of heads on one toss of a coin. For a random variable, you would always want to know what value to expect in repetitions of the underlying experiment. You're asking for a kind of "average value" for the random variable.

Recall Problem 1.1.9(d) where you toss an unfair coin which comes up heads 90% of the time and you're paid \$2.00 for a head and \$3.00 for a tail. How much would you expect to take in per toss? If the coin were fair, you'd expect on average \$2.50 per roll. Half the time \$2.00, half the time \$3.00, so you split the difference. But with the unfair coin, you get \$2.00 nine times as often as \$3.00 and on average you'd expect to take in only \$2.10 per roll. To see this, you must *weight* the values according to how often they occur:

$$\underbrace{2.00 \times 0.9}_{\substack{90\% \text{ of} \\ \text{the time} \\ X=2}} + \underbrace{3.00 \times 0.1}_{\substack{10\% \text{ of} \\ \text{the time} \\ X=3}} = 1.80 + 0.30$$

And so, on average you expect to receive only \$2.10, forty cents less than for a fair coin. Here, of course, X is the amount of money you receive on one toss of the coin—it's a random variable which takes on the values "two" or "three," in units of "one dollar."

This weighted average is called the ***expected value of X***, denoted by E(*X*). In symbols,

$$\mathrm{E}(X) = \sum X\mathrm{P}(X).$$

The expected value of a random variable is computed by adding the values *weighted* by their theoretical relative frequencies of occurrence, that is, weighted by their probabilities.

The computation can be very efficiently done by extending the probability distribution table of X to include a column for the weighted values of X, for the products $X\mathrm{P}(X)$. Then the sum of that column is $\mathrm{E}(X)$, the expected value:

X	$\mathrm{P}(X)$	$X\mathrm{P}(X)$
2	0.9	1.8
3	0.1	0.3
	1.0	2.1

so, $\mathrm{E}(X) = 2.1$.

You'll often hear the expected value referred to as the ***mean*** of the random variable. In general, a mean is a kind of average value. There are several types of mean as we'll see in the next chapter. What most people refer to as the average of a set of test scores, for example, is technically the "arithmetic mean"—add the scores and divide by how many you have. Because both terms "expected value" and "mean" are used interchangeably for random variables, we'll also use both terms.

When we refer to the mean of a random variable, it's customary to use the symbol μ_X (this is read: "mu sub X"), or simply μ if X is understood and there's no danger of confusion. Thus, $E(X)$ and μ_X are just two different symbols for the same number. If you are thinking in terms of what you "should expect" on average, you might prefer the term "expected value" and the symbol $E(X)$. All this will come clear as you . . .

Try Your Hand

Some general advice: In solving problems, before doing any calculations you should ATTEMPT TO GUESS THE ANSWER. An exact guess may be impossible, but at least you can guess a ballpark figure. This guessing is important for two reasons: First, it helps you to avoid errors—an erroneous calculation may be *obviously* wrong from a commonsense point of view. Second, by guessing on the basis of intuition and comparing your guess with the correctly calculated answer, you train your intuition. And that, in turn, deepens your understanding.

Here's some more advice: This is easy to do, but students often get in trouble when they don't to do it. Any time you're working with a random variable, be sure you're clear about the *possible values* of that random variable. For example, in Problem 1.2.1(b), when you toss two coins, what are the possible number of heads?

1.2.1 Calculate the expected value, $E(X)$, for the following random variables. Do this as we did above, by extending the table for the probability distribution to include a column for the products $XP(X)$.

(a) The number of dots on the uppermost face for one roll of a fair die.

(b) The number of heads on one toss of a pair of fair coins.

(c) The number of dots on the uppermost face for one roll of a die which comes up five half the time, with the other faces equally likely.

(d) Your "take" when you toss a fair coin and you receive two dollars for tossing a head and three dollars for a tail.

1.2.2 For the game described, if you want to break even in the long run, what should you pay for one play (for one roll of the die or one toss of the pair of coins)?

(a) You receive one dollar for each dot on the uppermost face for one roll of a fair die.

(**b**) You receive one dollar for each head on one toss of a pair of fair coins.

(**c**) You receive one dollar for each dot on the uppermost face on one roll of a die which comes up five half the time, with the other faces equally likely.

The Variance, Measuring the Accuracy of the Mean

We now come to a very important aspect of chance phenomena which is often overlooked. Initially, we'll consider only simple games of chance, but remember: These ideas are relevant not only to games of chance, but to all chance phenomena and so are really quite important!

The expected value of a random variable is far from telling the whole story of what we "should expect." To see why, consider a die loaded to have the following probability distribution. This die has the same expected value as a fair die because it's loaded *symmetrically*. But the die is far from fair even though the expected value is the same as for a fair die!

X	$\mathrm{P}(X)$	$X\mathrm{P}(X)$	
1	0.3	0.3	
2	0.1	0.2	
3	0.1	0.3	
4	0.1	0.4	
5	0.1	0.5	
6	0.3	1.8	
	1.0	3.5	$\mu_X = 3.5.$

As you can see, the smallest and largest of the possible values, $X = 1$ and $X = 6$, carry most of the probability. There's a 60% chance of getting one of those two values: $\mathrm{P}(X = 1 \text{ or } 6) = 0.6$. This means there's a high probability of a roll with an outcome quite far from the expected value. By contrast, on a fair die there's roughly a 30% chance of rolling a one or six.

This example shows very clearly how much information can be missed by a single parameter for a mathematical model. Here, the parameter is the expected value. Let's define this term: a ***parameter*** is a fixed number, such as the expected value, associated with a mathematical model. It's contrasted with the term "variable." In the table above you see two variable quantities, X and $\mathrm{P}(X)$, and one parameter, μ_X. The parameter by definition is just a number, in this case it's 3.5.

The idea is to try to capture the model as far as possible in its parameters. We'd like to have a list of parameters that will serve as a numeric summary of the model. Clearly, the expected value by itself will not suffice. As we see in the loaded die above, the expected value misses an important characteristic of the model for this die. It misses the fact that the probability is "spread" to the extreme values of X. We need a parameter to capture this characteristic of the model, the "spread" or "dispersion" of the values from the mean.

The notion of spread or dispersion for a probability distribution is well illustrated by the following graphs:

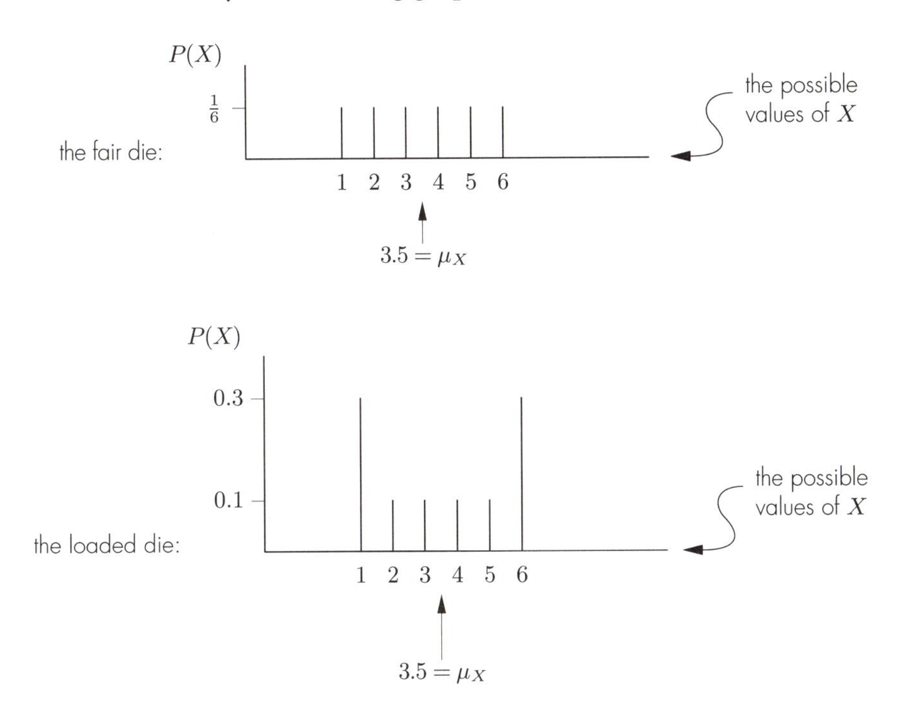

We'll discuss graphical presentations of distributions in more detail later, but as you can see, the graph for the loaded die shows at a glance that most of the probability for that die is concentrated at the extreme values of X. The distribution is "dispersed" away from its expected value, 3.5. We would like to capture this dispersion of the distribution in a numeric measure. In other words, we need a parameter to measure "spread about the mean" for a random variable.

The idea is to start with $\mathbf{X} - \mu$, the ***deviation of X from its mean.*** There's one deviation for each possible value of X. These deviations capture exactly what we wanted: the spread of X from the mean. But we wanted one number not many. Here, we have many numbers, one for each value of X. For the die, there are six deviations. One is -1.5.

Which value of X does it arise from? What are the other deviations for the die?

The natural thing is to combine these deviations from the mean into one number by just averaging. That means we would take the "average deviation from the mean" as our measure of spread. Don't forget that you must *weight* the numbers in an average according to the probability with which each occurs. So first weight the deviations: $(X - \mu)P(X)$. Then just add these weighted deviations to obtain $\Sigma(X - \mu)P(X)$, the average deviation. What is this for our weighted die? Well, this is the right idea alright, BUT IT DOESN'T WORK! WE'LL HAVE TO DO SOMETHING DIFFERENT. To see why it doesn't work, please . . .

Try Your Hand

1.2.3 (**a**) What's the probability of rolling a one or a six on a fair die?

(**b**) What's the probability of rolling a one or a six on the loaded die discussed in the text above?

(**c**) What's the point of this problem?

1.2.4 Compute the average deviation from the mean for

(**a**) the fair die,

(**b**) the loaded die given in the text above.

1.2.5 Suppose X is any random variable at all with mean μ.

(**a**) In constructing a parameter to measure "spread" or "dispersion" about the mean, what's the point of looking at the deviations from μ?

(**b**) What does it mean if $X - \mu$ is positive? Negative?

(**c**) Show that the average deviation from the mean for X is ZERO!

In Problem 1.2.5 you saw why the average deviation from the mean won't work as a measure of spread—you always get zero! The average of the deviations tells you nothing! What went wrong? The basic idea was right, but we inadvertently introduced a totally irrelevant consideration which vitiated our attempt to get a meaningful parameter.

This is typical of what may happen in trying to develop a mathematical model for real-world situations. The process of abstraction is a delicate one. In constructing your model, you want to "abstract" from the real-world situation everything which is relevant to your study and you want to omit everything irrelevant. But if you leave out too much detail, the model will be simplistic. If you include too much detail, your model will be too complex. You want to forget only what is IRRELEVANT to your question and keep what is relevant.

By looking at the average deviation from the mean, we've carried into the model an irrelevant consideration which, instead of making the model more complex, has introduced a trivialization: The number we thought would be a meaningful measure of dispersion about the mean turns out to be zero. What was the irrelevant consideration? It's just this: When we take the deviations $X - \mu$, we capture the idea of spread about the mean, but we also include the *direction* of spread. If a deviation $X - \mu$ is positive, that value of X is bigger than μ, and if the deviation is negative, that value of X is smaller than μ. We want to know only HOW FAR a value is from the mean. Whether it's above or below the mean is irrelevant to how far from the mean it may be.

There's more than one way to solve this problem. One NOT very common approach is to eliminate the irrelevant positive/negative consideration by taking absolute values. Then average the absolute values of the deviations. If you do that, you get the ***mean absolute deviation from*** $\boldsymbol{\mu}$: $\Sigma|X - \mu|\mathrm{P}(X)$.

However, the most common procedure is to square the deviations to remove the irrelevant positive/negative consideration. Instead of the average of the deviations, we take the average of the *squared* deviations from μ. This parameter is called the ***variance of the random variable*** $\mathbf{X}$ and is denoted by $\boldsymbol{\sigma^2_X}$. As with μ_X, the subscript can be dropped if the context makes it clear which random variable you're talking about. Here's the formula

$$\sigma^2_X = \Sigma(X - \mu)^2\mathrm{P}(X).$$

To calculate the variance of a random variable X, you extend the probability distribution table for X by putting in one more column, a column containing the weighted squared deviations from the mean, the products $(X - \mu)^2\mathrm{P}(X)$. Let's compute the variance for the loaded die we were discussing above:

X	$\mathrm{P}(X)$	$X\mathrm{P}(X)$	$(X - \mu)^2\mathrm{P}(X)$
1	0.3	0.3	1.875
2	0.1	0.2	0.225
3	0.1	0.3	0.025
4	0.1	0.4	0.025
5	0.1	0.5	0.225
6	0.3	1.8	1.875
	1.0	3.5	4.250

so, $\sigma^2 = 4.25$. As you see, this gives a variance of 4.25. When you compute the variance for the *fair* die, you'll find that it's 2.9167, reflecting the fact that the fair die is less spread about its mean than is the loaded die.

You may be wondering how we can justify squaring the deviations. After all, that significantly changes their values. True, but the variance is used only in a *comparative* way to see that one distribution is more dispersed or less dispersed about its mean than another one. Because we'll always use the same procedure—squaring the deviations—it's valid for comparative purposes.

So the variance by itself means nothing. We can't ask, for example, what it means intuitively for the fair die to have a variance of 2.9167! We CAN ask what it means that our loaded die has a variance bigger than that of the fair die. The larger variance for the loaded die means the distribution is more spread from its mean.

What does this mean in a practical sense? Simply this: The mean is the number of dots you expect on average. The loaded die is further from that, on average, than the fair die. To say it another way, because you would use μ to predict what "ought to happen," the loaded die is more unpredictable than the fair die. This unpredictability is reflected in the larger variance.

Finally, we need one more term: The ***standard deviation***, denoted by σ, is the square root of the variance. For the fair die, the variance is 2.9167 and so the standard deviation is $\sigma = 1.7078$. The standard deviation is not a different parameter; it measures exactly the same thing as the variance: spread, or dispersion about the mean. And it does it in exactly the same way as the variance. Given one of these two numbers, you can immediately calculate the other; so you learn nothing new.

Then why have two numbers at all? Purely for convenience. Square roots are algebraically a nuisance, and so, in computations, the variance is easier to work with. On the other hand, the *units* of the variance are squared. Therefore, in your final answer or in real-world discussions where the units may be mentioned, the standard deviation is better. After all, you don't usually talk about "squared dollars" or "squared cities"! For the fair die, you'd probably not feel completely comfortable talking about a variance of about three "squared dots"!

Well, now please . . .

Try Your Hand

1.2.6 In the text above we saw a loaded die that's less predictable than a fair die.

(a) Is it true that any loaded die will be less predictable than a fair die? Explain.

(b) Sketch a graph to illustrate part (a).

1.2.7 What about the unfair coin we looked at earlier which comes up heads 90% of the time? Compare it with a fair coin.

(a) Is the number of heads on the unfair coin more dispersed about its mean than on a fair coin, or less so? As always, first try to guess. Then compute the variance. Does this computed value confirm your guess?

(b) Draw and compare the graphs of the probability distributions for the number of heads on one toss of the fair coin and then of the unfair coin. Be sure to label the means.

(c) What's the standard deviation for each case in part (b)?

(d) Now let's be more general. Suppose we don't know the probability of heads on our coin. Denote that unknown probability by the symbol p. Let X be the number of heads for one toss of this coin. Derive a formula for the variance of X.

1.2.8 **(a)** Compute the variance for the random variables in Problem 1.1.5(a) and (b) and in Problem 1.1.6. Be sure to set up the complete probability distribution table.

(b) Compute the three mean absolute deviations for part (a).

(c) Compute the three standard deviations for part (a).

1.2.9 What are the mean and variance of a constant random variable? Be sure you first guess, then verify your guess with the formulas.

1.2.10 The variance may seem a bit abstract compared with the expected value. What's the PRACTICAL meaning of the variance? What does it mean in practical terms to say that a loaded die has a larger variance than a fair die, or that an unfair coin has a smaller variance than a fair coin?

The answer to this question may not be clear to you, but TRY ANYWAY! Think about the examples described in the text: What it would mean to you in practical terms that the loaded die has a probability distribution more dispersed about its mean than a fair die or that the unfair coin has a probability distribution less dispersed than that of a fair coin?

1.2.11 What's the difference between the variance and the standard deviation for a random variable?

1.2.12 We describe below three different dice, any one of which might be used for the following game: You'll be paid one dollar for each dot on the uppermost face after you roll the die. To roll the die once, you pay exactly your expected receipts, so that your expected gain/loss is zero. Here are the dice:

- all faces are equally likely;
- two dots come up 50% of the time and the other faces are equally likely;
- two dots come up 40% of the time, five dots 20% of the time, and the other faces are equally likely.

(**a**) Compare the predictability of your gain/loss for each of the three dice.

(**b**) In part (a), if the player breaks even in the long run, the gambling house won't be a profitable business! Suppose the house wants to make an average profit of 50 cents per play; what should they charge to play?

(**c**) In part (b), verify that the gain/loss is, indeed, a random variable.

1.3 Linear Functions of a Random Variable

It's very common to have a random variable determined through an equation involving another one which arises in a natural way from physical or real-world considerations. For example, the proportion of persons in a survey, all of whom share a common characteristic—who belong to a common ethnic group, for instance—is determined by the equation

$$\hat{p} = (1/n)X.$$

Here, $\hat{p}$ (it's called "p hat") is the standard notation for such a proportion and n represents the number of persons surveyed. X counts how many persons in the survey had the characteristic of interest.

To give another example, monetary amounts are often determined through an equation involving a random variable which describes some physical situation. To see in a very simple instance how this could occur, suppose you're betting on one roll of a die and you receive three dollars for each dot on the uppermost face. Then T, your "total receipts," is $T = 3X$, with X counting the number of dots on the uppermost face of the die. Note how X is determined physically and T is derived from that. Suppose you pay eight dollars per roll to play this game. Then Y, your total gain/loss, is given by the equation: $Y = 3X - 8$. So, for example, with one dot, you lose five dollars and with six dots you gain ten dollars. Would you play this game?

In a great variety of transactions like this one, your net result Y comes through a linear equation from a physically determined random variable X. A ***linear equation*** is an equation in which all variables appear added together with coefficients, but without any powers, square roots, and so on. For a linear equation, the standard notation is

$$Y = a + bX \qquad a,\ b \text{ constants,}$$

in the example

$$Y = -8 + 3X \qquad a = -8,\ b = 3.$$

Because linear functions play such an important role in statistics, we'll take a moment right now to see how they work. These fundamental equations will come up again and again in our work:

Fundamental equations for linearly related random variables:

If $Y = a + bX,\quad a, b$ constants then

$$\mu_Y = a + b\mu_X,$$
$$\sigma_Y^2 = b^2\sigma_X^2$$

To help you understand all this, please . . .

Try Your Hand

1.3.1 Think about the notation given in the text above for survey proportions. Show how that notation works: Suppose you surveyed 150 randomly chosen persons of whom 72% were Hispanic. Put those numbers into the equation

$$\hat{p} = (1/n)X.$$

1.3.2 You're paid one dollar for each dot on the top face of a die after one roll. To play—to roll the die once—you pay an amount equal to your expected receipts. Let X be the number of dots on the top face of the die after one roll and let G be the gain/loss random variable.

(**a**) Show that G is a linear function of X. That is, show that for some constants a and b, $G = a + bX$. Be sure you identify a and b clearly.

(**b**) Show that the variance of the gain/loss random variable is the same as the variance of X.

(**c**) Explain the result of part (b) on intuitive grounds.

(**d**) Now let's change the game and make it more realistic. Suppose the gambling house is making a profit of 50 cents per roll on average.

Express this new gain/loss random variable as a linear function of X and give its mean and the variance in terms of X.

1.3.3 Derive the "fundamental equations" for linear functions given in the text above. In other words, for any Y having the form $a + bX$, show that

(a) $\mu_Y = a + b\mu_X$.

(b) $\sigma_Y^2 = b^2\sigma_X^2$.

1.3.4 Based on data for the years 1919 to 1935, H.G. Wilm wanted to set up a model to predict April to July water yield (WY) in the Snake River watershed in Wyoming from the water content of snow (SC) on April 1. We'll study Wilm's data in Chapter 7. His data leads to some linear relationships. One analysis of the data suggests WY $= 0.7254 + 0.4981$ SC, measured in inches.

(a) The model gives WY as a linear function of SC. What are a and b for that model?

(b) Assuming the model, what was the average April to July water yield in the Snake River watershed over a ten-year period in which the water content of snow averaged 22.3 inches?

(c) What is the real-world meaning of a for the model?

(d) A more careful analysis of Wilm's data gives the model WY$= 0.52$ SC. Which model suggests more variability in April to July water yield?

(e) Why is the model in part (d) better than the original model?

1.3.5 Would you play the game described in the text where you receive three dollars for each dot on the uppermost face of the die after one roll and where you pay eight dollars per roll to play? Think about this carefully.

(a) Would you expect to win or lose in the long run? Would you play the game?

(b) Suppose on this die the face with one dot comes up half the time and all the other faces are equally likely. How risky is the game?

(c) In part (b), because half the probability for the die is concentrated on the single value "one dot," this game should be LESS risky than the same game with the fair die. But we showed that it's MORE risky! What's wrong?

1.4 The Fundamentals of Probability Theory

We've been using the intuitive notion of probability as captured in the idea of "long run relative frequency." A coin is fair if, "in the long run," heads should show uppermost half the time. This is justified if we think the coin is symmetric. Of course, no physical object is perfectly symmetrical, and for that reason, no physical coin is exactly fair. In fact, a physical coin does not even pretend to be symmetrical. The face is shaped differently from the tail. But such physical complications are beside the point! We know what we mean by fair and most coins, if they have not been damaged in some way, will be at least approximately fair.

In the previous paragraph, we've described two practically related but logically distinct ideas of probability: probability as "long-run relative frequency" and probability as determined by symmetry. There are other notions of probability. There is probability as "degree of rational belief," the degree of belief a rational person would invest in a given statement. For example, if storm clouds are gathering on the horizon, a rational person would invest little enthusiasm in a proposal for a picnic. Probability has also been thought of as a kind of continuous truth function, where every statement has a certain probabililty to be true. With perfect information, the probability is either zero or one, otherwise it's somewhere in between. In short, probability has been described in many diverse ways. Philosophically, it's a very thorny problem. Fortunately, this is not a philosophy course! Fortunately also, the development of statistics in an introductory course requires nothing of all this philosophical complexity.

Three Basic Rules of Probability

All definitions of probability agree on three rules which probabilities ought to obey. It's those rules which we need to know. To state the rules, we will think of probability in terms of "events." We've used this term before, but without giving a precise definition. An ***event*** is a set of possible outcomes of some random experiment. For example, think about event A, defined as "a number greater than two on one roll of a die." Here, A is the set of all outcomes for which a face with three or more dots comes uppermost. If the die is fair, then $\mathrm{P}(A) = 4/6 = 2/3$.

The probability rules are

1. $\mathrm{P}(\text{not } A) = 1 - \mathrm{P}(A)$,
2. $\mathrm{P}(A \text{ or } B) = \mathrm{P}(A) + \mathrm{P}(B) - \mathrm{P}(A \text{ and } B)$,
3. $\mathrm{P}(A \text{ and } B) = \mathrm{P}(A|B)\mathrm{P}(B)$.

In the third rule, you see ***the conditional probability of A given B***, denoted by the symbol **P(*A*|*B*)**. This is the probability that event A occurs, given that you know event B has already occurred. So the given condition B represents INFORMATION relevant to A. Because you now have more information, the conditional probability is often easier to understand than the unconditional probability. Note, by the way, because you know B has occurred, you also know that $P(B)$ is NOT zero! You cannot "condition" on an impossible event.

For example, suppose you draw two cards from a well-shuffled deck of 52 playing cards. Let H_2 be the event the second card is a heart. Then $P(H_2) = 1/4$. But that's not obvious! When you go for the second card, there are only 51 cards left and everything seems to depend on what the first card was. You can't calculate $P(H_2)$ because you don't know how many of the 51 cards are hearts. On the other hand, for event H_1 that the FIRST card is a heart, everything is obvious: $P(H_1) = 13/52 = 1/4$ and $P(H_2|H_1) = 12/51$. This last calculation, the conditional probability, is easy: You know the first card was a heart. When you go for the second card, there are 51 cards with 12 hearts left, giving a probability of 12/51.

From this, together with our three probability rules, we can calculate $P(H_2)$. First analyze H_2 into two separate events:

case I: H_2 and H_1,
case II: H_2 and "not H_1."

Note that H_2 is the event "case I or case II": Either both cards are hearts (case I) or only the second card is a heart, the first is not (case II). The event "not H_1" is called ***the complement of*** H_1, denoted by the symbol H_1^c. By Rule 3 we get

$$
\begin{aligned}
\text{P(case I)} &= P(H_2 \text{ and } H_1) = P(H_2|H_1)P(H_1) = (12/51) \times (13/52),\\
\text{P(case II)} &= P(H_2 \text{ and } H_1^c) = P(H_2|H_1^c)P(H_1^c) = (13/51) \times (39/52).
\end{aligned}
$$

Now compute $P(H_2)$. The trick here is that events "case I" and "case II" are ***mutually exclusive***; they cannot occur together. In terms of probability, to say A and B are mutually exclusive simply means their joint probability is 0: $P(A \text{ and } B) = 0$. So for us, P(I and II)= 0, and we get

$$
\begin{aligned}
P(H_2) &= \text{P(case I or case II)} \\
&= \text{P(case I)} + \text{P(case II)} && \text{by Rule 2}\\
&= 12/51 \times 13/52 + 13/51 \times 39/52 && \text{shown above}\\
&= 12/51 \times 1/4 + 1/51 \times 39/4 && \text{cancelling 13 into 52}
\end{aligned}
$$

$$= [(12/51) + (39/51)] \times (1/4)$$
$$= 1 \times (1/4)$$
$$= 1/4.$$

After all this, you may be surprised to see that $P(H_2) = P(H_1)$! It's an instance of a curious phenomenon in probability that allows you to "average across your ignorance"! That's possible if your "ignorance" can be broken down into several cases with known probability. So you're not completely ignorant. You don't know which case holds, but you do know the probability of each case. In our example, when we ask about the second card drawn with no information about the first card, the first card is your "ignoranace." First split H_2 into two cases according to your ignorance. That's (H_2 and H_1) and (H_2 and H_1^c). Then average, taking a WEIGHTED average, weighted according to your ignorance

$$\begin{aligned} P(H_2) &= P(\text{case I}) + P(\text{case II}) \\ &= P(H_2|H_1)P(H_1) + P(H_2|H_1^c)P(H_1^c). \end{aligned}$$

↑ ↑
weights, according to
your ignorance

Finally, sometimes the conditional probability simplifies. Suppose knowing that B has occurred does not affect the probability of A at all. In that case, $P(A|B) = P(A)$ and we say that A and B are ***independent events***. Note that on the draw of one card from a deck of 52, the event "ace" is independent of the event "club" because P(ace) is $4/52 = 1/13$ and so is P(ace|club). Because P(ace|club)=P(ace), the two events are independent.

Bayes' Theorem

A very important equation in probability theory is derived from our third probability rule. It's called Bayes' Theorem and is the beginning of "Bayesian Statistics," an entirely distinct and somewhat controversial approach to statistics. Thomas Bayes (d. 1761) was, in Stigler's words, ". . . a minor figure in the history of science whose published works show a spark of intelligence few of his contemporaries possessed." The theorem was not published until 1764, after Bayes' death, and did not receive any general recognition until about twenty years later. It's amazing that such a—to our present eyes—seemingly simple theorem could have had such a controversial history; not controversy about the theorem itself, but rather about the uses which have been made of it. That story goes far beyond the topic at hand into fascinating philosophical waters. We'll leave it to your investigation. But don't worry, in this text we'll keep away from any controversial uses! Recently, there has been a rebirth of Bayesian statistics with significant progress in understanding its proper use. It's becoming an important tool, for example, in business decision making. Here's the theorem

Bayes' Theorem:

$$P(A|B) = \frac{P(B|A)P(A)}{P(B)}$$

Note that Bayes' Theorem allows you to determine the conditional probability in the reverse order. For example, there's about a 24% chance on two draws from a deck of cards that the first card is a heart given that the second one is. The probability makes sense, but it can't be calculated directly from our rules. Try thinking about the second card as "affecting" the first draw! It's Bayes' Theorem that saves the day:

$$P(H_1|H_2) = \frac{P(H_2|H_1)P(H_1)}{P(H_2)} = \frac{(12/51) \times (1/4)}{(1/4)} = \frac{12}{51}.$$

Of course, it's very special to this example that $P(H_1) = P(H_2)$ so that they just cancel out. As you'll see in the exercises, it's not usually true that $P(A|B) = P(B|A)$.

Well, let's pause for a moment while you . . .

Try Your Hand

1.4.1 (a) What is P(ace|club) on one draw of a card from a well-shuffled deck of 52 playing cards? Do this from the definition of conditional probability.

(b) Now do part (a) using some of our three probability rules. Let A be the event that you draw an ace and C that you draw a club.

(c) How many outcomes are there for the event A of part (b)? For the event C?

(d) Give an example of two independent events. Verify your answer!

(e) Show that for independence, order does not matter. In other words, show that if A is independent of B, then B is independent of A.

(f) Show that the simple product rule P(A and B) $=$ P(A)P(B) is equivalent to independence.

(g) Does $\mathrm{P}(A|B) = \mathrm{P}(B|A)$ imply that A and B are independent?

(h) Show that $\mathrm{P}(A|B) = \mathrm{P}(B|A)$ is equivalent to $\mathrm{P}(A) = \mathrm{P}(B)$.

1.4.2 Before now we were using simple addition and multiplication rules for probabilities: "Or" means add. "And" means multiply. Of course, we warned you that these rules don't hold in complete generality. Let's explore these rules a bit. In what follows, let X be the number of dots on the top face of a fair die:

(a) Under what conditions do these simple rules hold?

(b) Which simple rule calculates $\mathrm{P}(2 \leq X \leq 5)$? Explain.

1.4.3 On one draw of a card from a well-shuffled deck of 52 playing cards, what's the probability that you do NOT draw an eight? Do this

(a) directly, by just counting,

(b) using one of our three probability rules.

1.4.4 Let D be the event a person has a certain disease. Let T be the event that a test for the disease is positive, indicating the person tested has the disease. Suppose studies of the test itself indicate that 99% of persons having the disease will test positive: $\mathrm{P}(T|D) = 0.99$. This is the "sensitivity" of the test. In addition, suppose at the same time, that 98% of unafflicted persons will test negative: $\mathrm{P}(T^c|D^c) = 0.98$. This is the "specificity" of the test. Finally, suppose this disease occurs in only 3 persons in 100, $\mathrm{P}(D) = 0.03$.

(a) What proportion of all persons tested would test positive?

(b) If you test postive, what are the chances you actually have the disease? This is the "predictive value of the test."

(c) What happens to the predictive value of the test as the disease becomes less common? Suppose, for example, $\mathrm{P}(D) = 0.003$.

(d) In 1987, a bill was introduced in the Senate of New York State stipulating that any screening test must "have a degree of accuracy of at least 95%" and "positive test results must then be confirmed by an independent test, using a fundamentally different method and having a degree of accuracy of 98%." Assume "accuracy" refers to both sensitivity and specificity. What would be the predictive value of the combined test, where a positive reading on the first test is confirmed on the second test? Assume $P(D) = 0.003$. (after [Finkelstein and Levin]).

(e) Show that the test and the disease are not independent. Of course that must be true, but can you show it analytically?

1.4.5 In 1980, after a series of airline hijackings in which the hijacker had passed through a magnetometer undetected with a plastic weapon, the federal government reinstituted a screening program that had been discontinued in 1973. The program attempted to identify potential airline highjackers on the basis of a behavioral "hijacker profile." One issue debated in the courts is whether such a profile gives "reasonable suspicion" to justify investigative detention. Does it? Let's look at an earlier case.

In a 1971 case (*United States* v. *Lopez*), the defendant was identified as fitting the hijacker profile and was subsequently arrested for possesion of heroin (no weapon, apparently). Lopez moved to suppress the evidence taken from his person. The court's decision reviewed a study of 500,000 passengers, 20 of whom were actually denied boarding.

Did the fact that Lopez fit the profile give "reasonable suspicion" to justify investigative detention? Suppose the sensitivity of the profile (see Problem 1.4.4) is actually 90% and the specificity 99.95%. Further, suppose we take the results of the court's study as valid in general. In other words, suppose 20 of any 500,000 passengers are carrying a weapon (after [Finkelstein and Levin]).

1.4.6 According to recent studies, lie detector tests have a sensitivity of 0.88 and a specificity of 0.86. How accurate is the test (after [Finkelstein and Levin])?

(a) Suppose one-fourth of all suspects will in fact lie.

(b) Suppose three-fourths of all suspects will in fact lie.

(c) What do parts (a) and (b) say about the predictive value of a screening test?

Random Experiments with Equally Likely Outcomes

To get some insight into our three probability rules, it's very helpful to take a careful look at a very special case: Distributions arising from experiments with equally likely outcomes. In that case, the probability of any event can be calculated just by counting. That was our procedure for analyzing the draw of a card from a deck of 52 playing cards. In fact, we've often drawn on our intuitive understanding of this principle of probability. Now we'll make it precise. Let's describe an outcome for the "draw one card" experiment as simply "one card." If the deck is well shuffled, each card is as likely to be drawn as any other and so the outcomes are "equally likely." Thus P(ace) = 4/52 because there are four aces among the 52 equally likely cards.

Suppose A is any event involving a random variable X for an experiment with equally likely outcomes. Recall that an event is a set of possible outcomes. Suppose there are N outcomes altogether and a of them comprise the event A, then $\mathrm{P}(A) = a/N$. Now you can see where our first probability rule comes from:

$$\mathrm{P}(\text{not } A) = (N - a)/N = N/N - a/N = 1 - a/N$$
$$= 1 - \mathrm{P}(A).$$

Note how this works for our "draw one card" example:

$$\mathrm{P}(\text{not ace}) = (52 - 4)/52 = 52/52 - 4/52 = 1 - \mathrm{P}(\text{ace})$$

The set of all possible outcomes of a random experiment is called the ***sample space*** of the experiment. We'll use the following notation: N is the number of all possible outcomes and $\#A$ is the number of outcomes which comprise an event A. Following normal usage in the English language, we say an event A "occurs" if we perform the experiment once and observe one of the outcomes comprising A. Otherwise, if we do the experiment and observe an outcome not among those comprising A, we say A did NOT occur.

Now for our second probability rule: Think how many outcomes comprise the event "A or B." In other words, what is $\#(A \text{ or } B)$? An outcome is part of this event if it's among the outcomes comprising A (A occurs) or if it's among those of B ("or" B occurs). So you might say, "Oh, it's easy, $\#(A \text{ or } B)$ is just $\#A + \#B$." Sorry, you fell into a trap! What about those outcomes which make up the event "A and B"? You counted them twice! Think of the event "ace or club." There are four aces and 13 clubs, but "ace or club" consists of 16 cards not $4+13 = 17$. It consists of 13 clubs (including one ace) and the three (not four) other aces. So $\#(A \text{ or } B)$ is 16, not 17.

In general, to get $\#(A \text{ or } B)$, count the number of outcomes in A, the number in B, and then SUBTRACT OUT the ones which were double counted, namely, the ones in "A and B":

$$\#(A \text{ or } B) = \#A + \#B - \#(A \text{ and } B).$$

That, in essence, is our second probability rule. Just divide by N,

$$\begin{aligned}
\mathrm{P}(A \text{ or } B) &= \frac{\#(A \text{ or } B)}{N} \\
&= \frac{\#A + \#B - \#(A \text{ and } B)}{N} \\
&= \frac{\#A}{N} + \frac{\#B}{N} - \frac{\#(A \text{ and } B)}{N} \\
&= \mathrm{P}(A) + \mathrm{P}(B) - \mathrm{P}(A \text{ and } B).
\end{aligned}$$

The third probability rule is obtained from $\mathrm{P}(A|B)$. With $\mathrm{P}(A|B)$, we know that B occurred. So the total number of possible outcomes has been reduced to $\#B$. Given that, what's the probability A occurred? Well, of course, we only think about those outcomes in A which are also in B because we know B occurred:

$$\mathrm{P}(A|B) = \frac{\#(A \text{ and } B)}{\#B}.$$

The rest is pure algebra. On the right-hand side, divide the numerator and denominator by N:

$$\begin{aligned}
\mathrm{P}(A|B) &= \frac{\#(A \text{ and } B)/N}{\#B/N} \\
&= \frac{\mathrm{P}(A \text{ and } B)}{\mathrm{P}(B)}.
\end{aligned}$$

Now multiply both sides by $\mathrm{P}(B)$ to get: $\mathrm{P}(A|B)\mathrm{P}(B) = \mathrm{P}(A \text{ and } B)$. This is nothing but our third probability rule!

So you see that our three rules hold in the special case of "equally likely outcomes." It's not so easy to prove that they hold in general, but they do. We'll ask you to believe that! Before we introduce more exercises, we'd like to tell you about a very surprising theorem.

Chebyshev's Theorem

We've said the standard deviation of a random variable is useful only as a comparative measure of spread, useful to compare the dispersion about the mean of one distribution with another. That statement cannot be maintained as unequivocally true. For instance, the standard deviation guarantees a minimum amount of spread in the sense that it's impossible for all the values of the random variable to fall strictly within one standard deviation of the mean.

For example, in a betting situation, if you have an expected loss of one dollar ($\mu = -1$) and a standard deviation for your gain/loss of two dollars, then to be "within one standard deviation of the mean" is to lose at most three dollars ($\mu - \sigma = -3$) and not more than one dollar ($\mu + \sigma$). But it's impossible your net gain/loss would stay within that range 100% of the time. It's impossible, in other words, that your net on any "go" of the game would always be within a loss of three dollars and a gain of one dollar. So knowing the standard deviation does, in fact, tell us a little something.

Chebyshev's Theorem helps the standard deviation to tell us a little more. It limits the chances for the values of the random variable to be more than k standard deviations from the mean:

Chebyshev's Theorem:

For any random variable X and positive integer k

$$P(|X - \mu| > k\sigma) \leq 1/k^2$$

or equivalently,

$$P(|X - \mu| \leq k\sigma) \geq 1 - 1/k^2$$

Well, now please . . .

Try Your Hand

1.4.7 Show that the condition in Chebyshev's Theorem, $|X - \mu| \leq k\sigma$, can be described verbally as "X is within k standard deviations of μ." Note how we use the phrase "within"—when we mean $|X - \mu| < k\sigma$, we'll say "STRICTLY within."

1.4.8 Chebyshev's Theorem is remarkable—we called it "surprising" earlier—because you do not have to know anything about the random variable. For any random variable at all:

(**a**) What's the probability of being within two standard deviations of the mean?

(**b**) What's the probability of being more than one and a half standard deviations away from the mean?

1.4.9 Suppose X counts the number of dots on the uppermost face of a fair die.

(a) What's the probability of being within two standard deviations of the mean?

(b) What's the probability of being more than one and a half standard deviations away from the mean?

1.4.10 Suppose X counts the number of dots on the uppermost face of a die for which the face with two dots comes uppermost half the time with all other faces equally likely.

(a) What's the probability of being within two standard deviations of the mean?

(b) What's the probability of being more than 1.1 standard deviations away from the mean?

(c) What does Chebyshev say about part (b)?

1.4.11 (a) Show that it's impossible for a random variable to have all its values strictly within one standard deviation of the mean.

(b) Show that if X counts the number of heads on one toss of a fair coin, then all the values of X are within one standard deviation of the mean.

(c) Why does Chebyshev's Theorem not apply to parts (a) and (b)?

1.4.12 Give an example of a random experiment for which the outcomes are NOT equally likely.

1.5 Some Review Exercises

1.5.1 (a) If a random variable is constant, what justifies the word "random"?

(b) Suppose X is any random variable whatsoever. Give a verbal description of ΣX, $\Sigma \mathrm{P}(X)$, $\Sigma X\mathrm{P}(X)$, $\Sigma(X-\mu)\mathrm{P}(X)$, $|X-\mu|$.

1.5.2 Suppose you toss a thumbtack over a table and, after it comes to rest on the top of the table, you assign U if the point of the tack points up and you assign D if it points down.

(a) Show that we have a random experiment here.

(b) Compute the mean and variance of

X	U	D
P(X)	0.3	0.7

(c) Suppose you receive 40 dollars for tossing the tack so that the point comes up and you lose 20 dollars if the point is down. Let $X = 1$ if the tack falls with the point up and let $X = 0$ otherwise. Write your gain/loss random variable as a function of X. Then compute the mean and variance of X and use that information to compute the mean and standard deviation of your gain/loss in this game.

(d) In part (c), you save a lot of work if you use the result of Problem 1.2.7(d). Show how.

(e) Would you play this game?

(f) Suppose you receive five dollars when the tack falls with the point up and lose five dollars otherwise. Which game would you prefer, this one or the one in part (c)?

1.5.3 The Scottish physicist James D. Forbes thought mountain climbers could avoid carrying the clumsy barometers of that day if they could determine altitude from the boiling point of water. He published data in 1857 which suggested the following model relating the boiling point (BP, degrees Fahrenheit) of water to barometric pressure (Pr, inches of mercury): $Pr = 0.5229\ BP - 81.0637$. We'll consider Forbes' data in more detail in Chapter 7.

(a) Describe this model.

(b) Suppose atmospheric pressure at a particular altitude over a period of three months as measured by a barometer averaged 26.7 inches of mercury. What does the model suggest as the average boiling point of water during that period of time?

(c) In part (b), would the boiling point of water be more or less variable than barometric pressure?

(d) The question in part (c) does not reflect an absolute characteristic of Pr and BP. It depends on the units of measurement. Explain.

1.5.4 (a) What are the two essential ingredients of any probability distribution?

(b) Name three modes of presentation for a probability distribution.

1.5.5 Consider the following probability distribution

X	P(X)
7	0.05
11	0.42
14	0.35
17	0.11
21	0.07
	1.00

(a) Complete the table with appropriate columns to compute the average of the deviations from the mean. Remember, it must turn out to be zero!

(b) Now set up a new table with appropriate columns to compute the mean and variance.

(c) What proportion of this distribution falls within one standard deviation of the mean? Think what "within" means: "Within one block of my house" means "one block in either direction."

(d) Make up a new random variable Y that's like X but has a smaller variance. By "like X" we mean Y has the same values and the same mean as X. Choose probabilities for Y so that σ_Y^2 is less than five. μ_Y need not be exactly the same as μ_X, but it should be close. Don't make any of the probabilities zero, that would amount to changing the possible values.

(e) What proportion of this distribution falls within one and a half standard deviations of the mean?

(f) This random variable is abstract in a very specific sense: Why do we say it's "abstract"? The word "abstract" is used here in the sense of "to abstract away from real-world complexities."

1.5.6 You are throwing darts at a dart board having a "bull's eye" within two concentric rings. The bull's eye is red, the innermost ring is blue, and the outer ring is white. The game is scored as follows: bull's eye ten points, blue ring five points, white ring three points. Further, there's a penalty of two points if the dart misses the dart board entirely.

(a) Show that "score" is a random variable.

(b) Set up an appropriate probability distribution and use it to compute the expected score and a measure of the predictability of the score.

(c) How would you measure an individual player's skill at this game?

(d) How would you measure an individual player's reliability at this game?

(e) If you want to compute the measures discussed in the previous two parts, what further information would you require?

(f) How many random variables are implicit in this game?

(g) Set up a probability distribution for your game and use it to compute your expected score and a measure of the predictability of your score. Assume you hit the bull's eye on average 65% of the time, the blue ring 23% of the time, the white ring 11% of the time, and that you miss the board 1% of the time.

(h) Now let's think about your opponent. We don't know her game. Make up probabilities for her game. Choose the probabilities so that her expected score is less than yours, but still her game is more exact. Make her "exactness" better by at least one point. Verify this.

(i) Who is more likely to be within one and a half standard deviations of their expected score, you or your opponent?

(j) Who is more likely to be within two standard deviations of their expected score, you or your opponent?

1.5.7 For a random variable X, suppose $P(X = 3) = 0.07$, what's the numeric value of $\Sigma_{X \neq 3} P(X)$?

1.5.8 Let X be the number of dots on the uppermost face of a die which shows four dots uppermost half the time with all other faces equally likely. Let Y be the number of dots on the hidden face of that die (the face on which the die comes to rest). Opposite faces of a die have seven dots total—did you know that?

(a) What's the random experiment for Y? Compare it with the experiment for X.

(b) Give the probability distribution of Y and use it to compute the mean and variance.

(c) What's the relationship between X and Y?

(d) Use the relationship in part (c) to obtain the mean and variance of X from that of Y.

1.5.9 Suppose one face of a die comes up half the time with all other faces equally likely. The mean and variance will depend, of course, on which face it is that comes up half the time. For both parts (a) and (b) below, try to guess first on intuitive grounds, then verify your guess by an appropriate calculation.

(a) Which face would have to come up half the time to yield the smallest variance?

(b) The largest variance?

(c) Is "the number of dots on the uppermost face" the same random variable for each of parts (a) and (b)? Or is it a different random variable each time?

1.5.10 For each of the following distributions, extend the distribution with appropriate columns and calculate the mean and standard deviation. Then specify what proportion of the distribution is within 1, 1.5, 2, and 2.8 standard deviations of the mean.

(a)

X	P(X)
22	0.13
23	0.62
24	0.09
25	0.16

(b)

X	P(X)
0.2	0.01
0.5	0.08
0.8	0.34
1.1	0.42
1.4	0.15

(c)

X	P(X)
1.7	0.22
1.8	0.17
1.9	0.14
2.0	0.17
2.1	0.12
2.2	0.18

1.5.11 For the distributions in parts (b) and (c) of the previous problem, you could avoid the decimal point nuisance if you multiply X by ten. To see how this would work, let $Y = 10X$. Set up distribution tables for the two Y's for parts (b) and (c) and use the table to calculate the mean and standard deviation of Y. Then use the fundamental equations for linearly related random variables to calculate the mean and variance of X from that of Y. Note that, of course, you get the same answer you calculated in the previous problem.

1.5.12 It's often possible to simplify calculations by a shift of the values such as you did in the previous problem. This is a common use of our "fundamental equations" for linearly related random variables. Define Y by the equation: $Y = 100X - 2{,}147{,}810$, where X is given in the table below:

X	$P(X)$
21,478.14	0.17
21,478.15	0.22
21,478.16	0.37
21,478.17	0.14
21,478.18	0.10

(a) Describe Y verbally in terms of X.

(b) With $Y = a + bX$, what are the numeric values of a and b?

(c) Set up a distribution table for Y and use it to calculate the mean and standard deviation of Y.

(d) Reason intuitively using parts (a) and (c) to get the mean and standard deviation of X.

(e) Write X as a linear function of Y and use that relationship to calculate the mean and variance of X.

Chapter

2 Understanding Observed Data

2.1 Observed Data from the Real World

Presenting Data Graphically

In Chapter 1, we concentrated on random variables, abstract models for the real world. By contrast, in this chapter, we'll consider observed data—observed in the real world of course, where else could you observe something? The contrast and interplay between real-world observations on the one hand and theoretical constructs on the other is one of the principal themes of this course. Before coming to the interplay of observation and theory, however, we need first to consider some of the ways of organizing, summarizing, and presenting observed data.

Graphs and charts are probably the easiest presentations of data to understand and for that reason they are found everywhere—in newspapers, magazines, advertisements, corporate reports, and so on. Because graphical displays have become such an important means of summarizing and communicating complex data, it's important to understand their use and misuse. Unfortunately, misuse of graphical presentations is not at all rare. Edward R. Tufte in his fascinating and informative book *The Visual Display of Quantitative Information* discusses this problem which he believes to arise in part from the lack of statistical experience among illustrators who are trained exclusively in the fine arts.

A great deal has been learned by psychologists about how the human eye perceives visual presentations of data and how the mind interprets these displays. We're all familiar with the optical illusions which result from the mind's interpretation of what is seen. For example, which of the following is the longer line:

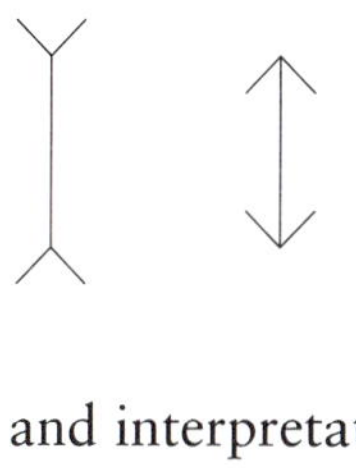

The process of perception and interpretation is largely subconscious and, consequently, the unwary reader is subject to deception by clever graphic manipulation. One of the most basic principles underlying graphic perception is the rule that

Numeric quantities are seen and interpreted
in terms of AREA not just height.

This principle is commonly abused in deceptive data displays:

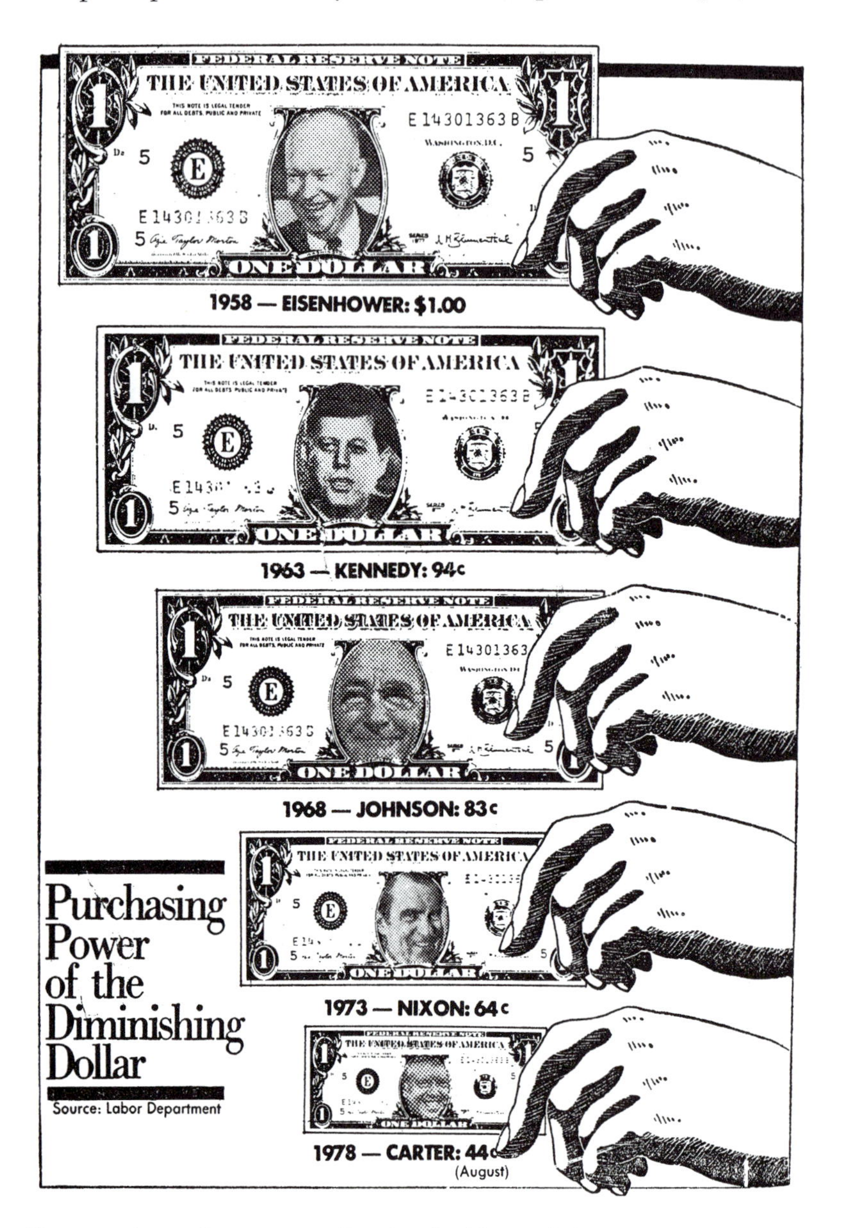

Here, the 1978 dollar is worth 44% of the 1958 dollar, but it's represented in the chart by less than 20% of the area of the 1958 dollar!

The height and width were both reduced to about 44% of the 1958 height and width, thus reducing the area to less than 20% of the 1958 area (measure them). You see the kind of exaggeration that results from ignoring this principle that "the eye judges by area." Sometimes this exaggeration is heightened by other deceptive devices. Note how the 1958 dollar at the top of the chart sits out in front of the top margin, giving it even further visual prominence and making the other dollars by contrast seem still smaller. Here's another chart on which you can . . .

Try Your Hand

2.1.1 When you look at the graph below, what initial impression do you get? Do you think "total budget expenditures and aid to localities" in New York State have increased significantly in "recent years" (that is, in the two or three years before 1977)?

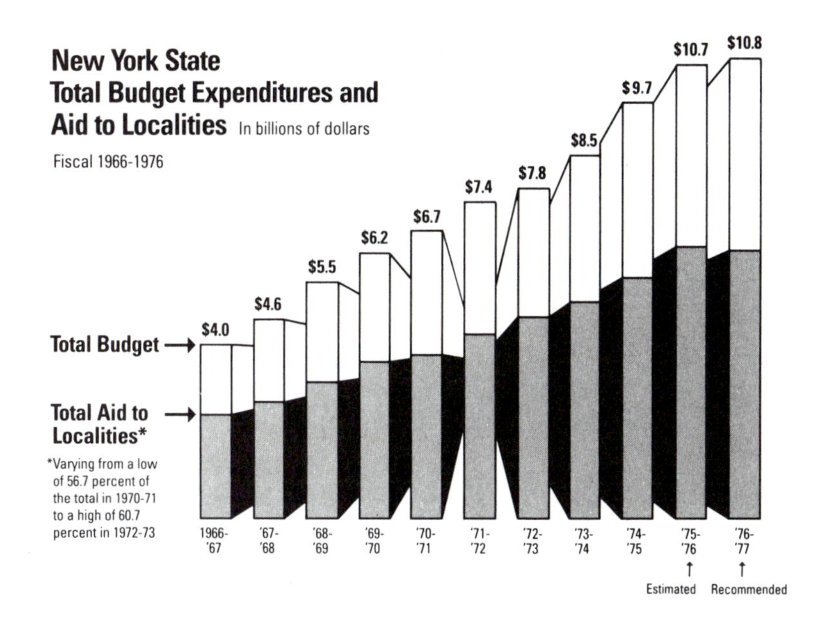

Collecting Data

The examples above should suffice to indicate the importance of clear and informative data presentations. In this chapter, we'll learn certain of the standard graphic and tabular methods for organizing and displaying data. So we leave aside theoretical considerations—random variables in particular—and concentrate our attention on observed data. First, we need to consider the question of data collection. Where does this data come from? Is it there, ready for your study? Or do you have the task of collecting the data yourself? Because the data-collection process can be very expensive and time-consuming, you may want to make use of existing data. In fact, as a practical, matter you may have no choice—funds may simply not be available for an expensive sample survey or other appropriate data-collection process. You may have to rely on data from some external source. Large databases from private and public agencies are coming more and more to be "on line" and available to anyone with access to a personal computer and modem. The federal government collects, organizes, and publishes an enormous amount of data through the census and other processes. For example, there's the *Statistical Abstract of the United States*, published annually by the Department of Commerce, and there are publications from the Department of Health, Education and Welfare such as *Vital and Health Statistics*. And many more. International organizations such as UNESCO and agencies of other governments also publish official data.

On the other hand, existing data relevant to the questions you're asking may not be available. Or it may not exist in a form relevant to your questions. Then you become involved in a major project—data collection. With the term "data collection," we're thinking, among other possibilities, of something like an opinion poll or some other kind of sample survey. Such a survey is an example of a statistical experiment. A ***statistical experiment*** is any random experiment which generates statistical data. A sample survey is just one example; there are others. To study the operation of a machine, or some more complex process, you may take a sample from a probability distribution which models the machine. To take a very simple example, you might take ten cups from a drink machine to study the operation of the fill mechanism of the machine. If you're trying to evaluate the effectiveness of several different fertilizers, you'll design a statistical experiment to test those "treatments" (fertilizers) under varying conditions. For instance, you may choose a number of different plots of ground across varying soil and climatic conditions and observe the yield of a number of different crops. These "crop yields" are then your data.

Note how complex the situation has become. The data will be useless if all you do is record numbers, the crop yields

$$12.3, \quad 317.4, \quad 3.7, \quad 0, \quad 63.2, \text{ and so on.}$$

Do these numbers refer to pounds, bushels, tons, . . . ? Which plot had ZERO yield? Why? If you got no yield at all, was it because of the soil? The climate? The fertilizer? These numbers just listed this way tell us nothing! You must organize the observations so that relevant aspects of the experiment are captured in your data presentation.

Furthermore, you must design the experiment carefully to make sure the questions you want answered can be analyzed on the basis of your data. In a sample survey, for example, if you don't frame the question for your respondents carefully, you'll have the right answer to the wrong question and the resulting data will be useless. Or worse, the data may seem quite clear in its implications and yet be WRONG!

Another problem is the failure to control for variables which might be confounded with the effect you want to study. This problem arises if your "effect" is not constant for some extraneous variable—if the groups or categories you're looking at are not homogeneous with regard to that "effect." When not controlled for, such a variable is said to be ***confounded*** with the effect under study. That means you can't separate out your effect from the effect of that other variable. Problems 2.1.2 and 2.1.3 provide a couple of simple but rather startling cases.

Try Your Hand

2.1.2 You are considering two treatments for a disease. You observed 390 patients, 160 of whom took the first treatment while the rest took the second treatment. Sixty of those who took the first treatment recovered, and 65 who took the second treatment recovered. Which treatment is better? [This is an example of Simpson's paradox.]

2.1.3 Your company has 100 employees, 50 men and 50 women. The average female employee earns $160,000 annually, while the average male employee earns only $140,000. Now this seems like a good company to work for; these are not bad salaries. But then, these figures suggest evidence of salary discrimination based on gender—men are making less on average than women. Comment.

2.1.4 Here's an example of how a statistical study can lead to gross misinterpretation if the authors of the study are ignorant of, or insensitive to, specific issues relevant to the study. It's one more example to illustrate that numbers out of context mean nothing. Statistical results require interpretation based on INFORMED judgment.

One widely reported and widely criticized conclusion of a 1991 federally funded study, the National Survey of Men (NSM),[1] claimed that 1.1% of men aged 20–39 in the United States were exclusively homosexual during the prior ten years. This is in stark contrast to the 10% figure—that 10% of the general population is homosexual—current since the Kinsey studies of the 1940s. In the NSM, respondents were asked a wide variety of questions concerning their sexual attitudes, behaviors, and relationships. The authors of the study claim that the "results presented here . . . can be generalized to the US population." Furthermore, the "privacy of the interview and the confidentiality of the information collected were stressed, and respondents were assured of anonymity." To facilitate follow-up studies at a later time, respondents were asked to provide full personal identification. In other words, for each respondent, the interviewer knew the home address, business or school address, social security number, and a reference to two persons—friends or relatives—who did not live with him.

(a) Of 3224 men who responded to this question on the survey, 2.3% reported some same-gender activity over the last ten years and 1.1% reported exclusively same-gender activity during that time. How would you interpret these percentages with respect to all men aged 20–39 in the United States?

(b) Suppose, in fact, that 10% of all gay men aged 20–39 in the United States are "out" enough to be willing to acknowledge their homosexuality to a stranger who knows their home address, work address, social security number, and so on. Based on the NSM, what percentage of U.S. males aged 20–39 would you think are homosexual?

(c) The NSM results break down by age as follows:

age	20–24	25–29	30–34	35–39
p	2.3%	1.2%	0.4%	0.7%

where p is the percentage who reported exclusively same-gender activity during the prior ten years. What do you make of these percentages?

(d) Of all men contacted for the NSM, 30% refused to participate. This

[1] The results of this study together with an account of the methodology is contained in five articles in the journal *Family Planning Perspectives*, March/April 1993. Our quotations are from these reports. The authors of the reports are John O.G. Billy, Koray Tanfer, William R. Grady, and Daniel H. Klepinger, all of the Battelle Human Affairs Research Centers in Seattle, Washington.

is the problem of "nonresponse" which plagues every statistical survey. How do you think this degree of nonresponse affects the validity of the findings?

(e) The nonresponse problem has been studied extensively and techniques have been devised to elicit truthful answers to sensitive questions. In 1965, S.L. Warner devised the "randomized response" technique whereby the respondent gets to choose between the survey question and a dummy question. Would you like to guess how this works?

The examples in Problems 2.1.2 and 2.1.3 pose a serious difficulty for any statistical study. Can you ever be sure you've taken ALL the relevant variables into consideration and controlled for them? Probably not. The best you can do is to control for those variables which *in your best judgment* you're able to identify as relevant. But there's always the possibility that time and further experience will turn up other variables which you overlooked, variables which, in fact, are confounded with the effect you're studying. This makes a very important point:

ONE CAN NEVER RELY ON NUMBERS TAKEN OUT OF CONTEXT

Numbers alone tell you nothing! Informed judgment is absolutely unavoidable in any statistical analysis. This goes contrary to the common ignorance which believes that statistics is "just a lot of numbers." It's impossible to say you've learned statistics if you haven't UNDERSTOOD what you've learned. Informed judgment, after all, can only be based on understanding.

Students are sometimes uneasy in a first statistics course, finding they always have to think their way through a problem. They feel something must be wrong because they can't "just solve" the problem. But "solving the problem without thinking" is a wrong approach to statistics. In real-world situations, it could lead to very wrong conclusions and expensive errors. Unfortunately, grossly misleading statistical "studies" are sometimes published. Often, it's not so much ignorance as an intent to mislead that produces such a smoke screen of disinformation. But without an understanding of fundamental principles, you'll never see through these smoke screens! The most fundamental goal of this course is to lay a foundation for that kind of understanding.

So we see in this section that designing a statistical experiment to generate data can be a major undertaking. Data collection can require a significant degree of expertise both in statistics and in the field under investigation. For this reason, the design and execution of a statistical experiment is often a team effort, involving persons with significant collective experience in statistics and in the field under study. It's a sophisticated process. Even a *first* course in experimental design requires a thorough grounding in the basics of statistics, a grounding such as you'll get from this text.

For that reason, the process of data collection is not treated in a first course such as ours. Throughout this course, we'll adopt the point of view that data appropriate to our problem is already available. When procedures for obtaining data go unmentioned, it's not because "Oh, you just go and get some data—everybody knows how to do that!," but rather because it's a topic too advanced for an introductory course. There's one exception to this rule: We will take a look at "simple random sampling." Still, simple random sampling, as opposed to more sophisticated sampling designs, is often not practical in real-world situations. We introduce it because it's a basic part of more complex experimental designs and because it provides some concrete experience with the process of sampling. We'll begin with simple random samples drawn from a probability distribution.

Simple Random Samples Drawn from a Probability Distribution

If you roll a die five times, let's say, and record the result as an ordered set of integers representing the number of dots on the top face of the die after each roll, for example, (4, 1, 4, 3, 3), then you've generated a simple random sample of size $n = 5$ from the probability distribution of the random variable X, where X is the number of dots on the uppermost face of the die after *one* roll. In general, suppose X is any random variable whatsoever, a ***simple random sample of size n from the probability distribution of X*** is an ordered set of n values of X obtained from n independent repetitions of the random experiment for X. The sample is "ordered" because you record the values of X in the order they're generated. We used parentheses for the sample instead of the usual set notation to indicate the set is "ordered."

Independence is a key assumption here and is often problematic. It means that any repetition of the experiment should be unaffected by previous executions. In particular, the probabilities for the distribution should not be affected by doing the experiment. In a sense, it shouldn't be necessary to say this. After all, if the probabilities change, you aren't repeating the "same" experiment. The independence condition holds for rolls of a die, of course, assuming you believe the die is not physically altered by rolling. Let's think about this kind of sampling.

Try Your Hand

2.1.5 We'll begin with the simplest example of all. Let X be the random variable which counts the number of heads for one toss of a coin. Just to be concrete, let's suppose heads comes up 70% of the time.

(a) Verify that three tosses of this coin generate a simple random sample from the distribution of X.

(b) Write out all the possible random samples of size $n = 3$ for simple random sampling from the distribution of X.

(c) How many samples of size 30 are there?

(d) What's the probability of a sample of size four for which $\Sigma X = 4$?

(e) What's the probability of a sample of size four for which $\Sigma X = 2$?

(f) What's the probability of a sample of size four having an average of "half a head" per toss? Note that one toss has either zero or one head, but the average for several tosses can be a fraction between zero and one.

2.1.6 Suppose you're rolling a die for which the face with five dots comes up half the time and all other faces are equally likely. As usual, X is the number of dots on the uppermost face after one roll.

(a) Show that a simple random sample from the distribution of X is generated by 10 rolls of this die.

(b) List all the possible simple random samples of size two from the distribution of X.

(c) How many simple random samples of size ten are there?

(d) What's the probability of a simple random sample of size three for which $\Sigma X = 4$?

(e) What's the probability of a sample of size three where you observe $1\frac{1}{3}$ dots on average per roll?

2.1.7 Suppose we have an industrial process that produces an item to specification. Let's say it's a machine part which is to be 2.5 cm in diameter. Because no physical process can be exact, each part will be off slightly from the exact specification. This is called "specification error."

(a) Show that specification error is a random variable.

(b) Show that the "next five parts" to be produced generate a simple random sample from the distribution of specification error.

(c) What might be a practical use of the sample in part (b)?

2.1.8 Another important kind of error arises in repeated measurement of some object or situation: the slight error that's always present when one makes repeated measurements. "Measurement error" is an important phenomenon which can be modeled by statistical techniques; there are entire books on the subject! Safeguards of radioactive material often involve periodic (repeated) measurement to assure that no material has been diverted, for example, to a hostile foreign power. The Bureau of Standards in Washington makes repeated measurements of its standard weights, measuring rods, and so on. Three different IQ tests taken on three different occasions could be regarded as repeated measurements of your IQ. In repeated measurement, measurement error is always present because of limitations in the accuracy of the measuring devices and because of limitations in the accuracy of observation or reading of those devices by the person doing the measurement.

(a) Show that measurement error is a random variable.

(b) Show that the "next five measurements" generate a simple random sample from the distribution of measurement error.

(c) How is "specification error" different from "measurement error"?

2.1.9 Here's another way simple random sampling from a probability distribution arises in industrial quality control. Suppose a manufacturing process turns out lots consisting of 500 silicon wafers. In addition, suppose 3% on average of all wafers being produced are defective. We're concerned with the percentage defective per lot. Let X denote the random variable of part (a).

(a) What's the appropriate random variable X for studying this manufacturing process?

(b) What are the mean and variance of the random variable X?

(c) Show that a lot generates a simple random sample from the distribution of X.

(d) Express the percent defective per lot in terms of X.

(e) Show that "percent defective per lot" is a random variable.

2.1.10 A disease with complex causes whose etiology is not well understood can be regarded as a random mechanism and studied with the powerful tools of statistical analysis. Suppose, let's say, a child is judged to have an 8% chance of contracting the disease. Then, trying to determine whether a particular child will contract the disease is like tossing a coin for which there's a probability $p = 0.08$ of heads.

(a) What's the appropriate random variable for studying this disease—the random variable that's comparable to counting the number of heads on one toss of a coin?

(b) Suppose you're interested in the incidence of this disease among children in a particular neighborhood of your city. Show that the children of that neighborhood can be thought of as a "simple random sample" from the distribution of the random variable of part (a).

(c) Show that the percentage of children in that particular neighborhood who contract the disease is one value of a random variable.

(d) How might you use the "sample" in part (b)?

(e) Give a symbol involving the random variable of part (a) that expresses "incidence of this disease" in that neighborhood.

Populations

Sampling from a probability distribution is just one form of simple random sampling. Another is to draw a sample from a population. This is familiar to you from survey sampling—opinion polls, for example, or surveys of voters, and so on. Before we discuss sampling from populations however, we need to look first at populations themselves.

Many statistical questions are questions or conjectures about some underlying population. Here are some examples of populations:

- registered voters in San Francisco
- a given day's output from a production line
- the San Francisco State University student body
- scores on the SAT test given on a particular date
- airplanes which are currently under the jurisdiction of the Federal Aviation Administration
- persons exhibiting a certain clinical symptom of glaucoma

As you can see, the term "population" does not necessarily refer to a population of persons. We can have any kind of objects whatsoever. Of the populations listed above, only three are populations of persons. Note that the fourth, SAT scores, may be a ***numeric population***, a population of numbers where it is the values of the numbers themselves that is of interest. That example would NOT be a numeric population if you were only interested, say, in scores above 1100 as compared with those below.

The exact specification of a population will depend on the question you want to ask. It's important that the population be precisely defined. A population is not well defined unless it's entirely and unambiguously clear which objects are and which are not members of the population. Thus, the first example is not well defined until you specify a particular time—voters registered by 5:00 P.M. on such and such a day, for instance. The population of registered voters, after all, changes from day to day up to the deadline for registration. Or should we say it changes from hour to hour, or even minute to minute? You must be very specific about such details.

Note also that a population may be well defined even though it's not easily accessible or known in any detail. In the last example given above, if the clinical symptom of glaucoma in question is clearly specified, we have a well-defined population—those persons who exhibit that symptom. Still, we may not know such details as their average age or weight or even how many persons are in the population. In fact, these might be exactly the questions we need to answer.

So, an exact specification of the population as determined by the question of interest is important. In the first population listed above, for example, you may be concerned only with registered voters for the upcoming election who do, in fact, vote. This partitions the population of registered voters into two categories—into a dichotomous population—depending on whether the person finally votes or not. A ***dichotomous population*** is a population each member of which either does or does not have some characteristic of interest—in our example, every registered voter either will or will not vote. Or, you may be interested in which candidate a person ultimately votes for. If so, you're not really interested in the population specified, but rather the subpopulation of those registered voters who do actually vote. This population, a subpopulation of the larger population of all registered voters, is dichotomous only if there are exactly two candidates. Otherwise, this subpopulation splits into several categories—one category for each candidate. A category for a particular candidate would consist of those registered voters who vote for that candidate.

In the second population above, suppose you're interested only in whether an item from the production line is defective. Again, you have a dichotomous population. If you're interested in the length of the items or in their diameters or some other numeric quantity, then you'll consider it to be a NUMERIC population with no question of categories. Unless, of course, you're not really interested in the numbers themselves but only, for example, in whether an item exceeds some limit, is "too long," say, or "too small." In that case, the population of numbers is again a DICHOTOMOUS population.

The third population listed above might restrict to the subpopulation of undergraduate, full-time, regularly enrolled students and split into four categories depending on the student's class standing. Or it might be dichotomous if you're concerned about "first year" versus "not first year." The fourth example is dichotomous if you are interested only in SAT scores above 1100 versus those below. The fifth example would become a numeric population if you're only interested in the age of an airplane as measured by the number of flight hours. It would become a dichotomous population if there is some standard which each plane may or may not meet, such as "less than a hundred hours of flight time."

Statistical Questions

Here are some examples of the types of statistical questions concerning populations which one might ask. We give one example for each of the populations given in the list of the previous section:

- What proportion of registered voters will vote for our candidate in the upcoming election?
- What's the average life of the electronic components which we manufacture?
- What's the average age of the student body at SFSU?
- What proportion of our students have SAT scores above 1000?
- What's the average age of airplanes under the jurisdiction of the Federal Aviation Administration?
- What proportion of patients exhibiting this clinical symptom of glaucoma will respond to treatment?

Each of these questions asks for a mean or a proportion. In other words, the question asks for the value of a population parameter. A ***population parameter*** is a fixed number associated with a population. This is parallel to the definition of the term "parameter" from the previous chapter as a fixed number for a mathematical model. Another class of typical questions asks for the difference in two means or two proportions: "What's the value of the parameter for this *pair* of populations?" The difference in two population means (or proportions) is a *fixed number* and so it's, indeed, a parameter for the pair of populations.

Note that to ask about a mean, you must have a numeric population, you must have numbers to be "averaged up." To ask about a proportion, you must have a dichotomous population. Only then can you make sense of the "proportion having the characteristic of interest." Be sure you're clear about this distinction between means of numeric populations on the one hand and proportions for dichotomous populations on the other—we'll come on it again and again.

There are many other types of statistical questions one might ask concerning a population. For example, how variable is the population? Here we'll be asking for the value of a parameter which measures variability. Is the quality of a manufacturer's product highly reliable, or is there a great deal of variation in quality? If you're considering two suppliers for an electronic component, where the mean life of the components of each supplier is the same, you might want to consider the variability of the lifetimes. Suppose the components from both suppliers have an average life of 1200 hours, but for one supplier 5% of the components burn out too early and for the other 15%. Which supplier would you prefer? Obviously the first, the "expected lifetime" (1200 hours) is the same, but the *reliability* of that expected lifetime is greater for the first supplier.

Statistics becomes relevant when you can't answer your question directly from the population. If you can afford to interview every voter or examine every item from a production line, you'll give an exact answer to your question with no recourse to statistics. However, populations are typically NOT accessible, either because of cost or for some other practical reason. Cost certainly prevents your interviewing every registered voter in an election. On the other hand, to take only one example, inspecting items from a production line often involves destruction of the item, you test the life of an electronic component by burning it until it burns out or you test the strength of a seal by putting stress on it until it breaks. So, to inspect every item means to destroy your entire inventory. You'll not get a promotion for that!

In cases such as these where the population is not accessible, we attempt to answer our question on the basis of a random sample of the population. On the basis of that one sample, we'll attempt to speak for the entire population. The naive idea is to obtain a sample which is "representative" so that the answer from our sample will be equally valid for the whole population. But how do we obtain a representative sample? And how can we be sure our sample is, in fact, representative? After all, to be sure the sample is representative, we must already know the population. Or at least it must be completely accessible so we CAN know it. Otherwise, how could we compare the sample with the population to say that it's representative? So, we seem to travel in a circle and come right back to our starting point

THE POPULATION IS UNKNOWN; we need to know it; we take a sample: Is the sample representative? To answer that we need to know the population. But THE POPULATION IS UNKNOWN!

Back to square one, the circle is complete.

Well, it's a *theory* of sampling that's required. It's the theory of random sampling which provides specific, very powerful techniques which break through this vicious circle. Using these techniques, we can obtain answers to many questions about unknown and inaccessible populations. But don't be misled. The answers are not as simple as the questions. A major part of our course will be focused on what this theory of random sampling says and its very important and powerful applications to concrete real-world problems. It will be interesting for you to see how such an abstract theory meets the challenge to say something valid about an entire population on the basis of a very much smaller sample, especially because any sample, no matter how carefully chosen, may fail to be representative of the parent population.

This last observation is a basic fact which is often forgotten, so let's highlight it now:

NO MATTER HOW YOU CHOOSE YOUR SAMPLE, IT COULD END UP BEING QUITE ATYPICAL OF THE PARENT POPULATION.

And furthermore . . .

YOU'LL NEVER KNOW WHETHER IT'S TYPICAL OR NOT!

If this makes the situation look entirely hopeless, GOOD! That means you see the problem. So you'll genuinely appreciate the power of the statistical theory which we develop in this text and which brings the situation under control.

In fact, the situation is not hopeless at all, but it does require an appropriate tool—the theory of random sampling. In keeping with the spirit of this text, we'll not develop the theory in a rigorous way. Rather we'll see what the theory says for a typical special case—simple random sampling—and then focus our attention on how the theory works in solving real-world problems. But for now, we need to learn the terminology and notation for samples and their populations. And we need to learn ways of organizing, summarizing, and presenting such data. The theory we'll leave to a later chapter. Let's pause for a moment while you . . .

Try Your Hand

2.1.11 Whether a particular population is "numeric" or "dichotomous" depends on the question being asked. For each of the six statistical questions listed in the text:

(a) Identify the underlying population as either dichotomous or numeric. If the population is dichotomous, identify the "characteristic of interest."

(b) Make up a different statistical question which would require thinking of the population as of the other type. If the question in the text required the population to be numeric, your question should require that it be dichotomous and vice versa.

2.1.12 In the example from the text where we were considering two potential suppliers for an electronic component, we said "suppose the components from both suppliers have an average life of 1200, hours but for one supplier 5% of the components burn out too early and for the other 15%." Explain why the criterion for "too early" cannot mean "burns out before 1200 hours."

Simple Random Samples Drawn from a Population

A ***sample of size n*** from a population is just a subset of the population, a selection of some n members of the population. A ***random sampling experiment*** is a random experiment—just as we defined that term in the previous chapter—which produces a sample as outcome. The sample is called a ***random sample*** because, as outcome of a random experiment, you cannot predict in advance which sample you'll get. More complex sampling designs allow for variable sample sizes, but we'll always assume a FIXED SAMPLE SIZE on repetitions of the experiment. So, n is not variable here, rather it's a parameter for the sampling experiment.

Let's verify that the deal of a five-card hand from a well-shuffled deck of 52 playing cards is, indeed, a random sampling experiment. The "doing" for the random experiment is to take the top five cards from the deck. Obviously, that's repeatable (be sure you replace the first hand you dealt and reshuffle the deck). This "doing" produces a five-card hand as outcome. Now, a "hand" is just a subset of the entire deck; it's a sample of size $n = 5$. Because the deck is well shuffled, you cannot predict in advance what hand you'll get, verifying that we have a *random* experiment. It's a random experiment which produces a sample as outcome. So, the deal is, indeed, a random *sampling* experiment from the whole deck of 52 cards as population, for which the parameter n takes on the value five.

The deal of a hand from a deck of playing cards is the prototypical example of simple random sampling from a population:

A simple random sample from a population . . .

> is a random sample selected so that whenever you select a sample element, each member of the population available for selection has an equal chance to be drawn next.

One can show that for simple random sampling from a population, any two samples have the same probability of being selected. By contrast, this is not generally true for simple random sampling from a probability distribution.

In dealing a five-card hand, once you have taken, say, the top two cards, there are 50 left. But if the deck is really well shuffled, all of those 50 cards have the same chance to be on top, namely, one chance in 50. In other words, as you prepare to draw the third card for your sample, each of the remaining cards has an equal chance to be drawn, satisfying the definition above for a simple random sample from a population. So, as you deal the cards one by one, you're selecting a simple random sample. Sampling in this manner is called ***sampling without replacement***.

By contrast, suppose each time you deal a card you record its value, then *replace* the card into the deck, shuffle the deck many times and deal again. When you repeat this process five times, you're still selecting a simple random sample of five cards (verify!). But this time you're doing ***sampling with replacement***: Each element of the sample is selected at random from the full original population. Sampling with replacement is often much more convenient than sampling without replacement. Imagine you're going to select a sample of 1000 voters. What a nuisance if you have to select names from a list one by one, making sure you don't select the same name twice. After all, the chances of selecting the same name twice are very small, so small as to be negligible. Of course, in this case, you'd want to sample WITH replacement. On the other hand, in many real-world situations sampling with replacement is absurd. If you want a sample of items from your production line to check quality, you certainly don't want to select the same item twice. In short, we need to allow for both types of simple random sampling.

Any random sampling experiment—simple or not—must be determined by some random mechanism to guarantee the randomness of the samples (the outcomes). In our card-playing example, the random mechanism is the shuffling of the deck. But shuffling is very artificial for most real-world situations. How do you shuffle registered voters?

Even for a deck of cards, shuffling may be an imperfect random mechanism. If you do it exactly—divide the deck exactly in half and reconstitute the deck card by card, alternately taking cards from each half so the top card goes to second place—the result is not random at all. Fifty-two such perfect shuffles returns the deck exactly to its original order (surprisingly, if the top card remains on top, only eight shuffles are required). If the deck is only approximately divided in half and recombined by cards falling randomly from each half—presumably that's how most card players shuffle—you'll get something like a random mix. But not until you've shuffled about seven times; fewer shuffles will probably leave you with a very nonrandom mix. These results are not obvious. The mathematics of card shuffling, which has been examined extensively by the Harvard University statistician Persi Diaconis, is not trivial. Most card players are accustomed to playing with poorly shuffled decks because they usually shuffle only a few times. This fact[2] resulted in much consternation among contract bridge players when computerized dealing was introduced (the computer "deals" from an idealized well-shuffled deck). Experienced bridge players were sure the odd-ball hands they got were somehow bogus and they blamed it on the computer. In fact, it's just that they had become accustomed to hands conditioned by the previous game. In all of their prior experience, decks were not properly shuffled and the cards reflected the order at the end of the previous game with cards of the same suit grouped together.

Possibly the most common random mechanism for generating samples is a random number table or its computerized equivalent, a random number generator. A ***random number generator*** is a random experiment for which an outcome is a number (with a fixed length, five digits long, say, or maybe 50 digits long) where all the numbers which can be generated are equally likely to occur. The numbers in a table or from a computer are technically called "pseudorandom" because they're generated by a deterministic rather than a random process. It's a very fundamental problem, still the subject of active research, to say exactly what the word "random" means and to specify a mechanism for generating truly random numbers. It may be impossible. Then we need some reasonable approximation. Certainly, randomness should imply the absence of any systematic pattern. But recently, sophisticated analysis has shown that many of the most frequently used random number generators (in fact, *pseudo*random number generators) produce sequences which contain subtly systematic patterns.[3]

[2] The discussion in this and the next three paragraphs is largely based on the very interesting *New York Times* article "The Quest for True Randomness Finally Appears Successful," 19 April 1988, p. 35.

[3] See the *New York Times* article referred to above which has interesting illustrations and a discussion of how these patterns have been discovered. Also see Ivars Peterson's article "Monte Carlo Physics: A Cautionary Lesson," *Science News*, December 1992.

One of the worst ways to generate "random" numbers is to ask somebody to write down some numbers "at random." It won't work, even if you make the process abstract by asking for a random series of zeros and ones which you could then interpret as a base two number. In a truly random sequence of, say, 100 zeros and ones, you're likely to find several strings five or six digits long of all zeros (or all ones). The term "random" here means the sequence of digits is like a sequence of zeros and ones from tosses of a fair coin where on each toss you record the number (zero or one) of heads. But psychologists have found that people rarely repeat the same digit more than four times in such sequences of zeros and ones, not five or six times as would be required. The human mind is built for patterns; it doesn't like boring repetitions.

Extensive experience has shown that any element of free human choice in situations where a *random* choice is required can result in important biases, seriously compromising the results of the study. In the previous chapter, we mentioned one instance of this, the 1970 draft lottery [see Problem 1.1.7(c), Level II]. There are numerous others. In the 1940 draft lottery, instead of the 366 capsules with birthdays as in the 1970 lottery, they had 9000 numbers in capsules which they attempted to stir into a random mix in a "fish bowl." It was chaos. In the 1948 presidential election, three major polls, Gallup, Roper for *Fortune* magazine, and Crossley for the Hearst newspaper group, predicted Thomas Dewey to be the winner over Harry Truman. They all were wrong by a significant margin. All of these polls used a sampling design called "quota sampling" which leaves a margin of choice for the interviewers. In ***quota sampling***, one determines the proportion of the population having various characteristics of importance to the question under study and then chooses a sample having those characteristics in the same proportion as the population. But the method of choice leaves room for human judgment; it's not random! For demographic and sociological reasons, this led to a bias in favor of Republican voters. For a very interesting and informative elementary account of the 1948 election and other issues which arise in survey sampling, see Chapter 19 of Freedman, Pisani, and Purves. Since the 1948 fiasco, quota sampling is no longer used by major polling organizations.

Because an element of human choice is involved in quota sampling, it's NOT a case of random sampling. Two very commonly used random sampling designs which we'll only mention here are stratified random sampling and cluster sampling. In ***stratified random sampling***, the population is divided into strata which are quite different with regard to some characteristic, whereas within a stratum there's relatively less variation with regard to that characteristic. For example, with a physical characteristic of persons such as height, you might want to stratify by gender, giving two strata, one for women and one for men. Then within each stratum, you choose a simple random sample. These simple random samples together, one for each stratum, make up your stratified random sample. This method of sampling can offer greater precision than the simple random sampling.

In ***cluster sampling***, you divide the population into many relatively small groups called "clusters"—for example, a city block—then you choose a simple random sample of clusters. Your cluster sample consists of all members of the population contained in the clusters you've chosen. In the example, you'll have randomly chosen, say, 100 city blocks across the city and then you interview all the households within those100 blocks.[4] This method of sampling can be much more economical than simple random sampling. After all, once you're on the block, you might as well interview everyone there. Both stratified random sampling and cluster sampling ARE cases of random sampling.

To get some idea of the complexity of sampling plans, look at the following description from the *New York Times* (1988) of a telephone survey:

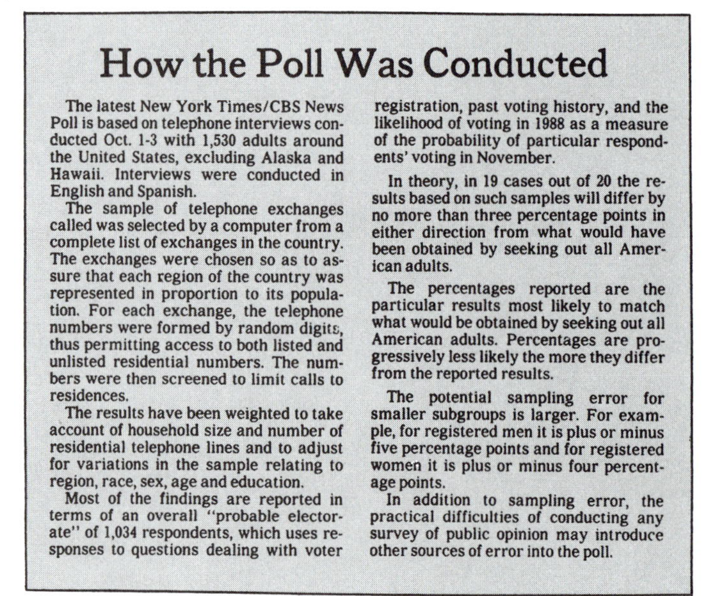

How the Poll Was Conducted

The latest New York Times/CBS News Poll is based on telephone interviews conducted Oct. 1-3 with 1,530 adults around the United States, excluding Alaska and Hawaii. Interviews were conducted in English and Spanish.

The sample of telephone exchanges called was selected by a computer from a complete list of exchanges in the country. The exchanges were chosen so as to assure that each region of the country was represented in proportion to its population. For each exchange, the telephone numbers were formed by random digits, thus permitting access to both listed and unlisted residential numbers. The numbers were then screened to limit calls to residences.

The results have been weighted to take account of household size and number of residential telephone lines and to adjust for variations in the sample relating to region, race, sex, age and education.

Most of the findings are reported in terms of an overall "probable electorate" of 1,034 respondents, which uses responses to questions dealing with voter registration, past voting history, and the likelihood of voting in 1988 as a measure of the probability of particular respondents' voting in November.

In theory, in 19 cases out of 20 the results based on such samples will differ by no more than three percentage points in either direction from what would have been obtained by seeking out all American adults.

The percentages reported are the particular results most likely to match what would be obtained by seeking out all American adults. Percentages are progressively less likely the more they differ from the reported results.

The potential sampling error for smaller subgroups is larger. For example, for registered men it is plus or minus five percentage points and for registered women it is plus or minus four percentage points.

In addition to sampling error, the practical difficulties of conducting any survey of public opinion may introduce other sources of error into the poll.

Now, think a bit about this discussion of random sampling . . .

[4] It must NOT be left up to the interviewer to decide which member of the household to interview; good experimental design requires that you give precise instructions about which person in the household is interviewed, and precise instructions about alternates if that person is not available. The principle: No element of human choice!

Try Your Hand

2.1.13 Let's explore "simple random sampling from a population."

(a) Sampling with replacement does, indeed, satisfy the definition of simple random sampling from a population. Let's take a concrete example: Show that drawing five cards from a deck of 52 with replacement is a case of "simple random sampling from a population."

(b) Show that sampling with replacement from a numeric population is a special case of simple random sampling from a probability distribution.

(c) Show that any two samples from a population have the same probability of being drawn.

(d) Let X be the number of dots on the top face of a die. By contrast with part (c), show that for simple random sampling from the distribution of X, the condition "any two samples have the same probability of being drawn" holds if and only if the die is fair.

2.1.14 In drawing a five-card hand from a deck of 52 playing cards, when you think of it as random sampling

(a) What's the random mechanism?

(b) Give a verbal description of the parameter n and give its value.

(c) Are you sampling with or without replacement?

2.1.15 If you could find a coin that was perfectly symmetric—either face *exactly* as likely as the other one—and if you could find a way to toss it so that even millions of tosses wouldn't disturb this symmetry (a physical impossibility), then you would have a perfect random number generator. In fact, this is precisely what designers of pseudorandom number generators try to emulate.

Well, just for purposes of instruction, suppose you've found such a coin and such a way of tossing it. Let's see how that random number generator would work. Suppose you want to generate a set of random numbers between, say, 0 and 125. If you toss your ideal coin repeatedly seven times, recording zero for "tails" and one for "heads", you'll generate a string of seven zeros and ones. Such a string can be interpreted as a binary number. For example, 1001011 would be the number

$$2^6 + 2^3 + 2^1 + 2^0 = 64 + 8 + 2 + 1 = 75.$$

Note that our binary number has a "one" digit in the zeroth, first, third, and sixth places (counting from the *right*). To evaluate the binary number, you simply add together powers of two, a power of two for each "one" digit in the binary number. The correct power is determined by the place number of the "one" digit.

(**a**) Evaluate the following binary numbers: 10110, 01101, 010001, and 0111.

(**b**) How many binary numbers are possible by tossing a coin seven times and interpreting the results as we've described?

(**c**) Why do we need seven coin tosses in this problem?

(**d**) Show that each binary number of part (b) is a simple random sample from the probability distribution of some random variable.

(**e**) Take out a coin and pretend it's truly fair. Use that coin as a random number generator to select a sample of size three from the numeric population consisting of the integers from 21 through 35 inclusive. There's not just one correct way to do this problem. There are some operational decisions you'll have to make and some imagination is required.

(**f**) Even if the coin were truly fair, part (e) actually is a very unrealistic, though instructive, problem. Why is it unrealistic?

(**g**) In part (e), contrary to the solution given, you might have thought to toss four coins once instead of one coin four times. This will work alright, but there's a slight hitch. Do you see what it is?

2.1.16 Suppose you're studying word frequency in the English language and have turned your attention to this very text. How would you determine the frequency of the word "the" as it's used in this text? Certainly, you would not just count. Why not? What would you do?

2.1.17 You're interested in the computer skills required of workers in the tourist industry in your city. Suggest a way to stratify this population for obtaining a stratified random sample. Remember that the point of stratification is to have the strata be as varied as possible for the characteristic in question, whereas a given stratum should be fairly homogeneous for that characteristic.

2.1.18 (**a**) Are cluster sampling and stratified random sampling examples of random experiments?

(**b**) Are cluster sampling and stratified random sampling special cases of simple random sampling?

(**c**) Suppose you have a given population and some random sampling plan. Name a random variable which would certainly be of interest in many situations. (Hint: First, what type of population are you sampling? Then, because you want to specify an example of a random variable, be sure you're clear about the underlying random experiment. In particular, what kinds of outcomes are you thinking about?)

2.2 Presenting and Summarizing Observed Numeric Data

Measures of Centrality for Observed Numeric Data

A simple example of observed numeric data familiar to every student is the list of scores on a test. Suppose the scores are

$$8,\ 4,\ 7,\ 7,\ 3,\ 5,\ 7,\ 9,\ 10,\ 2,\ 7,\ 3,\ 9,\ 8,\ 8,\ 5.$$

Even for such a small data set, this simple list is an inadequate presentation of the data. At the very least, the data should be ***ranked***:

$$2,\ 3,\ 3,\ 4,\ 5,\ 5,\ 7,\ 7,\ 7,\ 7,\ 8,\ 8,\ 8,\ 9,\ 9,\ 10.$$

Here we've ranked the data in ascending order. Descending order is also possible. But still, too many questions are left open: How many points were possible on this test? How many students were there? What was the average score on the test? A good data presentation should provide immediate answers to as many such obvious questions as possible. A good presentation of this data would be an appropriately labeled ***frequency distribution***. Note how the following frequency distribution answers all these questions at a glance:

Test scores (total of 10 points possible)

Possible score	Observed frequency
2	1
3	2
4	1
5	2
7	4
8	3
9	2
10	1
	16

mean = 6.375

Let X denote the test scores and f the frequency of occurrence of any particular value of X. So f is the number of times the value X was observed. Then Σf is just the size of the data set (the total number of observations). Here Σf is 16, the number of students who took the test. Note that 28 points come from the four students who scored seven. Those 28 points can be written: $Xf = 7 \times 4$. In general, the score weighted by its frequency of occurrence—the symbol is Xf—is the total number of points obtained by those students whose grade was X. So then, ΣXf is the total number of ALL points obtained by all the students,

$$\Sigma Xf = \text{the total of all the data.}$$

We get the average score on the test if we divide this by 16. But the word "average" is not exact here. There are many different kinds of average. This particular "average" is called the ***arithmetic mean***:

$$\begin{aligned}(1/16)\Sigma Xf &= \text{the mean of all the data}\\ &= 6.375.\end{aligned}$$

At this point we need to introduce some notation. For any statistical study, there are two parallel sets of notation depending on whether the notation refers to population data or sample data. For example, N refers to population size, n to sample size. It's important to learn the correct notation from the beginning. Statistical notation soon becomes quite complex and can pose a major problem for a beginning student who is inattentive or inconsistent in using it. So, it's important that you consistently use a capital letter N for the population size and a small letter n for the sample size. Here's some further notation: The arithmetic mean is denoted

$$\mu \quad \text{for the population mean,}$$
$$\overline{X} \quad \text{for the sample mean.}$$

The symbol $\overline{X}$ is read "X bar." With this notation, we obtain the formulas

$$N \text{ OR } n = \Sigma f \quad \text{(two notations for the same calculation depending on the type of data),}$$

$$\mu = (1/N)\Sigma X f \,,$$
$$\overline{X} = (1/n)\sigma X f \,.$$

A frequency distribution like the one we've given above for the test scores can be used efficiently to carry out calculations. This is just like what we did with probability distributions in Chapter 1:

X	f	Xf	
2	1	2	
3	2	6	
4	1	4	
5	2	10	
7	4	28	
8	3	24	
9	2	18	
10	1	10	
	16	102	mean = 6.375.

Note that the mean comes from dividing the third column sum by the second column sum. Another way of presenting this data is to give a ***relative frequency distribution***, where instead of frequencies, you record the relative frequencies

X	rf	X(rf)
2	1/16	2/16
3	2/16	6/16
4	1/16	4/16
5	2/16	10/16
7	4/16	28/16
8	3/16	24/16
9	2/16	18/16
10	1/16	10/16
	1	102/16

mean $= 6.375$.

Now in this relative frequency (rf) distribution, the mean is exactly the third column sum. Of course that's true because the division by 16 has already been done. The correct notation, if we have population data, is $\mu = 6.375$. Otherwise, if the scores are from a simple random sample, we write $\overline{X} = 6.375$.

For sample data obtained by some other sampling process—other than simple random sampling—this calculation would be wrong. For more complex sampling plans, calculations like this must be appropriately *weighted* according to the design of the sampling experiment. But we'll not be doing calculations for those more sophisticated sampling plans.

The arithmetic mean for a numeric data set is a ***measure of centrality***. That's exactly what any average is. An ***average*** is any number which indicates the "center" of the data set. But center in what sense? In addition to the arithmetic mean, we'll introduce two more averages, the median and the mode. The ***mode*** of a numeric data set is the most frequently occurring value (it's not necessarily unique). In our test scores above, seven is the mode. The terms ***bimodal*** and ***trimodal*** refer, obviously, to data sets with two and three modes, respectively. In other words, a bimodal data set has two values with the same frequency and no other values with the same or a larger frequency, so those two values are the two modes. There's no special notation for the mode.

The ***median*** of a numeric data set is the middle value after the data have been ranked. Of course, if there are an even number of observations, there's no "middle" value. This happens for the test scores given above. When there are an even number of observations, the median is simply the average of the two middle values. So, for our 16 test scores, the median is the average of the eighth and ninth scores. Because both of those values are seven, the median is just $(7 + 7)/2 = 7$.

For population data, the mean, median, and mode are just fixed numbers associated with the population, In other words, they're examples of population parameters. We defined the term "parameter" for a population just as we did in Chapter 1 for a mathematical model. It's a fixed number associated with the population. For sample data on the other hand, the mean, median, and mode vary from sample to sample. They're examples of what we call "statistics." A ***statistic*** is a number calculated from a random sample, usually for the purpose of estimating a corresponding population parameter.

Before we go any further, why don't you . . .

Try Your Hand

2.2.1 (a) Give a few integers with no repetitions (so, all the frequencies will be 1) for which the mean is larger than the median.

(b) What's the general condition under which the mean would be larger than the median?

(c) Under what conditions would the median be preferred to the mean as an average. Or, to say it differently, as a measure of centrality?

(d) For the data set you gave in part (a), give a verbal description and the numeric value of N, n, Σf, $\Sigma X f$, Σrf and $\Sigma X \text{rf}$.

2.2.2 (a) Identify the median and mode(s) of the following data:

$$5,\ 2,\ 8,\ 6,\ 2\ 5,\ 5,\ 7,\ 3,\ 7,\ 2,\ 8.$$

(b) For the data in part (a), set up a frequency distribution, assuming this to be population data, and use it to compute the mean.

(c) For the data in part (a), set up a relative frequency distribution, assuming this to be population data, and use it to compute the mean.

(d) Redo each of parts (b) and (c) assuming you have sample data.

(e) In part (a), assume you have sample data. Give a verbal description and the numeric value for each of: N, n, Σf, $\Sigma X f$, Σrf and $\Sigma X \text{rf}$.

2.2.3 Show that a statistic is a value of a random variable.

Measures of Spread for Observed Numeric Data

In the previous section, we defined three parameters which serve as averages for observed numeric data, the mean, median, and mode. Averages are measures of "centrality," but centrality is not the whole story. Just as with random variables, some numeric data sets will be tightly clustered about their centers and some will be widely dispersed. Two data sets with the *same* center could be quite different, depending on how they spread about that center.

We'll look at two parameters and their corresponding statistics which serve to measure the *spread* of observed numeric data. The simplest such measure is the range. The ***range*** of a numeric data set is the largest minus the smallest value. For the test scores given in the previous section, the range is eight. The range is easy to compute and carries a certain amount of information, but it's determined by only two values of the data set. It can't tell us much. The range is useful only as a quick and easy measure of spread. It has no special notation.

For a more informative measure of spread we require the variance, or its square root, the standard deviation. These are defined just as they were in Chapter 1 for random variables. But remember: Don't think of the variance and standard deviation as two different measures of spread. They're two different numbers which measure spread in exactly the same way. In other words, the variance and standard deviation are *one* measure of spread which can be expressed in two different numbers.

The ***variance*** of numeric data, just like the variance of a random variable, is the average of the squared deviations from the mean:

$$\sigma^2 = (1/N)\Sigma(X - \mu)^2 f$$

Here, we've used the notation for population data. If we have sample data, the variance is denoted by $\hat{\sigma}^2$. Of course, the ***standard deviation*** for population and sample data—the square root of the variance—are denoted by σ and $\hat{\sigma}$, respectively (the symbol $\hat{\sigma}$ is read "sigma hat"). The most efficient way to calculate the variance is by entering an appropriate column into the frequency distribution for the data, a column containing the squared deviations from the mean weighted according to their frequency of occurrence. For the test scores introduced at the beginning of the chapter

X	f	Xf	$(X-\mu)^2 f$
2	1	2	19.1406
3	2	6	22.7813
4	1	4	5.6406
5	2	10	3.7813
7	4	28	1.5625
8	3	24	7.9219
9	2	18	13.7813
10	1	10	13.1406
	16	102	87.7500

$\mu = 6.375$,
$\sigma^2 = 5.4844$.

Again, our notation implies we have population data. Here, to get σ^2, you divide the fourth column sum by the second column sum. If this were a relative frequency distribution, how would you calculate σ^2? Well now, please . . .

Try Your Hand

2.2.4 Construct a *relative* frequency distribution for the data given just above and use that relative frequency distribution to calculate the variance and standard deviation. To facilitate use of your calculator, give the relative frequencies as decimal numbers.

2.2.5 (a) We began by saying: "In the previous section, we defined three parameters which serve as averages for observed numeric data, the mean, median, and mode." But these three parameters refer to only one type of observed data. If our data was of the "other type," what word would be appropriate?

(b) Explain why it would be wrong to identify the range by saying that the test scores "range from two to ten."

(c) What's the formula for $\hat{\sigma}^2$?

2.2.6 Construct a bimodal data set. Identify the modes and the median and then compute the mean, variance, and standard deviation using a frequency distribution table.

2.2.7 If the variance of observed data is "just like the variance of a random variable," why is the formula different?

2.2.8 The variance measures spread about the mean. How does the *range* relate to the mean?

2.2.9 Identify the mode, median and range for the following population data. Then complete the table to compute the mean, variance, and standard deviation.

X	1.8	2.4	2.6	2.8	3.1	3.3
f	3	5	6	2	1	2

2.2.10 If you draw on your understanding of the relevant concepts, the parts of this problem can be done quickly without too much trial and error.

(a) Without changing the number of observations or the mean, change the frequencies for the following data to make the standard deviation larger than 1.1

X	3	4	5	6
f	7	9	4	2

(b) Replace the ?'s with numbers which make the median of this data 22 while making the mean more than 23

X	20	21	22	23	?
f	1	6	?	3	2

2.3 Grouped Data: Suppressing Irrelevant Detail

Grouped Distributions of Observed Real-World Data

In the previous section, we looked at a small data set, 16 test scores on a ten-point test. With ten possible points, we had at most 11 distinct data values and it was easy to construct a frequency distribution. This is not typical of all (or even most) data sets which often are very large with many possible values. For example, suppose you have a test taken by 1000 students where there are 100 possible points. So you have 1000 observed data points (the 1000 test scores) with 101 possible distinct values (zero to 100). Maybe the data looks like . . .

95 96 57 85 30 85 02 84 92 09 88 08 85 08 05
75 08 52 58 39 45 83 96 48 83 49 27 38 50 87
58 00 47 44 72 75 92 40 40 75 38 59 87 52 34
22 56 39 58 76 26 85 76 76 68 65 31 71 17 71
19 68 77 04 27 92 17 97 16 74 17 72 23 47 17
17 72 87 89 78 35 75 97 37 28 45 88 28 33 23
57 57 18 64 62 06 41 65 83 28 22 13 47 29 27
87 72 71 72 17 79 19 87 87 32 34 98 54 56 87
87 72 24 27 65 96 24 41 12 67 67 38 75 87 63
. . .
. . .

This is only 135 of the 1000 scores! It goes on and on. To bring some semblance of order to this chaos, you should, of course, rank the data and put it into a frequency distribution, just as we did in the previous section:

X	f	X	f	X	f
0	3	12	4	24	46
1	8	13	0	25	19
2	4	14	14	26	17
3	8	15	8	27	31
4	5	16	15	28	36
5	9	17	11	29	51
6	7	18	9	30	29
7	3	19	15	31	19
8	4	20	16	20	41
9	2	21	16	33	27
10	2	22	17	. . .	Hmmmmm!
11	6	23	35		

And this is only one-third of them.

This is not clarity! The cardinal rule of any data presentation is CLARITY!! You should be able to see the chief characteristics of the data at a glance.

What we need here is a ***grouped frequency distribution***, where the scores are grouped into classes:

Class	f	X	Xf	$(X-\mu)^2 f$
0–19	137	9.5	1301.5	191490.7780
20–39	202	29.5	5959.0	61061.9548
40–59	411	49.5	20344.5	2807.5019
60–79	157	69.5	10911.5	80285.8601
80–100	93	90.0	8370.0	172866.7730
	1000		46886.5	508512.8678

$$\mu \approx 46.8864, \sigma \approx 22.5502.$$

The symbol X refers to what's called the ***class mark***, the midpoint of the class. It's not hard to understand how a grouped frequency distribution like this one works. Instead of our boring you with the details, why don't you . . .

Try Your Hand

In the two problems given here, we're asking you to use your commonsense and guess how to construct a grouped distribution. Try to guess, even if you don't succeed. That will significantly help you in understanding the solutions. For this to work, you really must try for yourself to see what ought to be done.

2.3.1 For the grouped frequency distribution above, there are a number of points which require clarification. Try to identify them all on your own and explain them. That is, try to identify and explain anything in the table which—in any way whatsoever—is unlike what we've seen before. Be careful. There are things in the table which at first glance look the same as before but, in fact, require clarification. Give yourself some time to think about this. Assume this to be population data.

2.3.2 In the grade distribution of the previous problem, each class was 19 points wide. But sometimes it's more convenient to allow unequal class widths. Suppose you're looking at the employee salaries for a large corporation with 1000 employees. Can you see why it would be convenient to have unequal class widths?

2.3.3 Consider the following temperature readings:

Temperature (centigrade)	f
0–15	6
15–30	12
30–45	8
45–60	2

Let X be the class mark for each class.

(a) What does the 12 in the second column mean?

(b) Why do the endpoints overlap?

(c) Give a verbal description for the MEANING (not the calculation) of

$$f, \quad \Sigma f, \quad \Sigma Xf \text{ and } 1/\Sigma f \left[\Sigma X\right].$$

(d) Complete the table above and use it to estimate the mean, mode(s), variance, and standard deviation for these temperature readings.

(e) If we are to treat this data as a sample, what type of sampling is involved?

Histograms: Graphical Display of Grouped Relative Frequency Distributions

A histogram adheres to the principle stated at the beginning of this chapter:

> Numeric quantities are seen and interpreted
> by the human eye in terms of area.

We'll concern ourselves only with histograms for grouped *relative* frequency distributions for which the area, the relative frequency, can be interpreted as the percentage of the data falling in that class. If you're given a frequency distribution and asked for a histogram, first convert it to a grouped relative frequency distribution and give the histogram of that distribution. Here's a histogram for the relative frequency distribution of the test scores from the previous section

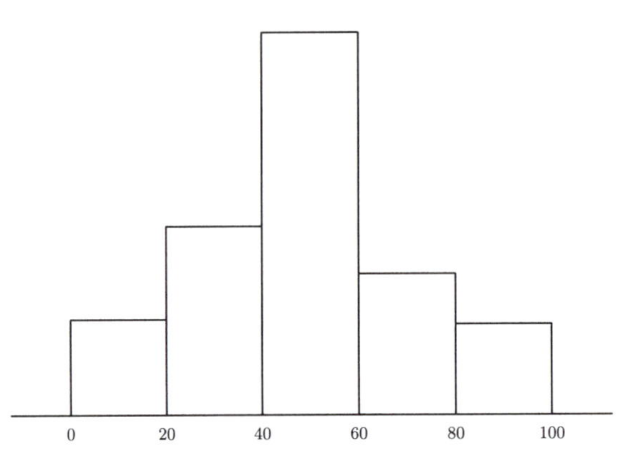

Note that we've not put in the vertical scale. The vertical scale can be confusing if the class widths are not the same. The point is that the area should represent the proportion or percentage of the data in each class. Thus, the first rectangle contains just slightly under 15% of the area, the second rectangle contains a little over 20% of the area, and so on.

Histograms can be very informative, revealing facts about the data which might not otherwise come to light. This is very well illustrated by the following example given by W. E. Deming. [5]

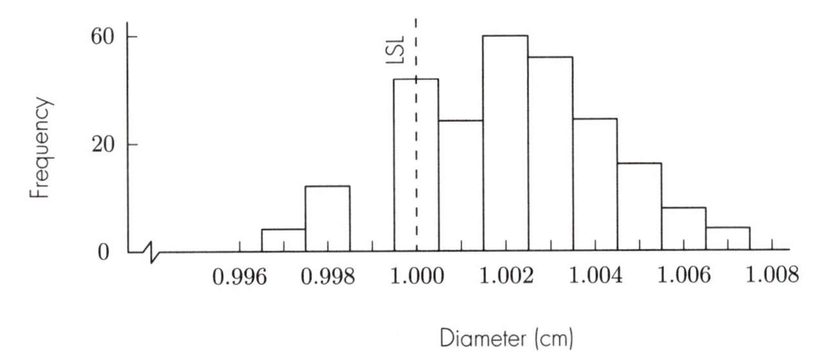

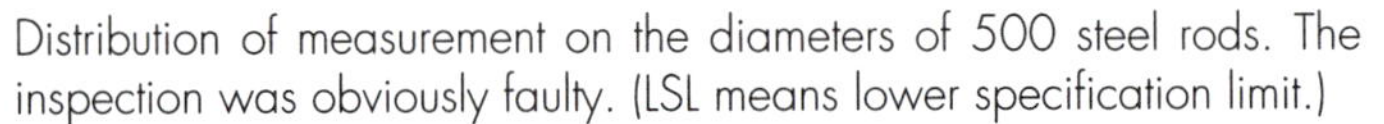

Distribution of measurement on the diameters of 500 steel rods. The inspection was obviously faulty. (LSL means lower specification limit.)

This histogram represents the diameters of 500 steel rods as measured at the time of quality control inspection. Rods smaller than 1 cm are too small; they would be too loose in their bearings and must be discarded. So now please . . .

[5] Judith H. Tanur et al., *Statistics, a Guide to the Unknown*, Holden-Day, San Francisco, 1972.

Try Your Hand

2.3.4 Suppose the test scores from the previous section had been grouped as in the following table. Note that the data is the same; we just happened to have broken the first class into two classes, but now the six resulting classes are of unequal width.

Class	f
0–9	68
10–19	69
20–39	202
40–59	411
60–79	157
80–100	93
	1000

Sketch the histogram for this data.

2.3.5 In Deming's histogram given in the text, how would you interpret the gap at 0.999 cm followed by a peak at 1 cm? These two characteristics of this histogram have a very important real-world significance. Can you guess what it is?

2.3.6 Draw a histogram for each of the following distributions, shading the area within one standard deviation of the mean

(1)

Class	f
0–10	10
10–20	15
20–30	10

(3)

Class	f
0–10	15
10–20	10
20–30	15

(2)

Class	f
0–10	5
10–20	30
20–30	5

(4)

Class	f
0–10	10
10–20	20
20–30	0
30–40	10

(5)

Class	f
0–10	10
10–20	20
20–60	0
60–70	10

2.4 Using the Computer

The advent of electronic computers capable of carrying out calculations and analyses of previously unimagined complexity has greatly increased the range of application of statistical techniques. You might like to investigate what resources are available for statistical analysis in the computer lab at your school. Minitab is a commonly available software package for statistical analysis. Three other widely used packages are SPSS, SAS, and BMDP. Because the focus of this text is not on data analysis and exploration—where computers are really essential—but rather on understanding certain basic and fundamental statistical concepts, we'll not require the use of any of these statistical packages.

The discussion in the rest of this section is intended just to whet your appetite for some independent explorations on your own. References will be in terms of Minitab, but you should have no difficulty carrying out the same exercises with other packages. The assistants at your computer lab will show you how to log onto your system and access Minitab or some other statistical program. They may also be able to provide some documentation for the package.

A good student reference for Minitab is the *MINITAB Handbook* [Ryan et al.]. It contains numerous exercises and suggestions for how you might use Minitab to explore your own data sets or the data sets included with the Minitab package. It also gives nine of the Minitab data sets along with descriptions of the variables and some background on the data.

Describing, Picturing, and Comparing Population and Sample Data

Minitab has a number of data sets which you can explore and from which you can take samples. It will be interesting for you to see how the samples which Minitab selects compare with the data set itself. Two commands which will make that comparison possible are DESCRIBE and HISTOGRAM. The command HISTOGRAM, of course, provides a histogram of your data, though not exactly in the format which was described in this text.

The DESCRIBE command will give you basic numeric summaries (parameters) for your data, including the population or sample size (using the symbol N for both cases), the mean, median, standard deviation, and the 5% "trimmed mean." The trimmed mean is the mean of that 90% of the data remaining after the largest 5% and smallest 5% have been removed. The point of the trimmed mean is to have a "resistant" version of the mean. A statistical procedure is ***resistant*** if it's not overly sensitive to a few extraordinary data values—values that are far out of the range of the rest of the data. Such values are called "outliers." The median is, of course, a resistant measure. Resistant measures protect against erroneous data values as well as against correct values that may distort the overall picture. For example, "median income" is usually reported to protect against the distortion resulting from a few exceptionally large income figures.

The DESCRIBE command also gives the smallest and largest observations and the first and third "quartiles." The first quartile is the point below which you find 25% of the data. The second quartile is the median because 50% of the data is below it. The third quartile is the point below which you find 75% of the data. Finally, the DESCRIBE command gives the "standard error of the mean" which we will not encounter until Chapter 5.

Among the data sets included with Minitab is data from the 1980 Wisconsin Restaurant survey, conducted by the University of Wisconsin Small Business Development Center. The data is saved in a Minitab file named "restrnt." That file contains 14 of the many variables in the actual survey. Some restaurants failed to answer one or more questions—missing responses are denoted with an asterisk (*). At the end of this section we have reproduced the screen output of our Minitab session in which we investigated this data. We will now describe that session in detail.

To explore the "restrnt" data, we first RETRIEVE the file (RETRIEVE is a Minitab command). Here's how we did that: After the Minitab prompt (the prompt is: MTB >), we typed

retrieve "restrnt".

In the screen output reproduced at the end of this section, you see how the RETRIEVE command brings that file "restrnt" into a Minitab worksheet. We then typed in the Minitab INFO command (after the prompt, we typed: info), which lists the variables in the file.

In the output at the end of the section, you see exactly what appeared on our computer screen during the Minitab session we are describing. Minitab gave most of what you see. We'll tell you exactly what part of that output we entered at the computer. Now we investigate the variable "wages" contained in column seven. "Wages" is a percentage: wages at the restaurant as a percentage of sales. We obtained a description of "wages" by typing: describe c7. Next, we obtained a histogram of "wages" by typing: hist c7. Note that although the file contains responses from 279 restaurants, only 235 gave information on wages. This is shown in the output by: $N = 235 \quad N^* = 44$.

Note that "wages" is skewed to the right. This is not just the usual problem which we expect with "a few large wages," because our variable is not actually wages at all; it's a percentage. The skew of the distribution is not terribly significant. It pulls the mean up only by a quarter of a percentage point. You might be suspicious of the accuracy of the two largest values. Are there really two restaurants which pay over 80% of sales in wages with less than 20% left for all other costs (including food and overhead)? Of course, we don't have enough information to answer that question. There are usually many conceivable explanations for such so-called "outliers."

We then drew two samples of size $n = 20$ from "wages" using the SAMPLE command, placing them in columns 15 and 16, respectively. After obtaining each sample, we looked at a histogram of that sample. It's interesting to compare those histograms with each other and with the histogram of the entire population of 235 values of the variable "wages." Note that only the first sample has picked up one of the outliers. Next, we described the two samples with one command by typing: describe c15 c16. You can see how much the samples are affected by the outliers. The first sample mean is more than five and a half percentage points above the second!

Finally, we looked in detail at the second sample by obtaining a "stem-and-leaf diagram" of the sample. This is a technique introduced the 1960s by John Tukey, a statistician who has made numerous important contributions to graphical displays of statistical data. Forget the first column for a moment. The second column gives the "stems." Here, our data consists of two-digit numbers; the stems are the leftmost digit. The leaves are the rightmost digit. A stem-and-leaf display is more informative than a histogram. It gives the histogram shape while at the same time displaying the actual values of your data. For example, here's how the second sample begins:

$$0,\ 8,\ 10,\ 15,\ 20,\ 20,\ 20,\ 21,\ 21,\ \ldots$$

That's interesting: One restaurant pays no wages! Wonder why?

The first column of the stem-and-leaf display is called the "depth" of a line. It tells how many leaves lie on that line or "beyond." "Beyond" means "beyond the middle." For example, there are nine observations on or before the fifth line. The parenthesis locates the line containing the median. Within the parenthesis is the number of observations on that line. If there are an even number of observations and the middle two are on separate lines, the parenthesis is omitted. Stem-and-leaf displays are very informative!

```
MTB >  sample 20  from c7 into c15
MTB > hist c15

Histogram of C15   N = 19   N* = 1

Midpoint       Count
      10           1  *
      20           7  *******
      30           7  *******
      40           3  ***
      50           0
      60           0
      70           0
      80           0
      90           1  *

MTB > sample 20 from c7 into c16
MTB > hist c16

Histogram of C16  N = 19   N* = 1

Midpoint       Count
       0           1
       5           0
      10           2  **
      15           1  *
      20           5  *****
      25           3  ***
      30           5  *****
      35           2  **

MTB > describe c15 c16

            N        N*      MEAN    MEDIAN    TRMEAN     STDEV    SEMEAN
C15        19         1     28.47     25.00     26.24     16.07      3.69
C16        19         1     22.79     25.00     23.41      9.42      2.16

          MIN       MAX        Q1        Q3
C15     10.00     85.00     19.00     34.00
C16      0.00     35.00     20.00     30.00

MTB >

MTB > retrieve 'restrnt'
  WORKSHEET SAVED 10/24/1989
```

```
Worksheet retrieved from file: restrnt.MTW
MTB  >  info

COLUMN    NAME        COUNT        MISSING
C1        ID           279
C2        OUTLOOK      279               1
C3        SALES        279              25
C4        NEWCAP       279              55
C5        VALUE        279              39
C6        COSTGOOD     279              42
C7        WAGES        279              44
C8        ADS          279              44
C9        TYPEFOOD     279              12
C10       SEATS        279              11
C11       OWNER        279              10
C12       FT.EMPL      279              14
C13       PT.EMPL      279              13
C14       SIZE         279              16

CONSTANTS USED: NONE

MTB > describe c7

              N        N*      MEAN    MEDIAN    TRMEAN     STDEV    SEMEAN
WAGES       235        44    25.251    25.000    24.972    10.886     0.710

            MIN       MAX        Q1        Q3
WAGES     0.000    85.000    20.000    30.000

MTB > hist c7

Histogram of WAGES  N = 235  N* = 44
Each * represents 2 obs.

Midpoint   Count
       0       6  ***
      10      23  ************
      20      73  *************************************
      30      98  *************************************************
      40      29  ***************
      50       2  *
      60       2  *
      70       0
      80       1  *
      90       1  *
```

```
MTB > stem-and-leaf c16

Stem-and-leaf of C16   N  = 19
Leaf Unit = 1.0        N* =  1

   1     0 0
   2     0 8
   3     1 0
   4     1 5
   9     2 00011
  (4)    2 5579
   6     3 0002
   2     3 55
```

Chapter 3 Discrete Probability Models

3.1 Introduction

In Chapter 1, we explored the idea of a random variable and saw some specific examples. Now it's time to see some of the standard "classes" of random variables. These classes serve as standard models for modeling real-world problems. There's an important advantage to having such standard models: The theory's already worked out! If you can match your problem with a standard model, you can draw on previous experience, intuition, and an established theory.

It's really a matter of classification. Many real-world problems fall into certain clearly defined types for which the models (and theory) have already been developed. That saves you a lot of work. When you're faced with such a problem, your first task is to "class"-ify the problem—to find an appropriate model. In this chapter, we introduce some of the standard classes of random variables and study the type of real-world problems which they appropriately model. First please . . .

Try Your Hand

3.1.1 There's a simple class of random variables for which we have already seen examples. This class could be characterized by the condition "the variance of X is zero." Can you identify this class in terms of the values of X?

3.2 The Discrete Uniform Distribution

We begin here with a particularly simple class of random variables. The ***discrete uniformly distributed random variable with parameter n*** is a random variable with n values, all of which are equally likely. We've already seen some examples which we'll recall in the exercises below. We say "uniform" because the probabilities are all "uniformly the same." "Discrete" simply means the values of the random variable are discrete points on the number line. You'll see the significance of that more clearly in Chapter 4 when we contrast "discrete" with "continuous." We'll not need that distinction in this chapter.

Do you see what we mean by a "class" of random variables? It's a group of random variables all having some common set of characteristics. For instance, the common characteristic "equally likely numeric values" defines the class we're studying in this section, the class of the uniform distribution. In Problem 3.1.1, we looked at the class of random variables characterized by the condition "the variance is zero," a rather trivial class, but instructive nevertheless. We think of a class of random variables as a "model": It models all those real-world situations which could be modeled by one of the random variables in the class. That's why we refer to such a class as a ***probability model.***

Before going any further, you'd better . . .

Try Your Hand

3.2.1 In Chapter 1, you worked with the "conceptual formula" for the variance of a random variable: $\sigma^2 = \Sigma(X - \mu)^2 \mathrm{P}(X)$. There's a much easier way to compute the variance, the so-called "computing formula":

$$\sigma^2 = \Sigma X^2 \mathrm{P}(X) - \mu^2.$$

(a) Derive the computing formula from the conceptual formula.

(b) Use the computing formula to calculate the variance for the number of dots on the hidden face of a fair four-sided die. Note it's a FOUR-sided die, just to make life easier for you.

(c) Why do you think the terms "computing" and "conceptual" have been chosen for these two formulas?

(d) Develop a computing formula for the variance of observed data.

(e) For any random variable X, what's the expected value of X^2?

3.2.2 Think back to the simple examples of random variables we studied in Chapter 1 and find two examples of uniformly distributed random variables. What's the parameter in each case?

3.2.3 Suppose we have a uniformly distributed random variable X which takes on the following values:

$$18.2, \quad 18.7, \quad 19.3, \quad 19.7, \quad 20.1.$$

What further information is required to compute the mean and standard deviation of X?

3.2.4 It's wrong to say: "Any real-world situation involving equally likely outcomes will be modeled by a uniformly distributed random variable." Why is it wrong? Hint: Look carefully at the definition.

3.2.5 Complete the following line graph for a random variable W, assuming W to be uniformly distributed

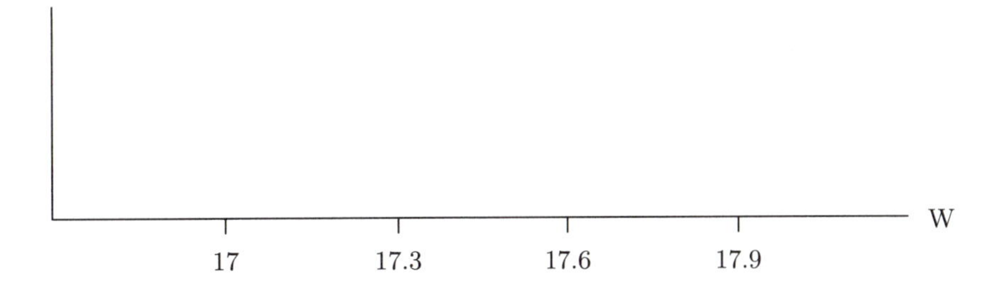

3.2.6 What information is required to specify one particular uniformly distributed random variable?

3.2.7 If you know X to be a uniformly distributed random variable, you can describe its mean without any reference to the probabilities. How?

3.3 The Hypergeometric Distribution

The next class we'll study is called the "hypergeometric random variable." Note how the usual terminology for such classes could be confusing. When we say "THE hypergeometric random variable," it sounds like just one random variable. In fact, it's many, an entire class. And why the term "hypergeometric"? The name, we're sorry to say, is not going to mean much to you. It derives from a connection between the probability formula we develop later and a very arcane creature called the "hypergeometric series." Don't worry about it.

Counting Rules

Before we turn to this new model, we need some counting techniques. All these techniques are derived from

the fundamental principle of counting:

> Suppose task #1 can be done in m ways and task #2 in n ways. Then you can accomplish task #1 followed by task #2 in mn ways.

This principle is reminiscent of the multiplication rule for probabilities which we learned in Chapter 1. Roughly stated, "and" means multiply. The probability rule, however, can't be applied blindly; it's only valid when the events in question are independent. But for counts, the multiplication rule—the fundamental principle of counting—always holds.

Here's how the fundamental principle of counting works: If there are three ways to go from my office to my favorite bar and two ways to go home from the bar, there are six ways to go home from the office by way of the bar. You can see this diagramatically . . .

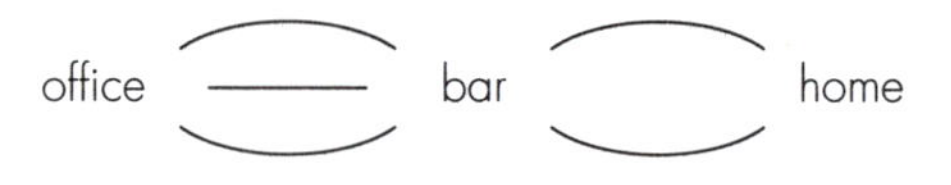

There are other counting rules based on the fundamental principle of counting, you'll derive them in Problem 3.3.1. They make use of the ***factorial notation***: The symbol is **n!**, read "n factorial." It's just the product of n and all integers less than n:

$$5! = 5 \times 4 \times 3 \times 2 \times 1 = 120.$$

Of course, $1! = 1$. We extend the definition by fixing the convention that $0! = 1$. This allows us to avoid always having to make a special case when zero shows up in a calculation. Now we're ready for you to . . .

Try Your Hand

3.3.1 (a) How many arrangements are there of n objects?

(b) Let $C(n, x)$ be the number of ways to choose X objects from a set of n objects. This is the ***combinations of n objects taken X at a time.*** Derive the formula

$$C(n, x) = \frac{n!}{x!(n-x)!}.$$

(c) Evaluate C(7, 3), C(7, 1), C(7, 7), C(120, 118).

(d) Evaluate $C(n, 0)$, $C(n, 1)$, $C(n, n)$, $C(n, n-1)$.

(e) How many ways can you arrange seven books on one bookshelf?

(f) How many ways can you choose three books to take on vacation from the 58 in your bookcase?

(g) How many ways can you seat six students in a class with six desks?

(h) How many ways can you seat 11 students in a class with 11 desks? How about 80 students with 80 desks?

(i) How many ways can you seat six students in a class with 14 desks?

(j) Suppose there are 96 workers in a machine shop. How many distinct committees of three could they form to represent them at an upcoming meeting with management?

What is the Hypergeometric Model?

The ***hypergeometric random variable*** arises from sampling without replacement from a dichotomous population. It counts how many in the sample have the characteristic of interest. Call it X; then

X = the number of observations in the sample having the characteristic of interest.

Usually, this model is required only when the population is small, less than 60 as a rule of thumb. In the next section, we'll see why.

Recall from Chapter 2 that a "statistic" is a number calculated from a sample. It will vary from sample to sample, of course, and so a statistic is a random variable. You showed this in Problem 2.2.3. The hypergeometric random variable is an example of such a statistic and its distribution is our first instance of a "sampling distribution." If the underlying random experiment for a random variable is random sampling, the probability distribution of that random variable is called a ***sampling distribution***.

If you're going to be successful in using our models, it's crucial that you be attentive to the common characteristics shared by all the random variables in the class. Only then will you be able to spot these characteristics in a real-world context. If you can do this, you'll be able to recognize which model is appropriate for a particular problem. This is probably the most difficult skill a beginning statistics student needs to develop—the skill of modeling, the skill to recognize the appropriate abstract model for a given real-world situation. So take careful note of the following description:

The hypergeometric random variable

- is a *count*
- is associated with *simple random sampling* with**out** replacement from a *dichotomous* population
- is used only for *small* populations
- tells *how many* in the sample have the characteristic of interest.

Suppose, for example, you're required to test the lifetimes of a batch of 50 electronic components, where the testing process is "destructive"—you put the component into operation and record the time to burnout. This is a common quality control situation which obviously requires sampling without replacement. Even when the batch is small, you're constrained to sample as opposed to testing every component, otherwise the entire batch is destroyed. And the sampling must necessarily be without replacement because you can't retest a component which you've already destroyed! Now let's explore the hypergeometric random variable . . .

Try Your Hand

3.3.2 The electronic components example mentioned just above is incompletely specified. What crucial piece of information is required before we can justify using the hypergeometric random variable as a model?

3.3.3 For the hypergeometric random variable:

(a) What's the underlying random experiment?

(b) What are the possible values?

3.3.4 From a pool of 40 candidates, the mayor has appointed a powerful committee of five. Because none of the committee members are women, there has been an accusation of prejudice. We might try to analyze this situation by determining the probability of no women on the committee if the committee had been chosen at random from among the 40 candidates. Discuss the suitability of the hypergeometric model here:

(a) What's the population?

(b) What's the underlying random experiment? Is it appropriate for the hypergeometric model?

(c) What is an outcome of the random experiment? Be very specific to the real-world situation.

(d) Exactly how many such committees are possible?

(e) What's the random variable?

(f) Can the random variable be regarded as belonging to the class of the hypergeometric random variable?

3.3.5 The neighborhood library wants to know how many books listed in the catalog are lost, nowhere to be found. To address this problem, you might take a sample—a simple random sample, without replacement—of the catalog listing and consider the number of listings which correspond to lost books. Suppose you find that two among a sample of 30 listings are for lost books. This situation can be modeled by the hypergeometric random variable:

(a) What's the population?

(b) What's the underlying random experiment for the random variable? Be sure you verify that it's indeed a random experiment.

(c) What are the possible values of the random variable?

(d) What value of the random variable are we asked about?

(e) In fact, we should not use the hypergeometric model here. Why not?

3.3.6 To make use of the probability models we're studying, it's important to recognize the *characteristics* of the model. In the box above this exercise set we gave a list of the characteristics of the hypergeometric random variable. Go back and give a similar list for the uniformly distributed random variable defined in the previous section.

Calculating the Probabilities

Now let's see how to calculate probabilities for the hypergeometric random variable. It will be a straightforward application of the "theoretical relative frequency" definition of probability. First some notation:

N = population size;

R = the number in the population which have the characteristic of interest;

n = sample size;

X = the random variable, the number in the sample having the characteristic;

p = R/N, the proportion of the population having the characteristic;

q = $1 - p$, the proportion of the population not having the characteristic.

In our quality control example (Problem 3.3.2), there were 50 electronic components. Suppose four of the 50 don't meet the specification for a lifetime of "at least 1000 hours." Of course, realistically, we would not know how many don't meet specification; that's exactly what we we're trying to study. But at the moment, we're just THINKING about it for purposes of developing the theory. Suppose we're going to test three components chosen at random. Then

$$N = 50, \quad R = 4, \quad n = 3, \quad \text{and } X = 0, 1, 2, \text{ or } 3.$$

[Hint: Always do this. Extract all numbers from the verbal description of a problem and write them down compactly in terms of the formalism of the model. You'll be amazed how much easier that makes the solution.]

To construct a probability distribution for X, we must first compute the probabilities. Because $X = 0$ or $X = 1$ may be less clear, let's start with the probability that $X = 2$. Using the "theoretical relative frequency" definition, the denominator will be the number of ways to get any sample (outcome) whatsoever and the numerator will be the number of ways to get a sample for which $X = 2$. So the denominator is

the number of ways to select a sample of n

and the numerator is

the number of ways to select a sample of n where exactly two have the characteristic.

So

$$P(\mathrm{X} = 2) = \frac{\text{ways to select a sample of } n \text{ where exactly two have the characteristic}}{\text{\# ways to select a sample of } n.}$$

Now, why don't you yourself just . . .

Try Your Hand

3.3.7 We're calculating the expression above for $\mathrm{P}(X = 2)$.

(**a**) First calculate the denominator.

(**b**) Then calculate the numerator.

(**c**) Then combine them to obtain $\mathrm{P}(X = 2)$.

(**d**) Make up a probability distribution table for X and compute the mean and variance.

3.3.8 In Problem 3.3.4:

(a) If the mayor chooses her committee randomly, what's the probability of having no women on the committee?

(b) What's the real-world meaning of the small probability in part (a)?

(c) Suppose the mayor's choice puts one woman on the committee. What's the probability of such a committee? What's the real-world meaning?

3.3.9 Now write down a formula for $P(X = x)$ by analogy with the case for $P(X = 2)$, where, instead of $X = 2$ we ask about $X = x$. Remember that the uppercase letter, X, is a name for the random variable, whereas the lowercase letter, x, is the symbol for one single unspecified value.

3.3.10 Take the case of a hypergeometric random variable X for which $N = 10$, $R = 4$, and $n = 4$.

(a) Describe the underlying random experiment as completely as possible.

(b) Guess the mean and variance of X.

(c) Give the probability distribution of X.

(d) Compute the mean and standard deviation of X.

The Formulas

Now, here are the formulas for the probabilities of the hypergeometric random variable together with its mean and variance:

the hypergeometric random variable:

If X is a hypergeometric random variable, then

$$P(X = x) = \frac{C(R, x)C(N - R, n - x)}{C(N, n)}$$

with

$$\mu = np,$$

$$\sigma^2 = \frac{npq(N - n)}{(N - 1)}.$$

The denominator of $P(X = x)$, which is just $C(N, n)$, is the total number of ways to draw a sample of n from this population of N members. It does not depend on the value of X. And the numerator is the product of

> the number of ways to choose x members of the sample from the R members of the population which have the characteristic of interest: this number is $C(R, x)$,

and

> the number of ways to choose $n - x$ members of the sample (the rest of the sample) from the $N - R$ members of the population which do NOT have the characteristic: it's $C(N - R, n - x)$.

Remember, the symbol x (the lowercase letter) refers to a particular but unspecified value of the random variable X. Here it refers to the particular value of X for which we want the probability.

The *parameters for the hypergeometric model* are:

N	$=$	the population size;
R	$=$	the number in the population which have the characteristic of interest;
n	$=$	the sample size.

It's not so easy to derive the formulas for the mean and variance. We'll not attempt it. In these formulas, we use the conventional notation for population proportions:

$p = R/N$, the proportion of the population having the characteristic of interest

and

$q = 1 - p$, the proportion of the population NOT having the characteristic of interest.

Note that p and q are not new parameters for the model because they are derived from the basic parameters R and N.

In Problem 3.3.10, you computed a mean of 1.6003. According to the formulas above, the mean for that example is $4 \times 0.4 = 1.6$. It's the formula that's exact; this 1.6 differs from your calculated value because of the rounding error in the calculation. The variance for Problem 3.3.10 as calculated from the formula above is

$$\sigma^2 = 4 \times 0.4 \times 0.6 \times (6/9) = 0.64,$$

agreeing exactly with your calculation.

In Problem 3.3.10, you saw that the formula for the mean is very intuitive. It just says that the proportion of a sample having the characteristic is, ON AVERAGE, the same as the proportion of the population having the characteristic. To give another example, suppose that 15% of the population has the characteristic of interest and you are taking samples of size $n = 20$. Then, on average, you should expect about three in the sample to have the characteristic—about 15% of your sample. That's the formula given above. Here it is again

$$\mu_X = np = 20 \times 0.15 = 3.$$

There are two observations we should make about the formula for the variance. Note first that the variance gets larger as the sample size increases. The variance has a factor of n in it, so n big means the variance is big. This is very reasonable because the larger the sample the more "room" there is for variability within the sample. More technically, when n is large, X has a larger range of values. The values range from zero up to n, inclusive. With a larger range, X has more potential variability.

Then there's that mysterious factor: $(N-n)/(N-1)$. This is called the ***finite population correction factor.*** We'll look at this again in the next section. The choice of name will become clear at that time (why "correction factor"?). The exact form of the finite population correction factor arises from very technical considerations. But, intuitively, the expression makes sense. To get that sense, think of the $N - 1$ in the denominator as if it were just N, the population size. It's one less than N for very technical reasons only. Now, if we use N instead of $N - 1$, the finite population correction factor would be

$$\frac{N-n}{N} = \frac{N}{N} - \frac{n}{N} = 1 - \frac{n}{N}.$$

So $(N-n)/N$ is just the proportion of the population not in the sample. Or, as $1-n/N$, it's just one minus the proportion of the population that is in the sample.

The variance will be large when $(N - n)/N$ is large, small when it's small. And that makes sense because when $(N - n)/N$ is large, much of the population is not in the sample. With much of the population not in the sample, there's less information in the sample. Less information means more uncertainty. More uncertainty reflects more variability. More variability means the variance should indeed be larger.

Now please . . .

Try Your Hand

3.3.11 Suppose you choose three light bulbs at random from a supply of 30, four of which are burnt out.

(a) What's the probability that at least one of the light bulbs you've chosen is burnt out?

(b) Give the mean and variance for this situation. Discuss the finite population correction factor.

(c) If only one of the 30 light bulbs is burnt out, what's the probability that the burnt-out light bulb will be among the three you choose?

(d) Now suppose two of the 30 light bulbs are burnt out. Make up a probability distribution for X, the number of burnt out light bulbs among your three. Give its mean and standard deviation (try to guess first and then calculate).

(e) How large a sample would be required to have a finite population correction factor of 90%? Of 80%?

3.3.12 In each part of this problem, try first to guess the answer on intuitive grounds and then verify your answer by precise reference to the model. Suppose you're drawing a six-card hand dealt from a well-shuffled deck of 52 playing cards . . .

(a) How many hearts would you expect?

(b) How many black cards would you expect?

(c) Which is more predictable, the color or the suit?

(d) Illustrate the previous parts of this problem by sketching possible line graphs for the two random variables. Do not attempt to compute any probabilities.

(e) Give a verbal description *in terms of this problem* for the finite population correction factor which appears in parts (a) and (b).

3.3.13 In the text just preceding this set of exercises, we have two analyses which work in opposing directions. We saw that when n is large, the variance of the hypergeometric random variable is large. But from another point of view, looking at the finite population correction factor, a SMALL n also makes the variance large because n small implies $(N - n)/(N - 1)$ is large. Which effect dominates the variance?

3.4 Sampling with Replacement from a Dichotomous Population

What is the Model?

Now let's look at simple random sampling WITH replacement. Just as before, we are considering a dichotomous population and our random variable will be a count, counting how many in the sample have the characteristic of interest. Of course, it's another instance of a "statistic" and its probability distribution another instance of a "sampling distribution." Note that the real-world situation we're discussing here is exactly the same as for the hypergeometric random variable except that the sampling is done WITH replacement instead of without. That does make a significant difference in the models, however, as we'll see. In particular, the model is less computationally cumbersome. It's an easier model!

This new model—for a reason which we explain later—doesn't have a name of its own. So we just refer to it as "sampling with replacement" (the random experiment) and speak of "how many in the sample have the characteristic" (the random variable). This is probably better than relying on a name anyway. You won't be able to forget so easily what it refers to.

We learned in the previous section that the hypergeometric random variable is not used as a model for sampling without replacement in the case of large populations. Instead, we use this new, simpler model, "sampling WITH replacement."—even though we're sampling without replacement. What justifies this? It's not the same model, after all, and will not give the same "answers." True. But if the population is large, it gives a very good approximation. It's clear why this should be the case. Think about it: Sampling with replacement allows the possibility of selecting the same population member twice. In sampling without replacement, this is impossible. But so what? If the population is very, very large, there's only a very, very small probability of selecting the same member twice. It doesn't happen, virtually speaking. So when the population is large, there's no significant difference between the two models. Use the simpler one!

Now let's think about constructing the probability distribution for this new, simpler model. You can do it yourself with the help of the following exercises. But here, in contrast to the argument for the hypergeometric random variable, you'll not use the relative frequency definition of probability. It's easier than that! The formulas are more naturally derived by analyzing the relevant events and using the following probability rules:

$$\begin{aligned} \mathrm{P}(A \text{ or } B) &= \mathrm{P}(A) + \mathrm{P}(B) \quad \text{if } A \text{ and } B \text{ are mutually exclusive} \\ \mathrm{P}(A \text{ and } B) &= \mathrm{P}(A)\mathrm{P}(B) \qquad\quad \text{if } A \text{ and } B \text{ are independent.} \end{aligned}$$

As you carry out the details, you'll need to justify the "mutually exclusive" and "independence" assumptions. Please . . .

Try Your Hand

3.4.1 Before we attempt to derive formulas for calculating the probability distribution for sampling with replacement, you should first:

(a) Describe the underlying random experiment for the random variable.

(b) Describe the outcomes of the experiment.

(c) Describe the possible values of the random variable.

3.4.2 Suppose we're drawing a sample of $n = 10$. Let's calculate $\mathrm{P}(X = 3)$. Recall the usual notation: p is the proportion of the population having the characteristic of interest and $1 - p = q$ is the proportion NOT having the characteristic.

(a) First compute the probability of the event "the first three selected for the sample have the characteristic, the rest do not."

(b) The probability in part (a) is NOT $\mathrm{P}(X = 3)$. Why not?

(c) How many different ways can we have a sample with three having the characteristic of interest and the rest not?

(d) What's the formula for $\mathrm{P}(X = 3)$?

(e) If 74% of the population have the characteristic of interest, what are the chances to get a sample with three having that characteristic?

(f) What role does the population size play in the analysis of part (e)?

3.4.3 (a) Give the formula for $P(X = x)$.

(b) What role does the population size play in part (a)?

The Formulas

Here are the formulas for the random variable which counts how many have the characteristic of interest when we are sampling with replacement from a dichotomous population:

sampling with replacement:

For sampling with replacement, if X is the number of observations in the sample having the characteristic of interest

$$P(X = x) = C(n, x)p^x q^{n-x}$$

with

$$\mu = np,$$
$$\sigma^2 = npq.$$

The ***parameters for sampling with replacement*** are just:

n = the sample size,
p = the proportion of the population having the characteristic of interest.

Note that the mean is the same as for the hypergeometric random variable and the formula for the variance is very similar. In fact, except for the finite population correction factor, the variance is the same also. Recall that the finite population correction factor was (approximately) just the proportion of the population not in the sample. So, if the sample was not large compared with the population—if n was not large compared with N—the finite population correction factor for the hypergeometric random variable was almost one. In that case, the variances for that model and the model of this section are approximately equal.

That's one reason why we said you don't need the hypergeometric random variable if you're sampling from a large population—your sample will not be a significant proportion of the population, the finite population correction factor will be approximately equal to one, and so the variance is essentially the same as with this new model. This explains the term "finite population correction factor"—you think of it as "correcting" the variance for the simpler model given in this section for large (or infinite) populations. Of course, the mean is also the same. And the probabilities will be approximately the same.

Here's a rule of thumb: The model for sampling with replacement is a good approximation for the hypergeometric distribution if

$$N \geq 60 \text{ and } N \geq 10n.$$

So, the population is large and the sample is small in comparison.

Now please . . .

Try Your Hand

3.4.4 Describe verbally the meaning of the condition $N \geq 10n$ given in the box just above.

3.4.5 Here we recall the situation of Problem 3.3.5 which was not suitably modeled by the hypergeometric random variable:

> The neighborhood library wants to know how many books listed in the catalog are lost, nowhere to be found. To address this problem, you might take a sample—a simple random sample, without replacement—of the catalog listing and consider the number of listings which correspond to lost books.

(a) What model is appropriate for this problem? Two explanations are possible. To see the situation more clearly, suppose we're going to take quite a large sample of a couple of hundred listings and use several staff people to carry out the work.

(b) Give a verbal description of the finite population correction factor.

(c) What's the probability that your sample of 200 listings contains three or more listings for lost books?

(d) Before doing part (b) you might have asked yourself how many lost books you should expect on average—just to get an idea of what kind of answer to look for. Well, how many?

(e) In the original problem (Problem 3.3.5) we said: "Suppose you find that two among a sample of 30 listings are for lost books." What would be the chances of that happening if, in fact, one percent of the listed books are lost? What are the chances that more than two of the books are lost?

(f) In part (e), how many lost books should you have expected to find among the 30 observed catalog listings?

(g) In part (e) what values of the parameters were specified?

3.5 The Bernoulli Trial

The very very simple model we're introducing in this section is not of interest in its own right. But it serves as the basic building block for three other models and those models ARE of great importance. A ***Bernoulli trial with parameter p*** is a random experiment with exactly two possible outcomes, where the outcome of interest has probability p. This "outcome of interest" is referred to as "success" and is denoted by the symbol S. Thus, $\mathrm{P}(S) = p$. The other possible outcome is denoted F for "failure." Because there are only two possibilities, we can conclude that $\mathrm{P}(F) = 1 - p$, which is also denoted by q. The ***Bernoulli random variable*** is the "number of successes" when you perform the trial once. So it's zero or one. You're already quite familiar with this very simple model, so please just . . .

Try Your Hand

3.5.1 (a) Give a simple example of a Bernoulli trial.

(b) Give a formula for the mean and variance of the Bernoulli random variable.

3.5.2 (a) Show that "select one at random" from a dichotomous population is an example of a Bernoulli trial.

(b) The mean and variance for the Bernoulli random variable in part (a) are p and pq, respectively. Suppose instead of modeling this as a Bernoulli trial, you modeled it as "sampling from a dichotomous population." Then what are the mean and variance?

3.6 The Geometric Distribution

What Is the Model?

Suppose you're attempting to locate someone with a rare blood type. This very natural situation gives rise immediately to a random variable: the number of persons you must test to find one who has the blood type you seek. The underlying random experiment is the process of testing individuals for blood type until you find someone with the desired blood type. It can be described as "independent repetitions of a Bernoulli trial." In our example, the Bernoulli trial is "test one person for blood type." We can think of the outcomes as "yes" or "no," depending on whether the person tested does or does not have the desired blood type. The outcome of interest is "has the blood type we seek." The probability p for this outcome is just the proportion of the population being tested which have that blood type.

Now, if we assume the repetitions of this Bernoulli trial to be independent—that blood type from one person to the next is independent—then the probabilities for our random variable will be easy to compute. Recall the definition: $X = \#$ persons you must test to find one who has the required blood type. Let's calculate the probability we would first find the desired blood type with the third person tested:

$$\begin{aligned} P(X = 3) &= P(\text{1st person is "no" AND 2nd person is "no" AND 3rd person is "yes"}) \\ &= P(\text{1st person "no"}) \times P(\text{2nd person "no"}) \times P(\text{3rd person "yes"}) \\ &= (1 - p) \times (1 - p) \times p \\ &= q^2 p \qquad \text{because } q = 1 - p. \end{aligned}$$

In general, the ***geometric random variable*** is the number of independent repetitions of a Bernoulli trial necessary to observe the first "success." The underlying random experiment for this random variable is

> independent repetitions of the Bernoulli trial,
> stopping once you observe a "success."

Now please . . .

Try Your Hand

3.6.1 Verify that the "very natural situation" of attempting to locate someone with a specific blood type is indeed a random experiment. If it were not, we could hardly claim to have a random variable in the number of persons required to be tested to find one having the desired blood type.

3.6.2 In the previous problem, what is the Bernoulli trial?

3.6.3 Is the independence assumption reasonable in the "blood type" example?

3.6.4 (**a**) What are the possible values of the geometric random variable?

(**b**) What are the parameters for the geometric random variable? Give the symbols and verbal descriptions.

3.6.5 Suppose eight percent of the population you're testing have the blood type in question and that this population has no persons who are blood relatives. What is the probability that

(**a**) you finally find a person of the desired blood type after 17 trials?

(**b**) the first person tested has the desired blood type?

(**c**) you never find anyone of the desired blood type?

Now, with X as the number of persons you must test before finding one with the desired blood type, what is the probability that

(**d**) $X = 3$?

(**e**) $X = 7$?

(**f**) $X = 0$?

3.6.6 What's the significance for the geometric model of the assumption in the previous problem about "no persons who are blood relatives"?

3.6.7 Why will it not be possible to summarize the geometric random variable by means of a distribution table?

3.6.8 (**a**) Is the geometric random variable a statistic?

(**b**) Is its probability distribution a sampling distribution?

The Formulas

From the exercises above, you may have guessed that there is a simple formula for the probabilities of the geometric random variable:

$$\begin{aligned} \mathrm{P}(X = x) &= \mathrm{P}\big(\ (x-1) \text{ failures AND one success }\big) \\ &= q^{x-1}p. \end{aligned}$$

Because the repetitions of the Bernoulli trial are independent, we use the multiplicative law in its simple special case—just multiply the probabilities. To see how the formula above is derived, you need only recall that the statement "the random variable X takes on the value x" just means that the first success occurred on the xth repetition of the Bernoulli trial. So the event that $X = x$ is just the outcome

$$F,\ F,\ F,\ F,\ F,\ F,\ F,\ F, \ldots,\ F,\ F,\ \underset{\substack{\uparrow \\ x\text{th position}}}{S}$$

where this means

"failure" AND "failure" AND "failure" AND . . .
AND "failure" AND "success."

Now, with *independence*, we can simply multiply probabilities.

We summarize the geometric random variable in the box below. Note, however, that it would be difficult for you to derive the formulas for the mean and variance because they involve infinite sums. For example,

$$\mu_X = \Sigma X \mathrm{P}(X),$$

where $X = 1, 2, 3, \ldots$ to infinity. We're not assuming you've studied the theory of such infinite sums, so-called "infinite series." We're just asking you to accept that there's an appropriate theory which produces the formulas below.

the geometric random variable:

> The geometric random variable X is the number of independent repetitions of a Bernoulli trial required to obtain exactly one success,
>
> OR EQUIVALENTLY X is the number of the repetition on which the first success appears.
>
> $$\begin{aligned} P(X = x) &= q^{x-1}p, \\ P(X \leq x) &= 1 - q^x. \end{aligned}$$
>
> Recursion formula:
>
> $$P(X = x + 1) = qP(X = x).$$
>
> Finally:
>
> $$\begin{aligned} \mu &= 1/p, \\ \sigma^2 &= q/p^2. \end{aligned}$$

The ***parameter for the geometric distribution*** is:

> p = the probability of success on one repetition of the Bernoulli trial

Because we will not have a complete table for the probabilities of X, you will find the cumulative distribution function for X given above to be useful. For any random variable, the ***cumulative probability distribution function*** is $P(X \leq x)$. For the geometric random variable, this is the probability that X takes the value 1 OR 2 OR . . . OR x. Because X cannot take on two different values at once—different values represent mutually exclusive events—the cumulative distribution function is just the sum (hence the word "cumulative") of all the probabilities up to and including x:

$$P(X \leq x) = P(X = 1) + P(X = 2) + P(X = 3) + \cdots + P(X = x).$$

For the geometric random variable, this is the probability for the first success to occur *on or before* the xth repetition of the Bernoulli trial. As you see above, this probability is just $1 - q^x$. To understand all of this better, please . . .

Try Your Hand

In the following problems, use the formulas in the box above.

3.6.9 You're attempting to locate a person of a particular blood type. Suppose eight percent of the population you're testing have the blood type in question and that this population has no persons who are blood relatives.

(a) How many persons would you expect to have to test before finally finding someone of the desired blood type?

(b) What's the probability you finally find a person of the desired blood type on the 15th trial?

(c) What's the probability you do not need to test more than 15 persons before finding the desired blood type?

(d) What's the probability you find a person of the desired blood type only after the 15th trial?

(e) What's the probability the first person tested has the desired blood type?

(f) What's the probability you never find anyone of the desired blood type?

(g) What's the probability you do not find a person of the desired blood type until sometime after the 10th person is tested?

Now, with X as the number of persons you must test before finding one with the desired blood type, what's the probability that

(h) $X \geq 3$?

(i) $X = 7$?

(j) $X \geq 0$?

(k) $X \leq 22$?

3.6.10 By Chebyshev's Theorem, for any random variable there is less than one chance in nine to have a value more than three standard deviations away from the mean. Show that for the specific case of the geometric random variable discussed in Problem 3.6.9, this probability is actually MUCH smaller than one in nine.

3.6.11 This problem will be difficult as we're stating it here. By analogy with Chapter 2, think how the labor of dealing with 50 values can be simplified. Check Level I of the answers before wasting too much time.

Make up a probability distribution for the geometric random variable of Problem 3.6.9, for all values of X through 50. Present this distribution as

(a) a table,

(b) a graph.

3.6.12 (a) You shouldn't think of the mean of a random variable as the "most likely value." Explain.

(b) For the geometric random variable of Problem 3.6.9, what proportion of the distribution falls on either side of the mean?

(c) What's the point of parts (a) and (b)?

3.6.13 Would the number of persons you must test be more predictable or less so if 12% instead of eight percent of the population have the required blood type?

3.6.14 Derive the cumulative probability distribution formula for the geometric random variable.

3.6.15 (a) Derive the recursion formula for the geometric random variable.

(b) Use the recursion formula to generate the probabilities for $X = 1$ through $X = 10$. Assume $p = 0.08$.

(c) Compare part (b) with Problem 3.6.11 (a) .

(d) For any $x > 10$, $P(X = x)$ is less than . . . ??

3.6.16 Various methods have been proposed for estimating the size of wildlife populations. One technique is to trap animals periodically in their home range. When an animal is caught, it's marked and then released. As the trapping proceeds, a record is kept of the number of times each animal is caught. W.R. Edwards and L.L. Eberhardt (1967) did an experiment with 135 cottontail rabbits in a protected enclosed 40-acre area where they repeated the trapping seven times. Here's their data:

Frequency of capture	0	1	2	3	4	5	6	7
Number of rabbits	59	43	16	8	6	0	2	1

For example, 16 of the 135 rabbits were caught on two occasions. The number 59 was not observed, of course; it was inferred from the rest of the data by assuming none of the 135 rabbits died or escaped the area or in some other way were lost from the population.

In a study of an extensive set of live-trapping data, Eberhardt together with T.J. Peterle and R. Schofield (1963) had given several arguments suggesting a geometric distribution as a good model for the capture–recapture experiment with X being "the number of times a particular animal is trapped." This proposes a purely abstract model because there's no Bernoulli trial here to be repeated. Because X starts at zero, the geometric random variable would be, let's say, $W = X + 1$. So $\mathrm{P}(W = w) = pq^{w-1}$ which means $\mathrm{P}(X = x) = pq^{x}$. Edwards and Eberhardt (1967) estimated p to be 0.4424. We'll see later (Problem 7.2.12) how they obtained this value and how they proposed using the model to estimate an unknown population size.

Compare Edwards and Eberhardt's data with what would be expected if indeed the geometric model is valid. One decimal place of accuracy is adequate for comparison purposes.

3.7 The Binomial Distribution

The Binomial Experiment

There are three models based on "independent repetitions of a Bernoulli trial." The geometric random variable of the previous section is one such model where the repetitions continue until we observe the first "success." In this section, we study the second of these models, the "binomial random variable." Its underlying random experiment, the ***binomial experiment***, consists of a FIXED number, n, of independent repetitions of the Bernoulli trial. Of course, with a fixed number of repetitions, we might have any number of successes. The binomial random variable will be that "number of successes." For the geometric model, the number of successes is fixed and the number of trials is the random variable. For the binomial model, it's just the reverse.

As always, before thinking about the random variable, we should make sure the random experiment is clear. The prototypical example of a Bernoulli trial is a coin toss, so you should think of the binomial experiment as a series of n coin tosses with n fixed. There are many real-world situations that look abstractly just like this. For example, an electronics manufacturer may want to model a day's output of 1500 electronic components where the manufacturing process is judged to have one chance in 200 to produce a defective component. Producing one component is like one toss of a coin where the probability for heads is $p = 0.005$ ($= 1/200$). If it's reasonable to assume defective components occur independently, then the 1500 components look abstractly like a series of 1500 coin tosses. Of course, the assumption of independence must be checked carefully—if it's not reasonable, the model is not valid. To verify independence of defects, you may have to consult an engineer who knows how defective components arise.

Note that our example assumes a day's output consists of 1500 "identical" components—identical from the point of view of defect rate. Otherwise, we're not repeating the SAME Bernoulli trial. This same consideration arises for the geometric random variable. But can we say "identical" components? No two physical objects are ever exactly identical. If the components are exactly identical, one component defective means they ALL are. But "exactly identical" is completely unrealistic! That's where our probability models come in. We make use of a probability model to account for the variability that inevitably exists among the components even when they're as identical as physically possible. That is, we assume the components are identical EXCEPT FOR RANDOM VARIATION and we account for the randomness by a probability model.

Here's another example of the binomial experiment: We could model a basketball player's skill, in part, by her probability of sinking a basket on one throw. If any one throw is unaffected by the success or failure of other throws—if the repetitions are independent—her throws are like repeated tosses of a coin.

For a binomial model, the actual value of the parameter p may be unknown. In fact, that may have been the very question you wanted to answer. Maybe the original question was "What's the probability of a defective component?" or "What's the rate of successful throws for this basketball player?" In other words, what's the value of p? If the assumptions seem appropriate, the model will be very useful for estimating the value of p. In fact, one important inferential technique of statistics (see Chapter 5) is for problems of just this type—to estimate the unknown value of some parameter. That requires a model for the underlying real-world situation and a theory for the model. Those models and that theory are what we're learning now.

One final caveat: See the difference between the Bernoulli trial—itself a random experiment—and the binomial experiment, a series of independent repetitions of the Bernoulli trial. It's the difference between one toss of a coin (the Bernoulli trial) and a series of n successive tosses (the binomial experiment).

To understand the binomial experiment better, please . . .

Try Your Hand

3.7.1 Let's explore the binomial experiment further:

(a) For the binomial experiment, what is the "doing" and what does it mean to repeat it?

(b) If $n = 15$ and you repeat the binomial experiment six times, how many repetitions of the Bernoulli trial have you made?

(c) What exactly does an outcome of the binomial experiment look like? Give a schematic description and a verbal description.

(d) Identify exactly where in the definition of the binomial experiment we see that the outcomes are unpredictable.

(e) Give a verbal description of the parameters for the binomial experiment in terms of a series of n coin tosses.

(f) How many of the outcomes of the binomial experiment have exactly k successes?

(g) List all possible outcomes for the binomial experiment where $n = 4$.

3.7.2 (a) Suppose we toss 20 coins into the air. Can this be modeled as a binomial experiment?

(b) Our example of the binomial model for 1500 electronic components is more like part (a) than it is like a series of 1500 coins tosses. Explain.

(c) The binomial model has one assumption that's automatically valid for both coin toss analogies—for both "1500 tosses of one coin" and "one toss of 1500 coins"—but which might fail for the electronics manufacturer's model. What is that assumption and is it reasonable for the electronic components model?

3.7.3 Describe the occurrences of stillbirths in a hospital maternity ward over the course of a week as the outcome of a binomial experiment.

3.7.4 Describe each of the following situations as the outcome of a binomial experiment. You may have to complete the example by supplying missing information. Discuss all aspects of the model with as much completeness as the problem allows:

(a) 112 drillings for oil on a large tract of land,

(b) 25 throws of a dart at a dart board,

(**c**) 15 telephone contacts by a telemarketing salesperson,

(**d**) truthful answers to a sensitive survey question (concerning drug use, sexuality, or some other such sensitive issue which often yields an untruthful answer) put to 1500 persons.

The Binomial Random Variable Itself

As you saw in Problem 3.7.1, an outcome for the binomial experiment consists of a string of n successes and failures:

$$S,\ F,\ S,\ F,\ F,\ F,\ S,\ S,\ S,\ S,\ F,\ F,\ S,\ F,\ F,\ F,\ S,\ F,\ F,\ F.$$

The ***binomial random variable***, assigns to each such outcome the number of successes. If we call the variable X, the outcome listed above is assigned the value $X = 8$. There are lots of other possible outcomes. For example,

$$F,\ S,\ S,\ F,\ F,\ F,\ S,\ S,\ S,\ S,\ F,\ F,\ S,\ F,\ F,\ F,\ S,\ F,\ F,\ F.$$

If you look carefully, you will see that this new outcome is identical to the first one except for the first two trials. Here also $X = 8$. Note that there were 20 trials and so $n = 20$.

The expression, "X is a binomial random variable with parameters n and p" is captured briefly in the conventional symbols $\boldsymbol{X \sim B(n,p)}$. This is often a convenient shorthand. So now we're ready for you to . . .

Try Your Hand

3.7.5 What are the possible values for a binomial random variable?

3.7.6 If X is a binomial random variable with $n = 20$ and $p = 0.3$, $X \sim B(20, 0.3)$, what is $P(X = 8)$? Here's some help:

(**a**) First calculate the probability of one outcome where $X = 8$. For example, what's the probability of

$$S,\ F,\ S,\ F,\ F,\ F,\ S,\ S,\ S,\ S,\ F,\ F,\ S,\ F,\ F,\ F,\ S,\ F,\ F,\ F.$$

(**b**) Now, what's the probability of

$$F,\ S,\ S,\ F,\ F,\ F,\ S,\ S,\ S,\ S,\ F,\ F,\ S,\ F,\ F,\ F,\ S,\ F,\ F,\ F.$$

(c) If A is the outcome in part (a) and B is the outcome in part (b), then P(A or B) =???

(d) How many ways can you have an outcome for which $X = 8$?

(e) What's the value of P($X = 8$)?

(f) If you had rounded at intermediate steps in the calculation of part (e), you would have obtained a very wrong answer! Explain.

3.7.7 In the previous problem, how many successes would you expect on average? Try to guess on intuitive grounds.

3.7.8 Let X_k be the Bernoulli random variable for the kth repetition of the Bernoulli trial (X_k ="the number of successes"). There are n of these X_k's , one for each repetition. Let's not be completely abstract, suppose $n = 20$ and $p = 0.3$ as in Problem 3.7.6.

(a) What's the relationship between the X_k's and the binomial random variable X.

(b) Use part (a) to derive a formula for the mean and variance of X.

3.7.9 For the binomial random variable:

(a) Derive a formula for P($X = x$).

(b) Derive a recursion formula.

3.7.10 (a) Show that the binomial random variable can be described as "the total or sum of a sample" for a certain kind of sampling.

(b) Show that X ="the number of observations in the sample having the characeristic of interest" for sampling with replacement from a dichotomous population is a binomial random variable. Give a verbal description of n and p.

(c) Part (b) answers a question we raised earlier. What was that question?

(d) For sampling with or without replacement from a large dichotomous population, we should use the binomial random variable for the number of observations in the sample having the characteristic of interest. Explain why.

From Problem 3.7.9, we obtain this summary of

the binomial random variable:

The binomial random variable X counts the number of successes on n independent repetitions of the same Bernoulli trial

$$X = \text{the number of } S\text{'s}$$
$$\mathrm{P}(X = x) = C(n, x)p^x q^{n-x}.$$

Recursion formula:

$$P(X = x + 1) = \frac{(n - x)p}{(x + 1)q}\mathrm{P}(X = x).$$

And:
$$\mu = np,$$
$$\sigma^2 = npq.$$

The ***parameters for the binomial distribution*** are:

n = the number of repetitions;
p = the probability for success on the Bernoulli trial.

Now you're ready to use the binomial model. Please . . .

Try Your Hand

3.7.11 We can model a basketball player's skill by the probability, p, of sinking a basket on one throw. For Shu Wen, $p = 0.17$ and for Juan, $p = 0.12$.

(a) Who is more predictable in basketball at sinking baskets, Shu Wen or Juan?

(b) How many baskets would Shu Wen have to attempt before sinking one?

(c) How likely is Juan to sink a basket on at least three of his first ten attempts?

(d) The further p is from one half, the more predictable that player is at sinking baskets. Explain this by reference to the appropriate formula.

(e) Explain why part (d) is reasonable from an intuitive point of view.

3.7.12 In 1953 (*Avery* v. *Georgia*), a black defendant was convicted by a jury selected from a panel of 60 "veniremen" (potential jurors). None of the 60 was black. They were chosen from the jury roll by drawing from a box containing tickets with the names of potential jurors, yellow tickets for blacks and white tickets for whites. Five percent of the tickets were yellow. Justice Frankfurter of the U.S. Supreme Court wrote that the "mind of justice, not merely its eyes, would have to be blind to attribute such an occasion to mere fortuity." Do you agree? (after [Finkelstein and Levin], p. 114).

3.7.13 Use the recursion formula to generate the binomial distribution with $n = 5$ for each of the following choices of the parameter p. Present the distribution in two ways: with a table AND with a line graph. Use the tables to compute the mean and variance and check that you get the answer given by the formulas (which answer is more accurate?).

(a) $p = 0.5$,

(b) $p = 0.25$,

(c) $p = 0.1$.

3.8 The Poisson Distribution

The Poisson distribution is a more abstract model than any we have seen so far. It can be derived through a mathematical limiting process from the binomial distribution by letting n get larger and larger as p gets smaller and smaller. There are other derivations as well. We'll not be concerned with the technical details of the derivation, but you will want to remember that the Poisson model was originally derived through an abstract process, not by any real-world considerations. This insight will clarify some of the discussion below.

The Poisson distribution seems to have first appeared in a treatise on probability and the law published in 1837 by Siméon Denis Poisson, a French academician and scientist. It only came into its own as a probability model 50 years later, with a publication in 1898 by Ladislaus von Bortkiewicz, a Russian-born Pole working in Germany. Bortkiewicz seems to have captured people's imagination with his use of the Poisson distribution as a model for the observed number of horsekick fatalities in the Prussian army from 1875 to 1895—a problem of more minor proportions in our own age! Since then, the Poisson distribution has proven useful in a wide variety of situations which we will illustrate as we go along.

So unlike all the previous models we've seen, we do not derive the abstract model from a real-world situation. Rather we "find" a model, derived independently from theoretical considerations, which happens to fit our data, a very common approach to model building. We'll explain this in more detail later when we take a look at Bortkiewicz' data to see how it "fits" the Poisson model, but first let's see the model itself. We're not going to give you a formal definition of the Poisson random variable—that would mean presenting the abstract derivation—rather we will give a "rule of thumb" description of the kinds of real-world situations in which the model has often been found appropriate.

The ***Poisson random variable*** is often appropriate for counting occurrences within some fixed interval of time (or space) for independent events such as accidents—Bortkiewicz' horsekick fatalities, for example—or arrivals of customers at a checkout counter or of telephone calls at a switchboard, and so on. It's commonly used in biostatistics as a model for the incidence of disease. The Poisson distribution also models such situations as the number of defects in a bolt of cloth or typographical errors in a magazine article. These last two examples involve intervals of space: A bolt of cloth is a two-dimensional "interval of space" as are the pages of a magazine article.

When might you expect a real-world situation to be appropriately modeled by a Poisson distribution? The model should be valid for any situation in which you are observing occurrences of something that looks like an "accidental" event, an event where

- simultaneous occurrences are impossible,
- any two occurrences are independent,
- the expected number of occurrences in any interval is proportional to the size of the interval (length, area, volume, depending on the type of interval).

These three conditions would often be reasonable in examples like those mentioned above. In a very broad sense, they capture what one means by "accidental" occurrences. For example, the incidence of noncontagious diseases or defects in a manufactured item are "accidents," broadly speaking. But certainly the model is not restricted to accidental occurrences. For example, these three conditions would often be satisfied for customer arrivals or arriving telephone calls and yet these are not "accidents," although in a certain sense they're like accidents.

The formulas for the Poisson distribution involve the ***natural exponential function,*** e^x. Here e is a constant, approximately equal to 2.7183 as you can see from your calculator using the e^x key with $x = 1$. The word "natural" refers to the especially simple character of this exponential function from the point of view of differential calculus. We will also have occasion later in this text to use the logarithm to the base e. It's called the ***natural logarithm, ln(x)*** (pronounced "lin" x). It has the same properties as the common logarithm (the logarithm to the base ten). In particular,

$$\ln(1) = 0, \ \ln(e) = 1, \ \ln(xy) = \ln(x) + \ln(y), \ \ln(x^y) = y \ln(x).$$

The last property is very convenient for solving equations where the unknown is in the exponent. To take a simple example, if $17 = 15^x$, what is x? Well, $\ln(17) = x\ln(15)$ and so, as your calculator will show you, $2.8332 = 2.7081x$. Solving for x, you find $x = 1.0462$.

As with the geometric distribution, there is, theoretically, no largest value for the Poisson random variable. There's one parameter for the model, denoted by λ (the Greek letter lambda). As you can see in the formulas below, it's the expected number of occurrences within the interval in question. It just so happens—for no reason that could be obvious to us—that the variance is also λ, the same as the mean. The probability distribution of the Poisson random variable—let's call it X—is determined by these equations:

the Poisson random variable:

$$P(X = x) = \frac{e^{-\lambda}\lambda^x}{x!}.$$

Recursion formula:

$$P(X = x + 1) = \frac{\lambda}{x+1} P(X = x).$$

And:

$$\mu = \lambda,$$

$$\sigma^2 = \lambda.$$

The ***parameter for the Poisson distribution*** is:

$\lambda =$ the expected number of occurrences in the interval in question.

Now, to understand the model better, here are some exercises for you to . . .

Try Your Hand

3.8.1 Think about the number of automobile accidents per year at a busy metropolitan intersection. Assume a Poisson model for this random variable. Compute the probabilities asked for below two ways:

(**i**) use the first formula given in the box above;

(**ii**) use the recursion formula.

(**a**) If you should expect 4.2 accidents per year at this intersection, what's the probability of less than two accidents?

(**b**) What's the probability of less than two accidents in a six-month period?

3.8.2 Consider the number of telephone calls arriving at a telephone switchboard in a five-minute interval. Again, assume the Poisson model for this situation.

(**a**) If you should expect 2.3 calls in a five-minute period, what's the probability of more than two calls?

(**b**) What's the probability of more than five calls? [Hint: Use part (a).]

(**c**) Do the calculation of part (b) again using the recursion formula.

(**d**) Give a verbal description of the Poisson recursion formula.

3.8.3 The Poisson distribution shares a characteristic in common with the geometric distribution which none of our other distibutions (so far!) exhibits. What is it?

3.8.4 The variance of the Poisson distribution is the same as the mean, namely, λ. Why is this fact explained "for no reason that could be obvious to us," as we said in the text?

3.8.5 Why do we say "noncontagious" when we talk about the incidence of disease being appropriately modeled by the Poisson distribution?

3.8.6 If observed data seems to fit the Poisson model, does that mean the data arose from a situation which satisfies the three rules of thumb?

3.8.7 Bortkiewicz studied 14 Prussian Army corps over a period of 20 years. We eliminate four of the corps as atypical; they were organized differently from the other ten. That leaves ten corps over 20 years, giving us 200 "corps-years." Let B be the number of fatalities in one year and let CY be the observed number of corps-years in which that number of fatalities occurred. Here's Borkiewicz' data:

B	CY
0	109
1	65
2	22
3	3
4	1
	200

(a) Give a verbal description of the number 109 from the table.

(b) Show that there were 122 fatalities over the 200 corps-years.

(c) What value of λ should you use for the Poisson model, assuming the model to be valid?

(d) Make up an "empirical" probability distribution for B, based on Bortkiewicz' observations. Compute the mean and $\hat{\sigma}^2$ for Bortkiewicz' observed data using this distribution.

(e) Make up a theoretical probability distribution for B, assuming B to be Poisson, using the value of λ in part (c). Use this distribution to approximate the mean and variance of B.

(f) Do you think Bortkiewicz' data fit the Poisson model?

3.8.8 Assume the Poisson model suggested by Bortkiewicz' data from the previous problem:

(a) In how many years over a ten-year period would a Prussian army corpsman have seen more than one of his comrades killed as a result of a horsekick? [answer: about 1.3]

(b) After how many years in the army would a Prussian army corpsman have first seen a year in which more than one of his comrades was killed as a result of a horsekick? [about eight]

(c) If you randomly chose three of the ten corps studied in the previous problem and looked at records for a five-year period, what are the chances you would observe more than four corps-years with no horsekick fatalities? [about 97%]

3.8.9 Solve the equation $e^{2x} = 14$ for x.

3.8.10 Because the Poisson model is obtained from the binomial by letting n get larger and larger, you won't be surprised to learn that the Poisson distribution provides a very good approximation to the binomial when n is large. For this to work, however, p must be small. There is a "rule of thumb" for the validity of the Poisson approximation

$$\begin{aligned} n &\geq 20, \\ p &\leq 0.05. \end{aligned}$$

This approximation can be very useful if you are relying on tables for your probabilities. A question involving binomial probabilities may take you out of the table if n is quite large. After all, any table is finite and will stop with some large value of n. Many tables stop with $n = 25$ or maybe $n = 100$. In these cases, the Poisson approximation will be helpful if p is not too large ($\leq 5\%$) and if your Poisson table contains entries for $\lambda = np$. On the other hand, in working binomial problems, if p is too small you will again find yourself out of the table. The Poisson approximation may help if n is large enough (≥ 20).

In other cases, you might prefer the Poisson approximation to the binomial if you're going to use the recursion formula—the Poisson recursion formula is much easier to use with a hand calculator than the binomial!

(a) For a given binomial distribution, which Poisson distribution should you select as the approximation? Try to make a reasonable guess. Think first about a specific case; for example, take the binomial random variable B(300, 0.01). Which Poisson distribution would you use?

(b) Show that a real-world situation which could be modeled by the binomial distribution with very large n and small p would satisfy the three rules of thumb which guide us in modeling real-world situations by the Poisson distribution. [Hint: Think of the binomial experiment as taking place over an interval of time.]

(c) Suppose $n = 300$ and $p = 0.01$. Compute the probability that the binomial X is equal to two and compare that with the approximation given by the Poisson distribution.

(d) To have the Poisson approximation be valid, we choose it so it has the same mean as the given binomial. Of course, we would also want the variances to be the same, but that's not possible! Why not?

(e) Show that the variance of the approximating Poisson will be *approximately* the same as the given binomial variance IF . . . ???

Here is a summary of Problem 3.8.10:

The Poisson approximation for the binomial distribution:

For a binomial random variable with . . .

$$n \geq 20,$$
$$p \leq 0.05,$$

the Poisson distribution with $\lambda = np$ will give a good approximation.

The parameter n must be large so that the probability formulas for the two models will give approximately the same values and p must be small so that the variance of the Poisson, λ, will be approximately the same as the variance of the binomial, npq.

3.9 The Negative Binomial Distribution

In this section we consider a generalization of the geometric distribution. First, recall our example of a geometrically distributed random variable:

> Suppose you're attempting to locate someone with a rare blood type. This very natural situation gives rise immediately to a random variable: the number of persons you must test to find one who has the blood type you seek. The underlying random experiment is the process of testing individuals for blood type until you find someone with the desired blood type.

But suppose instead of just one person having the desired blood type, you want, say, eight. How many persons must you test before finding EIGHT with the desired blood type? This requires the so-called "negative binomial" random variable with parameter k. Here, $k = 8$. Formally, the ***negative binomial random variable*** is the number of independent repetitions of a Bermoulli trial necessary to observe exactly k successes.

Compare the negative binomial model with the geometric model. The geometric distribution has only one parameter, p, the probability of success on one execution of the Bernoulli trial. For the negative binomial model there are two parameters, p and k, where k is the number of successes you must observe before you end the experiment. Note that for the special case $k = 1$, we have the geometric model. That's why we said the negative binomial model is a generalization of the geometric model.

Your experience in working with the geometric and binomial random variables will make it easy for you when you . . .

Try Your Hand

In the following problems, assume that X is a negative binomial random variable with $k = 8$.

3.9.1 (a) What does an outcome of the underlying negative binomial experiment look like? Give a verbal description.

(b) Give three specific outcomes.

(c) For the examples in part (b), what are the values of X?

(d) In general, what are all the possible values of X?

(e) What are the probabilities for the three examples you gave in part (b)?

3.9.2 How many ways can you get an outcome with

(a) $X = 10$?

(b) $X = 12$?

(c) $X = 9$?

(d) $X = 8$?

(e) $X = x$?

3.9.3 Now assume $p = 0.42$. What is

(a) $P(X = 10)$?

(b) $P(X = 12)$?

(c) $P(X = 9)$?

(d) $P(X = 8)$?

(e) $P(X = x)$?

3.9.4 Give the formula for $P(X = x)$ for general k.

Here's a summary of

the negative binomial random variable:

The negative binomial random variable X counts the number of independent repetitions of a Bernoulli trial required to observe k successes.

$$P(X = x) = C(x-1, k-1)p^k q^{x-k};$$

Recursion formula:

$$P(X = x+1) = \frac{xq}{x-(k-1)} P(X = x).$$

And:

$$\mu = \frac{k}{p}$$

$$\sigma^2 = \frac{kq}{p^2}.$$

The *parameters for the negative binomial distribution* are:

$p =$ the probability for success on the Bernoulli trial,
$k =$ the number of successes to be observed, after which you end the experiment.

The negative binomial model shows up in a wide variety of situations. In biology, it has been used as a model for insect counts. R.A. Fisher used it, for example, to model the number of ticks to be found on a sheep. For extensive applications in marketing, see Ehrenberg's book *Repeat Buying*. An alternative characterization of the negative binomial random variable as a sum of k independent geometric random variables with the same p is often useful. Although that description is not obvious in terms of "repetitions of a Bernoulli trial," it can be justified abstractly by showing it leads to the same probability formulas.

Well, now please . . .

Try Your Hand

3.9.5 Show that if $k = 1$ in the formulas above, you get the "correct result."

3.9.6 You're attempting to locate six persons of a particular blood type. Suppose eight percent of the population you are testing have the blood type in question and that this population has no persons who are blood relatives.

(a) How many persons would you expect to have to test before finally finding six people of the desired blood type?

(b) What is the probability that you finally have six persons of the desired blood type with the 15th trial?

(c) What is the probability that you do not need to test more than 15 persons before finding six persons with the desired blood type?

(d) What is the probability that you have found all six persons of the desired blood type only after the 15th trial?

(e) How many persons would you expect to test before finding someone having the desired blood type?

(f) How many persons would you expect to test before finding two having the desired blood type?

3.9.7 By Chebyshev's Theorem, for any random variable at all, there is less than one chance in nine to have a value more than three standard deviations away from the mean. Show that for the specific case of the negative binomial random variable discussed in Problem 3.9.6, this probability is much smaller.

3.9.8 Suppose the cloth which your company buys has about 0.62 serious defects per bolt. How likely is it that you would receive five or more bolts before getting three bolts with more than one serious defect?

3.9.9 In Problem 3.6.16, a negative binomial distribution for the frequency of capture of cottontail rabbits would be appropriate if, as you would expect, we have more than one trap per home range widely enough dispersed to assume independence of capture from one trap to another. Explain.

3.10 Some Review Problems

For each problem, identify the model clearly by name and verify that the model is appropriate for that situation. Most problems will involve a formula from the model; show clearly how you get your answer from that formula even though in a few cases the solution is obvious on intuitive grounds without reference to the model. If a problem requires an unstated assumption, work the problem under that assumption and be prepared to comment on the appropriateness of the assumption for that particular situation. We're leaving you partly on your own with all this. The solutions are rather sketchy. Sorry!

3.10.1 In a game of throwing darts at a dart board, your skill is such that you hit the bull's eye for a score of six points about ten percent of the time, you hit the second ring for three points about 60% of the time, the outer ring about 25% of the time for a score of two points, and you miss the board entirely with a penalty of two points (you lose two points) about five percent of the time. You're eliminated from the game if you miss the board more than three times. Suppose you play with an opponent who uniformly attains each of the possible scores on each throw. Suppose a "round" consists of one throw for each player, two throws altogether.

(a) Assuming you don't get eliminated before that, what's the probability you hit the bull's eye for the first time on your fifth throw? [6.56%]

(b) Assuming you don't get eliminated before that, what's the probability you don't hit the bull's eye before your fifth throw? [$\approx$ 66%]

(c) Assuming you don't get eliminated before that, what's the probability that you will hit the bull's eye at least four times on your first ten throws? [1.28%]

(d) Assuming you don't get eliminated before that, how many times would you expect to throw the dart to hit either the bull's eye or the second ring? [1.4286]

(e) Assuming you don't get eliminated, which is more predictable for you, the number of bull's eyes in ten throws or the number of times you actually miss the board entirely? [2nd]

(f) How many times would you expect to throw before being eliminated? [80]

(g) What score should you expect from your opponent on each throw? [2.25]

(h) From the point of view of score, who is the more consistent player, you or your opponent? [you]

(i) What's the probabilty your opponent is eliminated from the game on the 16th throw? [0.0563]

(j) How likely are we to see a bull's eye in one round? [0.325]

(k) Assuming no one's eliminated before that, how likely are you to see four rounds out of ten in which there is a bull's eye? [22.16%]

(l) Assuming you don't get eliminated before that, what's the probability of at least one bull's eye in your first six throws? [46.86%]

(m) How likely is it you will be eliminated before your sixth throw? This is possible but not likely, keep five decimal places in your answer so you won't report a probability of zero. [0.00003]

(n) What's the probability you get at least one bull's eye on your first six throws?

> [In part (n), being eliminated before you can make your sixth throw is NOT independent of the event in question! So this is not the same as part (l). But they're almost the same; to see the difference, use the 0.00003 answer from the previous part and give your answer with all the accuracy of your calculator: 0.468544943 compared with 0.468559 for part (l).]

3.10.2 **(a)** You're offering customers a gift of one box of a particular brand of tea. Suppose there are 80 boxes of that brand on your shelves in two flavors, one spicy and one not, with 30 boxes of spicy tea. You believe your customers prefer the spicy tea, but only one of the first five customers who arrive chooses spicy tea. If these customers really had no preference at all and they choose at random, what's the likelihood of a result such as you observed? Interpret the phrase "such as you observed" to mean "the result you observed or a result even more inconsistent with your observation." [0.3755]

(b) You are to select a committee of 50 from a group of which 18% are Hispanic. What's the probability of fewer than four Hispanics on the committee? [≈ 0.0137]

3.10.3 The city engineer's data suggests that, on average, in a quarter mile of city streets there are presently about two-tenths of a pothole requiring repair. Assume repair teams of three workers are assigned to two miles of streets.

(a) How many repairs should a team be prepared to make? [1.6]

(b) If the city engineer sends out ten teams, how likely is it that fewer than three of the teams will find more than two potholes to repair? [62.72%]

(c) Martha's team takes off for the beach on days when they find two or fewer potholes—they repair them quickly and leave. If her team goes out 20 days per month, what're the chances they have to work five days or more successively without going to the beach? [$\approx$ zero]

(d) What's the probability Martha's team enjoys no more than 16 days at the beach per month? [65.78%]

(e) How many beach days could Martha's team anticipate each month? [$\approx$ 16]

3.10.4 Suppose during any ten-minute period of the two hour lunch rush about three customers on average come to our service window.

(a) What is the probability of exactly five customers in one such ten-minute period?

(b) Which values of this random variable fall within two standard deviations of the mean?

3.10.5 Suppose you play a game in which you pay four dollars for each roll of a die and you receive one dollar for each dot which shows on the uppermost face when you roll. Suppose the die is loaded so the face with three dots comes uppermost 40% of the time with all other faces equally likely.

(a) What is your expected gain (loss) in this game?

(b) Is this game more or less predictable than playing with a fair die? What does this mean in terms of your cost to play the game?

(c) Make a probability distribution for your gain/loss and compute its mean and variance.

(d) Express G from part (c) in terms of X, the number of dots on the uppermost face; that is, give a formula for G in terms of X.

(e) Compare the mean and variance of G and X.

(f) How many rolls would you expect to make before sustaining a loss? [$\approx$ 1.5]

(g) What's the probability that your first loss would be on the third roll? [≈ 8%]

(h) What's the probability that your first loss would be only after the third roll? [≈ 5%]

(i) Suppose you've decided to quit playing once you have sustained a loss five times. What's the probability you will play for 15 rolls? [0.0039]

(j) In part (i), how many times would you expect to play before quitting? [≈ eight times]

3.10.6 Twelve identical machines operating independently produce defective parts randomly three percent of the time. A box of one dozen parts contains one part from each machine:

(a) What's the probability that in one box no more than one part will be defective? [≈ 95%]

(b) What's the average number defective in one box? [less than one]

(c) Suppose the machines produce defectives only eight tenths of a percent of the time. Is a box more or less reliable than before? [more]

(d) A quality control inspector passes a box only if none are defective. With a three percent defect rate, when should you find the first box that must be rejected? [about the third]

(e) With a three percent defect rate, what's the probability the quality control inspector inspects exactly five boxes before rejecting one? [≈ 7%]

3.10.7 Suppose you draw the top five cards from a well-shuffled deck of 52 cards.

(a) How many spades do you expect on average? [1.25]

(b) Which is more predictable, spade versus nonspade or black versus red?

(c) What's the predicted number of black cards among the five?

(d) What's the probability of two spades? [≈ 28%]

(e) What's the probability of two red cards? [≈ 33%]

(f) How many five-card hands would you expect to receive before getting one with two red cards? [≈ 3]

(g) What're the chances you get the first hand with two red cards only after the third deal? [$\approx$ 31%]

3.10.8 Suppose you and a friend each bring a ball to the tennis court and you find a third ball lying on the court. Suppose all three balls are identical in appearance and you leave one ball on the court when you go home. Thus, which ball each of you takes home is random.

(a) Make a probability distribution for the number of players, X, who take home the same ball they brought. Use your distribution to compute the mean and variance of X.

(b) What're the chances you take home the ball you brought?

(c) What're the chances at least one of you brings home the ball you brought? [fifty-fifty]

(d) If you repeat the same ritual tomorrow, what's the chance that on both days you bring home your own ball? [one chance in nine]

(e) Suppose you play under these conditions every day for a week. What are the chances you both bring home your own ball on four days or more? [$\approx$ 2%]

(f) Again, playing every day for a week, on how many days should you find yourselves both bringing home your own balls?

(g) If you're playing under these conditions, how many days should pass before you see yourselves both bringing home your own ball?

(h) Suppose there were two balls on the court instead of one. Now, is the number of players who go home with their own ball more or less predictable? Guess first and then do the appropriate calculation to verify that your intuition was correct.

(i) Repeat parts (b)–(g) under the assumptions of (h); but each time before calculating, guess whether the answer should be larger or smaller than in the original situation.

3.10.9 A laundromat has 17 washing machines which, according to the manager's estimate, have a 90% chance of trouble-free operation for the first month after her maintenance check.

(a) How much trouble should she expect during one such month?

(b) What's the probability of more than two such problems?

(c) If she brings in two more machines, would you guess that her operation would become more or less stable? Verify this guess.

(d) After several years of operation, the manager's new partner looks over the records and finds that, in fact, there had been an average of only 0.18 breakdowns within the first week after maintenance checks. With this more accurate information, she now believes it's unnecessary for either of them to be in attendance during that first week. If this plan is followed, what's the probability of at least one problem during such a week when no one is there to attend to it?

(e) If they make nine maintenance checks in one year, what is the probability of at least one problem in more than one of the weeks the two women are gone (that is, in the first week after each check)?

(f) If they make a maintenance check each month, what's the probability they get through the first six monthly checks without having any problem during the week following the inspection when they are gone? [$\approx$ 34%]

(g) If they make a maintenance check each month, how many months could they expect to go before having any problem during the week they are gone following the inspection? [about five, they should expect a problem in the sixth month after the sixth maintenance check.]

3.10.10 We know from records that we have about 1.73 serious defects in ten yards of high-grade cloth which we market. Our bolt consists of 30 yards of cloth.

(a) How many serious defects would we expect in one bolt?

(b) What's the probability of more than three such defects in one bolt?

(c) You buy ten yards of cloth each month for a special project which requires an unbroken stretch of ten yards of cloth with no serious defect. In one year of operation, how many times would you have to return your ten yards as unusable? [$\approx$ ten times]

(d) For part (c), when would you expect to see the first bolt with a serious defect? [in the first or second month]

3.10.11 For 120 employees in our company, 32% took no sick leave in the last six months, 41% took one day sick leave, 20% took two days, none took three, and the rest took four days.

(a) Find the median, mean, mode, range, and standard deviation for the number of days sick leave taken over these six months.

(b) How many days sick leave were taken all together by our work force?

(c) If some of those who took four days had taken only three, would that have increased or lowered the variability of this data? Guess, then compute for the modified situation.

(d) If you took a random sample (with replacement) of ten of the employee records, what is the probability that more than one of the ten would have been absent on sick leave for four days over the past six months? [$\approx$ 15%]

(e) How many employee records should you have to sample before finding one which showed four days of sick leave over the past four months? [$\approx$ 14]

(f) What's the probability you would have to sample more than 20 employee records before finding one which showed four days of sick leave over the past four months? [$\approx$ 23%]

3.10.12 Three defective light bulbs inadvertently got mixed up with six good ones. Suppose two bulbs are chosen at random for a ceiling lamp.

(a) What's the probability they're both good?

(b) How many good bulbs would you expect on average?

(c) Suppose one of the bulbs we thought to be good turns out to be defective also. Is this a more or less predictable situation? First guess the answer and then verify your guess by making the correct calculation. Be sure you make your calculations from an appropriate distribution.

3.10.13 In tossing a fair coin, we will assign zero to heads and one to tails. Let X be the assignment for the first of two tosses and Y the assignment for the second. Now considering the two tosses together, let $Z = X + Y$.

(a) Guess the mean and variance of each of these three random variables.

(b) Do you expect Z to be more or less variable than X?

(c) Guess how you could obtain the mean and variance of Z from that of X and Y without computing.

(d) Compute the mean and variance of each of the three random variables.

3.10.14 In Chapter 5, we will introduce the idea of a "confidence interval" for an unknown parameter. It's a range of possible values for the parameter together with the probability—the "confidence coefficient"—that the parameter actually falls within that range. Here, we'll look at a special case, a confidence interval for an unknown median.

Suppose we've taken a sample of size n from some probability distribution (or, as a special case, from a numeric population). Call the median M and let min and max refer, respectively, to the smallest and largest observations in the sample. Then the interval (min, max) can be thought of as a range of possible values, a confidence interval, for M. In this problem, we'll determine the confidence coefficient for this interval. That is, we'll determine the probability that this interval contains the median. For simplicity, assume a random observation has a zero chance to actually equal M.

(a) Identify M, min, max, and the confidence coefficient. Here's the sample:

X	1.2	1.3	1.4	1.5
f	3	7	11	2

(b) Let Y be the number of observations in the sample which are less than M. What's the model for Y?

(c) $P(M < \max) = ?$

(d) $P(M > \min) = ?$

(e) What's the confidence coefficient for the interval (min, max) as a confidence interval for M?

(f) What's the confidence coefficient for part (a)?

(g) What's the median weight of U.S. pennies? Here are the weights W of 100 newly minted pennies, reported to the nearest 0.02 gram (taken from W.J. Youden's National Bureau of Standards Publication 672, Experimentation and Measurement):

W	2.99	3.01	3.03	3.05	3.07	3.09	3.11	3.13	3.15	3.17	3.19	3.21
f	1	4	4	4	7	17	24	17	13	6	2	1

(h) In Chapter 4, we'll see that (3.11, 3.13) is a 99% confidence interval for the median weight, M, of U.S. pennies. How do you interpret this confidence interval? To what does the confidence coefficient refer?

(i) What's wrong with interpreting the confidence interval in part (h) by saying "ninty-nine percent of the time, M is between 3.11 and 3.13, the rest of the time it's not"?

(j) In part (g), what's the relationship between the sample median and M?

Chapter 4

Continuous Probability Models

4.1 Continuous Distributions and the Continuous Uniform Distribution

Continuous Distributions

Every probability model we've seen so far has been ***discrete***—the possible values are separated from each other as "discrete points" on a number line. With this section, we begin our study of the so-called "continuous" distributions. A ***continuously distributed random variable on an interval*** $[a, b]$ is a random variable which takes on any possible value in the interval $[a, b]$ of real numbers. Random variables whose values are measurements of time, weight, size, and so on, are typical examples of situations which may give rise to continuous distributions.

We very consciously say "may" give rise to continuous distributions because it is a question of interpretation. Weight measurements, for example, might or might not be best represented as continuous. If you measure only to within a quarter of a gram, let's say, then your measurements would be discrete:

$$0,\ 0.25,\ 0.50,\ 0.75,\ 1.00,\ 1.25,\ \ldots$$

0 1/4 1/2 3/4 1 . . . etc

But very often, one measures with a high degree of accuracy, getting numbers like

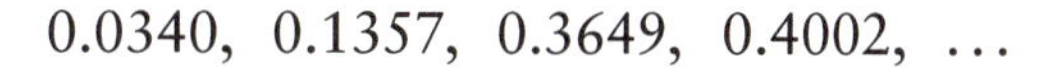

$$0.0340,\ 0.1357,\ 0.3649,\ 0.4002,\ \ldots$$

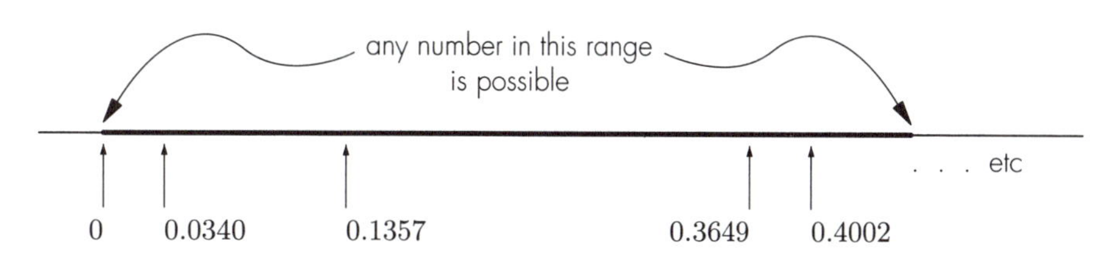

where all the values in between are possible—at least theoretically. In this case, a continuous interpretation of the measurements would be more appropriate. Note that in point of fact, any physical measurement is discrete because an actual measuring device has finite accuracy. So it's not a question of the actual physical measurements. When we say our model is continous and not discrete, it's an assertion about what measurements are THEORETICALLY possible.

Time measurements are often modeled as continuous. For example, in the next section, we introduce the exponential distribution which, among other applications, is typically used in industry to model lifetimes of electronic components. For the exponential distribution, as we'll see, any real number value from zero to infinity is possible—a component may go bad immediately, it may burn only a few seconds, or only a few minutes, or it may, in fact, burn several hours, or possibly many, many hours. The mean lifetime for the component in question may be several thousand hours, but any number less than that is conceivable. Furthermore, any number more than that is also conceivable. You cannot fix a number and say "no component will burn longer than this!" So, we choose a continuous model.

Before going further, please . . .

Try Your Hand

In these problems, assume X to be a continuously distributed random variable.

4.1.1 If c is any one of the possible values of X, then $P(X = c)$ is zero. Can you explain why on intuitive grounds?

4.1.2 Because for any c, $P(X = c)$ is zero, what DOES have nonzero probability?

The Probability Density Function

Because $P(X = x)$ is zero for any X when X is continuously distributed, the representation of probabilities for X cannot parallel the approach we have been taking for the discrete distributions. To avoid concepts from integral calculus, we will confine ourselves to representing these probabilities in graphical terms, in terms of pictures. In other words, for continuous distributions, we will think of probabilities in terms of areas in a specific graph.

The ***probability density function*** for a continuously distributed random variable X is the function—usually denoted by a symbol such as $f(x)$—whose graph determines the probabilities for X by means of "area under the curve," area under the graph of $f(x)$. Thus, in the following picture the shaded area represents the probability that X takes on a value in the interval (c, d)

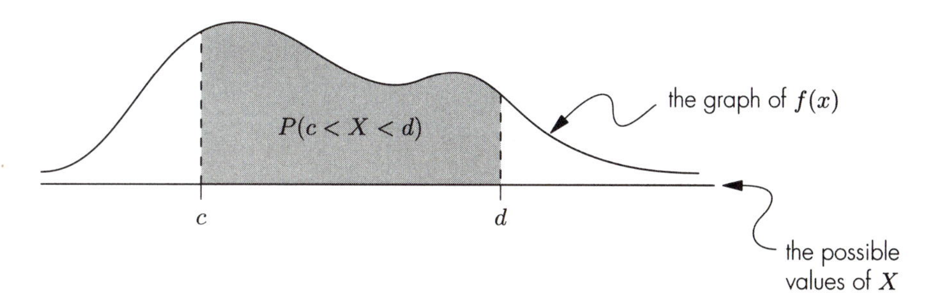

Now, please . . .

Try Your Hand

4.1.3 Assume X is continuously distributed with probability density function $f(x)$. Suppose X only takes on values between zero and 100; that is, $a = 0, b = 100$. What is the shaded area in each of the following pictures?

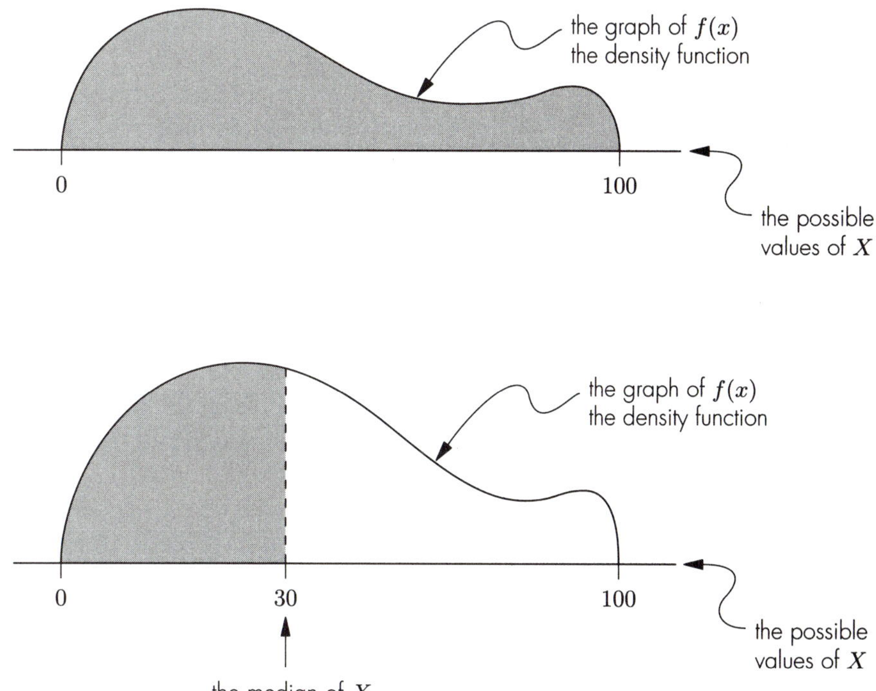

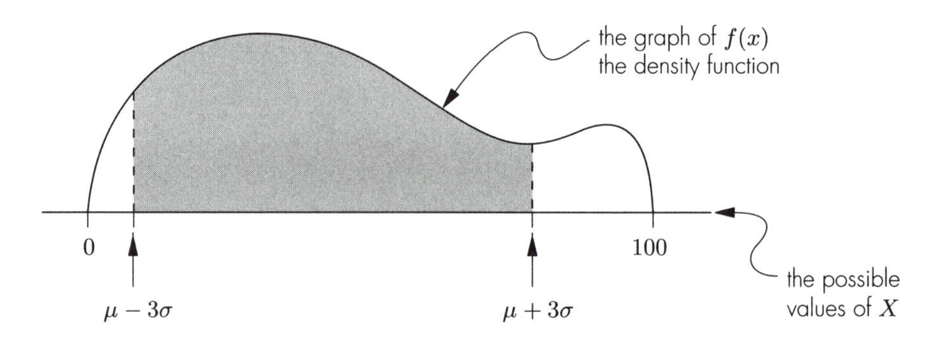

4.1.4 Suppose you have a sample of size n from a continuous distribution. Let x_i be the ith observation in the sample after having put the sample in ascending order. Show that the probability for a random future observation to fall in the interval between x_h and x_k is

$$\frac{h-k}{n+1}.$$

The Continuous Uniform Distribution

The uniform distribution is the simplest example of a continuous distribution. A ***continuous uniformly distributed random variable*** is a continuous random variable for which all intervals of a given length have the same probability. Be careful of the terminology being used here. When we speak of "the probability of an interval," we mean the probability that the random variable takes on a value within the interval. Sometimes for the phrase "the random variable X takes on a value in the interval $[c, d]$" we'll use the standard set notation: $X \in [c, d]$. For example, the probability of the interval [2, 3.5] is $P(2 \leq X \leq 3.5)$, or using set notation, $P(X \in [2, 3.5])$. If X is some kind of measurement, then we're talking about the probability of getting a measurement bigger than or equal to two and smaller than or equal to three and a half.

Although one can be more general, we will only consider uniform distributions defined on a fixed, finite interval whose endpoints we denote by a and b. This implies that X necessarily takes on a value somewhere between a and b. Consequently,

$$P(a \leq X \leq b) = 1.$$

As you'll see in the exercises below, the uniform distribution is completely determined by a and b, the ***parameters of the uniform distribution***.

Now, you will easily derive the basic properties of the uniform distribution as you . . .

Try Your Hand

In the following problems, assume X is continuously distributed with probability density function $f(x)$.

4.1.5 For a continuous distribution, the graphical representation should give the probabilities as area over the interval in question, like the histograms we studied in Chapter 2. Suppose you have a continuous and uniformly distributed random variable X on the interval [2, 3]. What would the graphical representation for $P(2.25 \leq X \leq 2.5)$ look like?

4.1.6 Suppose that X is uniformly distributed on the interval from two to seven. With the notation above, this means $a = 2$, $b = 7$. Thus, X only takes on values between two and seven. Draw some sketches and see if you can guess what the probability density function for X must be. This is not necessarily easy—you may see it or you may not. It requires a little experimenting and a little good luck. Try!

4.1.7 In the previous problem, you saw that if X is uniformly distributed on an interval $[a, b]$ then its probability density function must be constant:

$$f(x) = c.$$

What is the value of c?

4.1.8 Suppose your random variable X is uniformly distributed on $[a, b]$.

(a) Try to guess the mean for X .

(b) If c and d are between a and b , what is the probability that X is in the range $[c, d]$?

(c) Give a verbal description of the probability in part (b) . Describe it as a certain proportion.

Again, we remind you that our treatment of continuously distributed random variables cannot run parallel to the treatment of discrete random varibles because specific values now have probability zero. For a continuous distribution, we focus on the probability for intervals, for ranges of values.

Here is a summary of

the continuous uniform distribution:

A continuous random variable X taking on only values between a and b is uniformly distributed on $[a,\ b]$ with parameters a and b if the probability of any interval is proportional to the length of the interval.

$$f(x) = \frac{1}{b-a}, \quad \text{a constant function,}$$

$$\mathrm{P}(x_1 < X < x_2) = \frac{(x_2 - x_1)}{(b-a)}$$

whenever $a < x_1 < x_2 < b$.

Also:

$$\mu = \frac{(a+b)}{2}$$

$$\sigma^2 = \frac{(b-a)^2}{12}$$

So now, please . . .

Try Your Hand

In the following problems, suppose that X is uniformly distributed on the interval $[a,\ b]$.

4.1.9 (a) Show that the cumulative distribution function for X is

$$\mathrm{P}(X \le x) = (x-a)/(b-a).$$

(b) Illustrate part (a) with a picture.

(c) Show that $\mathrm{P}(|X-\mu| < 2\sigma) = 1$.

4.1.10 For each of the following, sketch the graph of $f(x)$ showing the desired probability and then evaluate the probability from the appropriate formula:

(a) $P(X < 2.5|a = 2, b = 4)$;

(b) $P(7 < X < 9.5|a = 6, b = 10)$;

(c) $P(X > \mu_X|a = 2, b = 4)$;

(d) $P(X < \mu_X)$ for any a, b.

(e) For the X of part (b), what is the probability that X is within one and a half standard deviations of its mean?

4.2 The Exponential Distribution

Modelling the Reliability of a System

We'll introduce the exponential distribution in terms of one of its most typical applications, reliability theory. In reliability theory, one is concerned with the "time to failure" of a system. The word "system" here is broadly defined. It can refer to some kind of mechanical or electronic device or component, it can refer to a piece of industrial equipment (made up of a number of components), or to an entire production or service process. It could refer to a computer system of considerable complexity or to something so simple as a single electronic component or even just a household fuse. The ***reliability*** of such a system is the probability of no failure in a specified time period under appropriate operating conditions. Note that "reliability" is a numeric quantity, a probability for satisfactory performance.

It's usual to divide the lifetime of such systems into three phases:

- The "burn-in" ("early failure" or "infant mortality") period in which failure may occur because of some defect in the system itself. A defective fuse, for example, may blow in the first few hours of operation.
- The "useful life" (or "random failure") period in which the system is functioning properly.
- The "wear-out" period in which failure can be expected when the system is used beyond its reasonable life.

You see that for the first and third of these phases, failure is the result of a problem with the system itself. We're not going to consider these two phases; instead, we will be concerned only with the second phase which can often be modeled by the exponential distribution. During this period of "useful life"—the period of random failure—the system may fail due to some external cause. For example, a perfectly good household fuse blows when there is a dangerous surge in the electrical current. Any system, even though operating properly, will fail when there is some unusual or unexpected demand which stresses the system beyond its capacity. Thus, failure in the second phase is thought of as randomly caused by independent factors external to the system itself.

This circle of ideas is really quite general, a series of checkout lines in a large supermarket is a system which may be considered to "fail" when a customer has to wait before checkout. Serving lines in a fast food restaurant provide another example of such a system.

Before we continue, it will be helpful if you give some thought to how we might model these systems. Here are some exercises for you to . . .

Try Your Hand

4.2.1 **(a)** How would you model the number of failures during the period of useful life for a system such as we have discussed above?

(b) What's the probability that the first failure occurs after some specific time t? In other words, what's $P(T > t)$? Here, T is the time elapsed from the beginning of the random failure period to the first failure.

4.2.2 For T as defined in the previous problem:

(a) Verify that T is indeed a random variable.

(b) Show that T is a continuous random variable.

(c) Give the cumulative probability distribution for T .

The Exponential Distribution

The random variable T of Problem 4.2.1 is an example of an exponentially distributed random variable. We need not have defined T as the time elapsed from the beginning of the period of useful life. T could be the time elapsed between any two failures. Then exactly the same analysis will carry through. We should then look at a period of time t

after the first of the two failures:

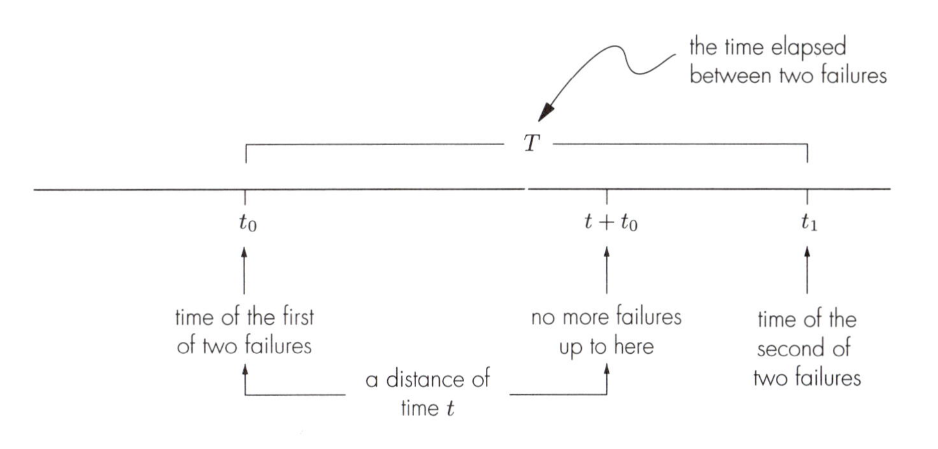

Saying there's no failure ($X = 0$) in the period of time t is to say the time to the next failure is greater than t (i.e. $T > t$). So t is the length of a time interval in which the number of failures, X , is zero. From this we obtain

$$P(T > t) = P(X = 0) \quad \text{with } X \text{ as the number of failures on the interval } t_0 \text{ to } t + t_0$$

$$= e^{-t\lambda} \quad \text{where } \lambda \text{ is the average number of failures per unit of time}$$

This gives us the cumulative probability function for T:

$$\begin{aligned} \mathrm{P}(T \leq t) &= 1 - \mathrm{P}(T > t) \\ &= 1 - \mathrm{P}(X = 0) \\ &= 1 - e^{-t\lambda} \end{aligned}$$

Now, because the probability for a continuously distributed random variable like T is represented as area under the graph of its probability density function, we have the following graph:

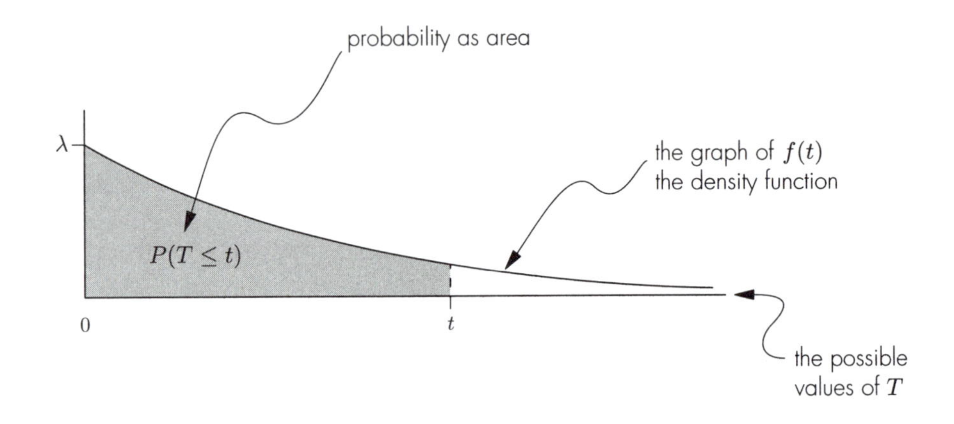

What is the function $f(t)$? Without reference to calculus, this is not obvious. But, believe it or not, the density function whose areas will generate the probabilities for T is just

$$f(t) = \lambda e^{-t\lambda}.$$

The probability distribution with this density function is called the ***exponential distribution***.

The discussion above gives us a rule of thumb for when to expect the exponential distribution to be appropriate: The exponential distribution models "time between two failures" in a system where the NUMBER of failures is appropriately modeled by the Poisson distribution. That is, where two simultaneous failures are impossible, any two failures are independent and the number of failures is proportional to the time in which they occur.

Furthermore, there is nothing special about looking for "failures of a system." We might be looking for any kind of occurrence where the number of occurrences follows the Poisson distribution. In that case, the time between two occurrences would follow the exponential distribution. In the exercises, you'll see some typical applications of the exponential distribution other than its application to reliability theory.

Here's a summary of

the exponential distribution:

> A random variable T has the exponential distribution if its density function is
>
> $$f(t) = \lambda e^{-t\lambda},$$
>
> in which case
>
> $$P(T < t) = 1 - e^{-t\lambda}, \quad t > 0.$$
>
> And
>
> $$\mu = 1/\lambda,$$
> $$\sigma^2 = 1/\lambda^2.$$
>
> With T as "time between two occurrences," λ is the "expected number of occurrences in a UNIT of time."

It's not important whether we write $P(T < t)$ or $P(T \leq t)$ because $T = t$ with probability zero. And it's not surprising that μ_X is $1/\lambda$ when you think of our reliability example. There, λ is the expected number of failures per unit time and so $1/\lambda$ should be the expected time per failure. If you expect five failures per hour, for example, then you should expect one-fifth of an hour between any two failures. Entirely understandable!

Here are some problems for you to . . .

Try Your Hand

4.2.3 Suppose one expects an average of 2.3 calls to be received at a telephone switchboard in a five minute period.

(a) How many minutes would you expect on average to elapse between any two calls?

(b) What's the probability of more than seven calls in a quarter of an hour?

(c) What's the probability of as much as ten minutes between calls?

4.2.4 A particular electronic component is judged to have an average life of about 2500 hours.

(a) What's the probability that one component will burn out before 1000 hours?

(**b**) How many components should we expect to purchase to run the system using this component for 10,000 hours?

(**c**) Suppose we have a system which uses these components in pairs. What's the probability that the system will run for more than 1000 hours without a failure due to these components?

4.2.5 Show that an exponentially distributed random variable T has no memory.

(**a**) That is, show that

$$\mathrm{P}(T > t + s | T > t) = P(T > s).$$

It can be shown that any random variable satisfying this equation—any memoryless random variable—must be exponentially distributed.

(**b**) Give a verbal description of the equation in part (a).

4.3 The Normal Distribution

The Normal Distribution as a Model for Measurement Error

The normal distribution which we introduce in this section is certainly the most important of the continuous distributions. Its discovery is a fascinating story in the history of statistical theory—a story of struggle, false starts, indulgent circular logic, and, finally, clarity, precision, and triumph.

The story begins with the self-taught Thomas Simpson, who began his career as a London weaver and part-time mathematics instructor. By 1755, he was a professor at the Royal Military Academy and Fellow of the Royal Society of London. In 1755, he read a paper before the Royal Society entitled "On the Advantage of Taking the Mean of a Number of Observations, in Practical Astronomy." Simpson attempted to justify taking the mean of several astronomical observations and to refute those who "have been of the opinion, and even publickly (sic) maintained, that one single observation, taken with due care, was as much to be relied on as the Mean of a great number." Simpson's first step toward the normal distribution "was his decision to focus, not on the observations themselves or on the astronomical body being observed, but on the errors made in the observations, on the differences between the recorded observations and the actual position of the body being observed." [Stigler]

The key idea here is that for repeated measurements, the measurement error is a random variable and we need to know its distribution: What is the appropriate probability distribution, the appropriate abstract model, of measurement error? Assuming no systematic source of error, either in the measuring instrument or the observer, it's clear the mean of the errors should be zero and that the distribution should be symmetric about that mean. In a 1757 revision of his paper, Simpson described the physical conditions which would imply a zero mean error and a symmetric distribution about that mean in these terms:

> That there is nothing in the construction, or position of the instrument whereby the errors are constantly made to tend the same way, but that the respective chances for their happening in excess, and in defect, are either accurately, or nearly, the same.

He also described the physical conditions which should determine the variability in the errors:

> That there are certain assignable limits between which all these errors may be supposed to fall; which limits depend on the goodness of the instrument and the skill of the observer.

Simpson chose—somewhat naively and purely for convenience—a triangular density function for his error distribution:

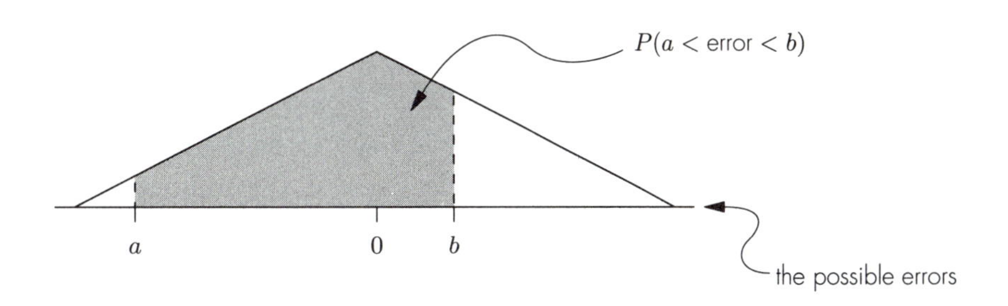

But any number of curves would satisfy the conditions described above. Laplace, looking at this problem expressed it this way: ". . . of an infinite number of possible functions, which choice is to be preferred?"

The appropriate choice for an error distribution, the curve which is now called the "normal density function," was first used in a work published in 1809 by Karl Friedrich Gauss. Gauss was one of the most brilliant mathematicians the world has ever known, but his derivation of the normal density as an error distribution was suspicious to say the least—his argument was essentially circular, giving no support at all for his particular choice!

Don't blame Gauss! Praise him. A common misconception holds that great scientists thrive only on precision and exactness, but just the opposite is true. Advances in knowledge are born from the chaos of ignorance—if you already knew the answer, we wouldn't call it an "advance." Therefore, success in scientific discovery requires a high degree of tolerance, prehaps relish, for being immersed in ignorance. And a willingness to risk error, even foolish error. In fact, at the very moment precision is obtained, the scientist abandons her problem, for it is exactly at that moment the problem is solved. She may afford herself some brief time to bask in the sunshine—the precision and exactness—of her success, but further progress demands she plunge back into the ocean of chaos and ignorance from which new discoveries will be born. This being the situation, it's inevitable that all sorts of crazy stuff will go on before knowledge is attained.

Well, in the 1770s, before Gauss had given any consideration to it, the French astronomer and mathematician Pierre Simon de Laplace (teacher of Napoleon at the École Militaire in Paris) had made two separate but unsuccessful attacks on this problem, the problem of determining from first principles a reasonable probability distribution for random error. And it proved to be very difficult! Laplace seems to have encountered Gauss" book for the first time in the early summer of 1810. In Stigler's words, ". . . it must have struck him like a bolt. Of course, Laplace may have said, Gauss's derivation was nonsense, but he, Laplace, already had an alternative in hand that was not—the Central Limit Theorem" (see [Stigler], p. 143). Laplace had presented this theorem in a paper delivered to the Academy of Sciences in April of 1810. By the time it was published, he had obviously seen Gauss' book, for he appended a supplement using his Central Limit Theorem to justify Gauss" choice of error distribution.

Laplace's Central Limit Theorem is a major focus of Chapter 5. This theorem, which Laplace had already proved in a completely different context, gave him a perfect argument for his choice of error distribution, what is now called the "normal distribution." This distribution would indeed be the right choice if one could think of the errors as compounded of many independent "elementary" errors (later known as the "hypothesis of elementary errors"). If that were true, Laplace's Central Limit Theorem would give the appropriate density function. You'll see why this is true in the next chapter.

What does the normal density function look like? Here's its graph, the so-called "bell-shaped" curve:

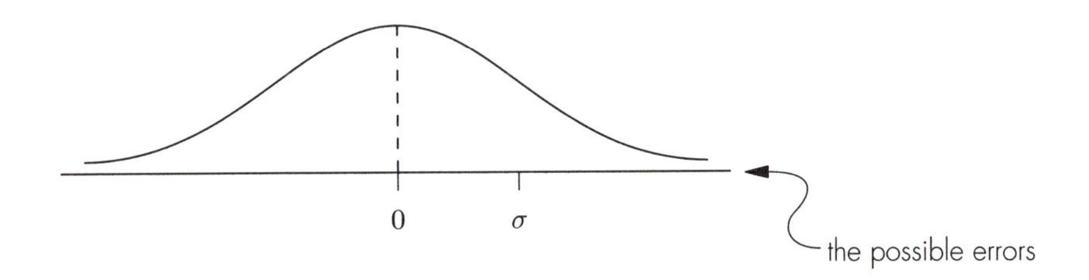

Because we're assuming an ERROR distribution, the mean is zero. A mean which is not zero is certainly also possible for a normal distribution, as we'll see later.

The location of σ in the picture is determined by the following rule:

A fundamental fact of the normal distribution:

> There's about a 68% chance for a value within one standard deviation of the mean of a normally distributed random variable. There's about a 95% chance of a value within two standard deviations of the mean and it's virtually certain for a value to fall within three standard deviations of the mean.

The probabilities given by this rule are rounded to the nearest whole percentage. We'll treat these as approximations. When we refer to an "exact" value, we'll usually mean a value obtained from the normal table in the appendix which gives percentages accurate to two decimal places (hardly exact!).

Thus, about 68% of the area under the curve should be centered between $\mu - \sigma$ and $\mu + \sigma$. The ***standard normally distributed random variable***, denoted by the symbol Z, is the normally distributed random variable for which $\mu = 0$ and $\sigma = 1$. So, if we're talking about the standard normal distribution, a little over two-thirds (about 68%) of the area under the curve should lie between Z = -1 and Z = 1:

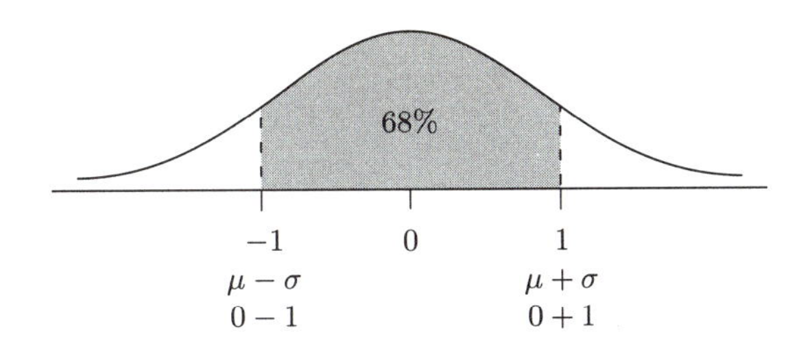

This curve, the standard normal density function, is given by

$$f(z) = (1/\sqrt{2\pi})\exp[-z^2/2].$$

It's called the "Gaussian curve" and is defined for all values of Z, although values outside of the range (-3, 3) are very unlikely (why?). The symbol exp refers to the exponential function to the base e (see page 111) and is introduced only for notational convenience. When the exponent is complicated, as it is here, $\exp[u]$ is easier to write than e^u.

The standard normal distribution is only one of the large ***family of normally distributed random variables***. If X is a member of this family, we write $\boldsymbol{X \sim N(\mu, \sigma^2)}$, which is read "$X$ is normally distributed with mean μ and variance σ^2." This family consists of all random variables having the density function given below. The notation $N(\mu,\ \sigma^2)$ identifies the two parameters for the model, namely the mean μ and the variance σ^2. Any one member of this family is determined by its mean and variance, which could be any real-numbers. Thus, Z is determined as the unique member of this family having mean zero and variance one. We think of the mean of a normally distributed random variable as the result of a systematic "effect", with any variation from that mean due to "random error," hence the "bell-shaped" curve. The density function for $N(\mu, \sigma^2)$ is

$$f(x) = (1/\sqrt{\sigma^2 2\pi})\exp[-(x-\mu)^2/2\sigma^2].$$

If this equation looks forbidding, don't worry! The model is mathematically too sophisticated for us to do direct computations. We'll use either a picture or the normal table in the appendix to determine normal distribution probabilities.

Here are some exercises for you to . . .

Try Your Hand

4.3.1 **(a)** Justify the statement made above for Z that ". . . values outside of the range (-3, 3) are very unlikely."

(b) Recall that the notation for the family of normally distributed random variables is $N(\mu,\ \sigma^2)$. Using this notation, Z is . . . ??

(c) Evaluate $P(Z < -2)$ using the relevant picture.

(d) Evaluate the following probabilities:

$$P(Z > 1),\;\; P(0 < Z < 1),\;\; P(-1 < Z < 2),\;\; P(1 < Z < 2).$$

4.3.2 Find the value of Z that will yield the required probability.

(a) $P(-1 < Z < ?) = 68\%$;

(b) $P(Z < ?) = 84\%$;

(c) $P(Z < ?) = 16\%$;

(d) $P(Z > ?) = 16\%$;

(e) $P(Z > ?) = 2.5\%$;

(f) $P(Z < ?) = 0$;

(g) $P(? < Z < ?) = 100\%$;

(h) $P(Z > ?) = 84\%$;

(i) $P(? < Z < 0) = 34\%$.

4.3.3 In the text we said "it's clear the mean of the errors should be zero." Why should an error distribution have mean zero?

4.3.4 **(a)** Sketch the graph for $X \sim N(2.5, 3.24)$.

(b) Locate zero in the picture for part (a).

4.3.5 **(a)** Sketch the graph for $X \sim N(2.5, 0.0064)$.

(b) Locate zero in the picture for part (a).

4.3.6 Which, if any, of the following pictures could be the graphs of the indicated density function? For those which obviously could not be, explain why not and redraw the curve to make it correct. Leave the axis and its labeling unchanged.

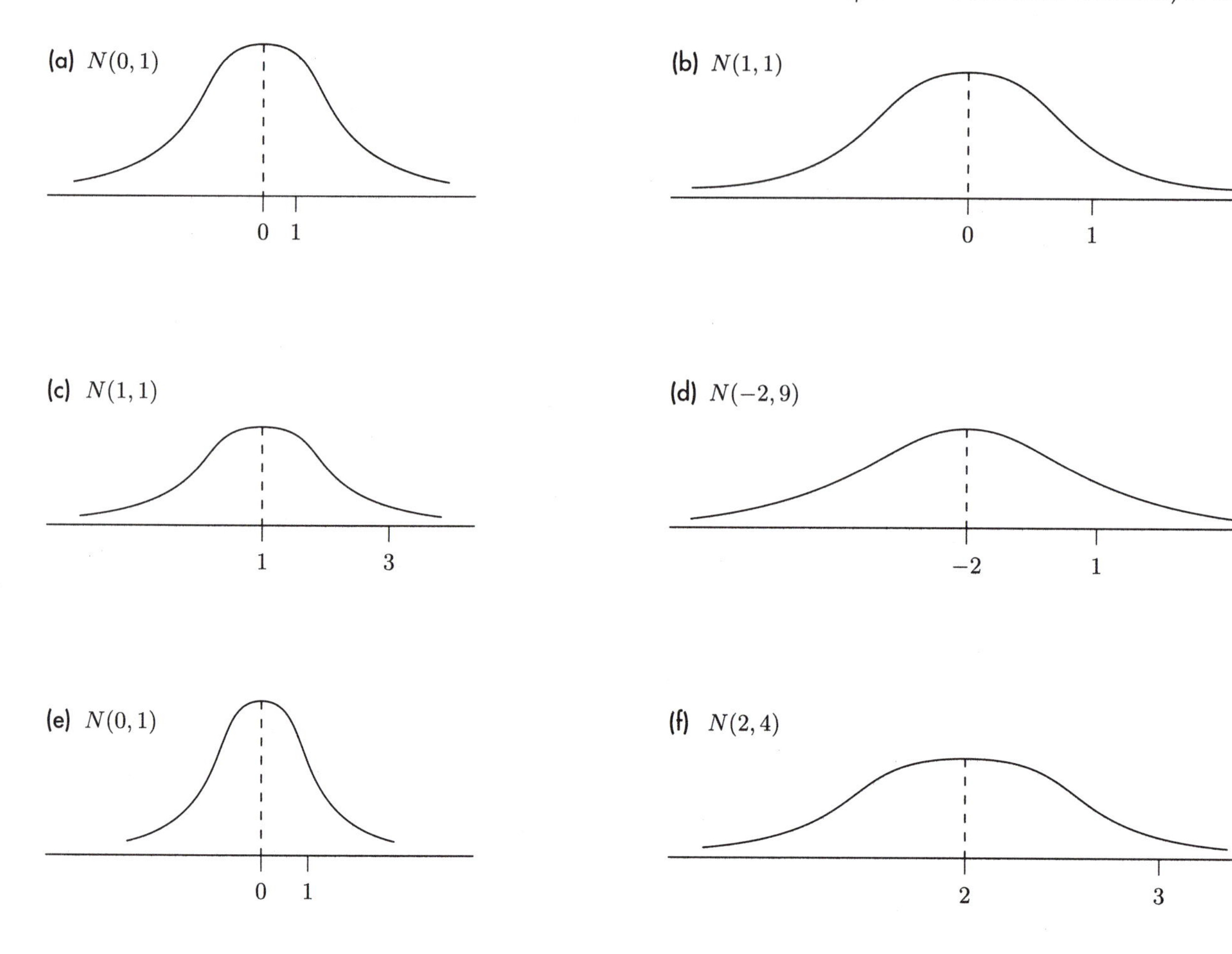

4.3.7 Show that the formula for the density function of a normal distribution . . .

$$f(x) = (1/\sqrt{\sigma^2 2\pi}) \exp[-(x-\mu)^2/2\sigma^2],$$

gives the "right" result when you specialize it to the standard normal distribution.

4.3.8 (a) In what sense is "measurement error" a random variable?

(b) If you make repeated measurements, what value would you expect to get on average?

(c) How variable is the measurement error?

(d) What determines the variability of the measurement error?

The Normal Distribution as an Abstract Model

As we have seen before, a probability distribution may serve to model many situations besides the one for which it was originally intended. This is emphatically true of the normal distribution. Under what real-world circumstances should you expect the normal distribution to be an appropriate model? This is not an easy question to answer. Historically, the normal distribution has often been used where it was not appropriate at all, as later experience (usually painful!) has shown. Nowadays, there are sophisticated "tests for normality." These are better guides than were formerly available, but even with the most sophisticated tests, the question is not easily resolved. We'll not study these "tests for normality." They're too sophisticated for an introductory statistics course. Later we'll show you a simple graphical technique—the "normal probability plot"—which is widely used as a rough "eyeball" test. But the normal probability plot gives no insight into the conditions that characterize normality. For that, we rely on the intuitive criterion presented below.

That rule-of-thumb criterion is closely related to the "measurement error" interpretation of the normal distribution. To focus on the error—this was Simpson's insight—you need to eliminate the systematic part of the measurement. The systematic part, of course, is the true value of the object being measured. By looking at the DIFFERENCE of two measurements, the systematic part gets eliminated: Suppose the measurement is M, with E being the error. If M_1 and M_2 are two such measurements,

$$M = \text{true value} + E,$$

and

$$\begin{aligned} M_1 - M_2 &= (\text{true value} + E_1) - (\text{true value} + E_2) \\ &= E_1 - E_2. \end{aligned}$$

Thus, the difference in two values of M is just the difference in two random errors, but the difference in two random errors looks like random error again. What do we mean, intuitively speaking, by "looks like random error"? Simply, the net result of many small and uncontrollable influences more or less independent of each other. So to say that a number like $M_1 - M_2$ "looks like random error" is just to say it's "due to many independent random factors." This last phrase—"due to many independent random factors"—is a somewhat more complete way of expressing our intuitive criterion for normality. It gives you an operational notion which you can actually apply to a situation to see if the assumption of normality seems reasonable.

Here's our criterion . . .

a rule of thumb for when to use the normal distribution:

> Any situation giving rise to numbers where the difference between any two values looks like random error (is due to many independent random factors) will often be modeled appropriately by a normal distribution.

When you're using this intuitive criterion as a justification for assuming normality, please remember that it's only a rough rule of thumb which might serve as a preliminary guide. The next step would be the "normal probability plot" which we introduce later—it's simple to use if you have a computer statistical package. But even this criterion is relatively weak. You would probably want to look into the matter more carefully at some point. In this course, the tools for doing that are not available. For a more subtle analysis, you should turn the question over to the experts. Or become an expert yourself!

In fact, a lot of criticism has been aimed at the naive use of our rule of thumb during the nineteenth century, especially in regard to social issues. For that reason, presumably, all mention of the rule has disappeared from current textbooks. Still, if taken in the proper spirit, it's helpful as a preliminary guide for understanding the normal distribution. We'll use it. Often.

Even in the nineteenth century this intuitive criterion was not used uncritically. It was understood that certain phenomena would follow other rules. Quantities such as income where changes are proportional to the quantity itself—when you get a raise, for example, it's determined as a percentage of your current income—will probably follow a "lognormal" distribution. That is, the quantities themselves are not normally distributed, rather their logarithms are. Much economic data follow the lognormal distribution. This distribution was well known in the nineteenth century. A full mathematical treatment was presented in 1879 by the Cambridge mathematician Donald McAlister. Similarly, it was recognized early on that certain quantities which are inherently quadratic—surface areas of organisms, for example—become normally distributed only after a square root transformation. Weights of organisms, being determined by volume (a cubic quantity), will often become normally distributed after a cube root transformation. Again, techniques for identifying these transformations and checking their appropriateness are beyond the scope of this course.

Let's stick to our rule of thumb! To explore the normal distribution a bit further, please . . .

Try Your Hand

4.3.9 (a) Using our intuitive criterion as a guide, show that you should expect repeated measurements of some object to be approximately normally distributed.

(b) You can think of the numbers in a numeric population as values of a random variable. What's the experiment? What's the random variable? How would you decide if it's a normally distributed random variable?

4.3.10 (a) Explain why "specification error" should be normally distributed. Recall that specification error for a manufactured object is the difference between the actual dimension of the object and the "ideal," the specified dimension (see page 47).

(b) In part (a), we said specification error "should be" normally distributed. What assumption does this make about the manufacturing process?

4.3.11 (a) The prototypical example of a "normally distributed population" is a set of scores for a test taken by a large homogeneous population of test takers, for example, the SAT test in a given year. Explain why test scores should be approximately normally distributed.

(b) Test scores will not always be normally distributed. How could it happen that a test given to a large class would have a "bimodal" distribution like

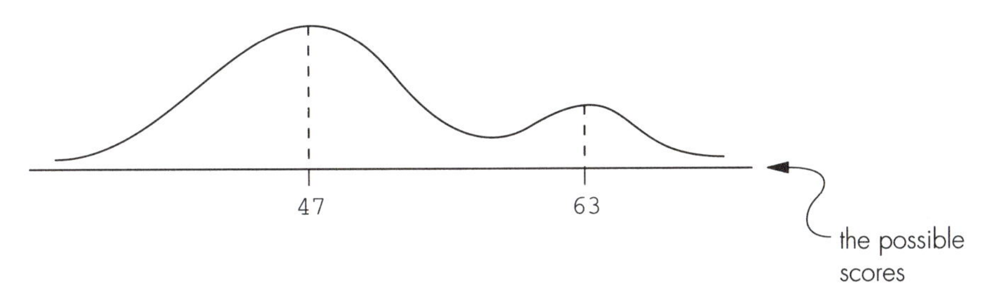

4.3.12 It's always possible to model a real-world situation in more than one way. What model you choose depends on your specific needs, the specific kinds of questions you want to address. Test scores can be thought of very simply as we did in Problem 4.3.11 or they can be modeled with more detail as we do in this problem.

The SAT test for a particular year is a device for measuring a student's "ability" for college. Let S ="SAT test score." Let T be the "true ability" for any student taking the SAT test that year and let E be the test's error in measuring that ability.

(a) How is S related to T and E?

Now explain the sense in which

(b) T should be thought of as a normally distributed random variable.

(c) E should be thought of as a normally distributed random variable.

(d) S should be thought of as a normally distributed random variable.

[Hint: Parts (b), (c), and (d) should be answered in different ways. Think about that and then look at the solutions level I.]

4.3.13 (a) Show that if X is normally distributed, then any linear function of X is normally distributed.

(b) Show that the rule in part (a) does not hold in general for all families of distributions. Do this by showing that a linear function of a binomial random variable is not necessarily binomial.

(c) Give the probability distribution of $Y = 2X+1$ where $X \sim B(1, p)$. Compute the mean and variance of Y in two ways.

(d) What's the point of this problem?

(e) Suppose X and Y are normally distributed. Show that $aX + bY$ is normally distributed, where a, b are any two constants.

4.3.14 Suppose X is any normally distributed random variable with mean μ and standard deviation σ; that is, $X \sim N(\mu, \sigma^2)$.

(a) Show that X can be expressed as a linear function of Z with positive slope.

(b) Show that Z can be expressed as a linear function of X with positive slope.

4.3.15 Let M be the observed measurement for some measurement process and let E be the measurement error.

(a) Show that M is a random variable.

(b) Identify the mean and variance of M.

(c) What is the distribution of M?

(**d**) How is the situation for M and E similar to the situation for X and Y where X and Y are, respectively, the number of dots on the top face and on the hidden face on one roll of a die?

The Standardizing Transformation

When you look for values of the standard normal distribution in the table of the appendix, you'll find probabilities of the form

$$\mathrm{P}(Z < z) \quad \text{for } z \geq 0.$$

For example, the table gives $\mathrm{P}(Z < 1.37) = 0.9147$:

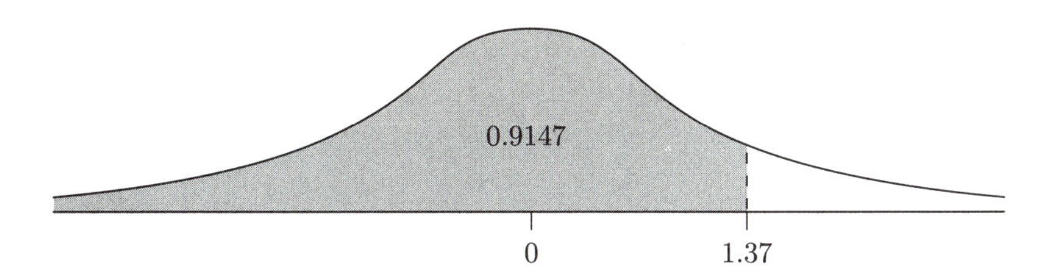

But, in fact, the standard normal table actually provides probabilities of any form whatsoever for any normally distributed random variable, not just for Z. Here's how: Suppose X is any normally distributed random variable with mean μ and standard deviation σ. Then Z is a linear function of X with positive slope (Problem 4.3.14): $Z = (X - \mu)/\sigma$. The slope is $1/\sigma$. But a linear function with positive slope "preserves inequalities," and so $X < x$ if and only if $Z < z$. In other words, the two conditions $X < x$ and $Z < z$ say exactly the same thing: one's true if and only if the other one is. Well if that's so, they must have the same probabilities: like $\mathrm{P}(X < x) = P(Z < z)$. This line of reasoning shows that

If you compute Z by standardizing the normally distributed random variable X:

If

$$Z = \frac{X - \mu}{\sigma},$$

then

$$\mathrm{P}(X < x) = \mathrm{P}(Z < z).$$

This is illustrated by the following:

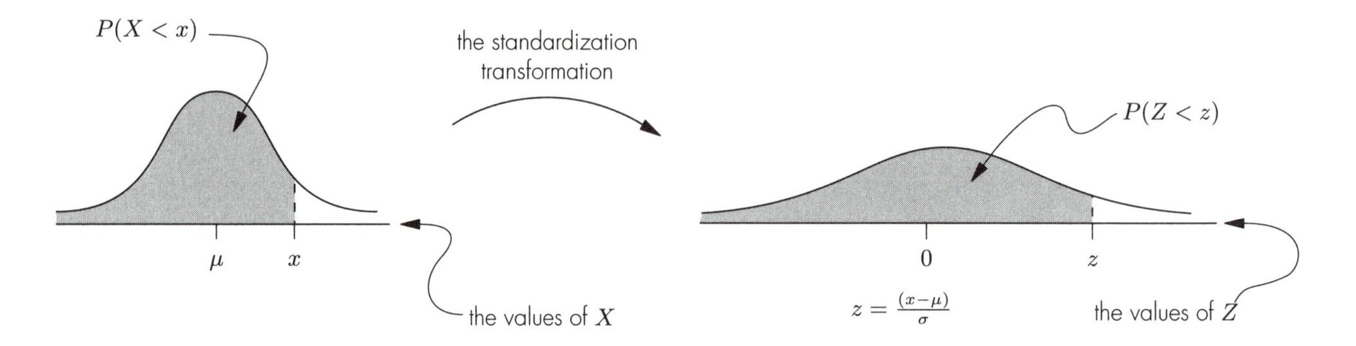

Note that Z is a linear function of $X : Z = a + bX$, with $a = -\mu/\sigma$ and $b = 1/\sigma$. This linear function is ***the standardizing transformation for*** X which converts X into Z where Z is the "standard" normally distributed random variable. This transformation allows you to find probabilities for X from the corresponding probabilities in the Z table of the appendix. If you understand the pictures properly and use the symmetry of the distribution, you'll be able to obtain any probability whatsoever!

To see this, please . . .

Try Your Hand

Hint for calculating: As we did above, draw both the picture for X and the corresponding picture for Z, then guess in advance what your answer ought to look like (you can avoid a lot of grief this way; you'll catch obvious errors!). Then after guessing, go to the Z table and find the exact value.

4.3.16 Suppose $X \sim N(2, 25)$. Compute the following probabilities.

(a) $P(X < 7.4)$;

(b) $P(X < 2)$;

(c) $P(X > 1)$;

(d) $P(X > -1)$;

(e) $P(X < -1)$;

(f) $P(X < 0)$;

(g) P(1 <X< 7.4).

4.3.17 Compute the probabilities:

(a) $X \sim N(3.8, 1.44)\ldots$ $P(X < 5)$, $P(X < 2.6)$, $P(X < 2)$;

(b) $X \sim N(12, 4)\ldots$ $P(X < 14)$, $P(X < 10)$, $P(X > 9)$;

(c) $X \sim N(0.18, 0.0001)\ldots$ $P(X < 0)$, $P(X < 0.17)$, $P(X > 0.16)$;

(d) $X \sim N(0, 1)\ldots$ $P(X < 1)$, $P(X < 2)$, $P(X > 1.5)$;

(e) $X \sim N(-2, 1.21)\ldots$ $P(X > 1)$, $P(X > 0)$, $P(X > -3)$;

(f) $X \sim N(4, 2.25)\ldots$ $P(4 < X < 9)$, $P(3 < X < 5)$, $P(X > 3)$.

(g) Now explain the logic which justifies the calculation in part (f) where you used the following equations:

$$P(4 < X < 9) = P(X < 9) - P(X < 4)$$

and

$$P(3 < X < 5) = P(X < 5) - P(X > 3).$$

4.3.18 In each of the following determine the value of X which yields the required probability. First make a rough guess and then find the true value from the table—if you don't find it exactly, take the closest value. [Hint: You'll gain maximum benefit and spend the least time on this exercise if you make a serious effort to answer the question BEFORE you look at the answers.]

(a) $X \sim N(1.3, 0.1764)\ldots$ $P(X < ?) = 0.05$;

(b) $X \sim N(1.3, 0.1764)\ldots$ $P(X > ?) = 0.18$;

(c) $X \sim N(5, 1)\ldots$ $P(X > ?) = 0.22$;

(d) $X \sim N(5, 1)\ldots$ $P(X < ?) = 0.22$;

(e) $X \sim N(-2, 4.84)\ldots$ $P(X < ?) = 0.73$;

(f) $X \sim N(-2, 4.84)\ldots$ $P(X < ?) = 0.51$;

(g) $X \sim N(14, 5.29)\ldots$ $P(10 < X < ?) = 0.38$;

(h) $X \sim N(14, 5.29)\ldots$ $P(? < X < 12) = 0.49$;

(i) $X \sim N(-36, 25)\ldots$ $P(-30 < X < ?) = 7\%$;

(j) $X \sim N(-36, 25)\ldots$ $P(\mu_X < X < ?) = 34\%$.

The Normal Probability Plot

Here we present a simple graphical technique, the "normal probability plot," customarily used to check on the assumption that a given sample was drawn from a normal distribution. The idea is simple, although the work involved is quite tedious for large samples—in practice, you would use a computer. Perhaps the most obvious graphical technique would be to check if a histogram for some grouping of the data seems to have the shape of a normal distribution. But, in fact, that doesn't work well at all. The shape of a histogram can change radically depending on exactly how you group the data. The "normal probability plot" presented below provides a more sensitive check on normality than does a histogram.

The ***normal probability plot*** for a sample of size n plots the observations in the sample against the corresponding percentiles of Z. We will explain this in more detail below. But first we need to know that a ***percentile*** for the probability distribution of a random variable is a value of the random variable which "cuts off" a given percentage of the distribution. For example, the term "median" can be defined as the 50th percentile because it cuts off half the distribution. In other words, half the distribution is below the median. A "tenth percentile" would be the value which cuts the distribution into 10% and 90%, so that ten percent of the distribution is below that point. Here's the picture for the tenth percentile of Z:

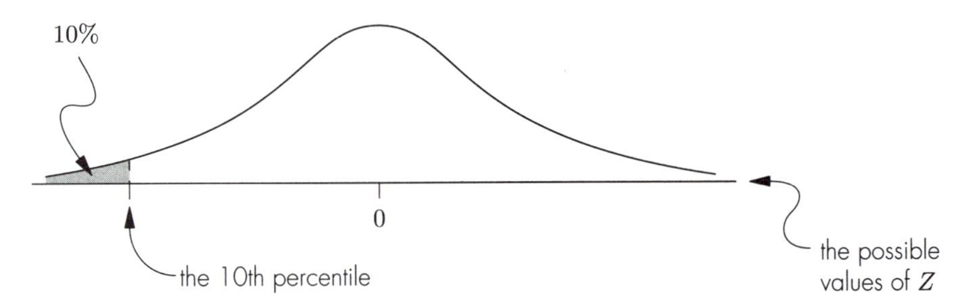

The idea behind the normal probability plot for a sample is to compare that sample with an ideal random sample from $N(\mu, \sigma^2)$. If you think your sample came from $N(\mu, \sigma^2)$, the comparison should be "favorable." Exactly what that means we'll see below. An ideal sample should have been evenly spread throughout the distribution it was drawn from. The phrase "evenly spread," however, does not refer to the observations in the sample but rather to the probabilities of the distribution. In other words, the sample should cut the distribution into equal probabilities. For example, four "ideal observations" from Z cut the distribution into five (FIVE notice, not four) equal probabilities:

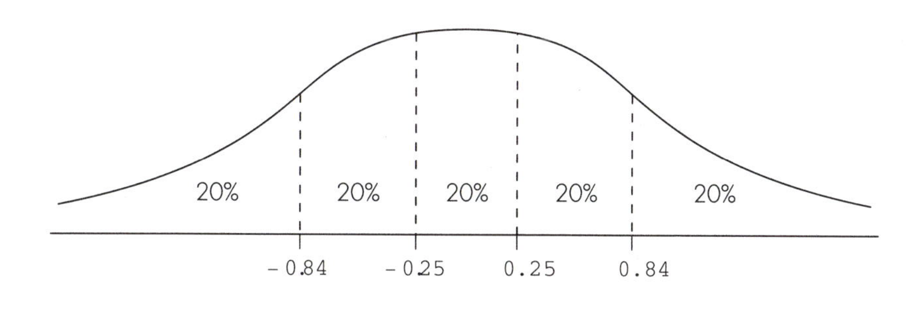

We obtained these four "ideal" observations of Z from the Z table by looking up the appropriate probabilities. Now, if you have a sample which you think came from the normal distribution $N(\mu, \sigma^2)$, the first thought is to compare it with the ideal sample from $N(\mu, \sigma^2)$. But there's no way to determine the values of the ideal sample because we don't know the parameters μ and σ of $N(\mu, \sigma^2)$. So, instead, we compare our sample to the standardized version of the ideal sample. In other words, we compare our sample to the ideal sample from the Z distribution (shown in the picture above for $n = 4$). You'll see the details of this comparison as you . . .

Try Your Hand

4.3.19 We'll lead you through the construction and interpretation of a normal probability plot for the following sample:

$$\{5.2,\ 0.4,\ 0.2,\ 2.5,\ 1.2,\ 3.5,\ 1.8,\ 4.8,\ 2.7\}.$$

(a) There are nine numbers in the sample, so we're talking about dividing the area under the curve of Z into ten equal probabilities. What values of Z will do that?

(b) The percentiles of Z which you determined in part (a) are not evenly spaced on the number line. Why not?

(c) What are the points we want to plot in our normal probability plot?

(d) Suppose our sample really is from a normal distribution, $N(\mu, \sigma^2)$ say, and that the sample is more or less like that distribution. What should the normal probability plot look like?

(e) Give the normal probability plot for our sample.

(f) Interpret the normal probability plot for our sample in the light of part (d).

(g) You should realize that there's some variation in the method of obtaining the Z percentiles corresponding to a given sample size. Our method determines them through the formula $k/(n+1)$. Explain.

4.3.20 The Stanford University geologist Kerry Sieh studied the occurrence of earthquakes at Pallett Creek northeast of Los Angeles on the San Andreas fault. Sieh estimates that earthquakes occurred in the years

$$1857,\ 1720,\ 1550,\ 1350,\ 1080,\ 1015,\ 935,\ 845,\ 735,\ 590.$$

Based on Sieh's estimates, do you think that "time between earthquakes" is normally distributed?

Continuous Approximations to Integer-Valued Random Variables

We have typically used line graphs to picture the probability distributions for discrete random variables:

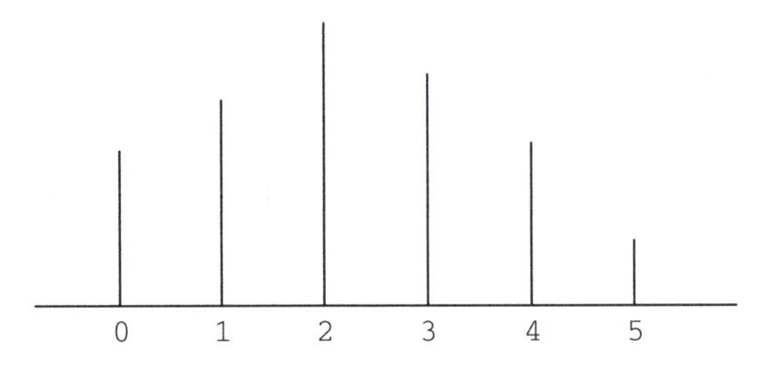

But there is nothing to keep us from replacing the vertical lines by rectangles of appropriate width so that area represents the probability of a given value:

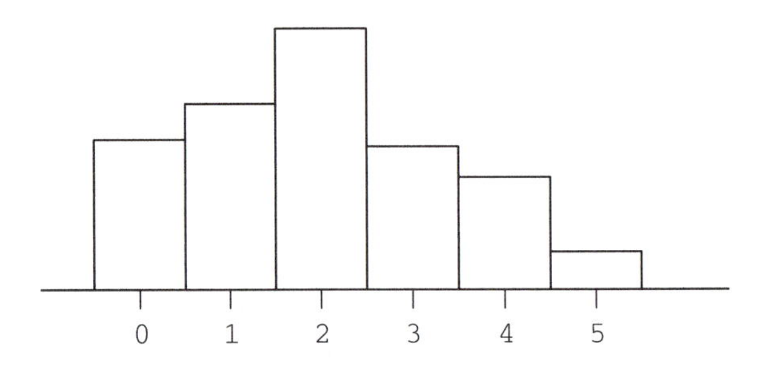

Now, imagine a discrete random variable with a very large number of possible values. Suppose it takes on any integer value from zero to

10,000. With probability represented by rectangles, the probability distribution will look like

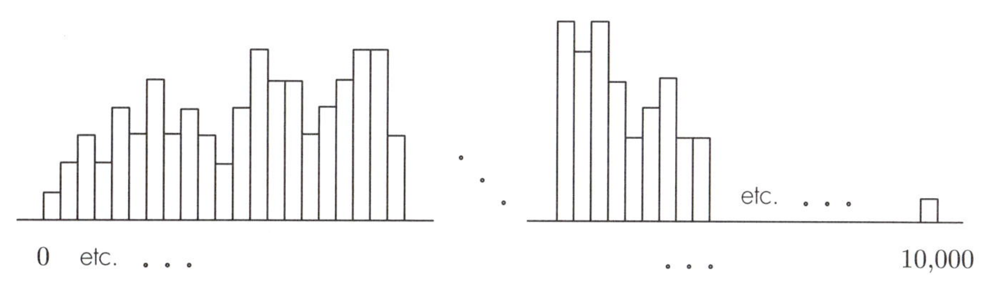

This picture would be visually easier to comprehend if we left out the edges of the rectangles and just drew the outline:

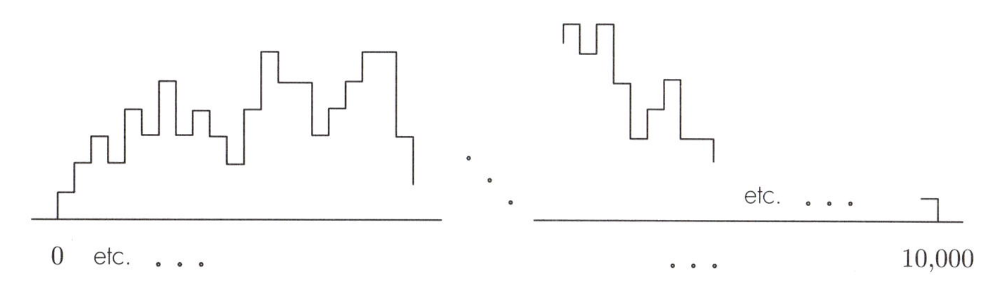

Or, by making the values much closer together and the rectangles proportionately smaller, you would get the whole picture on one page without having to break it with a lot of "dot-dot-dots"

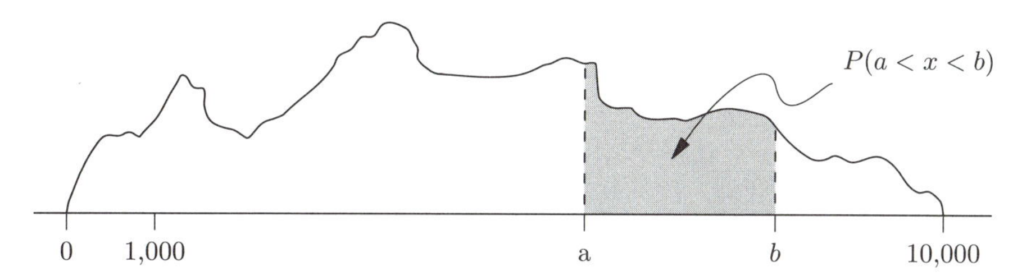

You may think this looks like a smooth curve, but it's not—it's just that the many horizontal and vertical lines have become so small you can't see them! We drew a smooth curve, of course, but the point is that with so many values, the probability distribution as a whole will look very much like a continuous distribution.

This purely heuristic line of reasoning based on the pictures suggests that we can approximate a discrete distribution by a continuous one if the discrete random variable has a large number of values. When we superimpose the continuous distribution over the discrete one, we get the following picture:

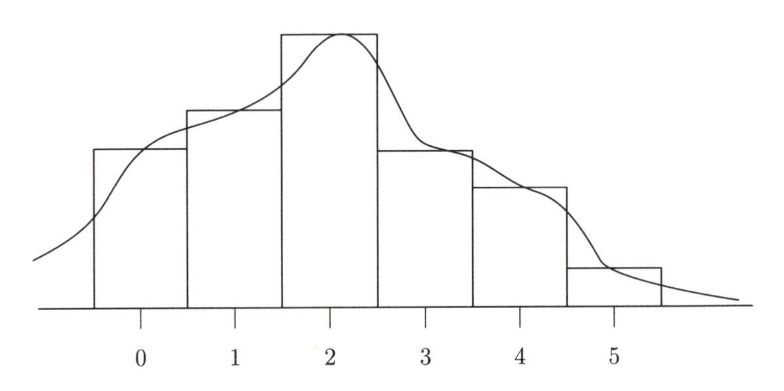

You can see that the area under the curve and the area enclosed by the rectangles is not exactly the same, but the "errors" tend to cancel out:

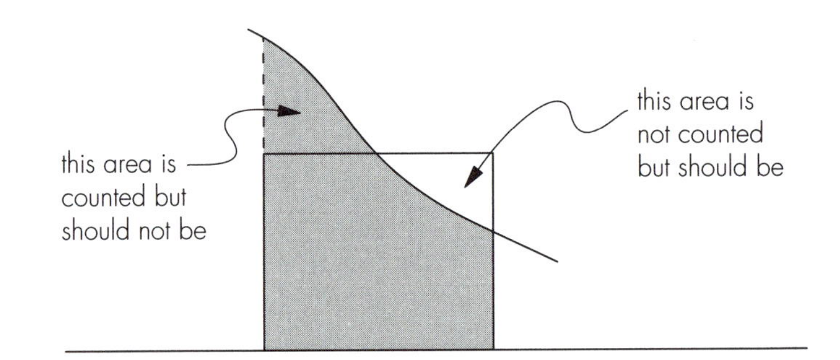

Note also that a probability of the form $P(X \geq 3)$ must be calculated by area under the curve starting at $\widetilde{X} > 2.5$, where $\widetilde{X}$ is the continuous approximation to X:

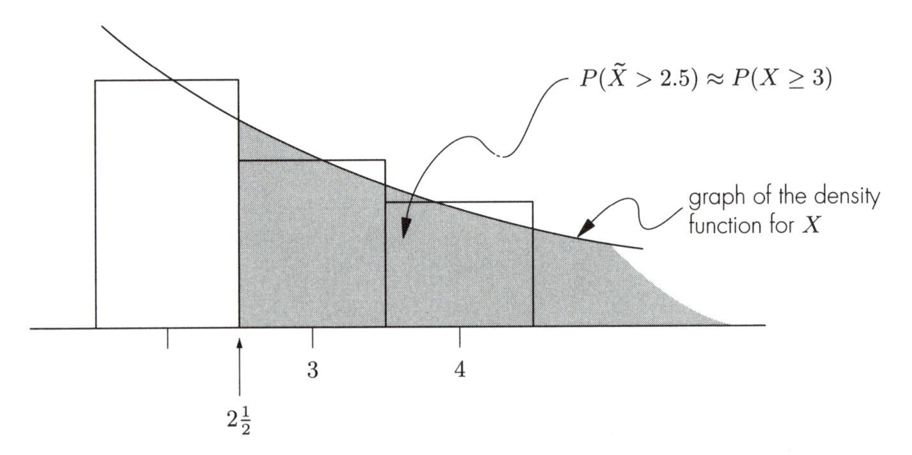

This adjustment of moving to the edge of the rectangle as we pass from X to its approximation $\widetilde{X}$ is called the ***continuity correction***. In many cases, the continuity correction will have little or no effect on the answer. But you'll never be wrong if you put it in—in some cases, it will make a significant difference. You should use the continuity correction unless specifically instructed to the contrary. But before we go any further, why don't you think about this a bit . . .

Try Your Hand

4.3.21 Each of the following conditions would be necessary for an integer-valued random variable to be approximated by the normal distribution. Explain why.

(a) It must have a large number of possible values.

(b) It must be approximately symmetric about its mean.

(c) It should have only one mode, and the mean, median, and mode should be approximately equal.

4.3.22 Suppose X is an integer-valued random variable which is approximately normally distributed with mean 32.8 and standard deviation 10.3. What's the probability that X is greater than 50?

4.3.23 Assuming $\widetilde{X}$ to be a continuous approximation to an integer valued random variable X, what should be the limits on $\widetilde{X}$ to give the correct approximation to the following probabilities:

(a) $P(X > 7)$;

(b) $P(7 \leq X)$;

(c) $P(2 < X \leq 8)$?

The Normal Approximation to the Binomial

Now let's take a look at the ***normal approximation to the binomial distribution.*** It's valid, as a rough rule, when both np and nq are greater than or equal to five.

> If X is binomially distributed, then X is approximately $N(np, npq)$ provided np and nq are at least five.

Note, if the smaller of np and nq is at least five, then, of course, the larger one is. This rule requires n to be large compared with p: If p is small, then n must be proportionately larger.

The normal approximation to the binomial is complementary to the Poisson approximation. The Poisson approximation (see Problem 3.8.10) is valid for cases in which "success" looks like the occurrence of a rare event. That means p should be quite small—our rule says "less than five percent." If $p = 0.001$ and $n = 100$, the Poisson approximation would be valid and the normal not because $np = 0.1$, much less than five. On the other hand, the normal approximation does not require p to be particularly small. It would be valid if $n = 50$ and $p = \frac{1}{2}$ or $p = \frac{1}{4}$.

To see how the normal approximation can be useful, please . . .

Try Your Hand

4.3.24 Suppose you want to approximate a binomial X by a normal distribution. How large must n be if p is

(a) $\frac{1}{2}$;

(b) 0.25;

(c) 5%

(d) 0.002

(e) 0.75?

How small can p be if n is

(f) 12;

(g) 100;

(h) 150;

(i) 1,000;

(j) 10,000?

4.3.25 Assume X is a binomial random variable. Evaluate the following probabilities. You should use the normal approximation where appropriate.

(a) $P(X > 4|n = 7, p = 0.4)$;

(b) $P(X \geq 137|n = 300, p = 0.4)$;

(c) $P(X < 8|n = 12, p = \frac{1}{2}$;

(d) $P(X = 24|n = 200, p = 0.12)$.

4.3.26 Suppose the machine parts which you manufacture have an average diameter of 2.3 cm with a standard deviation of 0.1 cm and an average life of about 6000 hours. Because this is a real-world question, BE SURE TO GIVE REAL-WORLD ANSWERS AT THE END.

(a) If you choose 12 machine parts at random, what's the probability at least five of them will be more than 2.35 cm in diameter? [$\approx$ 30%]

(b) If one of the customers you supply purchases a lot of 500 parts, what's the probability at least half of them will last six months? Assume the parts are in continual use. [$\approx$ 22%]

(c) In a lot of 500 machine parts, what are the chances more than five will be unusable? A part is unusable if its diameter exceeds 2.55 cm. [$\approx$ 9%]

4.4 The Chi-Squared Distribution

In Chapter 5, we'll begin to see, by Laplace's Central Limit Theorem, how the normal distribution is the key to certain types of questions concerning averages. Similar questions concerning variability involve the chi-squared distribution, the model of this section. The simplest version of this model is the chi-squared random variable with "one degree of freedom." It's the square of the standard normally distributed random variable: $\chi_1^2 = Z^2$.

Variability, of course, is not measured by just one squared quantity; the variance of a random variable or of observed data involves a sum of squares. We'll postpone looking at that for now, but the model we'll need, the model for a sum of d squared Z's , is

the chi-squared random variable with *d* degrees of freedom:

$$\chi_d^2 = Z_1^2 + Z_2^2 + Z_3^2 + \cdots + Z_d^2, \quad d \text{ independent } Z's.$$

The mean is d and the variance $2d$:

$$\mu = d,$$
$$\sigma^2 = 2d.$$

As with the normal distribution, the formulas for χ^2 are too complex to work with and so a table of probabilities is provided in the appendix. Note that the Z's must be independent of each other. This assumption is crucial: If it's not at least approximately valid, the model will give results which are also not valid. The term "degrees of freedom" is the number of Z^2's in the sum—they're "free" of each other in the sense of being independent.

The chi-squared distribution is really the most abstract of all our models. It arises from purely mathematical considerations; there's no "rule of thumb" for the type of real-world situation which it models. It doesn't model real-world situations. Rather, it's an important theoretical tool for addressing certain types of real-world questions concerning variability. We'll see all this in later chapters. For now, we just ask you to be familiar with the model and its terminology and learn how to use the table in the appendix.

Unlike the table for Z,

- the BODY of the chi-squared table lists the values of χ^2,
- the LEFT-HAND MARGIN gives d, the degrees of freedom (df),
- the TOP MARGIN gives the probabilities.

The probabilities are left-tail probabilities, as in the following picture ($d = 12$):

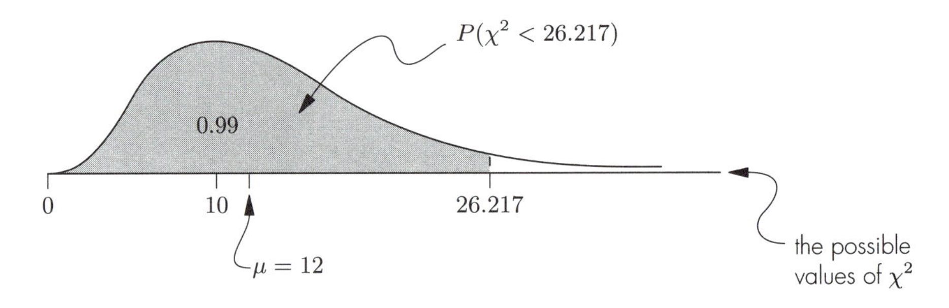

This picture shows there's a one percent chance for the sum of 12 independent squared Z's to take a value greater than 26.217. Note that the curve peaks out just to the left of the mean. In fact, the peak always occurs at $d - 2$. Because $d = 12$ in this picture, the peak is at ten. Also note that the curve starts out tangent to the axis. That's true provided d is bigger than four.

Well, now it's time for you to . . .

Try Your Hand

4.4.1 Give the following probabilities:

(a) With 12 degrees of freedom, $P(\chi^2 > 23.336) =??$

(b) $P(27.204 < \chi^2 < 32.852) =??$ Assume $d = 19$.

(c) Suppose you have a simple random sample of size 20 from the standard normal distribution. If you square all your observations, what's the probability those squares will total more than 40?

(d) In part (c), how do you know the twenty Z's are independent?

(e) Suppose you obtain a simple random sample of 23 quiz scores and you square each of those 23 scores. What's the probability the sum of those squared scores would exceed 35.17?

4.4.2 Find the value of χ^2 which gives the required probability.

(a) $P(\chi^2 >??) = 0.01$, with 19 degrees of freedom.

(b) $P(27.587 < \chi^2 <??) = 2.5\%$, with $d = 17$.

(c) $P(Z_1^2 + Z_2^2 + Z_3^2 + Z_4^2 + Z_5^2 + Z_6^2 >??) = 10\%$.

4.4.3 (a) In the pictures of χ^2, why are there no values to the left of zero?

(b) As the degrees of freedom increase for χ^2, should the mean increase or decrease? Answer this heuristically and in terms of the formulas.

(c) In the pictures of χ^2, why does the tail go off to the right with no limit but with smaller and smaller right-tailed probabilities? In other words, why is χ^2 skewed to the right?

(d) Draw a picture of the χ^2 distribution with 14 degrees of freedom. Be sure to label μ and $\mu + \sigma$.

4.5 A Few Review Problems

You should identify the model for each problem, even though in some cases the solution is obvious on intuitive grounds without reference to the model. Models from the previous chapter are fair game. If a problem requires an unstated assumption, work the problem under that assumption and be prepared to comment on the appropriateness of the assumption for that particular situation.

4.5.1 Records show that 30% of the screws produced by a particular machine are less than 1.3 cm and 40% more than 1.71 cm in diameter.

(a) What's the mean diameter and the standard deviation of diameters for these screws?

(b) What assumption are you making to solve part (a) and is that assumption justified?

(c) We want to report the maximum diameter for these screws. Of course, there's no meaningful absolute maximum (explain!), a few screws will be longer than our reported maximum. What value for the maximum should be reported if it's understood that we allow no more than one screw in 10,000 to be longer than the reported maximum? [3.6324 cm]

4.5.2 Reports from customers at your supermarket indicate that about 4.5% of them have to wait in line more than ten minutes during the weekday rush from 4:30 to 6:00 P.M. How many minutes must the average customer wait during that rush period? [≈ 3 min 14 sec]

4.5.3 Suppose 32,753 students nationwide take a standardized test for which the cutoff for passing is 1.2 standard deviations below the mean:

(a) About how many students will pass? [≈ 29,000]

(b) Suppose we choose 100 students at random, what is the probability exactly ten fail? [≈ 11%]

4.5.4 In a 1983 report for the U.S. Air Force (*New York Times* 2/11/86), R.K. Weatherwax estimated the probability of a catastrophic accident involving the space shuttle's solid-fuel booster rockets to be about one in 35 flights. On the other hand, in 1985 the National Aeronautics and Space Administration estimated the probability of booster failure to be one in 60,000 flights. On January 28, 1986 the shuttle exploded on its 25th flight. If Weatherwax's estimate was correct, what was the probability of at least one catastrophic accident of this sort in 25 flights? [≈ 52%]

4.5.5 You're thinking of maintaining a stock of semiprecious stones in your shop. The supplier of these stones has informed you that her stones weigh eight ounces on average with a standard deviation of 2.4 ounces. A carton consists of a dozen randomly chosen stones.

(a) Make a probability distribution for stone size for five sizes determined as follows:

Size	Weight
I	$\cdots$ –3.2
II	3.2–5.6
III	5.6–10.4
IV	10.4–12.8
V	12.8– $\cdots$

(b) How many stones of each size would you expect in each carton?

(c) How many stones in 110 cartons would be of size IV or V? [$\approx$ 209.5]

(d) Out of 110 cartons, what's the probability that more than 120 stones will be size IV or V?

(e) A customer needs 24 size I stones. What's the minimum number of cartons you would have to buy to guarantee at least a 90% chance of getting enough size I stones? [68]

4.5.6 The photocopying machine in your office seems to do about two and a half hours of more or less continuous copying before a malfunction.

(a) How many malfunctions should you expect in an eight hour day? [3.2]

(b) How likely is it you will get through an entire eight hour day of more or less continuous copying without such a malfunction? [$\approx$ 4%]

4.5.7 In discrimination cases, the question arises of what exactly is the pool of candidates you're choosing from. In *Hazelwood School District* v. *United States* (1977) only 15 of 405 teachers hired by the Hazelwood district were black. An accusation of discrimination was based on the proportion (15.4%) of blacks in St. Louis County which includes the city of St. Louis. The Hazelwood district claimed their pool of candidates should not include the city proper, in which case the proportion (5.7%) of blacks was much smaller. In each case, what's the probability that 15 or fewer of 405 teachers would be black? (after [Finkelstein and Levin])

4.5.8 Here we continue the discussion of Problem 3.10.14 by determining the endpoints of a confidence interval for an unknown median M. In Problem 3.10.14 we calculated the confidence coefficient for a specific confidence interval. That's the reverse of what one usually does. The true power of the confidence interval technique is that you get to choose whatever confidence coefficient provides adequate certainty for your purposes. Then you determine the confidence interval. Afterward, you have an "interval estimate" for your parameter with a predetermined degree of certainty.

Suppose you've taken a sample of size n from a probability distribution (or, as a special case, a numeric population) and arranged it in ascending order. So the sample looks like: $x_{(1)} \leq x_{(2)} \leq x_{(3)} \leq \cdots \leq x_{(n)}$. These $x_{(i)}$'s are called the ***sample order statistics***. Note that each one of these is a number calculated from the sample and so, indeed, is a "statistic." Thus, $x_{(1)}$ is the "first order statistic," $x_{(2)}$ the "second order statistic," and so on. In general, $x_{(k)}$ is the ***kth-order statistic***. We'll determine the endpoints of our confidence interval for the median in terms of the order statistics of a sample.

First, suppose we want a 90% confidence interval for the median M. If the endpoints of the interval are to be the order statistics $x_{(h)}$ and $x_{(k)}$, then just as in Problem 3.10.14(e)

$$0.9 = \mathrm{P}(x_{(h)} < M < x_{(k)}) \quad = \quad 1 - [\mathrm{P}(M < x_{(h)}) + \mathrm{P}(M > x_{(k)})].$$

We assume the sample is drawn from a continuous distribution (or very large population) so there's a zero chance for an observation in the sample to equal M. Thus, we can ignore equal signs in the square brackets.

(a) In Problem 3.10.14(a), show that the sample median is $x_{(12)}$. In that example, what's the value of $x_{(6)}$?

(b) Express the square bracket in part (a) in terms of Y = # observations in the sample which are less than M.

(c) Let $z_o = 1.645$. Show that if $n \geq 10$, the endpoints of the 90% confidence interval for M are $x_{(h)}$ and $x_{(k)}$, where h and k are determined by

$$h = \frac{n + 1 - z_o\sqrt{n}}{2} \quad \text{and} \quad k = \frac{n + 1 + z_o\sqrt{n}}{2}.$$

(d) Give a 90% confidence interval for the median fill of cups from the drink machine in the employee lounge. Here is the fill in ounces for ten cups: 6.7, 6.4, 6.3, 6.4, 6.3, 6.5, 6.4, 6.2, 6.8, 6.5 .

(e) What assumption are you making in the analysis of part (d)?

(f) Explain the meaning of the 90% confidence coefficient from part (d).

(g) Determine a 95% confidence interval for the median fill in part (d).

(h) Show how to calculate the 99% confidence interval for the median weight of U.S. pennies given the data in Problem 3.10.14(g).

(i) How do you interpret the confidence interval of part (h)?

Chapter 5 Estimation of Parameters

5.1 Parameters and Their Estimators

Introduction

Among the most fundamental questions arising in statistics are questions which concern one or more unknown parameters. If you think a particular population is normally distributed, for example, you'll need to determine μ and σ, the mean and variance for that population; or if you're looking at a Poisson process, you'll need λ, the average number of occurrences in a unit of time. If you have a problem that involves a Bernoulli trial (a geometric, binomial or negative binomial problem), you'll need a value for the parameter p, the probability of success on one repetition of the trial.

Typically, the parameter is not only unknown, it's *unknowable*! It would be impossible, in other words, to obtain information sufficient to calculate it. If that's the case, you have to estimate the parameter based on partial information; for us that means information in the form of a random sample. But this is not nearly so simple a solution as you might think! Suppose a random sample gives a sample mean of 2.37. What do you do with that information? Will you say the mean of the population is 2.37? You shouldn't, a sample isn't going to be exactly like the population it's drawn from. That's like saying 100 rolls of a fair die should yield exactly 350 dots total—not 348 or 354 but exactly 350!

It's usual to denote a population proportion by the symbol p, following the convention introduced in Chapter 3. For a sample from a dichotomous population, the ***sample proportion***, the proportion of the sample having the characteristic of interest, is denoted by the symbol $\hat{p}$ (read: "p-hat"). Sample proportions also arise in sampling from a Bernoulli distribution, where the parameter p is the probability of observing the outcome of particular interest ("success"). For instance, if you have a sample of 20 tosses of a coin, you may be interested in the proportion of heads observed, $\hat{p}$. Or you may be interested in the proportion, $\hat{p}$, of defectives among 40 silicon wafers produced in succession by your manufacturing process; or the proportion, again $\hat{p}$, of children in your neighborhood who contract a particular disease. In each case, your parameter is p, the probability of "success" for a Bernoulli random variable, and $\hat{p}$ is the sample proportion for a simple random sample from that probability distribution. You studied these three examples in Problems 2.1.5, 2.1.9, and 2.1.10. They are not random choices from a population, notice, but rather "n independent repetitions" of a Bernoulli trial. So, they are simple random samples from the distribution of that Bernoulli random variable.

Let's look more carefully at samples and what they might tell us about an unknown parameter. For simplicity, we'll confine our attention for the time being to the special case of population means and proportions. But there's nothing really special about this; everything we see here is applicable to other parameters as well, in particular to parameters for random variables and samples from their probability distributions. So now please . . .

Try Your Hand

In the following problems you will explore the relationship between the question "What's the value of this unknown parameter?" and the information for answering that question contained in one sample.

5.1.1 You need to know μ, the mean of a distribution you're studying. Maybe it's the mean of a population, maybe it's the mean of some random variable. Because it's impossible to actually compute the value of μ, you obtain a simple random sample from which you compute a sample mean of 2.37. Here's the situation . . .

$$\text{the question:} \quad \mu = ?,$$

$$\text{the information:} \quad \overline{X} = 2.37.$$

(a) If it's a population you're studying, what kind of population is it?

(b) What information would be required to actually compute μ?

(c) What assumption are you making if you say $\mu = 2.37$?

(d) What assumption are you making if you say $\mu \approx 2.37$?

(e) Which of the assumptions in parts (c) and (d) is valid?

(f) At what point will you know whether your sample is typical of the entire distribution you're sampling from?

(g) How do you make use of the information that $\overline{X} = 2.37$ to answer the question about μ?

5.1.2 Suppose you would like to know what proportion, p, of the population of a particular geographical area is over 50 years of age. Suppose further that you obtain a simple random sample for which 18% are over 50. Here's the situation . . .

$$\text{the question:} \quad p = ?,$$

$$\text{the information:} \quad \hat{p} = 0.18.$$

(a) What kind of population are you studying?

(b) What information would be required to actually compute p?

(c) Your simple random sample was chosen according to the strictest rules of sampling; the results were recorded carefully and doubled checked to eliminate any possible recording error. So, from the fact that $\hat{p} = 0.18$, we can safely conclude that 18% of this population is more than 50 years of age. Comment.

(d) How can you make use of the information that $\hat{p} = 0.18$ to answer the question about p?

5.1.3 Let's explore this idea of a "typical" sample in more detail. We'll take a very simple example so every detail will be completely obvious. Suppose you're sampling from a population which consists of the following numbers:

$$0, \ 1, \ 2, \ 3, \ 3, \ 4, \ 5.$$

Here, $N = 7$ and one value appears in the population twice. This is a very unrealistic example, of course. There's no need to talk of estimating the mean, in 15 seconds you compute it to be 2.5714. But the example will be very useful for exploring the "typicality" of random samples. Our example will be unrealistic in another sense also: We're going to take samples of size *two*. Nobody would ever take samples of size two because they don't contain enough information! Still this example is useful because it's easy and it allows us to see all the detail.

But don't forget the point of the example: You're supposed to imagine μ is not known even though in the example we do know it and you're supposed to imagine that computing its value would be impossible. In other words, you're to imagine that the population as a whole is not accessible and is to be studied through sampling. Occasionally, we'll make reference to the true value of μ (it's 2.5714) to compare what's suggested by a sample with what's really true. But in a real-world problem, we don't know "what's really true"; we only know the sample!

(a) Write out all the possible samples of size two chosen without replacement. Be sure you distinguish between the two different three's listed in the population.

(b) Compute the value of $\overline{X}$ for each of the samples in part (a).

(c) What kind of object is $\overline{X}$?

(d) Make a probability distribution of the sample means for all the possible samples of size two.

(e) Complete the distribution of part (d) to compute the mean, $\mu_{\overline{X}}$, and the variance, $\sigma^2_{\overline{X}}$, of the sample means. Compute the variance with the *conceptual* formula (so you can see exactly what's going on)!

(f) Identify the mean of the sample means which you computed in part (e) in terms of something simpler, something more immediate.

(g) Which is more variable, the various sample means or the various numbers in the population? Answer this in three ways:

(i) Guess! Look at the population and at the set of sample means and guess which is more variable.

(ii) Try to think of some general principle which would justify your guess in (i).

(iii) Look at the appropriate measures of variability.

(h) Which of the samples listed in part (a) are typical of the population? Which would you label as atypical?

(i) What could you say about the population mean based on one simple random sample?

5.1.4 Here we're thinking about the same population as in the previous problem, but we're considering p, the proportion of positive even numbers (of which there are two):

Here's the population: 0, 1, 2, 3, 3, 4, 5.

The sampling experiment will also be the same as in the previous problem—simple random sampling without replacement, $n = 2$. So, of course, the outcomes (the samples) are the same as those listed in part (a) of the previous problem.

(a) What's the value of p here?

(b) What kind of population are you studying?

(c) What kind of object is $\hat{p}$?

(d) What are the possible values of $\hat{p}$?

(e) Make up a probability distribution for $\hat{p}$ and use it to compute the mean, $\mu_{\hat{p}}$, and the variance, $\sigma^2_{\hat{p}}$, of the random variable $\hat{p}$. You should try to guess the mean of $\hat{p}$ before you compute it.

(f) Which samples are typical of the population? Which would you label as atypical?

(g) Suppose you drew the sample $\{2, 4\}$, what would that sample say about the proportion of the population which are positive even numbers?

(h) In this sampling experiment, what's the probability of drawing the very atypical sample $\{2, 4\}$?

Estimators—the Entire Context

For questions about unknown parameters, random variables like $\overline{X}$ or $\hat{p}$ are the key. Such a random variable is called an ***estimator*** for the parameter. The underlying random experiment is random sampling, with samples as outcomes. There's a very elaborate theory about such estimators and there's much to be said about which estimator is best for a particular parameter. What does "best" even mean?! But in this course, we'll not be concerned with such questions. It's enough for you to accept the sample mean as "estimator" for a population mean and the sample proportion for a population proportion.

In Problems 5.1.3 and 5.1.4, just to see what happens, we had you look at all possible samples from a very small population. That's very unrealistic, of course. In most real-world situations, the population is inaccessible (usually because it's too large) and looking at all possible samples would be absurd. It would be even more difficult than looking at the population as a whole because every member of the population is part of some sample. Beyond that, sampling is very expensive. A properly designed sampling experiment requires time, resources, and trained personnel. There's no question of taking many samples. One sample alone may stretch your resources to the limit. That's not to say someone else—or even you yourself at some later time—might not want to duplicate your study. Replication is essential for any study that's to have lasting significance. But in our course, we're learning techniques applicable to one single study. Therefore, throughout this text we'll assume that

ONE SAMPLE AND ONE SAMPLE ALONE IS THE MOST THAT'S PRACTICAL.

But that should cause you some disgust! You've seen how useless one sample alone would be for saying anything at all about the underlying population. Your one sample might just happen, unluckily, to be one of the atypical ones. After all, an atypical sample is just as likely to be drawn as a typical sample. In Problem 5.1.4, the sample $\{2, 4\}$ is atypical: Both numbers in the sample are even and positive ($\hat{p} = 100\%$) when, in fact, only about 30% of the population have that characteristic. Here the true value of the parameter ($p \approx 30\%$) is nowhere near the calculated value of the estimator ($\hat{p} = 100\%$). Yet, this very atypical sample has the same probability to be drawn as any other sample: one chance in 21.

In looking at a particular sample, how do you determine if it's typical or not? In fact, you have no way of knowing that! You would have to compare the sample with the entire population. But if that were possible, you'd have all the information required to actually calculate the parameter. There would be no need to estimate anything! Therefore, contrary to what the suggestive term "estimator" might lead you to expect, one value of an estimator determined by one sample is meaningless taken by itself because you have no way to know if that one sample is typical or not.

In Problem 5.1.1, you had the question $\mu = ?$ and the information $\overline{X} = 2.37$. It would be wrong to say μ is 2.37—that's the *sample* mean, it's not computed from the entire population. But to say the true value of μ is *approximately* 2.37 is also wrong. It's wrong not because of some problem with the sampling experiment nor because of possible human error. It's wrong because even though everything was properly done, a single random sample can be very misleading. The unknown true mean could be very far from the observed sample mean. The conclusion $\mu = 2.37$ is not justified and the conclusion $\mu \approx 2.37$ is also not justified—both conclusions are unjustified even though 2.37 is a correctly computed value of the proper estimator for μ!

Similarly, for Problem 5.1.2, we sought to answer the question $p = ?$, based on the information: $\hat{p} = 0.18$. You would not be justified in concluding that p is even approximately 18%. It could be quite far from that observed value.

To make use of the information contained in one random sample, more is required than just the one value of the estimator determined by that one sample. What's required is the ENTIRE CONTEXT of the estimator—the total picture for the estimator with our one sample seen in the context of that whole picture. What do we mean by the "total picture" or the "entire context" of an estimator?? To think about this, please . . .

Try Your Hand

5.1.5 Because the conclusion "μ is approximately 2.37" is not justified, it seems that a sample mean of 2.37 must imply μ is FAR from 2.37. But that's absurd! How could the only evidence you have (the sample), giving a mean of 2.37, suggest a true mean of some *other* value far away? What's wrong here?

5.1.6 Let's think about what an estimator really is:

(a) What kind of object is an estimator?

(b) What information would you routinely want to know about such an object?

5.1.7 (a) Give formulas for the mean and variance of the estimator $\hat{p}$.

(b) Sketch a picture of the probability distribution for the estimator $\hat{p}$, labeling it as completely as possible. Assume you're sampling from a very large population, taking samples of size $n = 100$. Assume also that n is large enough so that np, $nq \geq 5$.

(c) Show that the conclusions of parts (a) and (b) still hold if $\hat{p}$ is associated with simple random sampling from the distribution of a Bernoulli random variable.

5.1.8 a) What do we mean by "the entire context" or "the whole picture" for an estimator?

(b) What's the relationship between $\hat{p}$ and p? This is a crucial issue because just one value such as $\hat{p} = 0.18$ tells you nothing by itself.

5.1.9 As in Problem 5.1.2, suppose you would like to know what proportion, p, of the population of a particular geographical area is over 50 years of age and that you obtain a simple random sample for which 18% are over 50. Here's the situation:

$$\text{the question:} \quad p = ?,$$
$$\text{the information:} \quad \hat{p} = 0.18.$$

(a) Where does this particular value of the estimator fit into the total picture?

(b) Based on the available information, which of the following are possible pictures for the probability distribution of $\hat{p}$? [Hint: A picture is possible unless there is something *obviously* wrong with it.]

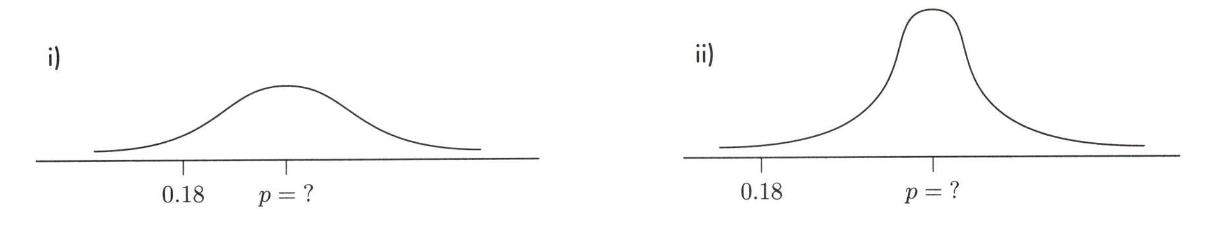

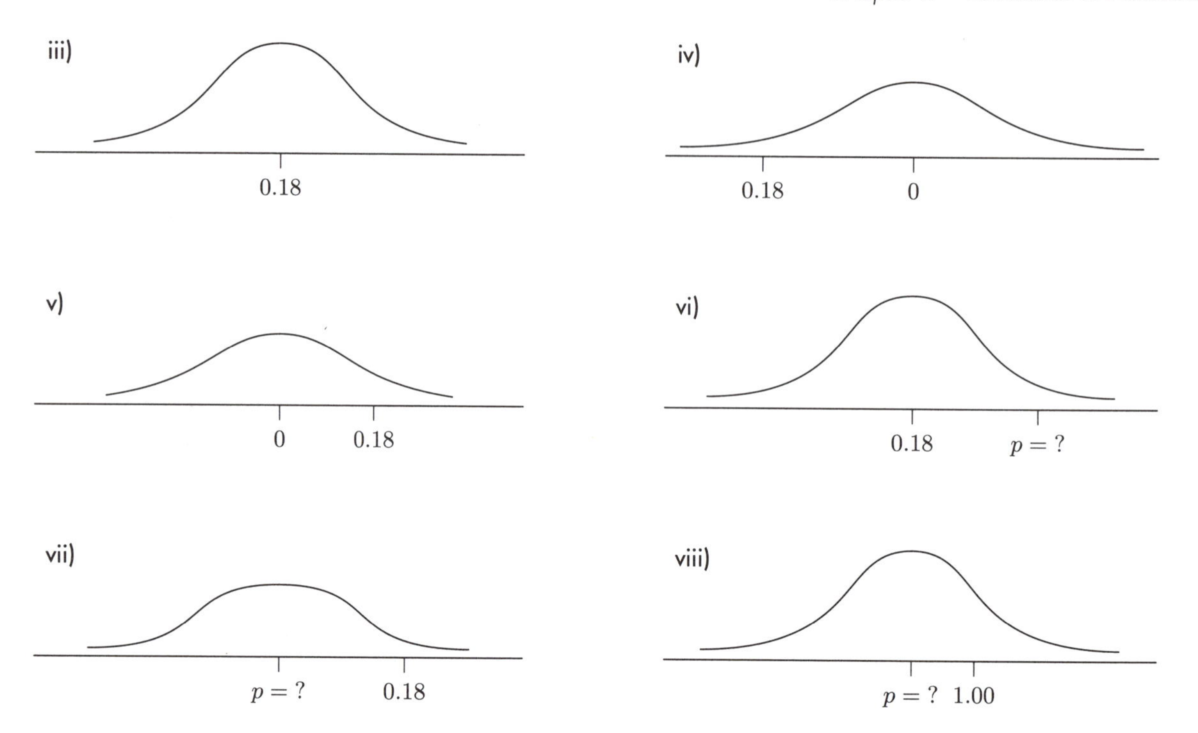

5.1.10 **(a)** In the situation of the previous problem, p cannot possibly be zero. Why not?

(b) When p is unknown, the population and the sample must be large to justify drawing a normal curve as the distribution of the estimator $\hat{p}$. Why?

5.2 Estimating an Unknown Proportion

The Sampling Distribution for $\hat{p}$

The probability distribution for an estimator like $\hat{p}$ is a sampling distribution. The term "sampling distribution," remember, refers to any probability distribution where the underlying random experiment is random sampling. An estimator for a parameter is called ***unbiased*** if its expected value (its "on average" value, the "center" of its sampling distribution) is the parameter in question. For example, $\mu_{\hat{p}} = p$, which means the estimator $\hat{p}$ is unbiased for the parameter p. You showed this in Problem 5.1.7(a) and (c). Later, we'll show that the sample mean is also an unbiased estimator for its parameter, μ. You saw a special case of this in Problem 5.1.3(f) where, for the particular population of that problem, you discovered that $\mu_{\overline{X}} = \mu$. The standard deviation of an estimator is usually referred to as the ***standard error*** (sometimes abbreviated as s.e.) because it measures the accuracy of the estimator—it measures the variability from one sample to another.

In the exercises at the end of the previous section, you explored the sampling distribution for $\hat{p}$ quite thoroughly. Here's a summary:

the sampling distribution for $\hat{p}$:

If you are sampling with replacement from a dichotomous population or taking a simple random sample from the distribution of a Bernoulli random variable and you're sure your sample size is large relative to p ($np, nq \geq 5$), then $\hat{p}$ is approximately *normally distributed*. It's also unbiased,

$$\mu_{\hat{p}} = p,$$

and its variance, its *squared* standard error, is

$$\sigma^2_{\hat{p}} = pq/n.$$

Here's the picture:

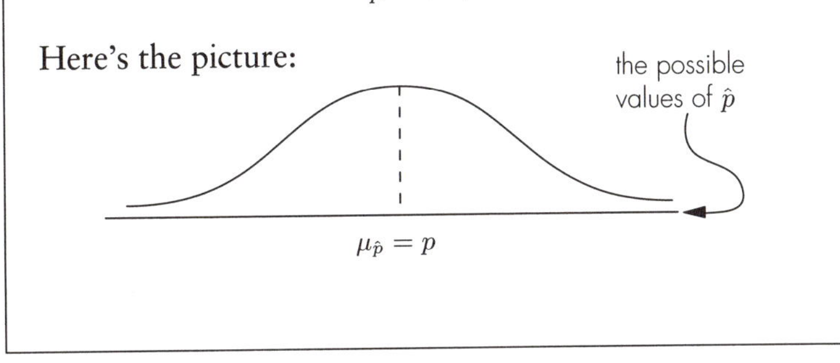

Although for ease of understanding we've described two sampling conditions in the box above, there's really only one. Sampling with replacement from a dichotomous population is modeled by the binomial random variable, so it's also a case of "simple random sampling from the distribution of a Bernoulli random variable." And remember: For very large populations—our typical case—the distinction between with and without replacement for the sampling experiment is insignificant. So even though in practice you may be sampling without replacement, you can use the simpler "with replacement" model of the box above.

In the box above, we've described the only situation where we'll use

$\hat{p}$, the situation where the normal approximation is valid. Certainly it's possible to deal with the other cases, but they introduce complications which are not particularly instructive. For the same reason, we'll ignore the continuity correction which in certain cases would become important. For a thorough discussion of these details together with references, please see [Larsen and Marx] or [Sachs].

It is helpful to remember, however, that $\hat{p}$ is $(1/n)X$ for a binomial X. This idea that $\hat{p}$ is essentially binomial is often the key to an easy view of certain facts. That's how you know $\hat{p}$ is approximately normally distributed. And that's also how you found formulas for its mean and variance. Let's review the derivation of those formulas:

$$\hat{p} = (1/n)X$$

↑ divide by the sample size

↑ count how many have the characteristic (X is binomial or hypergeometric)

So $\hat{p} = a + bX$ where $a = 0$ and $b = 1/n$. From this, you obtain the formulas for the mean and variance of $\hat{p}$:

$$\begin{aligned} \mu_{\hat{p}} &= (1/n)\mu_X &= (1/n)np &= p, \\ \sigma_{\hat{p}}^2 &= (1/n^2)\sigma_X^2 &= (1/n^2)npq &= pq/n. \end{aligned}$$

Now let's think a little more about our estimators. Please . . .

Try Your Hand

5.2.1 In this problem, we want to use the estimator $\hat{p}$ to address the question "What's the value of this unknown p?"

(a) The estimator $\hat{p}$ is unbiased. In fact, we would usually want any estimator to be unbiased. Why?

(b) What's the relevance of the standard error of $\hat{p}$ for our question about p?

(c) For the normal approximation of $\hat{p}$ to be valid, the condition $np, nq \geq 5$ is required. This means n must be large compared with the smaller of p and q. In fact, that's true even if we're not using the normal approximation. Here's an example to show you why: If $p = 0.01$ and $n = 10$, no sample can ever give a decent approximation to p. Why? [Hint: What are the possible values of $\hat{p}$?]

(d) In working with proportions, sometimes surprisingly large samples are required. Suppose $p = 50\%$ and you want the standard error to be no more than two percentage points. What size sample would be required?

(e) Complete the following: In the situation of part (d), there's about a 95% percent chance to observe a sample proportion within ___ percentage points of p.

5.2.2 Suppose that 42% of morning commuters take public transportation and suppose a telephone survey interviews 50 of these commuters, asking "Do you use public transportation?"

(a) What's the probability more than 60% of the persons interviewed will report that they use public transportation?

(b) For a larger sample, would you expect more or less variability in response?

(c) This problem is very unrealistic. How?

(d) What's the chance that of the persons interviewed between 15 and 20 will say they use public transportation?

(e) Would many of all morning commuters say they don't use public transportation?

(f) What's the probability that more than two-thirds of your sample say they do not use public transportation?

(g) What are the chances that somewhere between 20 and 23 of your sample say they do not use public transportation?

(h) Draw the picture of $\hat{p}$, the estimator for p, where p is the proportion of the population which say they use public transportation.

5.2.3 The height of all ten-year-olds in our geographic region is 129 cm with a standard deviation of 17 cm.

(a) "The height . . . is 129 cm" doesn't make sense! It can only mean . . . ?

(b) We're working with a class of 24 students, all ten-year-olds from our geographic region. What's the probability fewer than 15 of these students are less than 129 cm tall?

(c) In a group of 115 ten-year-olds from our region, what's the probability more than 25 of them are more than 145 cm tall?

(d) What assumptions did you make in parts (b) and (c)? Would they seem reasonable?

Estimating the Value of an Unknown p

A ***confidence interval*** for an unknown parameter is a range of possible values for the parameter together with the probability that the parameter actually falls within that range. At this point, you might like to review Problem 3.10.14 in which we anticipated the following discussion by looking at confidence intervals for a median. A confidence interval for any parameter serves to estimate the parameter. Note that although the parameter is just a number, the confidence interval is not—it's a much more sophisticated object. A confidence interval estimate for a parameter has two "aspects":

- a range of possible values,
- a probability.

What kind of solution does a confidence interval provide for the problem of estimating an unknown population proportion p? That's exactly the question we want to answer now. What we'll say here for the parameter p is just as valid for any other parameter whose estimator is unbiased and approximately normally distributed. We're restricting our attention to p in this section just for simplicity and because we already know the sampling distribution (the "entire context") for its estimator, $\hat{p}$. Once the case for p and $\hat{p}$ is clear, we'll turn to $\overline{X}$ as an estimator and easily see how to obtain a confidence interval for its parameter μ. Later in the course, we'll have estimators for other parameters. In each case, the idea of a confidence interval is the same. So . . .

If it's a confidence interval for p:

> The question we're asking is: $p =$?? and the answer will have the form
>
> "We can be about 95% sure that p is at least 15% and not more than 21%."

The 95% probability is called the ***confidence coefficient*** of the confidence interval. It's denoted by $1 - \alpha$. As we'll see later, you can choose the confidence coefficient in advance and set it at whatever value seems acceptable. But there are practical restrictions: A very high degree of confidence ($1 - \alpha = 99\%$ for example, or even say 99.9%) is going to be expensive! The symbol α is conventionally used to denote a small probability. Because the confidence coefficient should be large, we use the symbol $1 - \alpha$.

The endpoints for the interval—15% and 21% in the box above, just by way of example—are determined by an observed value of $\hat{p}$, the midpoint of the interval. In this example, the midpoint is 0.18, so you must have obtained a sample for which 18% had the characteristic of interest ($\hat{p} = 0.18$). Thus the endpoints of the interval, note, are of the form

$\hat{p}$ $\pm$ a small "margin of error."

In our example, the endpoints are 18%±3%. This three percent is called the ***maximum error of the estimate***, the ***error tolerance***, or the ***margin of error***. That means the width of the interval is twice the maximum error of the estimate.

We need to see how all this works of course, but before going further, think about the type of answer which the confidence interval provides . . .

Try Your Hand

5.2.4 Show that the answer given in the box above satisfies the definition of "confidence interval."

5.2.5 In the box above, the phrase "95% sure that . . . " is subject to misinterpretation. It's wrong to interpret it as saying

"95% of the time p is in the interval with endpoints 0.15 and 0.21."

Let's see why this is wrong:

(a) What is it that varies here?

(b) Can you discover anything wrong with the interpretation given above?

(c) Try to guess the correct interpretation of the 95% probability.

5.2.6 **(a)** What's the meaning of the word "confidence" for confidence intervals?

(b) Why do you think a high degree of confidence will be very "expensive," as we said above?

5.2.7 **(a)** In the box above we are asked for the unknown value of p. What are the sources of uncertainty in the answer?

(b) If you need an exact value of p for some computation or for some other purpose, what value would you use?

(c) In part (b), you used your observed value of $\hat{p}$, which involves two possible errors. How do you control those errors?

5.2.8 (a) We have said many times that no conclusion can be drawn from one sample alone. But a confidence interval is a conclusion, so it must draw on more information than just one sample. What is that more information?

(b) Suppose $\hat{p} = 0.18$. We've seen in several places that "based on this sample we believe $p \approx 18\%$," is an unacceptable conclusion. Exactly what does the confidence interval add to this conclusion? After all, a confidence interval is a valid conclusion.

The picture below summarizes the insight contained in Problem 5.2.5—it shows eight different confidence intervals for an unknown p obtained from eight different samples. Five of the eight intervals contain the true value of p—that's only $62\frac{1}{2}\%$. That proportion should *theoretically* be $1-\alpha$ (95% in our example). But $1-\alpha$, the confidence coefficient, is a theoretical, "on average" figure. So there's no surprise when a specific case of eight intervals exhibits a smaller (or larger) percentage of "good" intervals.

Even so, for a 95% confidence interval, there's a very small probability that no more than five out of eight would contain the true parameter value. We've chosen such an unlikely case only for purposes of illustration. Here's the picture:

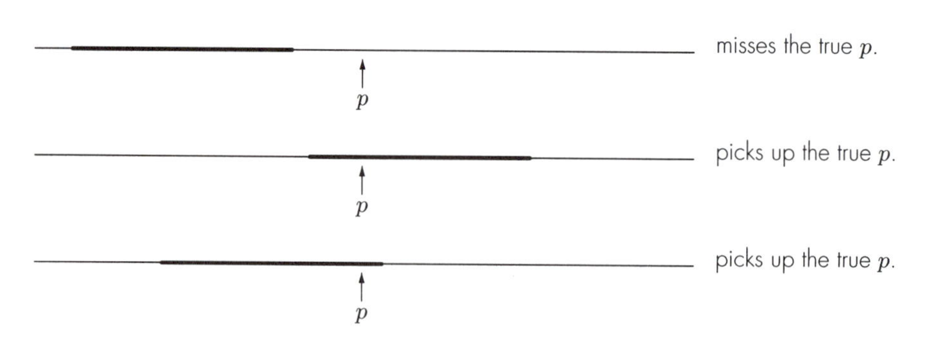

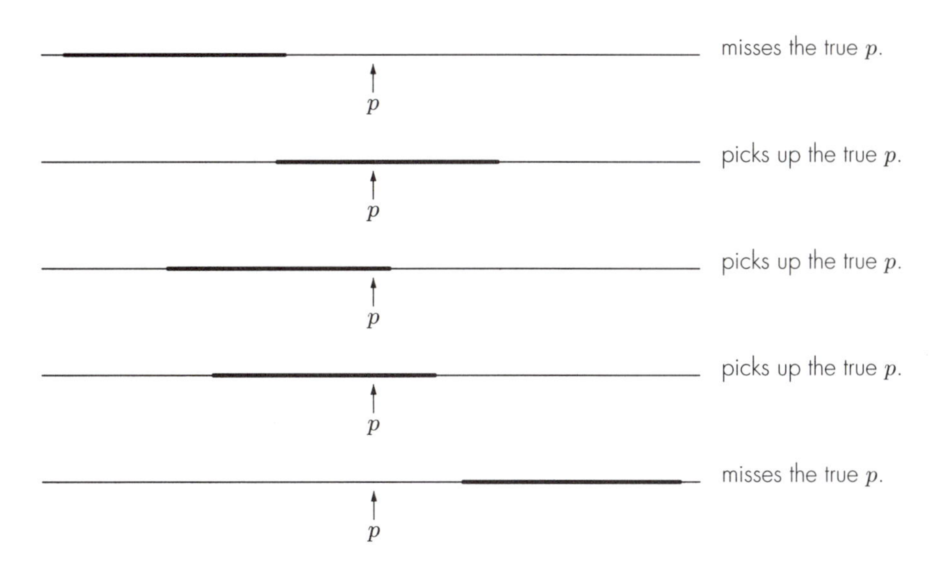

CAVEAT:

In giving your real-world conclusion for a confidence interval, you're "suppposed" to use the word "confident" as in, "We're 95% 'confident' that . . . ." This is supposed to mean you're not caught in the misinterpretation discussed in Problem 5.2.5, as if you could avoid a conceptual error by changing from a defined term (probable) to an undefined term (confident). Because the word "confident" is never defined in textbooks this is a useless distinction, it's inevitable a student will equate the two terms. So don't be surprised in the future if someone tries to catch you on your "misusage." If you understand Problems 5.2.5 and 5.2.6, you'll have no trouble defending yourself. Don't be too hard on them if they don't seem to understand. Be polite.

Here's a table summarizing Problem 5.2.7(c):

The possible error you make in estimating p from a value of $\hat{p}$	**How you exercise control over that error**
You're uncertain of the exact value of p within the confidence interval.	You know the maximum error of your estimate. [In our example, you're off by at most 3% in either direction. This assumes p is in the interval, it may not be. See below.]
The true value of p may not be within the interval at all.	You determine in advance the probability of this error. [In our example, 5%.]

Constructing the Confidence Interval for p

Now that we know what a confidence interval estimate for a parameter is—what kind of question it addresses and what kind of answer it provides—it's time to see how one actually constructs the interval. There are two aspects to a confidence interval:

- the range of possible values for the unknown parameter (the interval),
- the probability that that range of values actually contains the parameter (this is $1 - \alpha$, the confidence coefficient).

The confidence coefficient is determined in advance, balancing the desired degree of certainty against the cost of attaining that certainty. Greater certainty will require more information in the form of a larger sample. That, of course, costs money! The standard choices for confidence coefficient are 90%, 95%, and 99%. Rarely, under circumstances where a high degree of certainty is mandatory, you may see a choice of 99.5% or even 99.9% for the confidence coefficient. In our problems, just to secure a firm understanding, you'll sometimes be asked to construct intervals with nonstandard confidence coefficients.

In practice, the confidence coefficient is either chosen by you or the team you're working with or it's chosen in advance by some preset standard. There may be, for example, a procedures manual for your company which specifies the degree of confidence required under specific circumstances. Research journals may specify acceptable degrees of confidence (confidence coefficients) for any research to be published in their pages. For purposes of this course, unless otherwise specified, you should make your own choice of confidence coefficient.

The construction of a confidence interval begins by looking at the sampling distribution of the estimator, that "total context" for the data apart from which nothing at all can be said. The procedure is the same for all parameters and estimators as long as the sampling distribution is at least approximately normal.

Let's continue the example we've been discussing. We want to construct a 95% confidence interval for an unknown population proportion, p, based on a sample of which 18% had the characteristic of interest ($\hat{p} = 0.18$). Let's suppose we took a sample of size $n = 1000$ and let's assume the normal approximation for $\hat{p}$ is valid—it will be if we are sure both p and q are at least 0.005 (which implies they're both less than 0.995).

We determine the length of the confidence interval first. It should be a length which, centered in the distribution of $\hat{p}$, cuts off an area equal to $1 - \alpha$, the confidence coefficient. To see how to do this, please now . . .

Try Your Hand

5.2.9 In the picture below, there's a 95% chance for $\hat{p}$ to take a value in the interval between L and R. These are the left and right endpoints of an interval cutting off 95% of the area in the center of the distribution of $\hat{p}$. That center is located, of course, at $\mu_{\hat{p}} = p$. You cannot give a numeric value for L and R since the value of p is unknown. But you can identify L and R in terms of p and the standard error of $\hat{p}$.

(a) Express L as $p - \delta$ and R as $p + \delta$ for some appropriate δ. That puts L and R at the same distance (namely, δ) from the unknown p. The quantity δ will be expressed in terms of the standard error $\sigma_{\hat{p}}$.

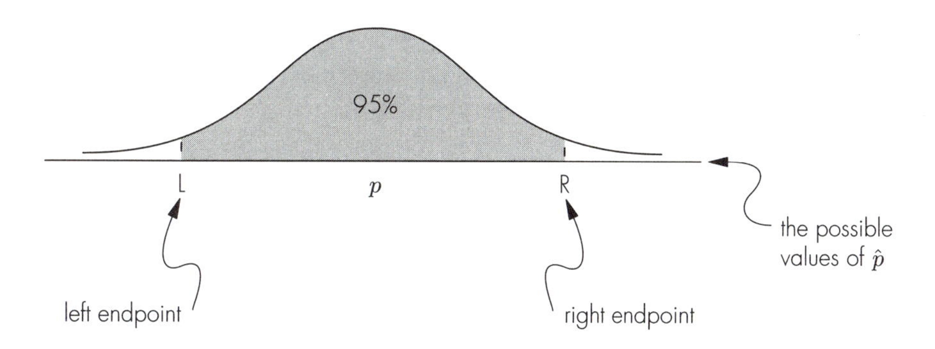

(b) You see from part (a) that we need to know the standard error of $\hat{p}$. But that raises a problem. What's the problem? [Hint: Try to evaluate the standard error for our example: $\hat{p} = 0.18$ and $n = 1000$.]

(c) Resolve the problem of the unknown standard error on a "worst case" basis. Does "worst case" mean the standard error would be made as large as possible or as small as possible? What value of p would accomplish this?

5.2.10 Explain the following statement: The solution provided by a confidence interval is much more complex than you would expect given the question it addresses.

Continuing our example, we need a confidence interval for p where we've obtained a sample of $n = 1000$, giving $\hat{p} = 0.18$. Drawing on Problem 5.2.9, we use the worst case estimate, 0.0158, for the standard error so that $1.96\sigma_{\hat{p}}$ is estimated by 0.0310. This gives the following picture:

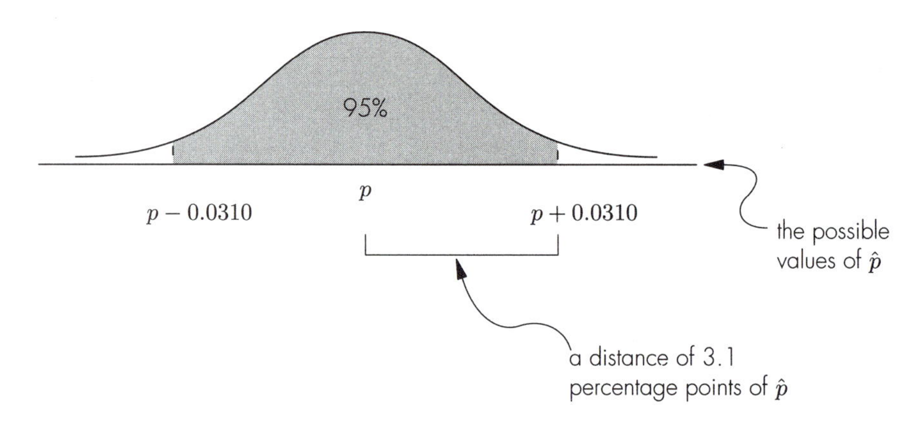

The values of $\hat{p}$ in the interval between these two endpoints are just those $\hat{p}$'s that fall within 3.1 percentage points of the unknown p. This means there's a 95% chance that our observed value of $\hat{p}$ is within 3.1 percentage points of the unknown p.

For the moment, let's assume our observed $\hat{p} = 0.18$ to be within this interval. Now shift the interval from its center at p (value unknown) and center it at $\hat{p}$ which is known to be 0.18:

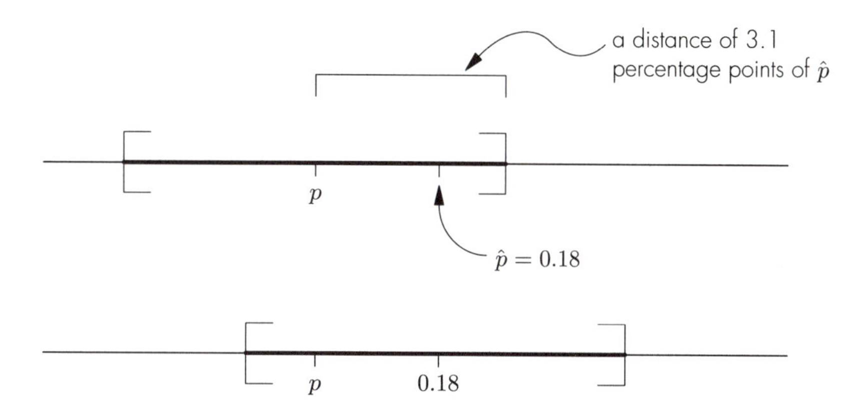

The first interval says "the observed 18% is within 3.1 percentage points of p," whereas the second interval says "p is within 3.1 percentage points of the observed 18%." Note that one of these statements is true if and only if the other one is. The logic of this is quite simple. It's like saying "If I live within 3.1 blocks of my friend, then my friend lives within 3.1 blocks of me and vice versa."

Note that the second interval is very concrete; it's the interval (0.1490, 0.2110). It has endpoints 18%±3.1%.

Now there's a 95% chance our sample gives rise to an interval like the first one, but if the first interval "happens," so does the second. So there's a 95% chance for an interval like the second one. That, at last, gives a completely concrete probability statement with no unknown quantities! Looking at the second interval, we get

Our confidence interval for p:

> There's a 95% chance that our unknown value of p is somewhere in the interval
>
> $$(0.1490,\ 0.2110).$$

This really is a confidence interval for p: We have (1) a range of possible values for the unknown p together with (2) the probability that p falls within that range!

What happens if our sample with $\hat{p} = 0.18$ happens to be one of the atypical samples? As the picture below shows, there's a five percent chance of such a sample. We don't know, of course, where our observed 18% falls compared with p, but suppose it's actually much larger than p:

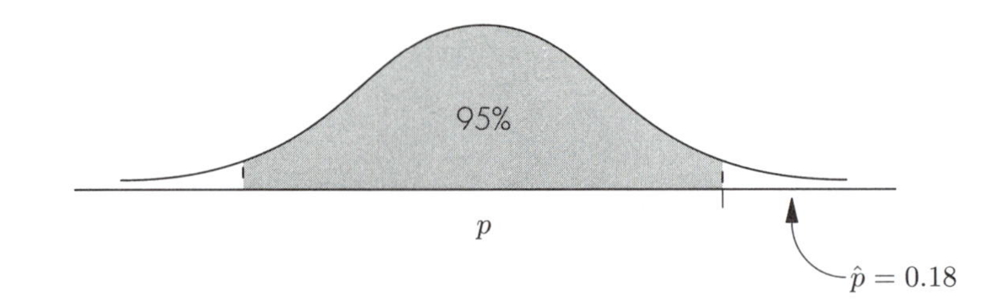

Here our observed 18% is NOT in the interval centered at p, and so p is not in the shifted interval

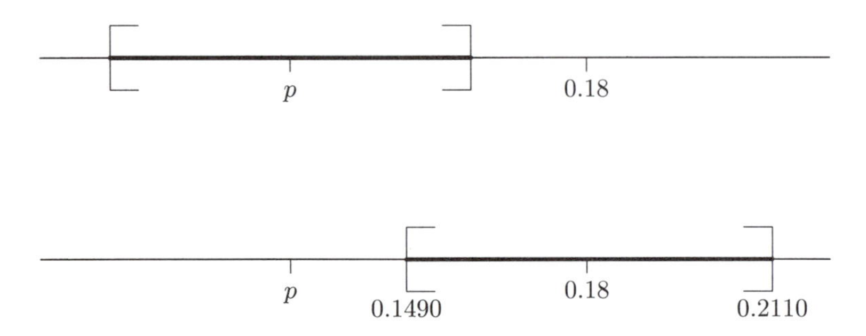

In 95% of the cases, the sample is more or less typical of the population as a whole, giving us an interval which really does contain p. In the other 5% of the cases, we obtain an interval which misses p. Which case are we in? Is our sample "typical" or not? WE DON'T KNOW! For the particular interval we obtain, we can't say whether it contains p or not. All we know is that MOST samples give intervals containing p. We can say nothing about our particular interval; we only know what happens on average with many intervals. Out of 100 such intervals, 95 on average do contain p, five on average miss p.

Note how we control the uncertainty: We don't know whether our interval contains p, but we *control* this uncertainty by specifying the confidence coefficient in advance. In this discussion, our confidence coefficient is 95% with a 5% risk of error. In general, of course, it may be 90% with a 10% risk of error, or 99% with a 1% risk or error, and so on.

Try Your Hand

5.2.11 Suppose you constructed ten confidence intervals for an unknown p. How many of those intervals would contain p?

5.2.12 In each of the following, assume p is an unknown parameter and use the given information together with the sampling distribution for $\hat{p}$ to construct the required confidence interval for p. In each case, use the worst case estimate for the standard error.

(a) $n = 600$, $\hat{p} = 43\%$, $1 - \alpha = 0.8$;

(b) $n = 450$, $\hat{p} = 0.71$, $1 - \alpha = 0.9$;

(c) $n = 280$, $\hat{p} = 45\%$, $1 - \alpha = 72\%$;

(d) $n = 40$, $\hat{p} = 43\%$, $1 - \alpha = 92\%$.

5.2.13 What is it about the normal distribution for $\hat{p}$ that allows us to say most samples are more or less typical of the population as a whole? This is a qualitative question—you should answer it without reference to any numbers or formulas.

The Exact (Almost) Endpoints of a Confidence Interval for p

When the value of p is unknown and we seek to construct a confidence interval, there's a problem: The standard error for the estimator $\hat{p}$ involves that unknown value of p. The worst case approach which we saw above approximates the term pq in the standard error by one-fourth because pq is maximum when $p = \frac{1}{2}$. That's a crude approach, however, and can give intervals so wide as to be useless. With a little more care, it's possible to derive an exact formula for the endpoints of a confidence interval without using an approximation for the standard error. We're saying "exact" here, but in fact even these endpoints are based on an approximation—the normal approximation for $\hat{p}$.

In the previous section, you saw that the endpoints of a confidence interval for p are $\hat{p} \pm z\sigma_{\hat{p}}$. Let's call these endpoints L and U for "lower" and "upper" confidence limits. Here the phrases ***lower confidence limit*** and ***upper confidence limit*** refer to the left and right endpoints of the confidence interval. If we continue with our example for which $\hat{p} = 0.18$, $1 - \alpha = 95\%$ and $n = 1000$, then

$$L = 0.18 - 1.96\sqrt{pq/n}$$

$$U = 0.18 + 1.96\sqrt{pq/n}.$$

These give a simple quadratic equation: L and U are those values of p which satisfy

$$p = 0.18 \pm 1.96\sqrt{pq/n},$$

that is,

$$p - 0.18 = \pm 1.96\sqrt{pq/n}.$$

Look at this equation in its more general form, using $\hat{p}$ and z instead of 0.18 and 1.96:

$$p - \hat{p} = \pm z\sqrt{pq/n}.$$

If we square both sides and simplify the algegra (try it if you like!), we get a quadratic equation, $ap^2 + bp + c = 0$, in p:

$$(1 + z^2/n)p^2 \;+\; (-1)(2\hat{p} + z^2/n)p \;+\; \hat{p}^2 = 0,$$

$$a \quad p^2 \;+\; b \quad p \;+\; c \;= 0.$$

Using the quadratic formula we get the two roots, L and U:

exact formula for the upper and lower confidence limits for p:

$$\frac{\hat{p} \;+\; z^2/2n \;\pm\; (z/\sqrt{n})\sqrt{\hat{p}\hat{q} + z^2/4n}}{1 + z^2/n}$$

In our example, we get

$$\frac{0.18 + (1.96)^2/2000 \pm (1.96/\sqrt{1000})\sqrt{0.18 \times 0.82 + (1.96)^2/4000}}{1 + (1.96^2)/1000}$$

which gives the confidence interval (0.1574, 0.2050). For the worst case estimate of the standard error, we obtained the confidence interval (0.1490, 0.2110).

A Less Conservative Approach to the Standard Error for $\hat{p}$

Now the amazing thing about the "exact endpoint formula" is how it gives us a simple approximation when the sample size is very large. In that case, we can just replace p by $\hat{p}$ in the standard error formula and write the endpoints as $\hat{p} \pm z\sigma_{\hat{p}}$! This is truly surprising because we know that $\hat{p}$ might be really quite far from p and, therefore, not a good approximation to p. Let's see how this simplification is possible.

Note that z^2/n is negligible in the exact endpoint formula if n is quite large. This is true because z is never more than three (for a 99% confidence interval, $z = 2.575$). If z^2/n is very small compared with $\hat{p}$, we can neglect it. By omitting the terms involving only z^2/n, we obtain the following formula for estimating the upper and lower confidence limits:

$$\hat{p} \pm z\sqrt{\hat{p}\hat{q}/n}.$$

Note that this formula seems to imply we're estimating the standard error using $\hat{p}$ instead of p. But now the estimate is justified because we know we can omit the negligible term z^2/n in the exact formula ($z^2/2n$ and $z^2/4n$ are even more negligible than z^2/n, because they are smaller still). Let's summarize:

The procedure for a confidence interval problem (large population and large sample assumed) for an unknown *p*

1. Choose an appropriate confidence coefficient.

2. Use the Z table to find the value of z which cuts off a probability of $1 - \alpha$, centrally located

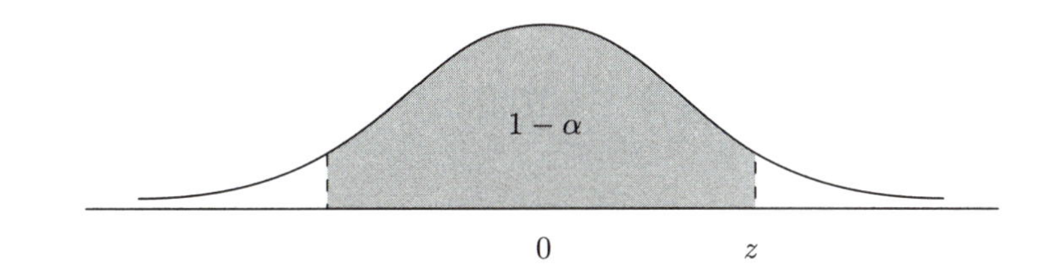

3. Compute the endpoints of the confidence interval one of three ways:

either

(i) use the exact formula

or use the formula $\hat{p} \pm z\sigma_{\hat{p}}$ and use

(ii) the worst case estimate of $\sigma_{\hat{p}}$ (use $p = \frac{1}{2}$)

or

(iii) the less conservative estimate of $\sigma_{\hat{p}}$ (use $p = \hat{p}$) (justified by neglecting z^2/n in the exact formula for the endpoints).

4. Give a real-world interpretation of the confidence interval with as much detail as the original statement of the problem allows. Be sure your real-world interpretation accounts for both aspects of the confidence interval: the range of possible values for p as well as the probability that this range of values really does contain the unknown value of p.

Our "less conservative" approach is the most commonly taught procedure and it is the one you should use unless instructed otherwise. Now please . . .

Try Your Hand

5.2.14 Suppose that $n = 240$ and $\hat{p} = 0.18$. Compute the upper and lower confidence limits of a 90% confidence interval for p

(a) exactly;

(b) using the worst case estimate of the standard error;

(c) using the less conservative estimate of the standard error.

(d) Compare the three intervals generated above.

(e) Why is the word "exact" not really valid in referring to our "exact formula for the endpoints of a confidence interval"?

5.2.15 Suppose that $n = 580$ and $\hat{p} = 0.22$. Compute the upper and lower confidence limits of a 92% confidence interval for p

(a) exactly;

(b) using the worst case estimate of the standard error;

(c) using the less conservative estimate of the standard error.

(d) Compare the three intervals generated above.

5.2.16 Show that when n is not particularly large, our three approaches to confidence intervals can give substantially different intervals. To do this, construct the three intervals which the three different approaches yield if $\hat{p} = 0.08$, $n = 100$ and $1 - \alpha = 0.95$.

5.2.17 A large corporation has installed in their printers 318 printheads manufactured by your company. Fourteen of those printheads have failed within the warranty period. What is the probability a printhead from your company will fail during the warranty period?

5.2.18 In the article "What Is Stuttering: Variations and Stereotypes" Richard E. Ham [1990] reports that 120 of 563 respondents to a telephone survey answered yes to the question "Have you ever stuttered?" The respondents were chosen randomly from the Tallahassee, Florida telephone directory.

(a) Based on this information, what proportion of persons in Tallahassee stutter?

(b) Sixty-eight of Ham's respondents answered affirmatively to the question "Do you stutter now?" How many people in Tallahassee stutter?

(c) Based on "qualifying remarks made by informants," in the judgment of the interviewers, 48 of the 68 respondents of part (b) had "normal dysfluencies" rather than "stuttered dysfluencies." Now how many people stutter?

Too Many Approximations: Are They Valid?

By now you may be feeling uneasy about the validity of so many approximations: We may have as many as three approximations:

(i) the normal approximation for the distribution of $\hat{p}$,

(ii) one of two possible approximations for the standard error, $\sigma_{\hat{p}}$,

(iii) the confidence interval itself, a "probable" approximation to p specified by a range of possible values.

For (i), the standard rule of thumb is that np and nq both be at least five. For the condition in (iii), the confidence coefficient and the maximum error of the estimate measure how good the approximation is.

But what about (ii)? Well, the conservative standard error approximation is ALWAYS valid—that's why it's called "conservative." The less conservative approximation depends upon z^2/n being small. It depends on having a large sample so that n will be large and z^2/n correspondingly small. But how large is large? This is not an easy question to answer! It depends on how small (or large) $\hat{p}$ is: The further $\hat{p}$ is from one half, the larger n must be. You approach these questions by studying the "exact" interval which we derived from the quadratic formula.

You could avoid (i) and (ii) entirely by taking a completely different approach: Forget the normal approximation and return to the binomial model. There are published tables of confidence intervals for an unknown p which have been computed using the binomial model. There is still another approximation, the so-called "F-distribution" approximation, which can be used instead of the normal approximation. For more details on all of these approaches, see [Sachs], [Bickel and Doksum], or [Larsen and Marx].

From this point on, unless otherwise stated, we will follow the usual approach of introductory courses, the less conservative approach: We approximate the standard error for $\hat{p}$ by just replacing p by the observed $\hat{p}$. So from now on, the endpoints of a confidence interval for p will be determined by the formula

$$\hat{p} \pm z\sqrt{\hat{p}\hat{q}/n}.$$

Try Your Hand

5.2.19 (**a**) How is it that the confidence coefficient and the maximum error of the estimate serve to "measure how good the approximation is" for a confidence interval?

(**b**) In our example for which $\hat{p} = 0.18$, $1 - \alpha = 0.95$, and $n = 1000$, give the numeric value of the maximum error of the estimate.

(**c**) Why do we say the conservative standard error approximation is ALWAYS valid?

5.2.20 Abbott Laboratories developed a test for the detection of antibodies to the HIV virus in human blood. This virus is generally considered to cause AIDS in humans. For such a test where a false positive can have serious consequences, a positive test result should be confirmed by a second test. But even then, false positives are possible. The control for this contingency is the "specificity" of the test (see Problem 1.4.4). To determine the specificity for persons who have already tested positive, Abbott tested the blood of 17,054 random blood donors assumed not to have the virus. Of those, 18 tested positive. It was later determined that one of those blood samples probably was infected with the virus after all and so that sample was omitted from the study. See Abbott Laboratories' 1992 report #83-7618/R3. Based on this information, what is the specificity of Abbott's test for persons who have already tested positive?

The Appropriate Sample Size for a Given Error Tolerance

In many situations, you want to control your estimate of a proportion by specifying in advance the maximum error you're willing to tolerate. Suppose, for example, you want to estimate the percentage of the vote which will be cast for your candidate in an election and you require the estimate to be correct to within four percentage points. This requirement can be met if you're willing to expend the neccessary resources to obtain a sufficiently large sample.

So the maximum error of the estimate—the "error tolerance"—is required to be 0.04:

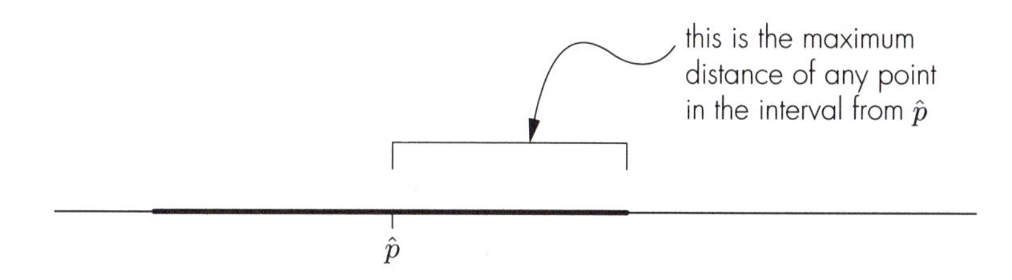

Because the endpoints of the interval are $\hat{p} \pm z\sigma_{\hat{p}}$, the equation for the maximum error of the estimate is

$$z\sqrt{pq/n} = 0.04.$$

This equation determines the sample size required to meet your four percentage point error tolerance. In this formula, you control z and n. But z is determined by your choice of confidence coefficient. That leaves only n as a free variable. If you want a 95% confidence coefficient, solving the equation for n, you obtain

$$(1.96)^2 pq/n = 0.0016,$$

and so

$$2401pq = n.$$

This means you must be prepared to obtain a sample of 601 voters. Finally, if you're going to use the normal approximation, you should check that $601p$ and $601q$ are both at least five. That means the smaller of p and q should be at least 0.0083.

Try Your Hand

5.2.21 (a) In the discussion above, where did the figure 601 come from? In other words, show how the equation $2401pq = n$ leads to the conclusion that $n = 601$.

(b) If you have more information, a smaller sample may be sufficient. For example, suppose you're SURE no fewer than 70% of the votes will be cast for your candidate. What sample size would you require to estimate the true percentage to within four percentage points?

5.2.22 You're trying to determine what proportion of the population of your state have a certain genetic defect and you're willing to have an error of no more than half of a percentage point. What sample size should you take if you want a

(a) 99% confidence coefficient?

(b) 92% confidence coefficient?

(c) Redo part (a) assuming you know for sure that at most one percent of the population have that defect.

(d) Redo part (a) assuming you know for sure that at most one tenth of one percent of the population have that defect.

5.3 Estimating an Unknown Mean

The Estimator $\overline{X}$

In the previous section, we have seen exactly how the sampling distribution of the estimator $\hat{p}$ together with the results of one sample allows us to construct a confidence interval estimate for an unknown proportion, p. We turn in this section to the corresponding problem for an unknown mean, μ. For that, we require the sampling distribution for the estimator $\overline{X}$. In most cases, but not all, $\overline{X}$ is at least approximately normally distributed. In the first problem below, before discussing normality, we ask you to derive the formulas for the mean and variance of $\overline{X}$. These formulas hold in all cases without exception. Then, in the next problem, we take one special case for which you can easily show that $\overline{X}$ is normally distributed.

In these problems, we have a simple random sample of size n from the probability distribution of a random variable X. That includes, as a special, case sampling with replacement from a numeric population. A simple random sample from the probability distribution of a random variable X, as you'll recall from Chapter 2, is simply an ordered set of n values of X obtained from n independent repetitions of the underlying random experiment. The sample looks like

$$X_1,\ X_2,\ X_3,\ \ldots,\ X_{n-1},\ X_n.$$

These numbers could be the result of rolling a die repeatedly n times. That would be a sample from the probability distribution of the random variable "number of dots on top face." The X_k's could be the measurements of n independently chosen objects (heights of n children, weight of n boxes, etc). That's a sample drawn from a numeric population. Or they could be n repeated measurements (presumably independent) of the same object, a sample from the probability distribution of the random variable "measurement." They could be the deviation from specification of n items off a production line, a sample from the distribution of "specification error."

Now please . . .

Try Your Hand

5.3.1 In this problem, you will show that $\overline{X}$ is an unbiased estimator for μ and derive a formula for the standard error. Let's employ the usual notation for the mean of a simple random sample:

$$\overline{X} = (1/n)(X_1 + X_2 + X_3 + \cdots + X_n) = (1/n)\Sigma X_k, \quad k = 1, 2, \ldots, n.$$

(a) Explain why for each X_k, the mean and variance is just μ and σ^2.

(b) Show that $\overline{X}$ is unbiased as an estimator for μ.

(c) Derive a formula for the standard error of $\overline{X}$.

5.3.2 Show that $\overline{X}$ is normally distributed if you're sampling from

(a) the distribution of a normally distributed random variable X;

(b) a normally distributed population.

The Central Limit Theorem

To make use of the estimator $\overline{X}$, we need a description of its probability distribution. In Problem 5.3.2, we saw that $\overline{X}$ is normally distributed provided the distribution we're sampling from is normal. But that special case is very limiting. We're trying to estimate the value of an unknown μ: If we don't know μ, we often wouldn't know enough about the underlying situation to say whether we're sampling from a normal distribution or not. We can do better *provided we're taking large samples*. That's not a serious restriction because the sample size is within our control (if our pockets are deep enough!). We're appealing here to the Central Limit Theorem of Laplace, mentioned in Chapter 4. It assures us that, without knowing anything at all about the distribution we're sampling from, $\overline{X}$ is approximately normally distributed for large samples. Here's the theorem . . .

The Central Limit Theorem:

> For simple random sampling from the distribution of any random variable,[1] as the sample size gets larger and larger the random variable $\overline{X}$ comes closer and closer to being normally distributed.

As a rule of thumb, $n \geq 30$ is taken as a sufficiently large sample size to guarantee that $\overline{X}$ is approximately normal. This is significantly smaller than the sample sizes (hundreds or even thousands) typically required for problems involving $\hat{p}$. A more precise rule and more exact interpretation of those terms in the theorem like "close to" requires a full mathematical treatment appropriate to a more advanced course.

Because sampling with replacement from a numeric population is a special case of sampling from a distribution, the Central Limit Theorem applies to that case as well. Furthermore, sampling *without* replacement from a numeric population, if the population is large, is essentially the same as sampling with replacement: The chances are tiny on the second draw (because the population is so large) of getting the same population element you got on the first draw even if you do replace the first one. So for sampling with or without replacement, if the sample size and the population are both large, $\overline{X}$ is approximately normally distributed. In summary, the Central Limit Theorem implies that $\overline{X}$ should be approximately normally distributed no matter what random sampling procedure we use, *if the sample size is large.*

In the exercises below you'll see that we can easily "prove" the Central Limit Theorem based on our heuristic criterion (see page 148) for a normal distribution. In fact, this criterion is just a practical interpretation of the Central Limit Theorem. All along, in using the criterion we've been tacitly relying on this major theorem. So it's not a matter of proving the theorem. Rather, we just want to see that our understanding derived from the criterion in Chapter 4 is consistent with the Central Limit Theorem. Before going on, let's see why this is true and explore the distribution of $\overline{X}$ further. Won't you please . . .

Try Your Hand

5.3.3 Use our intuitive criterion for a normal distribution to show that the Central Limit Theorem should be true. In other words, show that for

[1] Well, not just any random variable. There's a technical restriction that the first two moments must exist. We'll not try to explain what that means.

simple random sampling from a distribution, if the sample size is large, it's reasonable to expect $\overline{X}$ to be approximately normally distributed.

5.3.4 We derived our intuitive criterion for normality from the idea of "random error." Let's use the Central Limit Theorem to show that random error ought to be normally distributed.

(**a**) First, a preliminary fact. Use the Central Limit Theorem to show that the sum or total of the numbers in a large sample should be approximately normally distributed.

(**b**) Show that random error is "like" the sum or total of all the numbers in a large simple random sample.

(**c**) Use the Central Limit Theorem to show that random error should be approximately normally distributed.

5.3.5 Use the Central Limit Theorem to show that repeated measurements of the same object should be approximately normally distributed.

The Sampling Distribution for $\overline{X}$

The Central Limit Theorem together with Problems 5.3.1 and 5.3.2 gives the following summary of the sampling distribution for $\overline{X}$ . . .

The Sampling Distribution for $\overline{X}$:

The estimator $\overline{X}$ is unbiased. It's approximately normally distributed if the sample size is large ($n \geq 30$) and exactly normally distributed if the distribution you're sampling from is normally distributed:

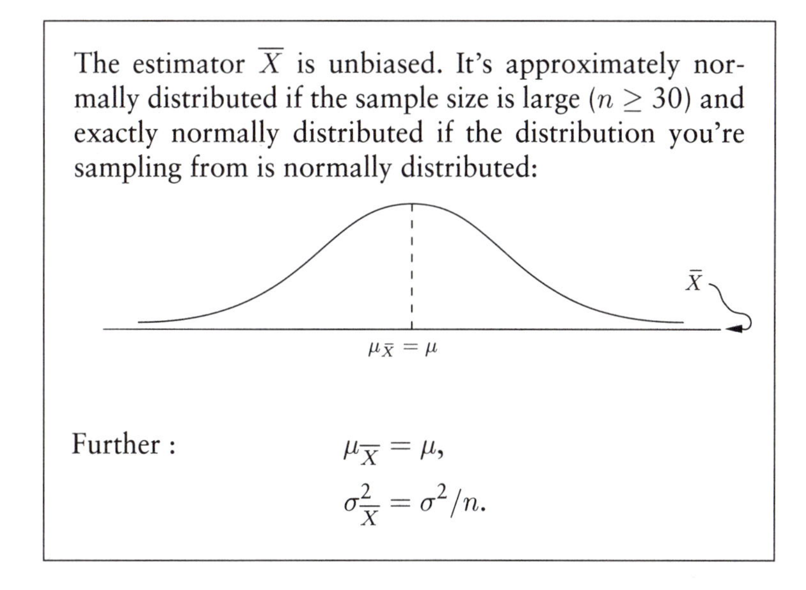

Further :

$$\mu_{\overline{X}} = \mu,$$
$$\sigma^2_{\overline{X}} = \sigma^2/n.$$

So there are two distinct situations where a normal distribution is appropriate for $\overline{X}$. First, the large sample case: This is the more useful case because it makes no assumption about the distribution from which the samples are being drawn. Second, the case in which your samples are known to come from a normal distribution: Then $\overline{X}$ is normally distributed no matter what sample size you're working with. This case is less common because detailed information about the distribution you're sampling from is not usually available.

Constructing a Confidence Interval for μ

Now let's return to the situation of Problem 5.1.1 where we wanted to estimate an unknown μ. There we had obtained a simple random sample giving a sample mean of 2.37. Here was the situation:

$$\text{the question:} \quad \mu = ?,$$

$$\text{the information:} \quad \overline{X} = 2.37.$$

With no other information to go on, our sample tells us nothing about the unknown μ, but taken in the total context of *all possible* samples, this one sample tells a lot. The "total context" is the sampling distribution of the estimator $\overline{X}$ as summarized in the box on the previous page.

Let's recall how we can make use of the sampling distribution to construct a confidence interval for this unknown μ. We say "recall" because the procedure is identical to the procedure for proportions. The questions are of the same type—what's the value of an unknown parameter—and the sampling distribution in each case is a normal distribution centered on the unknown parameter.

Suppose this time we require a 90% confidence interval estimate for the unknown μ. First, we turn to the Z distribution and find that value of Z which cuts off 90% of the probability in the center of the distribution:

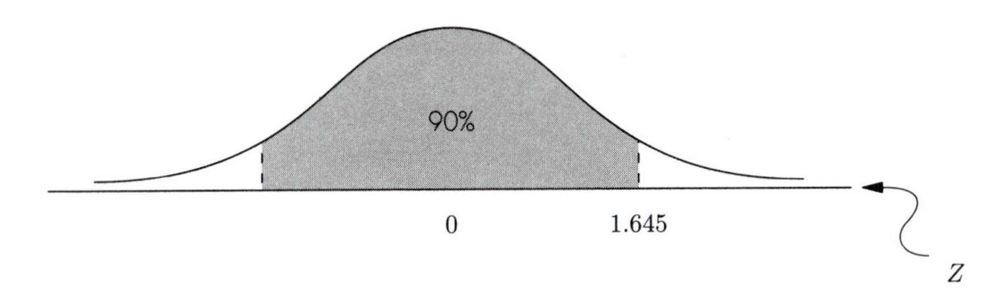

Now look at the corresponding picture for $\overline{X}$. If we go 1.645 standard errors above and below the center, we'll cut off 90% of the probability centrally located in this distribution:

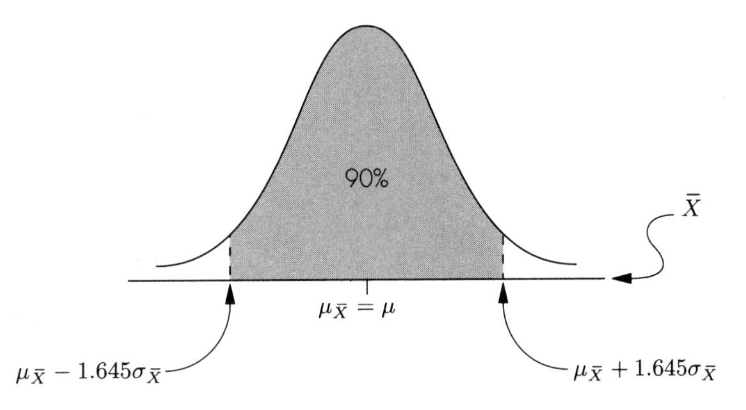

This means there is a 90% chance that our observed sample mean of 2.37 is somewhere between the endpoints $\mu_{\overline{X}} \pm 1.645\sigma_{\overline{X}}$. Now if $\overline{X} = 2.37$ is within 1.645 standard errors of the unknown value of μ, then, of course, μ is within 1.645 standard errors of 2.37. Here's a picture:

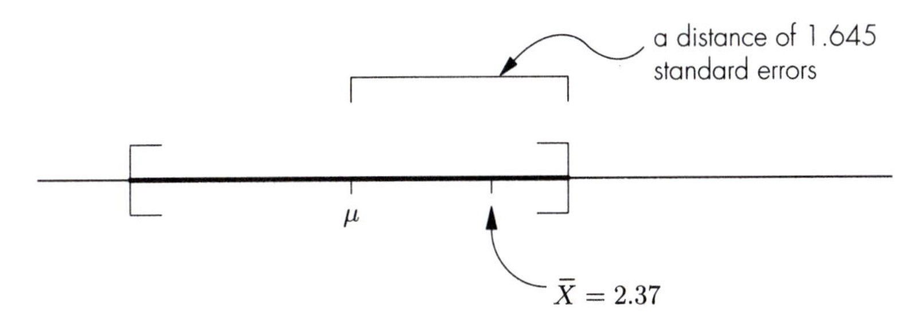

Now shift the interval from a center at μ to a center at 2.37, keeping the width of the interval the same:

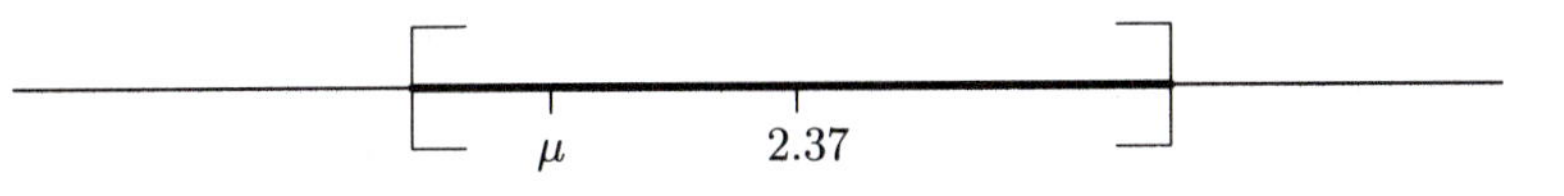

The first interval says "2.37 is within 1.645 standard errors of μ," whereas the second interval says, "μ is within 1.645 standard errors of 2.37." One of these statements is true if and only if the other one is true. So, of course, 2.37 is in the original interval if and only if μ is in the second interval. Recall our earlier analogy: This logic is like saying "my friend lives within 1.645 miles of me," or on the other hand, "I live within 1.645 miles of my friend." One of these statements is true if and only if the other one is.

The critical difference between the two situations is that you don't know the value of μ and so you don't know the endpoints of the first interval. You do know the endpoints of the second interval because you know $\overline{X}$ to be 2.37, those endpoints are just $2.37 \pm 1.645\sigma_{\overline{X}}$. If our observed 2.37 was in the original interval, then the final picture looks like

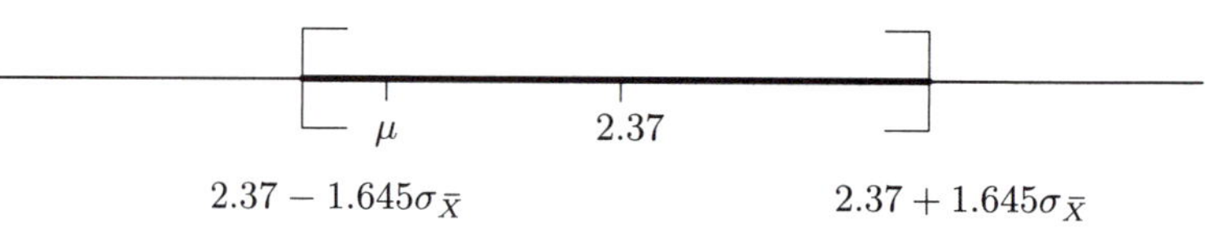

This is the situation we hope for—the "typical" situation—where the unknown μ and the observed sample mean of 2.37 are close to each other (within 1.645 standard errors of each other). The "bad" situation occurs when μ and the observed 2.37 are NOT close to each other. If $\overline{X} = 2.37$ is NOT in the original interval, we get

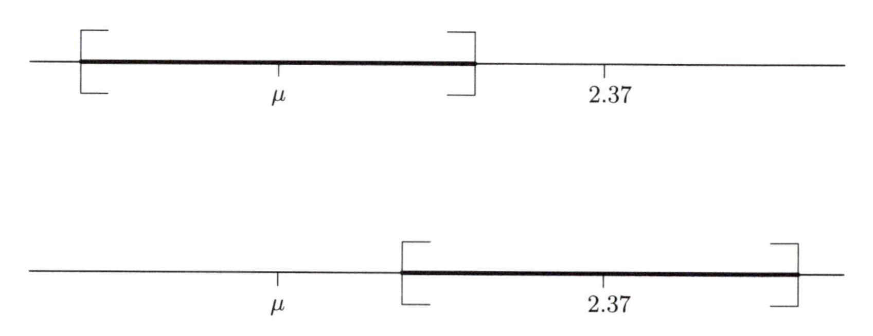

Here the first interval says "2.37 is NOT within 1.645 standard errors of μ" and the second interval says "μ is NOT within 1.645 standard errors of 2.37." As before, one of these statements is true if and only if the other one is. With a confidence coefficient of 90%, this "bad" or "atypical" situation happens only 10% of the time, on average. Ten out of 100 samples on average will generate values of $\overline{X}$ for which μ is not in the confidence interval. Those are the atypical samples: Their mean is too far from the true μ and so the interval centered on them misses the true μ.

The choice of confidence coefficient determines exactly what you want the words "typical" and "atypical" to mean. The confidence coefficient determines $Z = z$ (in our example, $Z = 1.645$) and that, in turn, determines the "typical" samples: those samples for which the mean is within z standard errors of the true value of μ. Then in most cases (here, in 90% of the cases), the sample is "typical" of the distribution you're sampling from and gives a confidence interval which really does contain the unknown value of μ. But don't forget: You never know whether μ is in your particular interval or not! All you ever know is that MOST intervals generated this way will contain μ. Your particular interval may or may not, AND YOU NEVER KNOW WHETHER IT DOES OR NOT! You control this inherent uncertainty by specifying the confidence coefficient in advance. So, although you don't know whether the interval contains μ or not, you are 90% sure that it does.

Note that the entire discussion here is identical to the corresponding discussion in the case of a confidence interval for an unknown proportion. The logic of confidence intervals is not dependent on exactly which parameter you're attempting to estimate as long as the sampling distribution is at least approximately normally distributed and as long as the estimator itself is unbiased so that the distribution is centered on the unknown parameter. In that case, the procedures for constructing confidence intervals are the same. By the end of this course, you'll have a long list of parameters whose estimators are unbiased and approximately normally distributed. That means you already know how to construct confidence intervals for all those parameters. You only need the formula for the standard error in each case. That, of course, is not known in advance: Standard error formulas do vary from one estimator to the next.

In fact, the standard error is going to pose a problem in the example we're working on here. We don't yet have a finished confidence interval; the endpoints are in the form $2.37 \pm 1.645\sigma_{\overline{X}}$. But what is $\sigma_{\overline{X}}$? The formula, of course, is $\sigma^2_{\overline{X}} = \sigma^2/n$, where σ^2 is the variance of the distribution you're sampling from. But if we don't know the mean of that distribution, it isn't likely we would know the variance. So, as in the case of proportions, there arises the problem of estimating the standard error because it involves an unknown parameter.

If the sample size is large, we can mimic our solution to the corresponding problem with proportions and use the *sample* standard deviation as an estimate for the unknown standard deviation, σ. However, we don't use $\hat{\sigma}$ as introduced in Chapter 2; for technical reasons, we require an unbiased estimator for σ^2, but $\hat{\sigma}^2$ is biased. Instead we use s^2 as the ***unbiased estimator for σ^2***. It's easy to obtain s^2 from $\hat{\sigma}^2$ using the relationship $s^2 = n/(n-1)\hat{\sigma}^2$. From this point on, unless otherwise stated, the phrase "sample variance" will refer to the unbiased estimator s^2 and the phrase "sample standard deviation" to its square root, s.

It's certainly not obvious that using s instead of σ in the standard error formula is acceptable. After all, because the sample could be quite atypical of the distribution you're sampling from, just replacing σ by s would seem very risky. But it works! How? Well, if you take a sufficiently sophisticated course in mathematical statistics you might see a proof. It's a difficult theorem. The idea is, intuitively, that a large n in the standard error will predominate over s or σ. For the sketch of a proof, see page 343 of [Mendenhall, Scheaffer and Wackerly].

So, let's suppose we took a sample of two hundred which gave a sample standard deviation of $s = 0.8742$. With that information our confidence interval will be

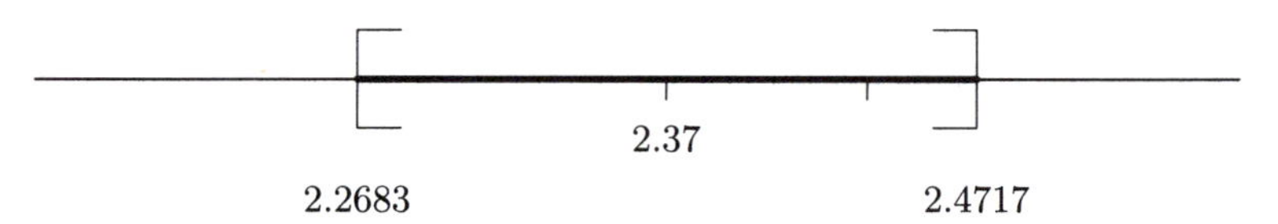

We don't have a real-world problem here, so we can't give a real-world conclusion. Still, the conclusion would have to take the form

> We can be about 90% sure that the range of values from 2.2683 to 2.4717 contains our unknown mean μ.

This conclusion provides "a range of possible values for an unknown parameter" together with "the probability that that range of values actually does contain the parameter." Therefore, the conclusion fits our description (see page 180) of what a confidence interval is supposed to be.

Now, why don't you . . .

Try Your Hand

5.3.6 In the text above, the endpoints for our confidence interval are 2.2683 and 2.4717. Show how we obtained these endpoints.

5.3.7 (a) Derive the conceptual formula:

$$s^2 = \frac{\Sigma(X - \overline{X})^2 f}{n-1}.$$

(b) Derive the computing formula:

$$s^2 = \frac{n\,\Sigma X^2 f - (\Sigma X f)^2}{n(n-1)}.$$

(c) Evaluate s using each of the two formulas above for the sample . . .

$$\{7, 5, 4, 5, 4, 7, 8, 7\}.$$

(d) Show that s is a random variable.

(e) Why do we use s instead of $\hat{\sigma}$ to estimate the standard error of the estimator $\overline{X}$?

5.3.8 (a) In the case of proportions, we have to rely on an estimate of the standard error because the true standard error formula involves an unknown parameter. What is that parameter?

(b) In the case of proportions, what estimate do you use for the standard error?

(c) We said ". . . we can mimic our solution to the corresponding problem with proportions" by using s instead of σ. Explain this comment.

5.3.9 Based on the information given below, what are the endpoints of the confidence interval for the parameter?

(a) $n = 250$, $\overline{X} = 6.4$, $s = 1.21$, and $1 - \alpha = 0.98$;

(b) $n = 400$, $\sigma^2 = 0.02$, $\overline{X} = 0.32$, $s = 0.01$, and $1 - \alpha = 0.98$;

(c) $n = 400$, $\overline{X} = 122.51$, $s = 12.00$, and $1 - \alpha = 0.92$;

(d) $n = 75$, $\overline{X} = 1.7$, $s = 0.1$, and $1 - \alpha = 0.89$;

(e) $n = 850$, $\hat{p} = 0.23$, and $1 - \alpha = 0.98$.

5.3.10 (a) The manager of a ski resort would like to encourage the parents of children too young to ski to take ski holidays at her resort. With this goal in mind, she is planning a daycare center for such children and would like to know their average age. You do a survey of visitors to the resort who have altogether 50 small children and you find the children's average age to be 34 months with a standard deviation of eight months. What do you report to the manager? Give your answer in months and years; for example, not as 26 months, but as two years and two months (round to the nearest month).

(**b**) The manager in part (a) comes to you several months later and asks how many of the children coming to the newly established daycare center are less than one year old. In looking at the data on the original 50 children you find that 11 were less than one year in age. What do you report to the manager?

(**c**) There is something suspicious in our treatment of the question in part (b), what is it?

5.3.11 Suppose you have reset the fill mechanism on a drink dispensing machine and need to estimate the average fill. Suppose, in resetting the mechanism, you don't change the variability of the fill as measured by the standard deviation of 0.0341 ounces. You obtain the following data from cups filled after resetting the fill mechanism:

Fill (ounces)	Number of cups with given fill
7.23	12
7.24	24
7.25	31
7.26	9

(**a**) Give a 90% confidence interval estimate of the new average fill.

(**b**) Give a 95% estimate of the new fill.

(**c**) How much drink is the machine dispensing per cup after you have reset the fill mechanism? Assume you're willing to run a two percent risk of error.

(**d**) How much drink is the machine dispensing per cup after you have reset the fill mechanism? Assume you want to avoid any risk of error.

(**e**) After resetting the fill mechanism, how many cups will overflow? Assume a cup overflows at 7.255 ounces.

(**f**) What percentage of cups were overflowing before you reset the fill mechanism?

(**g**) Is this data from a random sample?

(**h**) Verify that "fill" is a random variable.

(**i**) In part (e), you used the normal distribution. Was that appropriate?

5.3.12 Explain in what sense "... choosing your confidence coefficient determines exactly what you want the words 'typical' and 'atypical' to mean."

5.3.13 Later, we'll have many other parameters besides means and proportions for which we'll want to construct confidence interval estimates. Let's think about those situations:

(a) For questions concerning an unknown parameter, one sample by itself tells you nothing—what more is required?

(b) If the estimator for the parameter is unbiased and normally distributed, how will the procedure for constructing the confidence interval differ from what you've already learned?

5.3.14 Compute the endpoints of a confidence interval for the unknown mean of a normal distribution where

(a) $n = 8$, $\overline{X} = 3.78$, $\sigma = 1.01$, and $1 - \alpha = 0.90$.

(b) $n = 50$, $\overline{X} = 12.4$, $\sigma^2 = 3.6$, and $1 - \alpha = 0.95$.

5.3.15 **(a)** From a normally distributed population with σ about one, you randomly select a sample of 12 from which you compute a mean of 17 and a standard deviation of 2. What's the mean of this population? Assume you're willing to run about a five percent chance of being wrong.

(b) You have a normally distributed population whose variance is 3.6. You know the mean of the population has increased and you believe the variance has remained unchanged—what's the new mean? Assume you got a sample mean of 12.4 from 50 observations with $s = 1.7201$.

(c) Suppose in part (a) it has been decided that you need a more accurate estimate—your estimate should not be off by more than 0.4. What must you do?

(d) What are the endpoints of the new interval in part (c)?

(e) Suppose in part (b) you require a high degree of accuracy. You want your interval to be only half a point wide. You are willing to pay for this in part by lowering the confidence coefficient to 90%. How large a sample is required?

Estimating the Standard Error ($\sigma/\sqrt{n} \approx s/\sqrt{n}$), Using s Instead of σ Takes Us to Student's t-Distribution When the Distribution You're Sampling from Is Normal

In the previous section, when sampling from a distribution whose standard deviation was unknown, in the formula for the standard error we just substituted s in place of the unknown σ. That worked for us only because we assumed a large sample and because we were willing to accept a theorem from mathematical statistics which validates this in the large sample case. Unlike the situation for proportions, however, in the case of means there's a readily accessible theoretical tool for the small sample case, IF WE'RE SAMPLING FROM A NORMAL DISTRIBUTION. It's called "Student's t-distribution."

To understand Student's t-distribution, we should look more carefully at the standardizing transformation applied to $\overline{X}$:

$$\frac{\overline{X} - \mu}{\sigma/\sqrt{n}}.$$

This is a linear function of $\overline{X}$. If $\overline{X}$ is normally distributed, because linear functions "take normals to normals," this transformation of $\overline{X}$ is just Z. But σ, which appears in the denominator, is unknown. And replacing it by s introduces a major problem: The modified transformation will no longer be linear. For a linear function, variables should appear in their simplest form: multiplied by constants and combined by addition, nothing more complicated. But here, with σ replaced by s, the standardizing transformation has the VARIABLE s in the DENOMINATOR! When you look at the formula, we seem to have made a fairly innocuous change:

$$\frac{\overline{X} - \mu}{s/\sqrt{n}}.$$

But this modified transformation is radically problematic: Not only does it seem to draw on the fallacious assumption of a representative sample in replacing σ by s, but it involves two random variables, one of them in the denominator.

To think this new NONLINEAR, TWO-VARIABLE transformation of $\overline{X}$ is Z would be naive indeed! Still, this naive approach was exactly the practice of statisticians until the early 1900s. Of course, the science of statistics was still in its infancy at that time. For us, with 100 years or so of experience, such an assumption would be truly naive, but for statisticians in the nineteenth century, it was just one of many unresolved difficulties.

In any case, about the turn of the century a statistician with the Dublin brewery of Arthur Guinness and Son began to notice discrepencies when working with small samples—discrepencies which made clear the fallacy of just replacing σ by s in the standard error formula. This Guinness statistician, one William Gosset by name, investigated the situation in detail and worked out the correct probabilty distribution which he published in 1908 in a paper entitled "The Probable Error of a Mean." Because the Guinness brewery had a policy against the publication of in-house research, Gosset published his paper under the pseudonym of "A Student." The probability distribution which he discovered and published in this paper and which he denoted by the symbol t came to be known as ***Student's t-distribution with n – 1 degrees of freedom***. He computed a table of values for the t-distribution which you'll find in the appendix. So now, with Gosset's results in hand we can say

$$\frac{\overline{X} - \mu}{s/\sqrt{n}} = t_{n-1}.$$

In summary, here's the problem Gosset's research resolved: It's true for $\overline{X}$ normally distributed, the standardizing transformation takes $\overline{X}$ to Z:

$$\frac{\overline{X} - \mu}{\sigma/\sqrt{n}} = Z.$$

But if we replace σ by s in this formula, the resulting random variable is no longer normally distributed. In particular, it's NOT Z. Gosset really had some work to do to obtain the true distribution of the modified transformation of $\overline{X}$. But he successfully completed that project and we reap the benefits, a new and very useful probability distribution. Here it is:

Student's t-Distribution: The Sampling Distribution for $\overline{X}$ Using s Instead of σ, Sampling from a Normal Distribution:

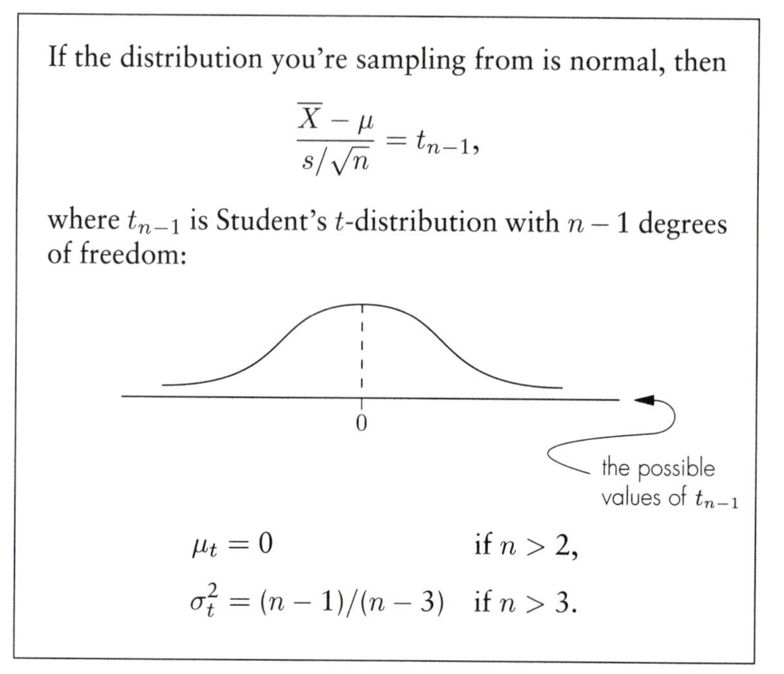

If the distribution you're sampling from is normal, then

$$\frac{\overline{X} - \mu}{s/\sqrt{n}} = t_{n-1},$$

where t_{n-1} is Student's t-distribution with $n - 1$ degrees of freedom:

$$\mu_t = 0 \qquad \text{if } n > 2,$$

$$\sigma_t^2 = (n-1)/(n-3) \qquad \text{if } n > 3.$$

Let's highlight an important point: This is not just one probability distribution. There's a different t-distribution for each of the possible "degrees of freedom." So we have a whole family of t-distributions, indexed by the degrees of freedom.

You may well wonder what the phrase "degrees of freedom" refers to. This is really a very technical matter—we'll not go into it in detail. Heuristically, you should imagine that the degrees of freedom for an estimator starts with n, the sample size, and then decreases by one for each unknown parameter in the standard error formula. Those unknown parameters must be estimated from the sample. In the case of the estimator $\overline{X}$, there's only one such parameter in the standard error formula. It's the parameter σ, estimated from the sample by s. For this reason, the degrees of freedom for $\overline{X}$ is $n - 1$, one less than the sample size. By the end of this course, you will have estimators with two such parameters in their standard error formulas resulting in $n - 2$ degrees of freedom. In other contexts, other degrees of freedom are also possible.

Note that the distribution you're sampling from must be normal; otherwise, the estimator s^2 won't have the appropriate distribution. Gosset's subtle analysis depends on the fact that s^2 has a particularly intimate relation to the chi-squared random variable, but that happens only if we're sampling from a normal distribution. Otherwise,

> If the distribution you're sampling from is NOT normal,
>
> $$\frac{\overline{X} - \mu}{s/\sqrt{n}} \quad \text{is NOT } t.$$

When the sample is small and you don't know if you're sampling from a normal distribution or not, we have no technique for dealing with the estimator $\overline{X}$. There are techniques available for this case, but they have a totally different flavor from the kinds of things we've been studying.

The t table given in the appendix gives the value of t which cuts off a given left tail area. With ten degrees of freedom, we get this picture:

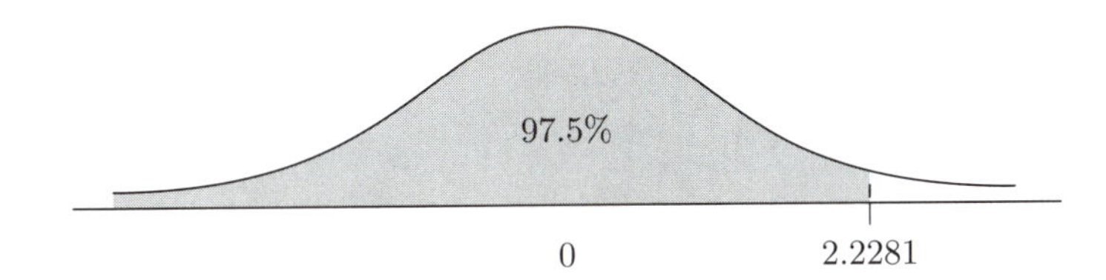

As you might expect, the larger the sample, the closer Student's t-distribution comes to Z. As a rough rule of thumb, if $n \geq 30$ we can approximate t by Z. For this reason, the t table is not required for large sample sizes. You can use the Z table instead. When you look at the t table, you'll see that after 30, the degrees of freedom are incomplete. The table begins to skip values because beyond 30 you can use Z instead. Values like $n - 1 = 50$ are given in the table only so you can see that t is getting closer and closer to Z. The values of Z are given in the last row as $n - 1 = \infty$. For example, with 97.5% in the left tail,

$n-1$	t
50	2.0086
100	1.9840
200	1.9719
∞	1.96

Note the familiar value 1.96. You already know that as the value which cuts off 97.5% of the area in the left tail of the Z-distribution:

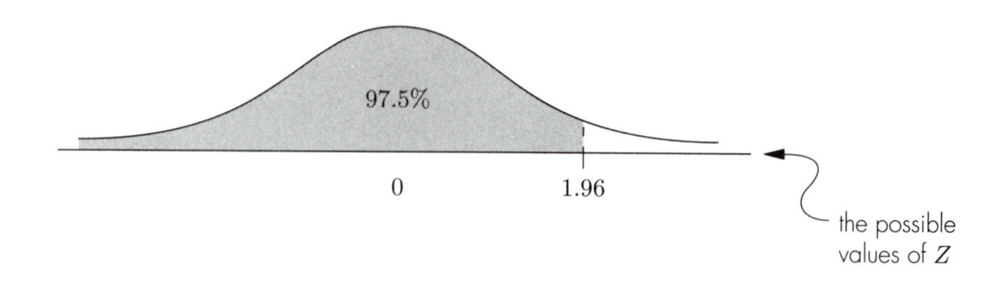

Since the logic of this situation has become rather complicated, it will be helpful to have a chart to summarize . . .

When to Use Z and When to Use Student's t-Distribution

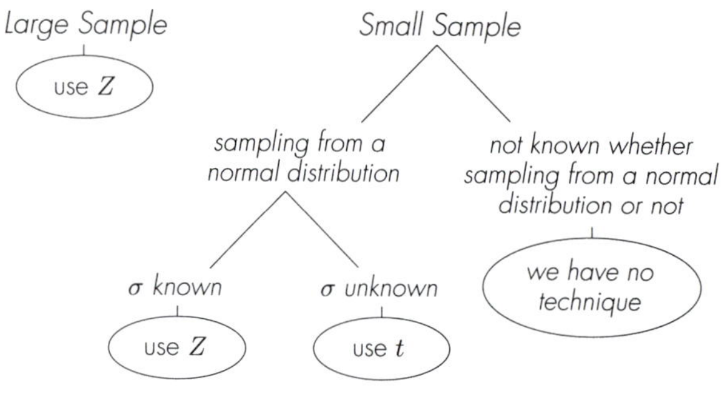

Try Your Hand

5.3.16 We said that Gosset provided "a readily accessible theoretical tool for the small sample case" to model the sample mean. That theoretical tool takes what form?

5.3.17 (**a**) Show that s^2 is a random variable.

(**b**) Why must we assume the population to be normally distributed to use Student's t-distribution?

5.3.18 In this problem, we will pin down some details about Student's t-distribution.

(**a**) To the "naked eye" (the eye looking at the picture, not at the formulas) Student's t-distribution looks just like Z. Our formulas reveal one way in which it's different from Z. What is that?

(**b**) Several times we've mentioned the "modified" standardizing transformation of $\overline{X}$. What modification are we talking about?

(c) Just above we said "As you might expect, the larger the sample, the closer Student's t-distribution comes to Z." Why might you reasonably expect that?

(d) So far in this third section of Chapter 5 we've discussed two distinct situations in which $\overline{X}$ would be at least approximately normally distributed. What are they?

5.3.19 Calculate the endpoints of a confidence interval for μ under the given conditions. State explicitly any unstated assumption which is required to solve the problem.

(a) $n = 15, \quad 1 - \alpha = 0.95, \quad \overline{X} = 1.2, s = 0.17;$

(b) $n = 9, \quad 1 - \alpha = 0.95, \quad \overline{X} = 1.2, s = 0.17;$

(c) $n = 45, \quad 1 - \alpha = 0.95, \quad \overline{X} = 1.2, s = 0.17;$

(d) $n = 9, \quad 1 - \alpha = 0.95, \quad \overline{X} = 1.2, \sigma = 0.17;$

(e) $n = 150, \quad 1 - \alpha = 0.95, \quad \overline{X} = 1.2, \sigma = 0.17;$

(f) $n = 12, \quad 1 - \alpha = 0.90, \quad \overline{X} = 3.4, s = 1.21;$

(g) $n = 8, \quad 1 - \alpha = 0.99, \quad \overline{X} = 0.21, \sigma = 0.03;$

(h) $n = 8, \quad 1 - \alpha = 0.99, \quad \overline{X} = 0.21, s = 0.03;$

(i) $n = 150, \quad 1 - \alpha = 0.95, \quad \overline{X} = 1.2, s = 0.17.$

5.3.20 In the chart which indicates when to use Z and when to use Student's t-distribution (page 212), justify the analysis of the

(a) small sample case;

(b) large sample case.

5.3.21 (a) You have a sample of size 40 from a population whose standard deviation is 5.23. In constructing a confidence interval for the unknown mean of this population, your result will be more accurate if you know for sure that the population is normally distributed. Explain why.

(b) The question in part (a) does not arise for small samples. Why?

5.3.22 Calculate the endpoints of an appropriate confidence interval under the given conditions. State explicitly any unstated assumption which is required to solve the problem. Finally, give a verbal conclusion for the estimation problem.

(a) $n = 5, 1 - \alpha = 0.90, \overline{X} = 0.42, s = 0.04;$

(b) $n = 110, \; 1 - \alpha = 0.95, \quad \hat{p} = 0.2;$

(c) $n = 35, \quad 1 - \alpha = 0.90, \; \overline{X} = 23, s = 2.4\,, \sigma = 2.1;$

(d) $n = 25, \quad 1 - \alpha = 0.99, \; \overline{X} = 1.27, s = 0.32;$

(e) $n = 14, \quad 1 - \alpha = 0.99, \; \overline{X} = 87, \sigma^2 = 16;$

(f) $n = 12, \quad 1 - \alpha = 0.90, \quad \hat{p} = 0.53.$

(g) Suppose in part (f) you recognized your mistake and supplemented the original sample by sampling 20 more items of the population obtaining $\hat{p} = 0.49$. What would be your estimate of the unknown p?

(h) $n = 19, 1 - \alpha = 90\%, \overline{X} = 7.2, s^2 = 2.73;$

(i) $n = 10, 1 - \alpha = 0.95, \overline{X} = 44, s = 0.03.$

5.4 A Confidence Interval Estimate for an Unknown σ

Using the fact that s^2 is intimately related to the chi-squared distribution if we're sampling from a normal distribution, we can construct confidence intervals for an unknown σ^2. Because the sampling distribution of the estimator—the distribution of s^2—is NOT normal, the analysis will be different from what we've seen up to now, different but similar.

Suppose we need to estimate the variability in the amount of a contaminant present in the chemical solution we receive from a supplier. For example, it may be that the effect of the contaminant is easily controlled, provided the amount of contaminant is accurately predictable. The "predictability," of course, is measured by the standard deviation σ of the amount of contaminant. We model this situation by the random variable X, the measured amount of contaminant in a container of solution of fixed size.

Now if X is at least approximately normally distributed, the following linear function of s^2 has a chi-squared distribution with $n - 1$ degrees of freedom:

$$\frac{(n-1)s^2}{\sigma^2} = \chi^2_{n-1}.$$

It's not at all easy to prove this fact, please accept that it's true. In Problem 5.3.17(b), you saw why X must be normally distributed.

Suppose we want a 95% confidence interval for σ. We begin by setting up the appropriate picture for the chi-squared distribution with $n-1$ degrees of freedom. We need to determine numbers L and U to cut off 95% of the area in the middle of the distribution with the remaining five percent equally divided into the two tails:

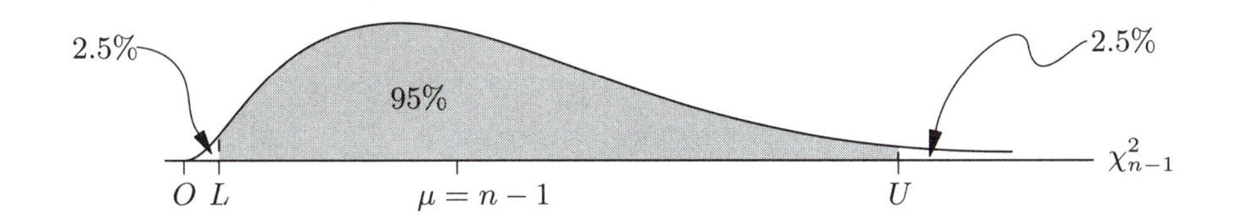

Suppose we obtain a simple random sample of size $n = 8$ from the distribution of X; that is, we make eight independent readings of the amount of contaminant in containers of the solution. From the chi-squared table in the appendix, with $n-1=7$ degrees of freedom, $L = 1.690$ and $U = 16.013$.

So there's a 95% chance that $(n-1)s^2/\sigma^2$ falls between these two numbers

$$1.690 < (n-1)s^2/\sigma^2 < 16.013.$$

Dividing by $(n-1)s^2$, we obtain an equivalent statement

$$\frac{1.690}{(n-1)s^2} < \frac{1}{\sigma^2} < \frac{16.013}{(n-1)s^2}.$$

The inequalities are preserved because you're dividing by a positive number. Then, inverting the fractions, we obtain

$$\frac{(n-1)s^2}{1.690} > \sigma^2 > \frac{(n-1)s^2}{16.013}.$$

Note that when you invert the terms of an inequality, if the terms are positive, the inequality is reversed. Because one of these three strings of inequalities holds if and only if the others do—they're logically equivalent—each one can be preceded by saying "there's a 95% chance that . . . ."

Now suppose our sample of eight readings on the amount of contaminant gave a sample variance of $s^2 = 1.0828$. Then we can say

> For the amount of contaminant in a container of the solution we receive from our supplier, we can be about 95% sure that the standard deviation σ is between 0.6880 and 2.1178.

Note the general rule:

the confidence interval for σ^2:

> Assuming we're sampling from a normal distribution, the endpoints of a confidence interval for σ^2, the variance of that distribution, are
>
> $$\frac{(n-1)s^2}{U}, \qquad \frac{(n-1)s^2}{L}.$$

Note here that the lower number, L, obtained from the chi-squared distribution table goes to the upper endpoint of the confidence interval, and the larger number from the table, U, goes to the lower endpoint of the confidence interval. This happens because of the reversing of inequalities when you pass from $1/\sigma^2$ to σ^2. Now please . . .

Try Your Hand

5.4.1 **(a)** Show how we calculated the endpoints, 0.6880 and 2.1178, of our confidence interval for σ.

(b) At the beginning of this section we said ". . . the sampling distribution of the estimator—the distribution of s^2—is NOT normal." Explain.

(c) If it's true that "the effect of the contaminant is easily controlled, provided the amount of contaminant is accurately predictable," which supplier would you prefer: the one with $\mu_X = 8.4$ and $\sigma_X = 0.8$ or the one with $\mu_X = 4.2$ and $\sigma_X = 1.6$?

(d) Verify that $(n-1)s^2/\sigma^2$ is a linear function of s^2.

(e) How did we find L and U from the χ^2 table?

5.4.2 Calculate the endpoints of a confidence interval for:

(a) σ^2 with $n = 15$, $1 - \alpha = 90\%$, and $s^2 = 0.9656$;

(b) σ with $n = 12$ and $s = 1.0266$;

(c) the variability in fill after resetting the fill mechanism for the drink machine in Problem 5.3.11.

5.5 One-Sided Intervals, Prediction Intervals, Tolerance Intervals

So far, we've considered only two-sided confidence interval estimates for an unknown parameter. But often ONE-sided interval estimates are required and often they're required to estimate something other than a parameter. We'll look at such estimates in this section.

One-Sided Confidence Intervals

A one-sided interval estimate is required if you're interested in a bound on a numeric quantity in one direction only. For example, you may be concerned to know if our drink machine is underfilling. You would then like to have an estimate giving a minimum for fill. This is a ***lower confidence interval***. The endpoint would be $\overline{X} - zs/\sqrt{n}$, the ***lower confidence limit***. Or perhaps you're concerned with the tensile strength of steel wire, again you would want an estimate giving a minimum (you don't care if the wire is "too strong"!). Or maybe you know μ_1 is greater than μ_2, but you want to know by how much. Then you need an estimate giving a maximum for $\mu_1 - \mu_2$, an ***upper confidence interval***, whose endpoint is called the ***upper confidence limit***. You can explore these ideas as you . . .

Try Your Hand

5.5.1 Your company markets a gourmet candy apple wrapped in nice tissue paper and packaged individually, with straw cushioning, into elegant, red cubical boxes. What's the largest dimension required for the boxes to accommodate the maximum diameter of the apples?

We're concerned with "maximum diameter" because the apples are not spherical. They have various diameters in various directions. We measure in the direction that gives the maximum diameter (MD, in centimeters). Because all apples are at least 3 cm, for simplicity we record only the "excess" diameter, $X = 10(MD - 3)$. Note that X is in millimeters. Here are the values of X for a random sample:

X	5	6	7	8	9	10	11	12
f	3	7	11	14	9	11	8	1

(a) Show how a confidence interval for μ_X will yield a confidence interval for μ_{MD}.

(b) Give a 95% one-sided confidence interval for the mean of MD.

(c) Give a 95% one-sided confidence interval for the median of MD.

(d) Neither of the intervals in parts (b) and (c) will be adequate to answer the question posed at the beginning of the problem. Why not?

(e) Give a 90% one-sided confidence interval for the proportion of apples with a maximum diameter below 3.7 cm.

5.5.2 Consider Youden's data on the weight of U.S. pennies given in Problem 3.10.14. Allow a 99% certainty for your answers.

(a) You work for the U.S. Mint and have to move steel carts carrying 100,000 pennies. What's the maximum weight of such a cart? The steel cart itself weighs 43 pounds.

(b) For a "Give a Penny" fund-raising drive, you're weighing bags of pennies which are supposed to contain ten dollars each. By weighing the bags you want to determine if the pennies seem to have been miscounted. If there has not been a miscount, how much should such a bag weigh? Ignore the weight of paper for rolls of pennies.

5.5.3 Sketch a picture of $\overline{X}$ which illustrates the confidence coefficient for a 95% one-sided confidence interval for μ.

Prediction Intervals for Observations from a Normal Distribution

A confidence interval is required if we're asked how much drink a machine dispenses into a cup—the "typical" cup. But suppose it's a question of how much drink the machine is going to give ME—right now. I'm not average or typical. I'm ME! I require a "prediction interval," an interval estimate for one particular numeric observation made at random from this drink machine. In general, a ***prediction interval*** is a range of possible values for one observation of a random process or of some population together with the probability that that range of values actually does contain the observation. More generally, a prediction interval can provide a range of values for the average of several observations. For example, there are three of us taking a drink break together. How much drink will the three of us get, on average? In the problems below, you'll see how to obtain prediction intervals.

As you'll see, the technique we introduce is only valid for observations from a normal distribution. It's very much a "distribution bound" technique, analogous to our small sample confidence interval for the unknown mean of normal distribution. In both cases, if the distribution you're sampling from is not normal, we have no technique. Of course, one would like to have techniques that make no extraneous assumptions beyond the actual data at hand. There's an entire branch of statistics known as ***non-parametric statistics*** whose goal is exactly that. It provides techniques not heavily dependent on a distributional assumption which would be difficult to verify. In the problems below, we'll develop a simple non-parametric prediction interval which can sometimes be useful. It's not hard to see how prediction intervals work. Please . . .

Try Your Hand

5.5.4 The drink machine in the employee lounge puts 6.3 ounces into a cup with a standard deviation of 0.26 ounces. How much drink am I going to get from this machine? Right now!

5.5.5 Three of us are taking a break in the employee lounge referred to in the previous problem. How much drink will the three of us get from this machine?

5.5.6 Consider any random observation of some process or population. Let this "random observation" be modeled by a normally distributed random variable X with mean μ and variance σ^2. Note that now, in contrast to the previous problem, we don't know the values of μ and σ. We'll get around this by taking a random sample from the distribution of X.

(a) What's the underlying random experiment for X?

(b) What was "the model" when we were generating a confidence interval for an unknown mean?

(c) For a prediction interval, the model is $X - \overline{X}$. Show that this model is normally distributed with mean zero and with variance given by

$$\sigma^2\left(1 + \frac{1}{n}\right).$$

(d) For part (c), show that there's a 95% chance for X to take a value within 1.96 standard errors of $\overline{X}$.

(e) Show that for a prediction interval, the endpoints are $\overline{X} \pm z$ s.e. with the standard error determined by the formula in part (c). Explain what to do if σ is unknown.

(f) Suppose the previous ten cups from the drink machine had a mean of 6.6 ounces with a standard deviation of 0.27 ounces. How much drink should I anticipate when I drop my coins in the machine?

(g) Now suppose you want a prediction interval for the mean of m independent future observations of X. Now the model is $A - \overline{X}$, where A is the average of m observations. Show that the squared standard error of the model is

$$\sigma^2\left(\frac{1}{m} + \frac{1}{n}\right).$$

(h) How much drink will three of us obtain from the machine given the information in part (f)?

(i) If n is large, $\overline{X}$ is approximately normally distributed even if X is not. That's the Central Limit Theorem. So why do we say our prediction interval is not valid unless X is normally distributed? Why couldn't we eliminate that assumption in the large sample case? Where have we used the normality of X in an essential way (even if n is large)?

(j) The confidence coefficient, let's say 95%, for a confidence interval says that, on average, 95 of 100 intervals obtained by the given technique will contain the parameter. What's the precise interpretation of the confidence level for a prediction interval to predict one future observation?

5.5.7 In Problem 5.5.1, what should be the dimensions of the cubical box for packaging your company's gourmet candy apples? Allow 5 mm for the tissue paper and cushioning.

5.5.8 We answered the question in Problem 5.5.2 in terms of a "typical" cart or bag of pennies. But now "typical" is not relevant:

(a) You've asked me to help you out so we can go to lunch early. You want me to push this cart into the next room. What's the most it's going to weigh?

(b) You want to weigh bags of pennies to see if they seem to have been miscounted. When would you want to check a bag by actually recounting?

(c) What assumption must we make about U.S. pennies which was not required in Problem 5.5.2?

5.5.9 A nonparametric prediction interval can be obtained from the order statistics of a sample (see Problem 4.5.8 for "order statistic"). Suppose you have taken a sample of size n from a continuous distribution.

(a) Show that the interval from $x_{(h)}$ to $x_{(k)}$ provides a $100(k - h)/(n + 1)$% prediction interval for the next observation from that distribution.

(b) Show that if the endpoints are to be symmetrically chosen within the sample, then $k = n - h + 1$.

(c) Show that the endpoints of a $1 - \alpha$ prediction interval symmetrically chosen within the sample are determined by $h = \frac{1}{2}(n+1)\alpha$.

(d) For a sample of 100, what would be a 90% prediction interval for the next observation?

Tolerance Intervals

We're concerned that employees using the drink machine in the employee lounge feel they've been cheated by getting too little drink. Of course, we can set the fill mechanism on the machine to put any amount we wish into a cup, but just how many cups are being adequately filled? Suppose we're not really sure. We can't say NO cup will have too little drink, but there is a technique whereby we can be "confident" that at least 93% of all cups will have adequate fill. We've chosen 93% because we're prepared to offer the other seven percent of employees some kind of compensation—four free cups—for being "cheated." Why are we only "confident" instead of sure? The word "confident" reflects the fact that we'll base our determination on a random sample which, if it happens to be atypical of all cups, will mean less than 93% of all cups having adequate fill. In other words, there's the possibility our 93% will not actually be within the range specified. We control for that unhappy possibility by choosing in advance an acceptable probabililty—let's say 10%—of such error. Then we can be "confident at the 90% level" that at least 93% of all cups will give adequate fill (fill within the range specified).

For this type of problem, a "tolerance interval" is required. A ***tolerance interval*** for the distribution of a random variable X is a range of values which encompasses a given percentage of all the possible values of X. In our situation, a "lower tolerance interval for 93% of the values of X" is called for. We need to be 90% certain—this is $1 - \alpha$, the "confidence coefficient"—that 93% of all cups have a fill greater than a given lower limit. So we take, let's say, 100 cups from our machine, measure the fill of each cup, and calculate a 90% ***lower tolerance limit***, a value for "fill" above which the fill of at least 93% of cups will fall. When that value turns out to be 6.4 ounces, we can conclude, with a ten percent chance of being wrong, that 93% of all cups dispensed will contain at least 6.4 ounces. Now we can offer four free cups to anyone who receives less than 6.4 ounces.

Note there are three elements for a tolerance interval:

1. **The confidence coefficient, $1 - \alpha$.** In the example, it is 90%. You choose this in advance. It's your "control of error." As usual, your conclusion is subject to error because your sample could be very atypical of what's going on—could be, but PROBABLY isn't. Through the probability distribution of the estimator, you control this uncertainty by specifying in advance an acceptable probability for the error.

2. **The proportion of all values of X which must be encompassed by the tolerance interval.** In the example, it is 93%. This is the "percentage" in the phrase "given percentage of all the possible values of X" from the definition of tolerance intervals.

3. **The range of values.** In the example, anything above 6.4 ounces.

In our analysis of tolerance intervals, we will use a binomial random variable just as we did in Problem 3.10.14 and then, to actually generate a formula, we'll use the normal approximation as we did in Problem 4.5.8. Thus, our technique will only be valid if np, $nq \geq 5$. We'll also need to assume that the relevant percentile of X has a zero chance to show up in random sampling. That would be true, for example, if X is continuously distributed.

In the example above, we've spoken of a one-sided tolerance interval. Two-sided tolerance intervals are also possible, but they involve considerations more technical than we're prepared to deal with, so we omit them from our study. One-sided intervals, by contrast, are quite accessible. To see how they work, please . . .

Try Your Hand

5.5.10 **(a)** One-sided tolerance intervals for the distribution of a random variable X are reasonably accessible because, in fact, they're just confidence intervals for the corresponding percentile. That's not true for TWO-sided intervals which require a more subtle analysis. Show that the 90% tolerance interval for 93% of the values of "fill" discussed in the text above is just a one-sided 90% confidence interval for the 93th percentile of the distribution of "fill." Do this in terms of a picture for the distribution of $X =$ "fill."

(b) The endpoint of a one-sided tolerance interval to encompass a proportion p of the values of any random variable X will be determined by an order statistic, $x_{(k)}$, from a sample (see Problem 4.5.8). We'll assume X has a negligible chance to take on any of the percentiles under discussion and that np, $nq \geq 5$.

Show that k is determined by

$$k = np + \tfrac{1}{2} + z\sqrt{npq} \qquad \text{for an upper tolerance limit,}$$
$$k = nq + \tfrac{1}{2} - z\sqrt{npq} \qquad \text{for a lower tolerance limit.}$$

where the value of Z is determined by $P(Z < z) = 1 - \alpha$.

(c) In the level II answer for part (b), why is the picture not a normal curve? And if it's not a normal curve, how do we end up with a value of Z in the formula?

(d) Show that the formula in part (b) gives the "right answer" if $p = \frac{1}{2}$.

(e) Show that the lower and upper tolerance limits taken together do NOT give the endpoints of a two-sided tolerance interval.

5.5.11 In the text we said the lower tolerance limit for X = "fill" from the drink machine in the employee lounge would be 6.4 ounces. Show how we got that value. Here's the result of our study of 100 cups:

$10(X - 6)$	3	4	5	6	7	8
f	2	18	37	33	9	1

5.5.12 Based on Youden's data given in Problem 3.10.14 (g):

(a) How much does a U.S. penny weigh?

(b) If I could weigh it, how much would this penny here in my hand right now weigh?

(c) Below what weight would we expect to find 90% of all U.S. pennies?

(d) In part (b), we gave a prediction interval according to the theory developed in Problem 5.5.6, assuming a normal distribution for the weight of U.S. pennies. But there's a nonparametric prediction interval possible (see Problem 5.5.9). Give the corresponding 95% nonparametric prediction interval for part (b).

(e) In parts (b) and (d), we answered the same question using two different techniques. Which technique is better?

(f) What's the smallest sample size which will allow a $1 - \alpha$ nonparametric prediction interval?

(g) Why might the nonparametric prediction interval be relatively weak?

Chapter 6
Introduction to Tests of Statistical Hypotheses

6.1 Introduction

Chapter 5 addressed real-world questions of the form, "Based on this sample data, what's the unknown value of our parameter?" A confidence interval is required as an answer. In this chapter, we consider two other types of real-world problem. Both types of problem involve a statistical hypothesis and ask us, in a sense which we must make precise, to "test" that hypothesis against sample data. First, we'll see exactly what a statistical hypothesis is and then give an overview of these two new problems.

Statistical Hypotheses

An hypothesis is simply a statement which might or might not be true. Sometimes it's believed to be true; sometimes not. Often, the hypothesis is simply set up as a straw man in the hope of showing it false. Here's a practical example of a situation where it could be said you were "testing an hypothesis": Suppose as you drive by your mother's house at night, you're hypothesizing that she's at home. You "test" your hypothesis by observing that no lights are on, except the one light in the living room which is always on. The evidence (no lights) seems to suggest your hypothesis is false. But the evidence is not complete enough to draw a *certain* conclusion. You might be more certain of your conclusion if you didn't see your mother's car in the driveway. To be absolutely certain your mother's not at home, you would have to actually stop, enter the house, and see if she's there or not. But if you do that, you're no longer "testing" the hypothesis; you're actually determining for certain whether it's true or not. One speaks of "testing" an hypothesis only in circumstances where a direct verification is not possible (or not practical) but where partial information is available against which you can, indeed, "test" the hypothesis.

A ***statistical hypothesis*** is an hypothesis which admits observations of a statistical nature. This covers a lot of ground. Here are some real-world examples:

(a) This is a fair die.

(b) Our parts supplier's claim is false.

(c) This new teaching method is superior to the old one.

(d) The air quality standard for our city is not being met.

(e) This employment data suggests a discriminatory hiring policy.

As statistical hypotheses, these can be formulated more precisely:

(a) All the probabilities for this die are equal. That means we have a uniform distribution for the random variable which counts the number of dots on the uppermost face for one roll of the die.

(b) The mean length of the chain links from our supplier is greater than claimed (or smaller than, or different than claimed, depending on your particular concern).

(c) The mean score on a standard test is greater for the population taught by the new teaching method than for the population taught by the old method.

(d) The parameter which measures air quality for our city (a percentage, a mean, etc.) is greater than the air-quality standard allows (or less than, depending on what the standard is).

(e) For these employees, the variable "was hired" fails to be independent of the variable "gender" (or "ethnic identity," or "religious affiliation," etc.—the variable which defines the group in question).

The hypothesis in (a) is an hypothesis about an *abstract model*, the probability distribution for a random variable which describes the die. The hypothesis might be tested by observations of, say, 100 rolls of the die. The next three hypotheses concern a *parameter*. They're tested by obtaining sample data. That data must be understood through the sampling distribution for the estimator of that particular parameter. The last hypothesis asserts the *independence* of two qualitative (non-numeric) variables. It's tested by comparing the observed data with what would be true if the variables were really independent. Note that only three of these five hypotheses involve a parameter; the other two are statistical statements of a different sort.

For each of these examples, it will be impractical if not impossible to determine the truth of the hypothesis directly. For example, the probability distribution for the die is a model for all possible rolls of the die. You cannot observe "all possible" rolls. Before you finish, you and the die will both be reduced to dust! Nor is it practical to measure exactly the length of each of several hundred chain links from the incoming shipments of your supplier. For hypothesis (c), you cannot test today all students to be taught in the coming 15 years by a new teaching method. Many of those students have not even been born! You can test the students over the years as you finish teaching them, of course, but the point is to evaluate the method BEFORE implementing it, not after. And for hypothesis (d), can you test every cubic foot of air in your city? Finally, in (e), in employment discrimination, what does "direct verification" even mean? Discrimination, after all, is not usually a deliberate policy so much as a result of subtle prejudices and negative attitudes which are difficult to verify in a direct way.

Because it's impossible to determine the truth of the hypothesis directly, you "test" the hypothesis. You collect data which you analyze through an appropriate model. Because the data may not be typical of the general situation (sampling error!), you will not be able to say for certain whether the hypothesis is true or not. There always remains the possibility you will be misled by atypical data. Still, you can control the possible error through your model by controlling the appropriate probability. Exactly what this means you'll see as we develop our testing procedures. In fact, understanding the sense in which one controls sampling error is at the heart of understanding a statistical testing procedure.

This is exactly parallel to the situation with confidence intervals. In a confidence interval problem, you are asked for the unknown value of a parameter in a situation where it's impossible or impractical to compute the value of that parameter exactly. So you do not answer the question by giving a simple number. Sampling error in your data must be taken into consideration. You do this by giving a range of possible values for the parameter together with the probability that this range of values actually contains the unknown parameter. Thus, you "control" the sampling error in a probabilistic sense. In similar but subtly different ways, one controls sampling error in statistical testing situations.

There are several procedures in statistics which are referred to as tests of a statistical hypothesis. We study two very widely used such procedures which we denote by the standard terms "test of significance" and "hypothesis test." However, the distinction between these two procedures is sometimes not clearly drawn in practical applications, with confusion as a result. Because of this, we will be especially careful in our initial presentation to distinguish between them. Once the procedures are clear in their distinct forms, we will discuss how they are combined in practice. There is a third procedure, sometimes called a "Bayesian test of significance," which requires a very different probabilistic approach. We will not attempt to deal with that procedure. And there are still other important procedures which we will not mention at all.

What Are These Two Testing Procedures?

Our two procedures, the "test of significance" on the one hand and the "hypothesis test" on the other, represent two distinct approaches to testing a statistical hypothesis. These two approaches were developed historically for two very different purposes. The procedure which we call a "test of significance" is conceptually straightforward and traces back to the very beginnings of statistical inference in the eighteenth century. The second procedure, the "hypothesis test," was only developed in the 1930s. It is most naturally conceived from a practical point of view as a monitoring procedure for such situations as industrial quality control. The procedure was developed originally by the English statistician Egon Pearson and his Polish colleague Jerzy Neyman who spent the last decades of his career at the University of California at Berkeley. In the next decade, the 1940s, Abraham Wald extended the procedure into a sophisticated method of analysis called Decision Theory.

Let's get an overview of the two types of procedure, with an example of each. For a ***test of significance***, the problem takes the form of a simple question[1]: Does our data seem to challenge the hypothesis? For example, we might ask, "Is it believable that this data came from a normally distributed population?" Contrary to what you might think, the solution to the problem is not a yes or no answer to the question. The solution, formally speaking, is a *number* called the p-value which is calculated from the data. We think of this number as measuring the consistency of the data with the hypothesis. We'll see later how to actually calculate the p-value.

Then, of course, one has to interpret that number to obtain a meaningful real-world conclusion. The usual interpretation is as follows (later we'll see other possibilities): If the p-value is small, the data seems inconsistent with the hypothesis and we'll believe the hypothesis false. If the p-value is not small, the data seems consistent with the hypothesis and we draw NO CONCLUSION because the data could be consistent with many other hypotheses as well! Again, we will see why this is true later. Note that the conclusion you finally come to, if there is a conclusion at all, is that the hypothesis is false. Thus, we think of a test of significance as "trying" to conclude that the hypothesis is false. In the example of the previous paragraph, the test of significance is "trying" to identify a non-normal distribution.

An ***hypothesis test*** is a monitoring procedure. It attempts to flag an "exceptional" situation calling for some alternative course of action and specifies, in advance, the probability of error in taking that action. We will see later how this is accomplished. Thus described, the hypothesis test provides an action-oriented *decision procedure* for a repeating decision, explicitly *exercising control over the possible error* which could arise from sampling error in the data.

[1] The exposition given here with the data as part of the question significantly clarifies the roles of the data, of the p-value, and so on.

Suppose, for example, we want to monitor incoming shipments of chain links as they arrive each month from our supplier, where we require a mean length of, say, at least 1.2 cm. With anything less than that, the chains we manufacture may be too short to function properly. If on the basis of a random sample it appears some month's shipment fails as a whole to meet specifications, it will be rejected. Note this is a "repeating decision," repeated once a month. The hypothesis test provides a procedure to "flag" a shipment which does not meet specifications. We determine in advance an acceptable probability, say five percent, of erroneously rejecting a shipment which, in fact, does meet specifications. That is, out of 100 shipments which do meet specifications, we could expect on average to erroneously reject five.

Contrasting the Two Testing Procedures

Now let's consider the various ways tests of significance and hypothesis tests compare and contrast one with the other.

A test of significance provides a numeric measure of consistency between the hypothesis and the data. It provides a number by way of answer. Of course, that number must be interpreted in real-world terms. An hypothesis test, by contrast, is focused on action. It provides a decision procedure controlling decision error. This is much more sophisticated than the simple number you get from a test of significance.

Note also that for a test of significance, the hypothesis and the data are on the same footing, they stare each other down (so to speak!). In particular, the data is part of the question. For the hypothesis test on the other hand, the data in a sense is secondary. The data is part of the solution alright, but it's not part of the original question at all. In fact, the hypothesis test must be set up without reference to the data. It's clear why: The data is collected periodically only after the test is set up. And the data changes from one repetition of the decision process to another. So, for the hypothesis test, the data is a tool for answering the question; it's not part of the question itself. For the test of significance, the data is intrinsic to the question.

Because the data is part of the original question, a test of significance can test for randomness of the data. We'll see later why this is true. For example, a test of significance is often used in employment discrimination cases to see if there appears to be a pattern of discrimination. Here "nonrandom" equates to evidence of a "pattern." By contrast, the secondary role of the data in an hypothesis test means an hypothesis test cannot reasonably be used as a test of randomness (explanation later!).

Further, for a test of significance, the hypothesis is either true or false. For example, a population is either normally distributed or it is not. For the hypothesis test, by contrast, the hypothesis is sometimes true and sometimes false. In the example, sometimes our shipment will meet specifications, sometimes not.

The distinction of the previous paragraph has important implications for what it means to repeat a test. A test of significance is a classic case of inductive inference in which you attempt to infer a general fact on the basis of specific cases. In the test of significance, you attempt to infer that the hypothesis is false ("general fact") based on our observed data (the "specific cases"). For this to make sense, to have a "general fact", the hypothesis must be either true or false. An hypothesis test, on the other hand, is a monitoring procedure; it's monitoring an hypothesis which is true on some repetitions and false on others. There is no "general fact." So a test of significance is, in principle, a one-time inductive inference; an hypothesis test is a repeating, action-oriented decision procedure. Note that this discussion also suggests that repetition is intrinsic to an hypothesis test, but not to a test of significance.

Because repetition is intrinsic to an hypothesis test, probabilities will have very concrete meaning as "theoretical relative frequency." The probability of error can be interpreted in terms, for example, of 100 runs of the test. Probabilities for a test of significance, because it is a one-time inference, have a more tenuous meaning.

Of course, for a test of significance, you certainly could repeat the entire argument with a new set of independently gathered data, accumulating more and more evidence relevant to that one unchanging hypothesis. Typically, that's exactly what does happen. Often the repetition is made by an independent team of investigators. This sort of exact replicability is required of a scientific experiment; it's essential to the "scientific method." For an hypothesis test, by contrast, replication shows up as part of the procedure. Because it's a procedure for a repeating decision made under *ever-changing circumstances*, the replication is NOT exact.

You can see that a test of significance might be preferred in testing a scientific hypothesis where you are concerned only with the hypothesis itself. Is it false? Is it challenged by our data? There is no "decision" or "decision error" in the sense of an hypothesis test. By contrast, in matters of public policy, business management, politics, and so on where one is constrained to take action under changing circumstances, an hypothesis test might be preferred. In such cases, the nature of decision error is often clear-cut and the consequences of error are of crucial importance.

But remember, the two distinct types of statistical test which we are introducing here tend to be combined in practice. We are introducing them in their original purity as distinct and separate procedures for purposes of clarity only. Once the logic of each procedure is clear to you, the common practice of combining them will seem less confusing. It fact, if one really understands the logic, there is no confusion.

So now, please . . .

Try Your Hand

6.1.1 Of the five examples of real-world hypotheses given at the beginning of the chapter ("this die is fair," etc.), which one(s) would be appropriately treated by a test of significance and which one(s) by an hypothesis test?

6.1.2 Why do we think of a test of significance as "trying" to show that the hypothesis is false?

6.1.3 (a) As in the discussion above, suppose you're monitoring incoming shipments of chain links as they arrive each month from your supplier. You require a mean length for the links of at least 1.2 cm and will reject a shipment if, on the basis of a random sample, it appears not to meet specifications. Suppose you decide in advance on a five percent probability of error, how many out of 500 shipments would you expect to be erroneously rejected?

(b) Suppose on second thought the five percent probability of error in part (a) seems unacceptably large. What would you do?

6.1.4 (a) Contrast in detail the test of significance with the hypothesis test.

(b) In terms of the examples we gave for each procedure, explain the contrasts you identified in part (a). Make specific reference to the examples, with as much detail as the examples allow! Omit the "test for randomness" contrast. It becomes clear for us only later.

6.2 Tests of Significance

When we were children we learned important things from the older kids, but then sometimes you had to be wary!

A Dialogue

Big Kid: What'd ya mean I'm cheating?

Little Kid: Your coin is not fair!

BK: Huh . . . ?

LK: It's not a fair coin! I *know* it's not a fair coin.

BK: Sure it is.

LK: Look—let's toss it 100 times and see how many times it comes up heads.

BK: Huh . . . ?

LK: Watch [tosses the coin repeatedly] . . . 57 . . . [toss, toss], . . . 58 . . . [toss, toss . . . toss]. There! 58 heads!

BK: So what?

LK: Well, if the coin comes up heads as often as tails, we should get about 50, not 58.

BK: On average. ON AVERAGE knucklehead! 58 heads out of a *specific* series of 100 doesn't say anything. Haven't you ever heard of random error?

LK: Yeah, . . . well . . . random error THAT big?

BK: Big, schmig—random error can be big!

LK: It's VERY unlikely to be THAT big!!

BK: Who says?

LK: ME!!

BK: Prove it!

LK: If the coin is fair, on 100 tosses the standard error is only five heads.

BK: How'd ya figure that?

LK: We're counting the number of heads in 100 tosses of a coin. That's a binomial random variable with $n = 100$. If the coin is fair, $p = 0.5$. So $npq = 25$ and the standard error is five.

BK: So what? Just because the standard error is five doesn't mean anything! 58 heads is perfectly possible!

LK: Possible, but not likely. The *p*-value is going to be VERY small!

BK: *P*-value, *p*-value!!?? What's this *p*-value?? I never heard o'no p-value!

LK: The *p*-value's a measure of how consistent your data is with your hypothesis. For us it's the probability of 58 or more heads out of 100 tosses, assuming your hypothesis true. You said the coin is fair right? That's your hypothesis. If that's true, $p = 0.5$ and . . .

$$\begin{aligned} p\text{-value} &= \mathrm{P}[X \geq 58 | n = 100, p = 0.5] \\ &\approx \mathrm{P}[Z > (57.5 - 50)/5] \\ &= \mathrm{P}[Z > 1.5] \\ &= 0.0668. \end{aligned}$$

Less than seven percent!!

BK: (weakly) Seven percent's not so small.

LK: Yeah, well . . . it's too small for ME! I'm not going to play with this coin, I think it's biased.

BK: Awww . . . I didn't know it wasn't fair! Seven percent's not THAT small.

LK: Okay, okay . . . just gimme my marbles back! You didn't win them fair and square.

BK: Even the BLUE one?

In this dialogue we see exactly how to do a test of significance. It's really easy. You just assume the hypothesis true and then compute the p-value of your data. Don't be misled by the light-hearted tone of our example! We chose it precisely because it really does illustrate the various situations that arise in serious applications. You'll see how that's true as we come to more realistic problems.

If the p-value is too small, the hypothesis doesn't stand. It seems inconsistent with your observed data. Of course, there is a question of what is meant by small here. The dialogue illustrates well what happens in actual practice. If the test is just for your information, you decide subjectively for yourself what is small. In the dialogue, the little kid felt a seven percent p-value was too small and refused to play with the coin. The big kid wasn't so sure but evidently didn't want to push the issue.

If it's a question of making some public statement, then a more objective criterion is required. For example, if the kids wanted to publish their results in a scientific research journal, they would have to see what the requirements of that journal are. Some journals require a p-value of no more than one percent others allow up to five percent. By either of those criteria, the little kid would have been in trouble. Still other criteria are sometimes imposed. If the criterion had been that p-values of no more than ten percent are considered small, then the little kid would have been on firm ground.

But it was not a question of any public statement and apparently even the big kid felt the data casts doubt on the fairness of the coin. So the issue was resolved on the basis of a clear-cut, objective analysis of the observed data (7% p-value) with everyone involved knowing and accepting the interpretation of that analysis.

Whether a p-value is to be considered small or not can depend very much on the context. A p-value would have to be VERY small indeed to convince most scientists that you had observed an instance of mental telepathy. Scientists tend to be VERY sceptical about the possibility of mental telepathy because it seems inconsistent with almost everything else they know. In fact, there has been a lot of heated discussion about certain controversial statistical studies of paranormal phenomena.

The p-value

Now let's look more carefully at the p-value. The ***p-value*** is the probability of data as extreme as yours or worse *if the hypothesis is true.* It's the probability of data that could cause you as much or more doubt about the hypothesis as YOUR data. To compute the p-value you must think what kind of data would be even worse than yours. Then you compute the probability of those values together with your value ("as extreme as yours or worse"). So the p-value has the form

$$p\text{–value} \quad = \quad \text{P}(\text{ data like yours or worse} \mid \text{hypothesis})$$

For the kids, if the hypothesis of "fair coin" is true, they would expect 50 heads on 100 tosses. They observed 58 heads. What would be even worse? Obviously, 59, 60, 61, ... Any number of heads more than 58 would be even more unbelievable (worse) IF THE COIN IS FAIR. So the p-value, just as the little kid said, is the probability of getting 58 or more heads assuming a fair coin:

$$\text{P}(X \geq 58 | p = 0.5, n = 100) = 0.0668.$$

A small p-value suggests two possible explanations for your data:

- The hypothesis might be true and the data extreme just because of sampling error. All heads on 100 tosses of a fair coin is *possible* however unlikely it may be!
- The hypothesis might be wrong and the data not really extreme at all. Suppose the coin really is biased with a 60% probability of heads. Then 58 heads out of 100 tosses is certainly not extreme.

The p-value is calculated assuming the hypothesis correct and so it is like taking a "look" at the first explanation. If the p-value is very small, the first explanation seems unlikely. Not because atypical data is impossible, but because there seems to be a more plausible explanation: It seems more believable that the hypothesis is wrong. So we accept the second explanation, thinking the hypothesis makes the data seem so extreme that the hypothesis itself becomes unbelievable. Thus, the p-value serves as a numeric measure of the consistency of our data with the hypothesis. A small p-value suggests data inconsistent with the hypothesis.

The most unambiguous use of tests of significance are those for which the p-value is very small indeed. Like one chance in 100,000. If the kids had observed 100 heads in 100 tosses, we would not think the situation in any way equivocal (the p-value has 30 zeros after the decimal before the first nonzero digit!). Even less so if they had observed a billion heads on a billion tosses! In actual practice, tests of significance often do lead to unequivocal results. You might object that the hypothesis would be *obviously* false in such a situation and that a test of significance would be a futile exercise in proving the obvious. For something as simple as a coin you're right. But for a highly complex problem, trying to get any kind of intuitive conclusion from a large data set may be impossible.

Thus, for the kids, there is only about a seven percent chance of data like theirs or worse if the coin is really fair. Are you going to believe the coin fair or not? That's strictly up to you. The test of significance only provides you with a measure (the p-value) of the "statistical significance" of the difference between the observed 58 heads and the 50 heads you should theoretically expect if the coin is fair.

What happens if the p-value is not small? That means the difference between the observed data and the hypothesized value (58 versus 50 heads) can be explained as "just due to chance," due to random error. In that case, we can't conclude the coin is fair! It may or may not be fair. Problem 6.2.4 below shows why: An observed 53 heads gives the same p-value for a biased coin as for a fair coin. In other words, the observed data is just as consistent with the "fair coin" hypothesis as it is with the "biased coin" hypothesis. It's true, when the p-value is not small the data seems consistent with your hypothesis. But because the data will always be consistent with many other hypotheses as well, that says nothing at all! So admit it: When the p-value is not small, the data is inconclusive.

Now why don't you please . . .

Try Your Hand

6.2.1 In each part of this problem compute the p-value and state what you think about the kids' hypothesis on the basis of their observed data. For simplicity, use the *proportion* of heads instead of the count of heads and treat it as the normally distributed sample proportion, $\hat{p}$ (our convention is to ignore the continuity correction with $\hat{p}$).

(a) The kids observed 58 heads on 100 tosses to see if the coin seems fair, as in the dialogue. [Hint: Recalculate the little kid's p-value using $\hat{p}$. Your p-value will be slightly less accurate than hers because you're ignoring the continuity correction.]

(b) The kids observe 62 heads on 100 tosses, wanting to check if the coin seems fair.

(c) The kids observe 58 heads on 100 tosses just as in the dialogue, but here they think the coin comes up heads only 40% of the time.

(d) The kids observe 44 heads on 100 tosses for a coin they think to be fair.

6.2.2 In a test of significance, if the p-value is small, we conclude that the data seems to challenge the hypothesis. In other words, the data seems to *support* . . . ? [Hint: Think about Problem 6.2.1(c).]

6.2.3 Here are some general questions about tests of significance to help you clarify your understanding:

(a) When in the light of the hypothesis the data seems extreme, there are two possible explanations. What are they?

(b) If the data yields a very small p-value, which explanation from part (a) do we accept?

(c) How do you justify the comment in the answer to part (b) that random data might be atypical, but it *probably* is not?

(d) Give the meaning of the kids' seven percent p-value in terms specific to their situation. Be as detailed as the situation allows.

(e) The kids' seven percent p-value means there's a seven percent chance the die is actually fair. No, no, it means there's a seven percent chance it is NOT fair. Comment.

6.2.4 Show that an observed 53 heads on 100 tosses is consistent with any one of the following hypotheses for the probability p of heads:

$$p = 0.5, \qquad p = 0.56, \qquad p = 0.53.$$

What does this fact tell you about tests of significance?

6.2.5 Parts (a) and (b) of this problem are very similar; be sure you see the difference.

(a) You know a certain population of numbers has variance 6.2 and you thought the mean was 70.18. You take a sample of 50 and find a sample mean of 69.3. What do you think now?

(b) You have good reason to think the mean of a certain population is 70.18. The variance is unknown. You take a sample of ten and obtain a sample mean of 69.3 and sample variance of 6.2. Now what do you think about the mean?

(c) Which yields a more certain conclusion, part (a) or part (b)?

6.2.6 Suppose the kids are going to play a coin-tossing game almost everyday during the summer. Making specific reference to this example, discuss the issues of "repetition" and "accumulation of evidence" as they relate to tests of significance and hypothesis tests.

Comparing Means and Comparing Proportions (Large Samples): Two New Parameters and Their Estimators

Many problems which arise in practical applications seek to compare two means or two proportions by looking at their difference. The problem may simply ask, "What is the difference?" (confidence interval) or it may involve an hypothesis about the difference (test of significance or hypothesis test). For example, you may think there's no difference in academic preparation for the student bodies at two schools. If you measure "academic preparation" in terms of the mean score on some standard test—the SAT test, for example—your hypothesis of "no difference in academic preparation" becomes a statistical hypothesis by asserting that $\mu_1 - \mu_2 = 0$, where μ_i is for the mean score for the ith school. Of course, academic preparation might be measured in a number of other ways as well. Maybe "academically prepared" is determined by the proportion of students with an SAT score above 800. If so, your hypothesis of "no difference" becomes $p_1 - p_2 = 0$, where p_i is the proportion of students at the ith school with an SAT score above 800.

As you might guess, the estimator for the difference between two parameters is just the difference between their estimators. For example, for the parameter $p_1 - p_2$, the estimator is $\hat{p}_1 - \hat{p}_2$. In the problems, you'll see that these estimators are unbiased and, in the large sample case, approximately normally distributed. For the condition "large sample," the same rules hold as before: For means, a sample is large enough if $n \geq 30$; for proportions we require np, $nq \geq 5$. We'll not introduce the small sample case for differences because they involve technicalities that are beyond the scope of this course.

You'll also see in the problems how easily one derives the standard error formulas if the samples are chosen independently. All of this analysis depends on the following not so trivial facts about sums of random variables. We'll not attempt a proof:

for the sum of two random variables:

$$\mu_{X+Y} = \mu_X + \mu_Y;$$

and if the random variables are independent

$$\sigma^2_{X+Y} = \sigma^2_X + \sigma^2_Y$$

If the samples are chosen from two different populations, they would usually be independent. We'll deal only with that case:

for large, independent samples:

$\overline{X}_1 - \overline{X}_2$ is approximately normally distributed and is an unbiased estimator for the parameter $\mu_1 - \mu_2$

and

$\hat{p}_1 - \hat{p}_2$ is approximately normally distributed and is is an unbiased estimator for the parameter $p_1 - p_2$.

Here's a summary of our parameters and their estimators up to this point. We'll have several more before the end of the course.

Parameter	Estimator	Standard error
p	$\hat{p}$	$\sqrt{pq/n}$
μ	$\overline{X}$	$\sqrt{\sigma^2/n}$
σ^2	s^2	We don't need[2]
$p_1 - p_2$	$\hat{p}_1 - \hat{p}_2$	$\sqrt{p_1q_1/n_1 + p_2q_2/n_2}$
$\mu_1 - \mu_2$	$\overline{X}_1 - \overline{X}_2$	$\sqrt{\sigma_1^2/n_1 + \sigma_2^2/n_2}$

In the large sample case—the only case we treat—the standard errors can be estimated if necessary from the samples, replacing p's and q's by $\hat{p}$'s and $\hat{q}$'s and replacing σ's by s's. Of course, if the true values are known, this is not necessary. For an hypothesis of "no difference" in two proportions, we can get a more sensitive estimate of the standard error by "pooling" the samples. Because you're assuming the hypothesis true, the two unknown proportions become equal. That means the two populations look exactly alike from the point of view of the characteristic of interest. If so, both samples can be thought of as coming from the same population with unknown population proportion, p. By pooling the samples, you obtain one sample of size $n_1 + n_2$ from which you calculate $\hat{p}$, the proportion of the pooled sample having the characteristic. That value of $\hat{p}$ is used in the standard error in place of the unknown p_1 and p_2. So you estimate the standard error by the square root of

$$\frac{\hat{p}\hat{q}}{n_1} + \frac{\hat{p}\hat{q}}{n_2} = \hat{p}\hat{q}(\frac{1}{n_1} + \frac{1}{n_2}) = \frac{\hat{p}\hat{q}(n_1 + n_2)}{n_1 n_2}.$$

This "pooling" of the samples is not relevant in the case of $\mu_1 - \mu_2$ because the standard error involves σ_1 and σ_2 which may still be different even when $\mu_1 = \mu_2$.

Just to give you some idea of the complications that arise in the small sample case, using s's instead of σ's in the standard error for the $\overline{X}_1 - \overline{X}_2$ may not lead to a t-distribution! As a consequence the small sample theory becomes quite complex. An enormous amount of effort has been devoted to this problem. If the samples are drawn from two normal distributions having the same variance, life's not too hard. You'll get

[2] Note that s^2 is the only one of our estimators that is not normally distributed. Its distribution is given by χ^2 and does not require "standardizing." So we don't need the standard error.

a t-distritution if you estimate the standard error by pooling the samples. But the solution is very sensitive, indeed, to that assumption of equal variances, an assumption that's virtually unverifiable.[3] So you shouldn't assume the variances equal. But if the variances are not equal, the problem is more complex still. It even has a name. It's called the Behrens–Fisher problem.[4] So please excuse us, we're omitting the small sample case for differences. Such technicalities belong to a more advanced course!

Well now please . . .

Try Your Hand

6.2.7 In the text above, we said, ". . . the standard errors can be estimated if necessary from the samples, replacing p's and q's by $\hat{p}$'s and $\hat{q}$'s and replacing σ's by s's . Of course, if the true values are known, this is not necessary." Under what circumstances would they be known so that estimating the standard error would be unnecessary?

6.2.8 **(a)** What's the difference in height between second graders in the suburbs of your city compared with second graders in the inner city? Suppose a sample of 50 second graders from the inner city gave a mean height of 92.5 cm with a standard deviation of 4.8 cm and a sample of 42 second graders from the suburbs gave a mean height of 97.3 cm with a standard deviation of 5.2 cm.

(b) Is the observed difference in height in part (a) significant?

6.2.9 We thought that both populations we're interested in had the same proportion for some characteristic of interest. A pair of samples each of size 100 yield proportions of 0.22 and 0.31. Now what do you think?

6.2.10 Let $W = X + Y$ where X and Y are the number of dots on the top and bottom faces, respectively, of a die. So W is the constant random variable which always takes the value seven. First, guess the mean and variance of W. Then calculate the mean and variance of W from those of X and Y using the appropriate formulas.

6.2.11 Show that each estimator described below is unbiased and approximately normally distributed with standard error as given in the table of the text. Assume you have chosen independent samples.

(a) The estimator for the parameter $\mu_1 - \mu_2$;

[3] See Hoaglin and Moore, p. 14 and 15.

[4] See Bickel and Doksum, p. 219 for a discussion of this.

(b) The estimator for the parameter $p_1 - p_2$.

6.2.12 A legal challenge was brought against an employer alleging that a job promotion test was racially biased. Of 48 blacks who took the test, only 26 passed, whereas of 259 whites, 206 passed (Connecticut v. Teal [457 US 440 (1982)] (after [Finkelstein and Levin]).

(a) Is it believable this difference is due just to chance and that racial bias is not a factor?

(b) What exactly is the difference in pass rates for blacks and whites on this test?

Practical Versus Statistical Significance

In the dialogue, if the kids were not playing for such high stakes (beautiful marbles, even the blue one!), if they were playing for match sticks or pebbles say, then the little kid might be convinced by a 7% p-value to go ahead and play. This raises yet another question. If the kids go ahead with the game, is it because they think the coin is fair or is it because they think, "Well . . . with a p-value of seven percent, even if the coin is biased, it doesn't seem to be VERY biased!"? If this is how they think, it means they accept that the coin may be biased, but the bias doesn't seem significant in a practical sense.

They admit the data is "statistically significant"—that the data seems to have detected a real bias—but they question whether the bias that's been detected is of any "practical significance." This is an important distinction. The p-value measures the significance of your data for challenging the truth of the hypothesis. It says something about the data and the hypothesis; it doesn't say anything about the hypothesis and the real world. It says nothing about the practical significance of the hypothesis being false. These two types of "significance" are totally unrelated!

So when a small p-value suggests your hypothesis is false, the question remains, "How false?" Once the kids decide there's more than a 50/50 chance for heads on this coin, they may wonder just how serious that bias really is. Suppose the probability of heads is 0.5001. That's bigger than a 50/50 chance, but few people would think it meant the coin was seriously biased. Such a bias has no practical significance. On the other hand, suppose there's an 85% chance of heads—THAT is of practical significance! As you can see, the determination of what is or is not of practical significance has nothing to do with statistics.

Of course, if we ask what's the probability of heads on a coin—for example, to see if there's any practical significance to a bias which we have detected (or for any other reason)—it's time to compute a confidence interval. This is what happens in general for statistical tests involving an hypothesis about a parameter: If you decide the hypothesis is false, you may want to compute a confidence interval to see what is a probable range of values for the parameter (because your hypothesized value seems wrong). Once you have that interval, you can decide whether the difference between the true value and the hypothesized value of the parameter is of any practical significance. That practical decision has nothing to do with statistics—it uses the information provided by statistics through the confidence interval, but the decision itself is a matter of practical, not statistical, considerations.

So now, please . . .

Try Your Hand

6.2.13 On January 7, 1993, the *San Francisco Examiner* reported on California job losses for 1993 (projected) and the previous three years. The information comes from a year-by-year report on job losses from the Commission on State Finance. For example, in 1991, there were 12.43 million unemployed in the state, and in 1992, 12.13 million unemployed, a loss of 300,000 jobs. Is that a significant drop?

The Test of Significance as an Argument by Contradiction

The basic logic of a test of significance is really quite familiar. It's a form of logic we all use constantly in our daily lives—argument by contradiction. It's the "monkey's uncle" reasoning: "If that's true, I'm a monkey's uncle!" But I'M NOT! Contradiction, therefore "that's not true."

The test of significance is a probabilistic version of such an argument by contradiction. The only difference is that instead of arriving at a contradiction you arrive at a statement with small probability. You might call that a "probabilistic contradiction." Instead of, "If this die's fair, I'm a monkey's uncle," we get, "If this die's fair, I'm PROBABLY a monkey's uncle." Well, it's highly unlikely I'm a monkey's uncle, so I don't BELIEVE the die's fair. But maybe it is, after all I don't seem to be completely sure whether I'm a monkey's uncle or not!

More exactly, "If this die's fair, there's only a seven percent chance of seeing 58 heads or more in 100 tosses." Well, that's pretty unlikely; therefore, I don't BELIEVE the die's fair. But maybe it is. After all, seven percent is not so terribly small. You'd feel more secure if it went like this: "If this die's fair, there's only one chance in a billion to get what we got (or worse) on our 100 tosses." One chance in a billion is pretty slim odds. Again, I don't believe the die's fair and I'm much more secure in that belief than I was before with the seven percent probability.

What If the p-value Is Not Small?

What happens if the p-value is not small? Such a p-value suggests the data is entirely consistent with your hypothesis. But the data will be consistent with many other hypotheses as well. This is what you saw in Problem 6.2.4 where 53 heads on 100 tosses was consistent with having a fair coin and equally consistent with having a coin which comes up heads 56% of the time. In fact, 53 heads is consistent with a probability for heads of anywhere between 50% and 56%! It's consistent with a probability of 52.078%. Or 54.29402%. Well, that's what we said: 53 heads is consistent with many different hypotheses!

So, to say the data is consistent with the hypothesis (p-value not small) really says nothing about the hypothesis. In such a case, the test of significance should be considered inconclusive. In actual practice, one of two situations usually arises:

- you had reason to believe the hypothesis and were using the test of significance only as a double check;
- you had reason to doubt the hypothesis and were using the test in an attempt to challenge the hypothesis.

Only rarely would you be totally indifferent about the hypothesis with no reason either to believe or disbelieve.

Consider the first instance where you had reason to believe the hypothesis. If the data seems consistent with your hypothesis (p-value not small), then, of course, your original reason for believing it stands unchallenged and you will proceed, confident that it's true. But only because your original reason looks good. The hypothesis is accepted as true, not because the data shows it to be true but because your original reason for believing it stands unchallenged by the data.

Now consider the second instance where you had reason to doubt the hypothesis even though the data seems consistent with it. This simply means you went in search of evidence against the hypothesis and failed to find it. So you don't know whether the hypothesis is true or not; you doubted it but you have no evidence against it.

Thus, in both cases, when the p-value cannot be considered small enough to say the data challenges the hypothesis, the test itself is inconclusive. But YOU are not necessarily inconclusive with regard to your original question. Your conclusion or lack thereof depends on the prior information you have concerning the hypothesis. Information supporting the hypothesis remains unchallenged; information causing doubt about the hypothesis has not been corroborated by the data. The first case is stronger: You believed and will continue to believe the hypothesis. But not based on the data, the data has become irrelevant because it's inconclusive. In the second case, not only is the data inconclusive, so are you. You started out with reason to doubt the hypothesis but that doubt is not supported by the data. You're back to square one.

Here's one final point about tests where the p-value is not small, rendering the data inconclusive. A very important point: It's a FUNDAMENTAL FALLACY to repeat the test hoping on the second try to get data that does finally challenge the hypothesis. Why is this a fallacy? Well, even if the hypothesis is true, eventually you'll get an atypical sample which seems to challenge it—erroneously seems to challenge it! After all, atypical samples are possible, however unlikely they may be. The moral to this story: Any true hypothesis can be deceptively rejected with repeated testing if you test it often enough.

Of course, if your test is inconclusive, it's entirely reasonable you might want to gather more data. That's fine, put that new data together with the data you've already gathered. Now you have a larger sample. See what the larger data set tells you. That's legitimate and you'll get a more accurate result because the new, expanded sample contains more information.

This discussion may suggest an idea which was introduced in the 1940s by Abraham Wald. Wald created a branch of statistics called Sequential Analysis based on the idea of stopping after each sample element is selected and testing the sample accumulated up to that point. Sequential Analysis was such an efficient technique for industrial quality control it was held by the government as a military secret during World War II. Such statistical techniques were more or less ignored by American industry after the war, even though they were no longer secret. Ironically, with the encouragement of American advisors, such techniques of statistical quality control were adopted enthusiastically and to great advantage after the war by the Japanese! American industry has recently begun to play catch-up.

A Case Where "Not Small p-value" Is Conclusive and "Small" Is Not

It's important to keep in mind the distinction between the test of significance itself and the problem which gave rise to the test. Thinking only of the test itself, "NOT small p-value" means the data seems consistent with the hypothesis. "Small p-value" means the data seems INconsistent with the hypothesis. But then so what? What exactly does that p-value mean for the original problem?

This is well illustrated by Problem 3.3.8 which is typical of a kind of anyalysis often employed in cases of purported discriminatory selection. Let's look at part (c) of that problem. The mayor put only one woman on a committee of five chosen from a pool of 40 candidates where 15 of the candidates were women. Usually with a test of significance the population is what's in question; the data is given. Here, that usual situation is reversed. Here, the population is completely known ($N = 40$, $R = 15$) and it's the data that's in question. Specifically, the *choice mechanism* for the data is what's in question: Was it a random choice? So "randomness" is the hypothesis to be challenged by the data.

In this case, the data is the specific committee chosen by the mayor, a committee with only one woman on it. The p-value for this data is the probability of a committee with as few or fewer women than we observed, $P(X \leq 1)$. In Problem 3.3.8, you calculated $P(X \leq 1) \approx 37\%$, which is NOT small, so the data is entirely consistent with the hypothesis. Up to this point the analysis is standard. But note what happens when we interpret our analysis in the real-world terms of the problem. A NOT small p-value gives a very definite conclusion: If the mayor's choice is "entirely consistent with a random choice," the accusation of discrimination cannot be maintained. This is exactly the reverse of what usually happens. Usually to get a conclusion we look for a small p-value, but here, "not small p-value" is the more conclusive situation.

By contrast, in the situation of the mayor's committee, if the p-value is SMALL—so the data seems inconsistent with the hypothesis—we're led to the "conclusion" that the mayor's choice does not seem random. But this is NOT genuinely conclusive. Of course she didn't choose randomly! Everybody knows that. The only question was whether her criteria of choice were free of gender considerations, as they should have been. It's true the choice doesn't look like a random choice and that certainly puts the mayor on the defensive. But the argument is not over; those who suspect bias must engage the mayor on the issue of her criteria of choice. Were those criteria gender neutral? A small p-value is only the first step which establishes that the accusation has some basis on which to proceed.[5]

Is this always the pattern when the hypothesis is "randomness"? No, not exactly, that's why we keep telling you there's no substitute for genuine understanding. Look at Problem 3.7.12 which again was a p-value calculation (although you didn't know it at the time). Or Problem 6.2.22 below. In these problems the hypothesis is again "random choice," this time for the panel of "veniremen" from which a jury will be selected.[6] In these cases, the selection process SHOULD be random (the mayor's choice presumably was not). Now, just as with the mayor, a NOT small p-value is very conclusive. The accusation of bias in selection cannot be maintained if the choice is consistent with a random choice. The discussion is over!

But a SMALL p-value carries a different implication than in the mayor's case. Here, a small p-value is stronger evidence of bias. Nobody can say, "Well, of course, the panel of veniremen wasn't chosen randomly." It should have been! As Justice Frankfurter said in *Avery* v. *Georgia*, "the mind of justice . . . would have to be blind" to believe the hypothesis of unbiased selection (see Problem 3.7.12, p-value < 0.05). We should note that in this early case (1953) Justice Frankfurter apparently did not rely on a p-value calculation. Such statistical analysis became common in jury discrimination challenges only after 1967 with *Whitus* v. *Georgia* (Problem 6.2.22).

There are other situations where one wants to test "randomness." For example, the so-called "efficient market hypothesis" for the stock market assumes that information is dispersed virtually instantaneously into the market so that attempts by the ordinary investor to profit from published information will necessarily fail. This hypothesis can be made precise by asserting "randomness" for price movements in the market after a major announcement is published.

Finally, in Problem 6.2.27, you'll see a very curious historical case of "p-value only TOO, TOO small"! For now, we'll leave it to you to think what THAT could mean.

With this discussion you see once again what we've emphasized often in this course: Statistical techniques are not blind routines to be applied with no understanding of the logic. The logic of p-values is that "small"

[5] As you may imagine, there has been much discussion in the legal literature of the use and misuse of statistical analysis. See Finkelstein and Levin, *Statistics for Lawyers*.

[6] The jury itself is not chosen randomly; it's chosen after intensive questioning by both parties to the litigation. Both sides can, without any justification, eliminate prospective jurors they don't like. Here, we're discussing the panel of "veniremen," the panel of *prospective* jurors from which the jury is selected. In the cases we're discussing, the panel of veniremen was supposed to have been chosen randomly.

means the data seems inconsistent with the hypothesis (there's a small probability of such data given the hypothesis) and "not small" means it seems consistent. But the interpretation of this in a given real-world context cannot be described in advance. Therefore, you have to understand the logic and the context!

Well, now we're ready for you to . . .

Try Your Hand

For any real-world question involving a p-value calculation, you should be able to do four things: (1) calculate the p-value, (2) identify it as "small" or "not small," (3) state the real-world meaning of the p-value for that problem, and (4) interpret the p-value in the real-world terms of the problem with as much detail as the problem allows.

6.2.14 **(a)** Do you think the little kid would be likely to say, "Well . . . with a p-value of seven percent, even if the coin is biased, it doesn't seem to be VERY biased!"?

(b) If the little kid is willing to play with a coin that's slightly biased in favor of the big kid, how should they test the coin?

(c) In the situation of part (b), would the little kid be willing to play once they have observed 58 heads on 100 trials? Recall, there are two distinct situations [part (b), level II answer].

(d) What exactly is the difference between the two situations of part (b)?

6.2.15 Our soft drink dispensing machine has a fill variance of 0.0324 ounces. Resetting the machine does not appreciably change the variance. After resetting the machine, it is now supposed to dispense about 7.4 ounces of soft drink into the cups. We obtain a random sample of 35 filled cups and determine the mean fill to be 7.53 ounces. Does it appear the resetting device is working properly?

6.2.16 Over the past year there have been 12 cancers of a certain type in the southern part of your city. The expected number of such cancers in a population of that size is 8.4. Could this difference be due just to chance or does it appear there may be some environmental cause of cancer present in your city?

6.2.17 **(a)** You observe 56 boarding passengers for your airline's New York to San Francisco flight and find that four requested a vegetarian lunch. On the New York to Chicago flight, only one passenger of 72 requested a vegetarian lunch. Does this suggest a pattern of any significant difference in the number of such requests on the two flights?

(b) Your solution in part (a) fails on one important criterion. What is it?

(c) How might you get around the problem identified in part (b)?

6.2.18 Suppose you have an hypothesis and some data and carry out a test of significance. Technically, what's wrong with the following conclusion? "Our calculations from this data show that the hypothesis is probably correct (we calculated a big probability)."

6.2.19 When the kids decided their p-value was small, we used the phrase "the data seems to have detected a real bias." What would be "illusory" bias?

6.2.20 **(a)** You work for a tire manufacturer. A random sample has given a mean life of 28,204 miles. The head of marketing says, "That's no good! We can't advertise a life of only 28,000 miles—it's got to be at least 35,000 to 40,000! Go get another sample!" Comment.

(b) What has the question in part (a) got to do with tests of significance?

6.2.21 **(a)** You are considering a coin which supposedly is fair. You're told that it was tossed 12 times showing three heads, the third head occurring on the last toss. This data could be modeled two ways. Compute the p-value both ways. Assume a 5% criterion for "small p-value." Is it a fair coin?

(b) What does the example in part (a) tell you about tests of significance?

6.2.22 **(a)** In *Whitus* v. *Georgia* (1967), a black defendant was convicted by an all white jury. In considering the challenge of this jury, the U.S. Supreme Court concluded, "Assuming that 27% of the list was made up of the names of qualified Negroes, the mathematical probabilty of having seven Negroes on a venire of 90 is 0.000006." Comment.

(b) To complete the level II discussion of part (a), show that even the most probable value of a binomial distribution will have small probability. For example, if $n = 90$ and $p = 0.5$, what's the probability of the most probable value?

6.2.23 On March 6, 1978, *Newsweek* reported on a study by Brigham Young University sociologist Philip R. Kunz. He found that for a random sample of 747 persons whose obituaries were published in one year in Salt Lake City, only eight percent had died in the three months prior to their birthdays. From this we can see that dying people succeed in holding out until after their birthday before giving up the ghost.

(**a**) Carry out an appropriate test of significance for this data to see if the claim seems believable.

(**b**) In what ways would this seem to be more appropriate as a test of significance than as an hypothesis test?

6.2.24 An accounting firm was taken to court because none, of 17 fraudulent invoices, showed up in their sample of 100 taken in the course of an audit. After the company failed, the accounting firm was sued by a creditor who had relied on their certification of the company's financial statements. Was the accounting firm negligent? Assume there were 1000 invoices total. (*Ultramares Corporation* v. *Touche*, 1931, after [Finkelstein and Levin]).

Chi-Squared Tests for Goodness of Fit, Homogeneity, and Independence

Suppose we observe 100 rolls of a die to see if the die seems fair. This is a version of the kids' problem, only more complicated—their coin had only two faces, the die has six faces. Saying the die is fair means there's a uniform distribution for the random variable which counts the number of dots on the uppermost face after one roll. Because we're asking if the data seems to "fit" a uniform distribution, this type of test is called a "goodness of fit" test. Of course, there's nothing sacred about the uniform distribution. Maybe we think the die is loaded so that the face with two dots comes up half the time with all other faces equally likely. Then we're asking if the data seems to fit that distribution.

The trick for this test is an ingenious application of the χ^2 distribution. It was first developed in 1900 by Karl Pearson, the father of Egon Pearson. It was Egon, you may recall, who in the 1930s with Jerzy Neymann developed the procedure of hypothesis tests for industrial quality control. Karl Pearson is not only the father of Egon, he's often referred to as the father of modern statistics. Karl Pearson's χ^2 test is very widely used. It won't be difficult for us to understand because it follows exactly the logic of any test of significance: Compare your data with what you would expect to observe if the hypothesis were true.

In the general pattern of the test, we have a fixed number, k, of "cells" and an hypothesis which specifies the probability, p_i, to obtain an observation in cell number i. In our example, a cell is a "face of the die." Because there are six cells, $k = 6$. If you thought the die fair, each p_i would be $1/6$. But if it's loaded the way we described earlier, $p_2 = 0.5$ and each of the other p_i's is 0.1. With n observations, np_i is the expected number of observations which should fall in the ith cell. So, on 600 rolls of a fair die, you expect 100 in each "cell"; that is, you expect each face to come up 100 times. The expected number of observations for cell i is usually denoted by E_i, a theoretical number. The corresponding observed number is denoted by O_i. It's the number of observations which fall in cell i. Then $O_i - E_i$ captures the discrepency between your observation and what should be expected if the hypothesis is true. Pearson showed that the following statistic based on the squares of the various $(O_i - E_i)$'s has approximately a chi-squared distribution with $k - 1$ degrees of freedom:

$$\chi^2 \approx \Sigma\big[(O - E)^2/E\big].$$

Note that χ^2 will be large if your observations are mostly quite far from what you expect, suggesting that what you expect, and so the hypothesized model, is wrong. Of course, if E is quite large, we should be ready to allow the corresponding $(O - E)^2$ to be proportionately large. That's why E enters the formula in the denominator. At the other extreme, if each of your observations hit the mark exactly, if they were each *equal* to what you expect, then $\chi^2 = 0$. Of course, that's highly unlikely for random data. If such a suspicious set of data shows up, fitting the model only too, too well with χ^2 close to zero, you might suspect the data has been "massaged"—manipulated to fit the model. That happens!

Let's continue with our example of the die which we think is loaded so the face with two dots comes up half the time with all other faces equally likely. Suppose we roll that die 100 times. Then 50 of the 100 rolls ought to show two dots on the top face. The other faces should come up ten times each. That specifies the E_i's. Suppose we recorded the results of our 100 rolls in a table like the one below. We can extend the table with two more columns, recording the expected frequencies and the terms of Pearson's χ^2 statistic.

Face	O_i	E_i	$(O_i - E_i)^2/E_i$
1	6	10	1.6
2	61	50	2.42
3	9	10	0.1
4	6	10	1.6
5	10	10	0.0
6	8	10	0.4
	100	100	6.12

The sum of the last column is the value of Pearson's χ^2 statistic for this data. From the χ^2 table with five degrees of freedom ($k - 1 = 5$), the p-value for this data is greater than 10%. In fact, this observed $\chi^2 = 6.12$ is not far above the mean of the distribution. The mean of χ^2, recall, is the degrees of freedom. So our observations ("not small" p-value) provide no reason to doubt our description for the loading of this die.

Pearson's χ^2 test is very flexible. It can be used to test not only for goodness of fit but also for "homogeneity" and for independence. You will see how all this works if you'll just . . .

Try Your Hand

6.2.25 In the table of the text above, we recorded observations for what we believed to be a loaded die. Suppose, instead, we had thought the die fair. Would the data have suggested that, in fact, it was not fair?

6.2.26 With χ^2 we have a way to test fit to the Poisson distribution for Bortkiewicz' data on horsekick fatalities. Bortkiewicz, you will recall, was the first person to realize that Poisson's abstractly derived distribution would model real-world situations. In the leftmost table below, we give Bortkiewicz' data (see Problem 3.8.7). Recall that we estimated λ, the average number of horsekick fatalities per corps-year, from Bortkiewicz' data. That gave $\lambda \approx 0.61$. In the table on the right, we give that Poisson distribution through $X = 4$.

B	CY
0	109
1	65
2	22
3	3
4	1
	200

X	P(X)
0	0.5434
1	0.3314
2	0.1011
3	0.0206
4	0.0031
	0.9996

(a) Calculate the expected number of corps-years for each of the five values of B, assuming the Poisson distribution with $\lambda = 0.61$.

(b) Evaluate Bortkiewicz' observed value of χ^2.

(c) Does the Poisson model seem to fit Bortkiewicz' data?

(d) Give a verbal description of the p-value calculated in part (b).

6.2.27 In 1865, at the February and March meetings of the Natural History Society of Brno, Gregor Mendel read his now famous paper on the transmission of genetic characteristics of garden peas from one generation to the next. In the table below, we present one set of Mendel's observations (see [Mendel] p. 23) together with the values which are to be expected in the light of his theory. Does this data seem consistent with his theory?

Type of pea	Actually observed	Theoretically expected
Smooth yellow	315	313
Wrinkled yellow	101	104
Smooth green	108	104
Wrinkled green	32	35

6.2.28 A test of "homogeneity." You believe that in each of three voting precincts the proportion of voters for your candidate is more or less the same. In other words, you believe the three precincts taken together are "homogeneous." Suppose for the three precincts you observed,

$$\hat{p}_1 = 253/587, \quad \hat{p}_2 = 127/319, \quad \hat{p}_3 = 296/647,$$

where $\hat{p}_i$ is the proportion of voters in precinct number i who support your candidate.

(a) Calculate the E_i's .

(b) Evaluate χ^2 for this data.

(c) Does the data challenge your hypothesis?

6.2.29 A test for "independence." You want to know if death within five years from a certain type of cancer is independent of gender. If the patient survives the first five years, she is considered "recovered." Consider the data:

	Recovered	Died within five years	
Male	127	17	144
Female	42	8	50
	169	25	194

The logic for this test follows the pattern of the goodness-of-fit test. Here's how you do it:

(**a**) Suppose that 84% of persons suffering from this type of cancer are still living after five years and that half are male and half female. This is our first approach to this data; the usual "test of independence" would not make these two assumptions [see part (f)]. Determine the expected number, E_i, for each category under the assumption that recovery is independent of gender.

(**b**) Evaluate Pearson's χ^2 statistic for part (a).

(**c**) Based on part (b), does the data suggest that death rate for this cancer is independent of gender?

(**d**) In part (b), what else does the data suggest beyond just independence?

(**e**) Suppose you had not known the recovery rate for this type of cancer. What would you do?

(**f**) Now we'll do the usual "test of independence," where the recovery rate and the proportion of male victims are estimated from the data.

Suppose the recovery rate for patients stricken with this type of cancer is not known and you suspect the population of victims is not divided 50/50 between men and women. Would the data still suggest that recovery is independent of gender?

(**g**) Does the data really seem to suggest, as we noted in part (f), that this type of cancer is dependent on gender?

6.2.30 In Problem 6.2.12, you tested "racial bias" for the *Connecticut* v. *Teal* case by testing the difference in pass rates for blacks and whites. A more complete analysis would test to see if passing the job promotion test seems to be independent of race. Do that test (after [Finkelstein and Levin]).

6.2.31 The χ^2 goodness-of-fit test cannot show that your data fits the hypothesized distribution. What can it show?

6.2.32 In Problem 6.2.23 we reported a study by the sociologist Phillip R. Kunz who found that only 8% of a sample of 747 persons had died in the three months prior to their birthday. In fact, his data was more complete than that; he also found that 46% died within three months after their birthday and 31% within the next three months. Now what do you think?

6.2.33 In Problem 3.6.16, you gave a theoretical model for the capture-recapture data of cottontail rabbits given by Edwards and Eberhardt.

(a) Does that data seem to fit the model?

(b) Edwards and Eberhardt also tried estimating p for the geometric model using the "maximum likelihood estimate" (MLE), an important technique for estimating unknown parameters. In many cases, MLE just gives the obvious estimate. For example, MLE for the mean of a normal or a Poisson model is $\overline{X}$. In the geometric model, because $\mu = 1/p$, you'd expect to estimate p by $1/\overline{X}$. Well, that's the maximum likelihood estimate! How well does Edwards and Eberhardt's data fit the geometric distribution if you use the MLE estimate for p?

6.3 Hypothesis Tests

Introduction

We turn now to the Neyman–Pearson hypothesis test. Before you begin this discussion, review Section 6.1 where we compared and contrasted tests of significance with hypothesis tests. You need a clear map of the territory before entering new jungles!

The prototypical example of an hypothesis test arises in industrial quality control. For instance, in a manufacturing process, it may be impossible or impractical to insist on absolutely no defectives. An hypothesis test can be set up to monitor the proportion of defective items. This situation arises for certain types of highly complex electronic components, to name just one example. Or again, in the manufacture of products which contain an undesirable contaminant—chemical compounds, for example—it may be impractical to strive for a zero level of contaminant. So you monitor the contaminant level. Such situations are legion.

Let's think about the first of these examples. Of course, any real-world situation will be more complex than our example; industrial quality control is not a matter of one simple hypothesis test! Because "zero defects" in the manufacturing process is often an unattainable goal, the proportion of defects must be monitored. Suppose you monitor the output on a daily basis by obtaining a random sample of that day's output. If the sample provides evidence of an unacceptably high proportion of defects, you will, let us say, stop production and take corrective action. So you require a decision procedure to obtain and analyze daily a sample of the day's output and to decide if corrective action is required. The hypothesis test provides that decision procedure.

To be specific, suppose the decision criterion is "with more than 1% defects, take corrective action." The criterion, note, is formulated in terms of a condition on the parameter p, the proportion of defective items from the day's output. The condition is $p > 0.01$. If the condition holds, corrective action is required.

There are a number of elements in this situation:

- TWO COURSES OF ACTIONS: "stop production and take corrective action" or "stay in production."
- TWO ERRORS, correspondingly, which may occur: "stop production unnecessarily" or "stay in production even though you're producing too many defects."
- a CONDITION ON THE PARAMETER, $p > 0.01$, which signals a problem calling for corrective action.
- a PERIODIC PROCEDURE of sampling day by day from that day's output, analyzing the data, and deciding between the two possible courses of action.

Of course, it's not enough to just say "two errors" are possible. We'll have to think about controlling those errors. But first let's take a moment for you to . . .

Try Your Hand

6.3.1 (**a**) Why would you ever make an error in the situation of the text above? Why would you "stop production unnecessarily" or "stay in production even though you're producing too many defects"?

(**b**) How might you control the error in part (a)?

(**c**) Although possible, it's not likely you would be misled into a wrong decision by an atypical sample. Why is it not likely?

Setting Up the Hypothesis Test

Before looking further at the possible errors, let's learn the standard format for setting up an hypothesis test along with some standard terminology and notation. In the example, one of the actions is routine ("stay in production"), the other exceptional ("take corrective action," if it's not exceptional, you'd better shut down production right now!). The point of the test is to flag that "exceptional" or alternative action. The condition on the parameter which flags the exceptional action is called the ***alternative hypothesis***, denoted H_A. In the example, you'll "stop production and take corrective action" only if there's evidence the proportion of defective items exceeds one percent. That condition on p is the alternative hypothesis: $p > 0.01$. The term "alternative hypothesis" could be confusing. It's "alternative" only because it flags an exceptional action. From the point of view of the test, the alternative is, in fact, the main hypothesis. After all, it's precisely that "exceptional action" that the test wants to flag.

The ***null hypothesis***, H_o, simply asserts that the parameter takes on the borderline value from the alternative hypothesis. It plays a purely logical role in the hypothesis test by giving us a value of the parameter to work with. By contrast, the alternative hypothesis plays a very practical role. As we'll see later, the analysis of the data will be carried out by assuming the null hypothesis. Here's the standard format:

$$H_o: \; p = 0.01 \qquad (p < 0.01 \text{ irrelevant}),$$

$$H_A: \; p > 0.01.$$

The test given here is called a ***right-tailed test*** because evidence for H_A would be a large value of the estimator $\hat{p}$, a value in the right tail of the distribution of the estimator. At the side of the hypotheses we note the irrelevant values of the parameter. In the example, fewer than one percent defective items is irrelevant. Certainly, you're not going to stop production because "not enough defectives" are being produced. That's absurd!

Note how you set up the hypotheses. First, you identify the exceptional action the test is trying to flag and write down the alternative hypothesis. It's the condition on your parameter which flags the exceptional (alternative) action. Then you write down the null hypothesis, following the standard format by writing the null hypothesis *above* the alternative. All other values of the parameter should be irrelevant. One caveat: We're not following the standard practice exactly. Standard practice confuses the logic of the test somewhat—excuse our saying so—by combining the irrelevant values of the parameter into the null hypothesis. We'll not do that. It will be much clearer if you follow the format given above which writes the irrelevant values out to the side. And of course, you should check that those values really are irrelevant.

There are other possibilities for the hypotheses. In "acceptance sampling" for quality control, you inspect a sample from an incoming shipment of items before you accept the shipment. If there's evidence from the sample that too many items are defective, you refuse to accept the shipment. By way of example, let's say you're receiving machine parts from a supplier under a contract that specifies a diameter of 3.2 mm for the parts. The contract may also specify that you will reject a shipment if a sample provides evidence that the mean diameter is too small. "Too small" might, for example, mean $\mu < 3.15$ mm. Then we set up the hypotheses as follows:

$$H_o: \mu = 3.15 \qquad (\mu > 3.15 \text{ irrelevant}),$$
$$H_A: \mu < 3.15.$$

This is call a ***left-tailed test*** because small values of the estimator will be evidence for H_A. Here, it's implicit in the statement of the problem that we're looking for small values of μ, large values are irrelevant. In another situation that might not be true.

We should note here that "acceptance sampling," although it illustrates well the logic of hypothesis tests, is much less common in modern quality control than formerly. It has been largely replaced by the more efficient (and more sophisticated) methods of *statistical process control*. Instead of waiting until the end of the process, after you've already produced a number of defective items, you monitor the process itself. When you see a problem developing, correct the situation BEFORE you produce any defects.

Finally, suppose the contract with your supplier says the diameters are to be 3.2 mm and you intend to reject a shipment if the average diameter is NOT 3.2 mm. Then your hypotheses are

$$H_o: \mu = 3.2,$$
$$H_A: \mu \neq 3.2.$$

This is a ***two-tailed test***. In real-world problems, two-tailed tests will not be common for the simple reason that you would usually contemplate a different action depending on whether $\mu < 3.2$ or $\mu > 3.2$. They are common for the *difference* of two parameters, if you're looking for evidence they are NOT EQUAL. Now please . . .

Try Your Hand

6.3.2 Think about the left-tailed test in the text just above.

(a) What kind of "evidence" would indicate that $\mu < 3.15$?

(b) The contract was written for a mean diameter of 3.2 mm. Why didn't we set the alternative hypothesis at $\mu < 3.2$ (instead of $\mu < 3.15$)?

(c) It's unrealistic that you would want to monitor only the mean diameter of incoming machine parts. Why?

(d) Instead of monitoring the mean diameter, you might want to monitor the diameters through a proportion. What proportion?

6.3.3 Suppose the contract with your parts supplier allows you to reject a shipment if there's evidence of too much uncertainty in the diameters of the parts. Let's explore this situation a bit.

(a) Assuming an hypothesis test is intended, the word "evidence" in the phrase ". . . evidence of too much uncertainty" refers to what?

(b) What parameter will you be monitoring?

(c) If the mean diameter is exactly 3.2 mm and the standard deviation exactly 0.025 mm, what proportion of the parts would be useless? Assume a part is useless if it has a diameter less than 3.15 mm.

(d) Redo part (c) assuming the standard deviation is 0.05.

(e) Suppose the contract allows for rejection of a shipment if there's evidence the standard deviation of diameters is greater than 0.025 mm. Set up the hypotheses for this test.

(f) Why is $\sigma^2 < 0.000625$ irrelevant in part (e)?

(g) Does a sample with $s^2 > 0.000625$ provide evidence to support H_A?

6.3.4 The two hypotheses of an hypothesis test do not play symmetric roles. The roles of the two hypotheses can each be characterized by one simple word—what are those two words?

6.3.5 Set up the hypotheses for each of the situations in parts (a)–(d). Keep your notes for this problem, we'll return to it in several other problems.

(a) Suppose that any machine part from your manufacturing process with a diameter less than 3.15 mm is entirely useless, but otherwise the part is functional. You want to monitor the process to avoid too many useless parts. Assume you're willing to discard up to one in 50 parts.

(b) You are campaign consultant for a presidential candidate. One particular district seems certain, with some 84% of voters supporting your candidate. Nevertheless, over the coming year before the election you will monitor support in that district on a monthly basis. If it ever appears that support in that district has fallen more than three percentage points below the present level, you will step up the campaign effort in that district.

(c) You're monitoring incoming shipments of chain links as they arrive each month from your supplier. The contract with your supplier guarantees a mean length for the links of at least 1.2 cm and permits you to reject a shipment if on the basis of a random sample it appears not to meet specifications.

(d) The chains which you're manufacturing in part (c) contain 92 links. You want to write a contract with your supplier of chain links that guarantees no more than one percent of the chains vary in length by more than a centimeter. So in the contract with the supplier you've guaranteed that the standard deviation for the lengths of chains is at most one-third of a centimeter. Why "one-third"? Set up your hypothesis test to reject a shipment if it appears on the basis of a random sample not to meet specifications.

(e) In part (d), the contract will not contain an "at most one-third of a centimeter" condition. What condition will it contain?

The Possible Errors

Now let's take a more careful look at the possible errors involved in an hypothesis test. For any real-world context, you'll find a bewildering array of possible errors—human error, machine error, careless error, intentional error (sabotage). Customer errors, billing errors, specification errors. Errors of omission, errors of commission. Some trivial, some serious. Some recognized, some not. Cruel errors, funny errors . . . !

But from a purely formal point of view, from the point of view of the logic of an hypothesis test, there are only two possible errors, called ***type I error*** and ***type II error***, as displayed in the table below. The table shows the four combinations of possible actions given the possible "states of the world." There are two "states of the world," either H_A is true or it's false . . .

		State of the world (unknown to us)	
		H_A false ("H_o true")	H_A true
Our decision (based on the data)	Act on H_o	Correct decision	Type II error
	Act on H_A	Type I error	Correct decision

We use the term "act on H_o" as shorthand to mean "take the routine action" as opposed to the alternative "exceptional" action. Similarly, "act on H_A" means "take the alternative action." Note that this table is partly theoretical. After all, in real life we never know the true state of the world. If we could *know* that, there would be no need of a statistical analysis based on such partial and possibly misleading information as a random sample!

Note the phrase "H_o is true." It has two very distinct meanings, one logical and the other practical. Think of our original example with H_o : $p = 0.01$ and $H_A : p > 0.01$. Logically, H_o means p EQUALS one percent; it means we have EXACTLY one percent defectives. But from a practical point of view, p won't be exactly one percent. It's either more or less. If it's more, H_A is true. If it's less, p takes one of the "irrelevant" values written out to the side of H_o and we say—in its *practical* sense—that "H_o is true". Those values are irrelevant only from the point of view of the *logic* of the test; we do not claim they are irrelevant beyond the test itself. Note how the logical meaning for "H_o is true" is the worst case of the practical meaning. Taken together, they're the negation of H_A. In other words, H_A says p is greater than 0.01, H_o says it isn't.

Let's summarize the two types of error as described in the table. Type I error is wrongly acting on H_A—acting on H_A when, in fact, H_o is true (although you didn't know that). Type II error is wrongly acting on H_o—acting on H_o when H_A is true. These errors are conditional events, conditioned on the unknown state of the world. This understanding is important for interpreting the probability of error; it's a *conditional* probability. Recall Problem 6.1.3 and you'll see the difference. The conditional probabilities for the two types of error are denoted respectively by α and β. In the example,

$$\begin{aligned} \alpha &= \text{P(type I error)} \\ &= \text{P(act on } H_A | H_o \text{ is true)} \\ &= \text{P(act on } H_A | p = 0.01), \end{aligned}$$

$$\begin{aligned} \beta &= \text{P(type II error)} \\ &= \text{P(act on H}_\text{o}|\text{H}_\text{A}\text{ is true)} \\ &= \text{P(act on H}_\text{o}|p = ?). \end{aligned}$$

Here the lack of symmetry of the hypotheses surfaces again (see Problem 6.3.4). Note that α is completely specified, but β is not. For α, we have a specific value for p given by the null hypothesis (in its logical meaning). On the other hand, β remains indeterminant because the alternative hypothesis does NOT give a specific value for p.

Because the null hypothesis gives a precise value of p, we can control type I error by specifying α in advance. This means we're specifying an acceptable probability of such error. We'll see later how it's possible to do that. This is analogous to control of error in a confidence interval problem by specifying the confidence coefficient in advance. There, even though the interval may NOT contain the parameter, we believe it does. If it doesn't, we're in error. We "control" this error because we avoid it 95% of the time (assuming a confidence coefficient of 95%).

By contrast, we cannot control type II error so easily because we can't compute β exactly. That would require a specific value of p which we don't have because now it's H_A that's true and H_A does NOT give a specific value for the parameter. For most of our discussion in the rest of this chapter, we take the point of view that type II error is not controlled at all. At the end of the chapter we'll see how, in fact, it's possible to exercise some rather hypothetical control over type II error. Hypothetical, yes, but important from a theoretical point of view.

So the testing procedure is more conclusive when it decides in favor of the alternative hypothesis. In that case, the probability of error is completely under our control. We cannot eliminate that error, but we can specify its probability in advance. This is why the alternative hypothesis is the principal hypothesis from a practical point of view. It's important to bear this in mind when setting up an hypothesis test and in interpreting the results of the test. We'll comment further on this in the text below and in the exercises. In view of this privileged position of the alternative hypothesis, the hypothesis test is thought of as "trying" to decide in favor of H_A.

This discussion, remembering that type I error is the error we control, suggests a second method of determining the direction of the test (left, right, or two tailed). Obviously, if one error is more serious than the other, the test should control for the serious error. For that to happen, the serious error must be type I error. So you choose the alternative hypothesis to force the more serious of the two errors to be type I error. This is a different way of thinking about the hypothesis test. Now it's not a matter of focusing on an "alternative action" to be flagged by the data; it's rather a question of setting up a decision procedure which controls for a particular error.

Finally, two related technical terms: The ***significance level*** of the test is the probability we're misled by the data to act on H_A even though it's false. Note that the significance level is just α, the probability of type I error. Significance levels of 1%, 5%, and 10% are considered standard, although we'll ask you to set up tests with other significance levels to assure that you master the procedure.

There is also a special name for $1 - \beta$. It's called the ***power of the test***. It is NOT the probability of an error; it's the probability of a correct decision. The hypothesis test is trying to decide in favor of H_A, so the power of the test is the probability to succeed in what you were trying to do. It's the probability the data *correctly* leads you to act on H_A:

$$\begin{aligned} 1 - \beta &= \mathrm{P}(\text{act on } \mathrm{H_A} | \mathrm{H_A} \text{ is true}) \\ &= \mathrm{P}(\text{the test “succeeds”}). \end{aligned}$$

Because we do not, for now, exercise control over type II error—over β—we also do not exercise control over the power.

Now you're ready to . . .

Try Your Hand

6.3.6 Let's explore a bit.

(a) Why must the alternative hypothesis be the "principal" hypothesis from a practical point of view?

(b) At what point in the monitoring procedure do we determine the significance level?

(c) Why is the term "power of the test" very reasonable heuristically speaking? That is, why does it make sense from a practical, real-world point of view?

(d) What is an hypothesis test "trying" to do?

(e) By contrast, what is a test of significance "trying" to do?

(f) The table which defines the types of error at the beginning of this subsection is unrealistic. What's unrealistic about that table?

(g) Why, from a practical point of view, should the proportion of defective items from a production process be either more than one percent or less, but not EQUAL to one percent?

(h) In speaking about the "irrelevant" values of p, we said "those values are irrelevant only from the point of view of the *logic* of the test, we do not claim they are irrelevant beyond the test itself." Explain this.

(i) Which decision is more conclusive, "act on H_o" or "act on H_A"?

(j) What's wrong with this: $H_A : \hat{p} > 0.01$?

6.3.7 Let's think again about how to set up an hypothesis test.

(a) Given a real-world problem, how do you determine the direction of an hypothesis test?

(b) Suppose you are testing a new medical procedure where you have a choice between the two errors: "using the new procedure when it's not really more effective than the old" and "using the old procedure when, in fact, the new procedure is more effective." Describe a situation for which the second of these errors would probably be considered more crucial than the first.

(c) Suppose under the old treatment for a very serious disease the recovery rate is one in 1000. Set up the hypotheses for testing a new treatment to determine if it should be accepted for use. Suppose there is little risk associated with the new treatment.

6.3.8 Let's return to the original quality control example, where we were monitoring the proportion of defects, with $H_A : p > 0.01$. If we "control error" by specifying in advance a five percent chance of type I error, then we can expect about five times out of 100 to halt production for corrective action when it was not really necessary. True or false?

Real-World Interpretation of the Conclusions and Errors

As we've seen, there are two possible conclusions for an hypothesis test. Formally, either you ***reject the null hypothesis*** ("reject H_o") or you ***fail to reject the null hypothesis*** ("fail to reject H_o"). In any real-world context, the meaning of each of these two conclusions and of the corresponding errors must be clear. For the quality control example, we had these hypotheses:

$H_o : p = 0.01$ ($p < 0.01$ irrelevant)

$H_A : p > 0.01$ p is the proportion of defective items from our production line

Assuming a 5% significance level, here are the possible conclusions and errors in their real-world interpretations:

Reject H_o: Our evidence suggests quality control has weakened, stop production and take corrective action.

Fail to reject H_o: The test is inconclusive, there's no evidence that quality control has weakened, stay in production.

Type I error: With a 5% risk of error, we stop production unnecessarily. The proportion of defective items does not exceed our criterion even though we believe it does based on misleading evidence.

Type II error: With an unknown risk of error, we continue in production when, although we didn't know it, we're producing more defective items than our quality control criterion permits.

Note how the formal conclusions lead to action in the real world. "Reject H_o" means "act on H_A," "fail to reject H_o" means "act on H_o."

Why does "failure to reject H_o" mean the test is inconclusive? Well, we don't control the error (type II error) in that case. Because the error is not controlled—because the probability of that error could be quite large—we should draw no conclusion. But it's not that WE are inconclusive! Even when the test is inconclusive, some action is required. In the real world, you're always constrained to take action in some form or other. Because the hypothesis test is inconclusive, our action is not based on the test but rather on some other, prior information.

In the example, if there's no evidence of poor quality, you "stay in production." But you've not proven quality is in control, it's just that if there's a problem, you don't know about it! Your action ("stay in production") is based on the fact ("prior information") that you have a production process which is well designed, free of problems, and run by well-trained workers. With no evidence to the contrary, it makes sense to let the process go forward.

Now look at the acceptance sampling example:

$$H_o : \mu = 3.15 \quad (\mu > 3.15 \text{ irrelevant}),$$

$$H_A : \mu < 3.15 \quad \mu \text{ is the mean diameter of parts in the present shipment.}$$

Reject H_o: It appears the machine parts we just received from our supplier are unacceptably small (mean diameter less than 3.15 mm). There's a five percent chance for error here.

Fail to reject H_0: The test is inconclusive. There's no evidence of a problem with the present shipment of parts. Accept the shipment.

Type I error: There's a 5% risk that we reject this shipment of parts even though, contrary to our misleading evidence, the mean diameter is not below 3.15 mm.

Type II error: We accept this shipment of parts even though on average they are too small. We found no evidence of this problem. There's an unknown risk of this error.

Try Your Hand

6.3.9 For the situations of Problem 6.3.5(a)–(d), you have already given the parameters and hypotheses. Now, in real-world terms with as much detail as the problem allows, state the possible conclusions and errors for those situations.

6.3.10 Contrast the formal meaning and the real-world meaning for

(a) the two possible conclusions for an hypothesis test;

(b) the two possible errors;

(c) the statement "H_0 is true";

(d) the statement "H_A is true";

(e) the statement "the test is inconclusive."

6.3.11 In the acceptance sampling example, you're checking only the average diameter. That says nothing about individual parts and yet it's the individual parts you finally have to work with. Doesn't that leave open the possibility that an individual part will be much too big or much too small?

Moving in the Direction of Commmon Practice

It is very common in actual practice to use an hypothesis test for a one-time decision. So then we don't have a monitoring procedure at all and there's no question of an "error rate" among many decisions, some of which may be wrong. The justification for the error analysis is to think of what WOULD happen if the decision were made many times.

In such a context, "control of error" must be based on a willingness to consider the long-term odds as a relevant guide in a one-time situation. Is it? Suppose you're going to place a bet on one toss of a biased coin.

You'll never see this coin again, no bets at a later date. Would you consider it relevant to know that the coin came up heads 824 times on the last 1000 tosses? Some would say no—after all, either it comes up heads on your one toss or it does not. Others would say yes. They would say the past history of this coin seems to suggest about an 80% chance of heads. But there's a problem. The word "chance" referring to your one bet does not refer to a theoretical relative frequency. You're using some other definition of probability. What definition? At this point, we turn you over to your local philosopher! A subjective interpretation of probability is often used in such problems.

Well, we're not among the "some," we're among the "others." In other words, from here on we'll assume the long-run odds are informative even for a one-time decision. To see some examples please . . .

Try Your Hand

In these problems, identify the parameter and its estimator in real-world terms, set up the hypotheses, and then give the real-world conclusions and errors. To identify H_A, maybe you can recast the problem into the form "follow routine action unless alternative action seems appropriate." That would pick out H_A. Or maybe you can identify H_A by seeing what type I error must be.

6.3.12 The purchasing department of your company must decide whether to continue with the same supplier for a certain machine part or possibly to switch to a new supplier. After some questioning, you find they're prepared to stay with the present supplier in the absence of strong evidence to back the new supplier's advertising claim that their part has a mean life of greater than 310 hours. The present supplier provides process control documentation showing a mean life of 285 hours for their part. You suggest to the purchasing department that they might want to control the risk of unnecessarily switching to a new supplier. Together, you decide to allow a one percent risk.

6.3.13 You are campaign manager for one of the mayoral candidates in the upcoming election. Your candidate is under pressure by her supporters to launch an expensive series of television spots. They claim that more than 30% of the voters would see the spots. You have decided to go ahead with the series of spots, but you want to do a preliminary study at the five percent significance level to see if there is evidence that fewer than 25% of the registered voters will see the spots, in which case you will not launch the series.

6.3.14 You are campaign manager for one of the mayoral candidates in the upcoming election. Your candidate is under pressure by her supporters to launch an expensive series of television spots. They claim that more than 30% of the voters would see the spots. You have decided to do a preliminary study. You decide to take a sample of the registered voters and allow no more than a five percent risk of launching this series if fewer than 25% of the registered voters will see the spots.

6.3.15 By taking appropriate random samples, you want to see if the students at your school have higher SAT scores than at Bad U. If your study reveals a significant difference, that is, a difference of more than 15 points, you will publish the good news in the school newspaper. Allow a ten percent risk of error.

6.3.16 By taking appropriate random samples, you want to see if the students at your school have higher SAT scores by at least 15 points than at Bad U. You have decided to allow a ten percent risk of missing a chance to boast.

6.3.17 By taking appropriate random samples, you are going to attempt to determine if the students at your school have higher SAT scores by at least 15 points than at Bad U. You have decided to allow a ten percent risk of falsely publishing your school's superiority.

6.3.18 The Internal Revenue Service is considering auditing all income tax returns which show a certain set of common characteristics on the assumption that those characteristics in combination flag an attempt at tax evasion. They want to test this hypothesis and want to run a risk of no more than one percent of implementing this policy when the proportion of returns which indeed reveal attempts at evasion is less than ten percent.

6.3.19 The Internal Revenue Service is considering auditing all income tax returns which show a certain set of common characteristics on the assumption that those characteristics in combination flag an attempt at tax evasion. They have decided not to implement this new policy unless there is evidence at the ten percent significance level that more than 20% of such returns reveal attempts at evasion.

6.3.20 We need to maintain strict control over the variability of the thickness of the coating on the coated paper which we manufacture. Our criterion for the variability is that the variance of the thickness should not exceed 0.3 mm. You are requested to set up a quality control procedure for this situation with five percent significance level.

The Rejection Region

You've learned how to set up an hypothesis test and how to interpret the possible conclusions and errors. Now it's time to see exactly how you decide—based on the data—whether to reject H_o or not. To see this, we take up once more our quality control example:

$$H_o : p = 0.01 \quad (p < 0.01 \text{ irrelevant}),$$

$$H_A : p > 0.01 \quad p \text{ is the proportion of defective parts from the production line.}$$

The probability of type I error is α, the significance level of the test. Suppose we've specified it to be five percent. Here's the picture for $\hat{p}$:

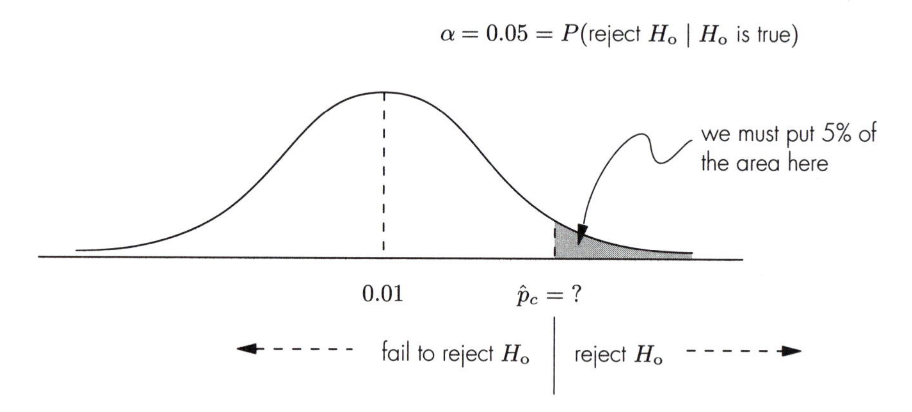

The cutoff value of $\hat{p}$ which separates "fail to reject H_o" from "reject H_o" is called the ***critical*** $\boldsymbol{\hat{p}}$, denoted $\boldsymbol{\hat{p}_c}$. When $\hat{p} > \hat{p}_c$, we reject H_o, otherwise we fail to reject H_o. Those $\hat{p}$ values which reject H_o are called the ***rejection region*** for the test,

$$\text{rejection region} : \quad \{\hat{p} \mid \hat{p} > \hat{p}_c\}.$$

You determine the specific value of $\hat{p}_c$ this way: Because $Z = 1.645$ puts five percent of the area into the right tail of the Z distribution, $\hat{p}_c$ must standardize to 1.645:

$$1.645 = \frac{\hat{p}_c - 0.01}{\text{s.e.}}.$$

The standard error for $\hat{p}$ is calculated using $p = 0.01$ because we assume the null hypothesis true. To complete the calculation we need to know the value of n. Suppose we have been given $n = 700$. Solving the equation above, we find $\hat{p}_c = 0.0162$, and so the rejection region is

$$\text{rejection region}: \quad \{\hat{p} \mid \hat{p} > 0.0162\}.$$

Note how we guarantee $\alpha = 5\%$. We guarantee a 5% chance to reject H_o given that it's true. First, we assume the null hypothesis true. Then the condition "given H_o true" is seen in the picture this way:

- the picture is centered on p because $\hat{p}$ is unbiased,
- $p = 0.01$ since we assume H_o true.

This is the logical role of the null hypothesis in an hypothesis test. It provides a specific value of the parameter and so a specific center for the distribution of the estimator because the estimator is unbiased.

The Decision Rule and Test Statistic

In an hypothesis test, the "decision rule" is for someone like a production line supervisor. For the quality control example above, the decision rule would be

> On any day in which more than eleven items in the sample are defective, stop production and take corrective action. [Hint: 1.62% of 700.]

In general, the decision rule takes the form

> If the value of the estimator computed from the sample falls in the rejection region, then act on H_A.

Sometimes it's convenient, as we just did, to express the decision rule in terms of the estimator. But sometimes it's more conveniently expressed in terms of the ***test statistic***, the computed value of Z derived from our sample. In the quality control example, the test statistic is this value of Z:

$$\frac{\hat{p} - 0.01}{s.e.}.$$

For a mean, the test statistic would be

$$\frac{\overline{X} - \mu_0}{\sigma/\sqrt{n}} = Z \quad \text{or} \quad \frac{\overline{X} - \mu_0}{s/\sqrt{n}} = t.$$

Here, μ_0 is the value of μ specified in the null hypothesis. For a variance, the test statistic would be

$$\frac{(n-1)s^2}{\sigma_0^2} = \chi^2,$$

where σ_0^2 is the value of the variance given in the null hypothesis.

In an hypothesis test for μ, if we must estimate σ by s, the rejection region will change from one sample to another because the standard error now depends on s. In this case, you may want to express the rejection region and associated decision rule in terms of the test statistic instead of the sample mean. If the test statistic is t,

$$\text{rejection region} = \{t \mid t > t_c\},$$

where t_c is the value of t which puts α into the right tail of the t-distribution with $n-1$ degrees of freedom. If $\alpha = 0.05$ and $n = 10$, say, then $t_c = 1.8331$. In this case, the decision rule will be:

From the sample data calculate

$$\frac{\overline{X} - \mu_0}{s/\sqrt{n}} = t,$$

If it's greater than 1.8331, act on H_A.

So the rejection region and the decision rule can be given in two ways, either in terms of the value of the estimator, or in terms of the test statistic and its standard model (Z, t or χ^2). In more advanced courses, you would see still other models for the test statistic.

Now please . . .

Try Your Hand

6.3.21 In each of the situations of Problem 6.3.5, set up the rejection region and state the decision rule. Do this first with a 5% significance level and then with a 10% significance level. Save your results for the continuation in later problems.

(a) Each week, you'll be taking samples of 300 parts.

(b) Each month, you'll interview 50 registered voters randomly chosen from the registration list for that district.

(c) You'll take a sample of ten links each month from the incoming shipment.

(d) Follow the same procedure as in part (c).

6.3.22 Here we provide periodically gathered data for Problem 6.3.5. What conclusion would you draw in each case for the 5% significance level test of that problem?

(a) In each of eight successive weeks, you observed the following number of useless parts: 2, 11, 0, 0, 1, 14, 3, 7.

(b) Among the 50 voters interviewed in each of eight successive months you found the following number who support your candidate: 41, 31, 44, 42, 38, 27, 29, 46.

c) In each of eight successive months, you observed the following results for the ten chain links which you sampled from that month's shipment:

$\overline{X}$	1.28	1.24	1.17	1.26	1.22	1.02	1.11	1.21
s^2	0.09	0.11	0.10	0.09	0.08	0.09	0.09	0.10

(d) Same data as part (c).

6.3.23 For part (c) of the previous problem:

(a) Suppose in one month you had observed a sample mean of 1.02 cm with $s^2 = 0.22$. What conclusion would you have drawn?

(b) From a *practical* point of view, what's the problem with the shipment described in part (a)?

Now we continue Problems 6.3.13 through 6.3.20 by providing data. The problems below are numbered c6.3.13, and so on to show that we're continuing previous problems. You'll find the hypotheses for each situation stated in the level I answers to the original problems.

For each problem, you're given a summary of several sets of data. In Problem c6.3.13(a), for example, 43 of the 130 registered voters would see your television spot. In part (b), 30 of the 130 registered voters would see the spot.

For each problem, first determine the rejection region and state the decision rule in real-world terms. Then using the given data, determine for each part of each problem whether you would reject H_0. Finally, recall what that conclusion means in real-world terms.

c6.3.12 You obtain a sample of 15 parts from the new supplier for which the mean and standard deviation, respectively, are

(**a**) 287, 35; (**b**) 326, 27; (**c**) 326, 21; (**d**) 352, 22; (**e**) 364, 82.

(**f**) What assumption are you making in this problem? Is it reasonable?

c6.3.13 You interview a randomly selected sample of 130 registered voters and find that the number who would see your television spot is

(**a**) 43; (**b**) 30; (**c**) 27; (**d**) 22; (**e**) 16; (**f**) 11; (**g**) 8.

c6.3.14 Use the same data as in Problem c6.3.13.

c6.3.15 Suppose the standard deviation for SAT scores at your school is 157 and at Bad U, 112. You obtain the SAT scores of 42 randomly selected students from your school and 31 randomly selected Bad U students from the Student Records Office of each school and find the mean score of the samples from your school and from Bad U to be respectively

(**a**) 1116, 1143; (**b**) 1132, 1108; (**c**) 1148, 1093; (**d**) 1152, 1015.

c6.3.16 Use the same data as in Problem c6.3.15.

c6.3.17 Use the same data as in Problem c6.3.15.

c6.3.18 The Internal Revenue Service obtains a random sample of 1215 tax returns and finds that the number which show evidence of attempts at tax evasion is

(**a**) 280; (**b**) 260; (**c**) 215; (**d**) 170; (**e**) 145; (**f**) 130.

c6.3.19 Use the same data as in Problem c6.3.18.

c6.3.20 Suppose on a weekly basis you measure the thickness of each of a sample of seven pieces of paper from your manufacturing process. Here are the observed standard deviations for the thicknesses:

(**a**) 0.28; (**b**) 0.54; (**c**) 0.98; (**d**) 1.05.

6.3.24 Aside from the parameter, Problem 6.3.20 is different logically from Problems 6.3.12 through 6.3.19 in another important way. What is it? What justifies that difference for the other problems?

p-values for Hypothesis Tests

Now that you're clear about the logic of tests of significance and hypothesis tests, let's look at the question which we have alluded to so often: How do those two logically distinct procedures come to be combined in actual practice?

It's just a question of computing p-values for an hypothesis test, and that's easy to do. Let's look again at the quality control example. Suppose we obtain a sample for which $\hat{p}$ is 0.0391. Because this is bigger than the critical $\hat{p}$ of 0.0162, we will "reject H_o." The picture of the sampling distribution is

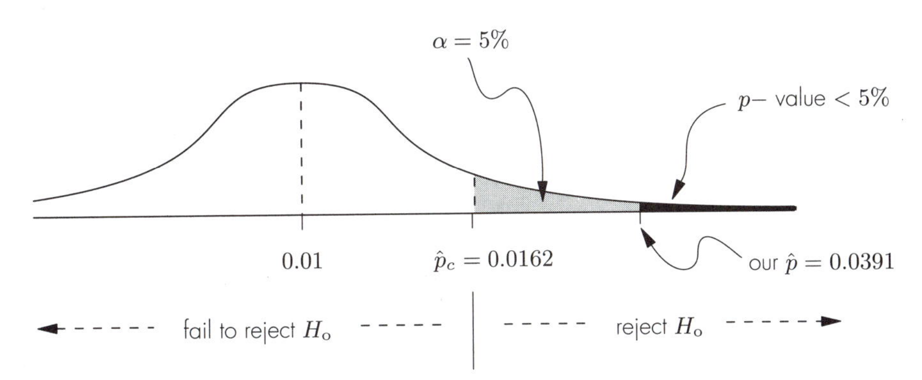

Note that the p-value for our observed 0.0391 is going to be less than five percent because $\hat{p}$ falls in the rejection region. If it did not fall in the rejection region, the p-value would be *greater* than five percent. So the p-value provides a new way of expressing the decision rule: "If the p-value is less than α, reject H_o." But this works only if the observed $\hat{p}$ falls in the right tail! On those days when you get a $\hat{p}$ less than one percent, you do not need to compute a p-value at all—it's obvious that such data does not support H_A!

So we get . . .

THE p-VALUE DECISION RULE (for a one-tailed test): If the data does not fall in the wrong tail, compute the p-value. If this computed p-value is less than α, reject H_o.

Here, the "wrong tail" is the left tail of a right-tailed test or the right tail of a left-tailed test.

Finally, if you're doing a two-tailed test, you compare the p-value with $\alpha/2$ instead of α because only half of α is in the tail with your observed data. However, for two-tailed tests, it's customary to report a ***two-sided***

***p*-value**, the usual p-value multiplied by two, and compare that with α. So we get

> THE p-VALUE DECISION RULE (for a two-tailed test): Compute the two-sided p-value. If this computed p-value is less that α, reject H_o.

Once more, please . . .

Try Your Hand

6.3.25 (a) For the data in Problems c6.3.13–c6.3.20, calculate p-values and note how the p-value decision rule obtains the same conclusion without reference to the rejection region. Well, OK, you don't have to do it for ALL that data—just do enough to see how the p-value decision rule works!

(b) When is the rejection region and its decision rule more appropriate than the p-value decision rule?

Controlling Power and Type II Error

The power of an hypothesis test is the probability your data correctly leads you to act on H_A:

$$\begin{aligned} 1-\beta \;=\; \text{the power} &= P(\text{act on } H_A | H_A \text{ is true}) \\ &= P(\text{the test "succeeds"}). \end{aligned}$$

This concept has both practical and theoretical importance. From a theoretical point of view, the power is an important criterion for choosing among possible tests. There's an elaborate theory of optimal tests in mathematical statistics that attempts to identify, for example, "uniformly most powerful" tests.

To see the practical importance of the concept of "power" for a test and to see how we actually work with it, recall our original quality control example:

$$\begin{aligned} &H_o : p = 0.01 \quad (p < 0.01 \text{ irrelevant}) \\ &H_A : p > 0.01. \end{aligned}$$

Here, the power is P(act on $H_A \mid p > 0.01$), the probability we succeed in catching a bad situation. It's the probability we stop production and take corrective action when we really were producing too many defectives.

In the quality control example, the rejection region, assuming a 5% significance level, is $\{\hat{p} \mid \hat{p} > 0.0162\}$. Because we "act on H_A" exactly when the data gives a value of $\hat{p}$ in the rejection region, the power for this test is

$$1 - \beta \;=\; P(\hat{p} > 0.0162 \mid p > 0.01).$$

Note that we don't know the value of p here; we only know it's greater than one percent. To sketch a picture for the power, we need to pick a value of p. Here's a picture centered on a value of p greater than the critical 0.0162:

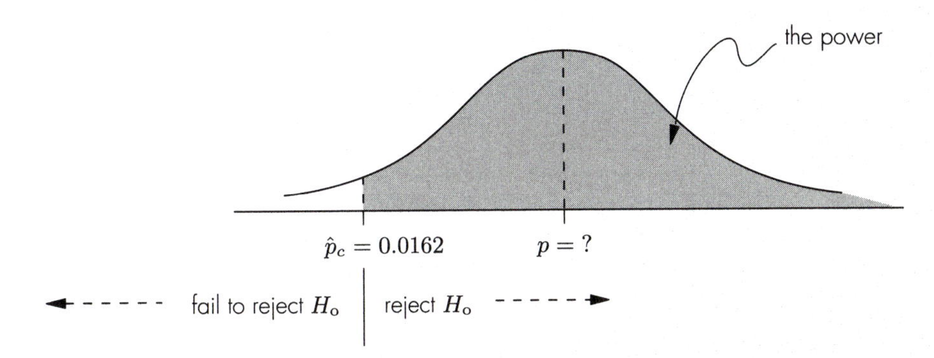

In this picture, clearly the power is greater than 50%. That's because we centered the picture on a value of p in the rejection region. Of course, it could be that p is only slightly bigger than 0.01, not big enough to be in the rejection region. In that case, the power is less than 50%:

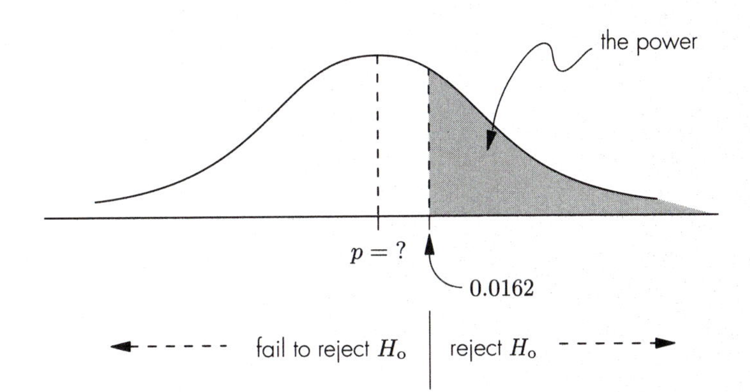

Now let's compare the picture for the power with the picture we used to determine the rejection region. To determine the rejection region, we assumed H_o true, centering the picture at $p = 0.01$. Here are the two pictures, superimposed. The dotted picture is the one from which we determined the rejection region:

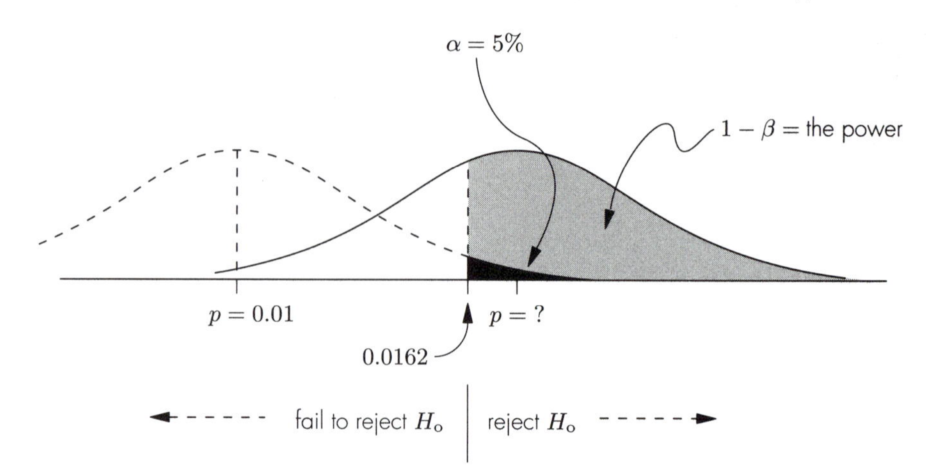

The difficulty in controlling the power—or equivalently, controlling type II error—is that H_A, unlike H_o, does NOT give a specific value to work with for the parameter. So we control the power only hypothetically by *supposing* values for the parameter. Suppose $p = 0.04$ in the quality control example, then with $n = 700$, the power is

$$\begin{aligned} P(\hat{p} > 0.0162 \mid p = 0.04) &= P(Z > -3.21) \\ &= 0.9993. \end{aligned}$$

That's fairly strong power! If p is 0.025, the power is reduced to about 93%. Well, you should do some power calculations. Please just . . .

Try Your Hand

6.3.26 You're interested in the "power" a test has to flag a situation in which H_A is true. For the quality control example in the text above, compute the power for the following values of p. Note that we've only chosen values of p for which H_A is true.

(**a**) $p = 0.015$; (**b**) $p = 0.02$; (**c**) $p = 0.035$; (**d**) $p = 0.01$;

(**e**) $p = 1$.

6.3.27 (**a**) In Problem 6.3.26, why did we only chose values of p for which H_A is true?

(**b**) Because it's the probability the test "succeeds," the power should certainly be reasonably large. In the quality control example discussed above, what's the smallest value of p which would give a power of at least 75%? Give your conclusion in real-world terms.

6.3.28 Calculate the power for each of the tests in Problem 6.3.5 (continued in Problem 6.3.21) at the following values of the parameter:

(a) $p = 0.025, p = 0.035, p = 0.05$ (recall: $n = 300$ and $\hat{p}_c = 0.0333$);

(b) $p = 0.75, p = 0.65$ (recall: $n = 50$ and $\hat{p}_c = 0.7187$);

(c) $\mu = 1.1, \mu = 1.0$ assume $\sigma^2 = 0.1$ (recall: $n = 10$ and $\overline{X}_c = 1.0167$);

(d) $\sigma^2 = 0.0065$ (recall: $n = 10$ and $s_c^2 = (0.0475^2) = 0.0023$).

The power of an hypothesis test depends on the true value of the parameter. But we don't know that true value; it could be any of the values encompassed by H_A. So we can plot the graph of the power. The graph of the power as a function of p is called the ***operating characteristic curve.***

For the quality control example, the power can be computed for any value of p above one percent. We've already computed the following values:

p	0.01	0.015	0.02	0.025	0.035	0.04
$1 - \beta$	5%	39.74%	76.42%	93%	99.66%	99.93%

This information gives the following operating characteristic curve for our quality control example:

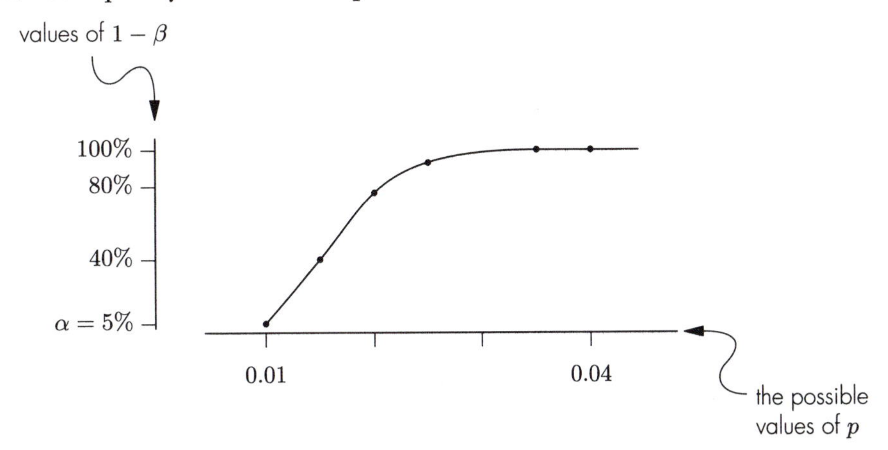

Note that the power is not defined for values less than one percent because then H_A is no longer true. It's clear this test has excellent power to detect a defect rate of 2.5% or more. Below that, the power weakens rapidly. (You can control this situation, if it's a cause for concern, by rethinking the design of your test (increase n).)

Considerations of power play an important role in the design of a statistical study. There's much to be said on this topic, mostly far beyond the scope of this text. But one simple fact is clear: a LARGER SAMPLE will narrow the distribution of the estimator, clearly giving greater power. Look again at the pictures of power for our quality control example:

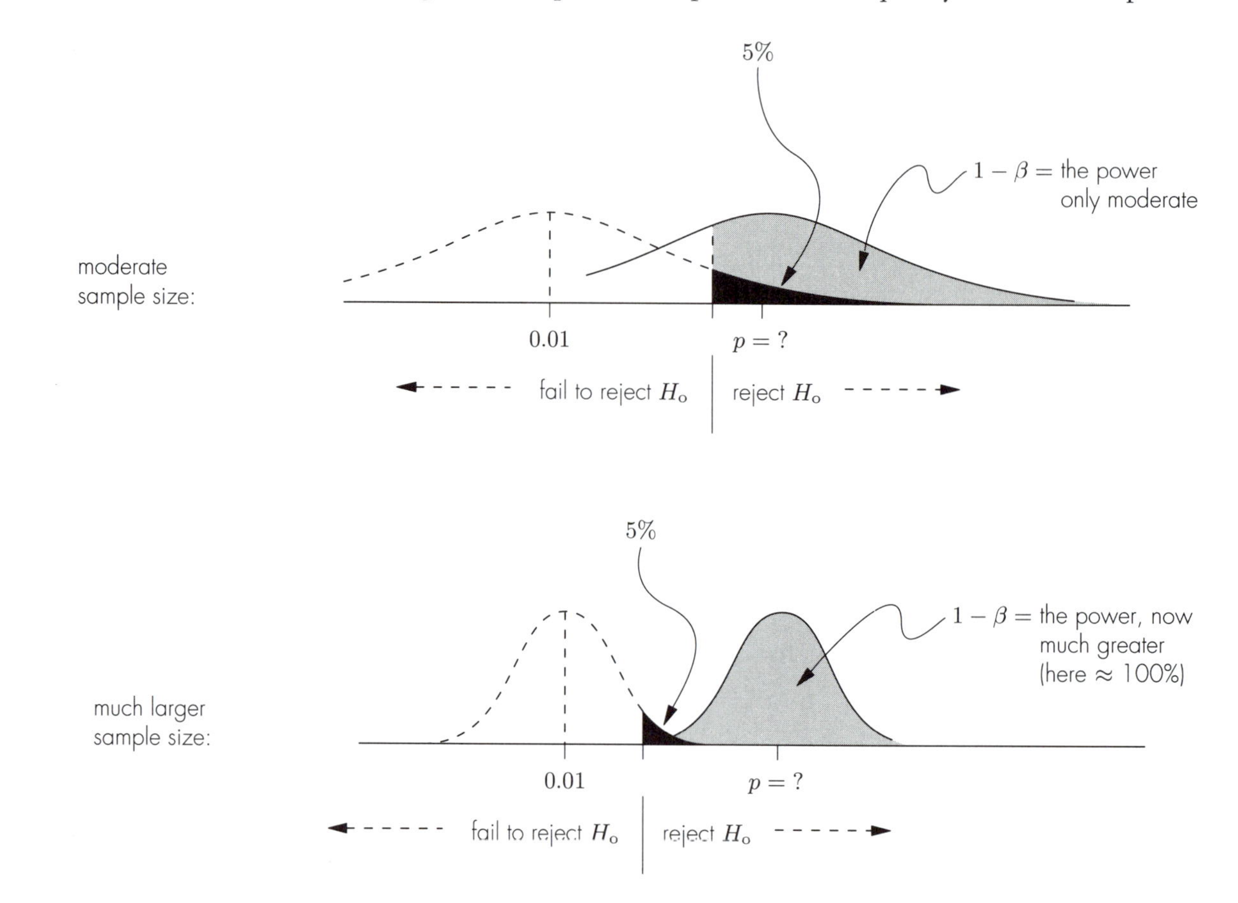

To see how this permits some control of power, you should . . .

Try Your Hand

6.3.29 Sketch the operating characteristic curve for each of the tests in Problem 6.3.5.

6.3.30 In the quality control example, suppose you want to guarantee at least a 60% chance to detect that you are producing too many defectives once the proportion of defectives has reached as much as one and a half percent. To achieve this without increasing the sample size beyond what is practical, you're willing to relax the control of type I error by setting the significance level at 10%.

(a) What sample size is required for your monitoring procedure?

(b) What is the decision rule now?

6.4 A Somewhat Comprehensive Review

Before attempting these review problems, go back and study the types of question which you've learned to deal with. It's not enough to study the solutions to problems—it's the QUESTIONS you must study. Note the distinction between the question and the information on the basis of which you will answer that question. The same question may require several distinct types of answer depending on the available information. The first problem below illustrates this very concretely.

6.4.1 *Question*: How many cups from our drink machine will overflow in one day of operation? Answer this on the basis of the following:

(a) *Information*: For X = fill in ounces per cup for this machine, $\mu = 7$ and $\sigma = 0.12$. A cup overflows at 7.3 ounces or more.

(b) *Information*: Let X = fill in ounces per cup for this machine. A cup overflows at 7.3 ounces or more. You observed

X	6.9	7.0	7.1	7.2	7.3	7.4	7.5
f	4	9	14	11	7	3	1

(c) *Information*: For Y = # cups overflowing in one day, you observed

Y	0	1	2	3	4	5
# days	17	8	5	2	1	1

(d) With less than 30 observations in part (c), you could not answer the question. Why not?

6.4.2 How much does the baggage in the storage compartment of today's flight weigh? There are 118 bags stored on today's flight. Stored baggage for our airline weighs 37.2 pounds per bag with a standard deviation of 8.3 pounds.

6.4.3 How much does a stored bag for one of our airline's passengers weigh? There are 214 bags in the storage compartment for today's flight weighing a total of 6425 pounds with a standard deviation per bag of 5.8 pounds.

6.4.4 How likely is the typical bag stored on today's flight to weigh less than 36.5 pounds? There are 713 bags stored on today's flight. Stored baggage for our airline weighs 37.2 pounds per bag with a standard deviation of 8.3 pounds.

6.4.5 Cost considerations prohibit more than 6000 pounds in the storage compartment for any flight of our airline. Based on today's flight, do you think we are, at least on average, within the cost constraints? There are 214 bags in the storage compartment on today's flight weighing a total of 6425 pounds with a standard deviation per bag of 5.8 pounds.

6.4.6 The Canadian province of New Brunswick and the state of Maine share a common border and a similar economic base in fishing and forestry. Being in different countries, however, they were subject to different government policies in response to the energy crisis of 1973. The United States provided funding for programs of energy education; Canada did not. In 1984, a test consisting of 74 questions assessing attitudes and knowledge on energy issues was given to ninth graders in 19 randomly chosen schools. The average score for the 429 students in Maine was 17.489 with a standard deviation of 4.269. The average for the 452 students of New Brunswick was 13.966 with a standard deviation of 4.761 (see [Barrow and Morrisey]). Was the energy education program in Maine effective?

6.4.7 For each of Problems 6.4.1—6.4.6:

(**a**) What assumption(s) did you have to make? Is the assumption reasonable? How might it fail?

(**b**) Which are statistical questions and which are exact calculations? A statistical question draws a conclusion on the basis of partial information in the form of data from a statistical experiment, such as a sampling experiment.

(**c**) Describe in real-world terms all parameters and estimators.

6.4.8 You must be able to identify the PATTERN of questions. For any numeric population, you can ask the following pattern of questions: Suppose a chain consists of 81 links where the mean length of a link is 0.7 cm with a standard deviation of 0.06 cm. What's the probability that

(a) a link is more than 0.71 cm?

(b) a chain is more than 57.51 ($\approx 81 \times 0.71$) cm?

(c) at least ten links in a chain are more than 0.71 cm?

(d) at least ten of 100 chains are more than 57.51 cm?

(e) at least 20% of the links in a chain are too long (more than 0.71 cm)?

Or if you did NOT know that $\mu = 0.7$, you might get these questions:

(f) Based on the fact that one chain which you happen to have is 57.51 cm long, what's the length of a link (from among ALL links)?

(g) We believe the links are about 0.7 cm long. Does that seem to be inconsistent with the fact that this chain is 57.51 cm long?

(h) We will be monitoring the lengths of the links to be sure they meet specifications: The mean length of a link should be 0.7 cm with a standard deviation of 0.06 cm.

6.4.9 An alphabet of problems:

You are hired as consultant for a fleet of small fishing boats which hire out to amateur fishermen. You will consider one day's haul for the entire fleet to be a simple random sample from an appropriate probability distribution. Unless otherwise indicated, let $\alpha = 0.05$.

Suppose the average weight of fish in the area of your fishing operation is 7.5202 pounds with a standard deviation of 2.5311 pounds. Further, suppose there were 1756 fish in today's haul for the fleet with 547 flounder and with 368 fish weighing more than ten pounds.

(a) How likely is it a haul like today's of 1756 fish would weigh more than 13,000 pounds total? [answer: 0.9738]

(b) Justify the distribution you used in part (a). Why do you not need to assume the weight of fish to be normally distributed?

(c) What should today's haul weigh? Justify this precisely in terms of our models. [a range around 13205.4712]

(d) What is the standard deviation for the total weight of a haul of 1756 fish? [106.0649]

(e) What is the random variable whose probability distribution is being sampled in part (a)? Verify that it is, indeed, a random variable and that a "haul" is a simple random sample from its distribution.

(**f**) Explain how in part (e) "schools of fish" might cause the independence assumption in the definition "simple random sampling from a probability distribution" to fail and why for your operation it's reasonable to think the assumption is approximately valid.

(**g**) The weight of a haul is ΣX, where X is the weight of one fish in the haul (sample). You used this in parts (a) and (d). Verify that ΣX is a random variable. What is its distribution? Give formulas for its mean and its variance. [Hint: ΣX is a linear function of the sample mean.]

(**h**) A teenage girl goes out into the fleet's fishing waters with her new boat. Her day's haul of four fish weighs a total of 23.2 pounds. The total squared deviations from the mean for her sample is 2.2707 squared pounds. She wonders how much the fish in these waters weigh on average. Based on her catch, what might she conclude? [Hint: 5.8 is wrong!]

(**i**) The girl in part (h) returns to her fishing area the next day and decides to stay until she gets one fish. How much will that fish weigh?

(**j**) What two assumptions must the girl in part (h) make? Are they reasonable? How might they fail?

(**k**) What is it that tells us the girl in part (h) seems to be fishing in an area where the fish are smaller than the fish of your fleet? Explain how such a situation might actually arise.

(**l**) What proportion of the fish in the fleet's area of operations weigh more than ten pounds? [20.96% is wrong. It's 16.35%.]

(**m**) What proportion of the fish in the fleet's area of operations are flounder? [31.15% is wrong.]

(**n**) How large a haul would you have to get to estimate the proportion of flounder in the fleet's waters to within 0.14 of a percentage point? Do a worst case analysis because today's haul might be very unrepresentative. [4900]

(**o**) The teenage girl in part (h) rented her boat from an old man who said, "Waal, de fish 'round here don' weigh much more'n about five pounds." Does the girl's catch call the old man's information into question?

(**p**) Each month, you examine a sample of 400 fish from the month's haul for evidence of toxic contamination. How many of the fish in your sample showing evidence of such contamination would cause you to take action to identify and correct the contamination? According to the director of operations, fewer than one contaminated fish in 1000, on average, would be insignificant. [anything more than 3.6 per 1000]

(q) What are the chances an amateur fisherman on one of your boats will haul in a fish weighing more than 15 pounds? [answer 15 to 10,000 chance]

(r) For your fleet, how much would a haul of 2000 fish weigh on average? [about 15,000 pounds]

(s) What's the probability the typical fish in a haul of 2000 fish would weigh less than seven and a half pounds? [about 36%]

(t) For several days, a very large school of fish are in the vicinity of one of your boats. How large a haul would be required to estimate the size of these fish to within one pound? As a rough estimate, assume these fish vary in size pretty much as all the fish in the area of your fleet's operation. [25]

(u) In the previous part, why might one think about the t-distribution and why in the last analysis would it not be required?

(v) You are watching the weight of the fish in your fleet's waters from season to season to assure satisfaction of the fleet's customers with the quality of fish they catch. If there is ever evidence that the average weight of all fish in the fleet's area of operation has fallen below seven pounds, you will suggest a study to identify any possible cause for a decrease in weight. What should be the criterion for initiating such a study? Assume you monitor the fish each season by weighing a sample of 3500 fish. [sample mean below 6.9296 pounds]

(w) A year ago there were 601 flounder in a haul of 2144 fish from these waters. Does today's haul support the contention there were fewer flounder in these waters at that time? [yes]

(x) Each month you examine a sample of 400 fish from the month's haul for evidence of toxic contamination. How much contamination in your sample would cause you to take action to identify and correct the contamination? "Toxic contamination" in this case means more than 0.23 g of the contaminant per fish. [0.4382 g]

(y) How many fish would you have to catch before getting one that weighed more than eight pounds? [1.3546 on average—one less than 2.3546]

(**z**) On one day, one of your boats made a haul of 52 fish. Seven of those fish weighed more than eight pounds. Were there a surprising number of large fish in that day's haul? [yes-what's the meaning of this?]

6.4.10 At the end of World War II, the Germans initiated the "flying bombs" attack on London. It was strategically important for the British to know the aiming accuracy of these bombs. If they were highly accurate, it would be best to spread key administrative and engineering sites widely over the city. On the other hand, if the bombs were falling more or less randomly, it would be best to maintain efficient operation with closely clustered units.

For purposes of analysis, the city was divided into 576 regions of equal area and the number of hits per region recorded as shown in the following table (from [Clarke]). For example, 229 regions were not hit at all, 211 suffered one hit, and so on.

No. of hits	Observed frequency
0	229
1	211
2	93
3	35
≥ 4	8

(**a**) What's the probability more than two bombs would fall into one region if the bombs were falling more or less randomly? [$\approx 93\%$]

(**b**) Thinking of the solution to part (a), you might expect to answer a question like "How much time would pass on average between the falling of two bombs?" But that question cannot be answered from this data. Explain.

(**c**) Should the British maintain "closely clustered units"?

6.4.11 The label on bags of frozen green peas from the distributor for your supermarket chain indicates the weight to be 1.2 pounds. Be sure you give real-world conclusions to these questions.

(a) How many of the distributor's bags of peas weigh less than one pound? Suppose the standard deviation of weight is 0.14 pounds.

(b) How many of the distributor's bags of peas weigh less than one pound? Here you have no information about the standard deviation of weight, but you do know that 13 of 124 bags were less than a pound.

(c) Bags of peas are shipped to a store in cartons of 50. How many bags per carton weigh less than one pound? Here you have the following observations where X = # bags in a carton that weigh less than one pound.

X	0	1	2	3	4
f	8	3	3	2	1

(d) Using the observations in part (c), estimate the proportion of bags of peas from this distributor which weigh less than one pound.

(e) Identify the assumptions you made in each part of this problem.

6.4.12 Treat each of the following as hypothesis tests and set up the hypotheses. Discuss the appropriateness of an hypothesis test here as opposed to a test of significance.

(a) In trying to promote the city, the Chamber of Commerce funds a study of air pollution. They hope to show the mean level of a certain toxin in the downtown core is less than 4.9 parts per million.

(b) An ecologist wants to have pollution control devices imposed on certain manufacturers. The question: Is air pollution too high, more than 4.9 parts per million of a certain toxin in the downtown core?

(c) The local newspaper wants to do a study of air pollution in the downtown core in view of the critical 4.9 parts per million which is considered the maximum acceptable average level of a certain toxin.

6.4.13 All drugs must be approved by the Food and Drug Administration (FDA) before a drug manufacturer can market it. The FDA must weigh the error of allowing an ineffective drug on the market with the risks of side effects versus the consequences of blocking an effective drug. Their analysis of a new drug might might well include an hypothesis test for r, where r is the mortality rate under treatment with the new drug. Set up such a test assuming

(a) the mortality rate without the new drug is 95%;

(b) the mortality rate without the new drug is 5%.

6.4.14 We turn once again to the "birthday/deathday" problem which we saw in Problems 6.2.23 and 6.2.32. David Phillips published an interesting article on this subject in *Statistics, a Guide to the Unknown* [Tanur et al., 1972], entitled "Deathday and Birthday: An Unexpected Connection." Phillips studies four samples taken from sources listing famous persons. Here's a summary of his four samples. Does it seem to support the contention that some people, famous people at least, have an ability to postpone their death until after their birthday?

Here, N is the number of months from the birth month (where a negative sign means "before birth month") and X is the number of deaths observed for that value of N.

N	−6	−5	−4	−3	−2	−1	0	1	2	3	4	5
X	90	100	87	96	101	86	119	118	121	114	113	106

(a) Graph this data and comment on the "delayed death" hypothesis.

(b) Do a goodness of fit test.

(c) Test the contention that there are no fewer deaths in the month prior to the birth month than would be expected just through chance.

(d) In what other ways might you test the "delayed death" hypothesis with this data?

6.4.15 We have an infinite numeric population which is NOT normally distributed and which has a mean of 2.7. Match each of the following with the appropriate picture (given below), taking that picture which represents the most probable answer:

(a) the population distribution;

(b) the distribution of a large sample;

(c) the sampling distribution of sample means for large samples;

(d) the standard normal distribution;

(e) the distribution of sample proportions for the proportion of the population below two;

(f) Student's t-distribution.

Here are 12 pictures as possible answers . . .

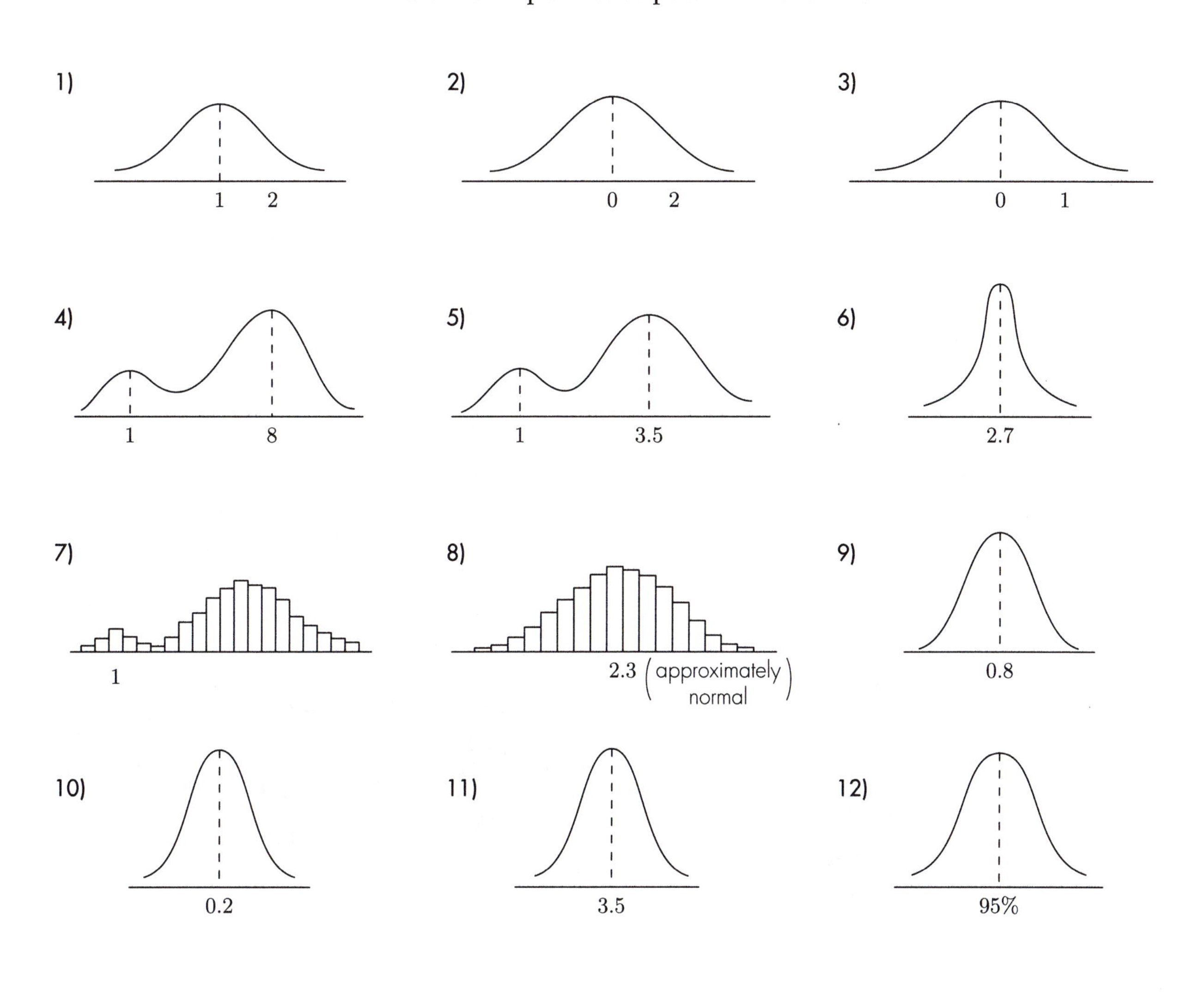

Chapter 7

Introduction to Simple Linear Regression

Warning signs

BY TIM REDMOND

I THINK MY DERANGED obsession with the transit-demand function started back in 1982, when Dianne Feinstein raised the bus fare from 50 cents to 60 cents and I bought my first counterfeit Fast Pass.

Two bucks, right off the color-photocopying machine. A little touch-up with a pastel crayon, and the drivers never knew the difference. I wasn't the only one stiffing the transit system, either.

Times were tough; people were broke. Sixty cents was too much money. I watched crafty passengers pour nickels and pennies into the slot, too many coins for the driver to count, too few to cover the fare. Everyone I knew asked for a transfer, even if we didn't need it, and when we got off the bus, we'd hand someone at the bus stop a free ride.

When all else failed (or we managed to get caught), we walked. A lot of law-abiding people just skipped the tricks and walked anyway. A few decided to go back to their cars.

I knew what was going on, and so did every other marginal member of San Francisco society. I almost laughed out loud when I heard the Muni finance staff had announced to the Public Utilities Commission one afternoon that the bus system had an "unexpected shortfall" of a few million dollars and nobody could figure out why.

After all, the fares had just gone up 20 percent, and service hadn't changed a bit. The way the Muni economists saw it, the system should have been taking in a whole lot more money.

THE AWFUL TRUTH

I think something horrible happened to my mind that afternoon. Flashbacks from my sordid past began haunting me on the Hayes Street bus: images of Stanley Lebergott's early-morning economics lectures, attacking my poor defenseless brain ... multivariable matrices ... the slope of the curve, the function of change over time, autocorrelation ...

This stuff can get ugly fast. When you're hooked and you have a big computer, you can sit in your swivel chair all night long, easing reams of hard, raw data into smooth, gentle arcs. If the program really starts to hum, you can grab one hot little factor, isolate its coefficient, and then see how fast it changes when you jack the input up a few notches.

You could jack it up, say, about 10 cents on a 50-cent San Francisco transit fare.... With a decent program, you'd find out that a lot of people who rode the bus in San Francisco were worried about every dime — and that, when the price went up, some of them would just stop paying.

You see why I get so excited.

HOT STUFF

Economics is a dismal game, but it's not always entirely abstract. Back in 1982, we were talking about several million dollars, but the city budget was healthy enough to absorb the blow. Now we're talking about even more — and we can't afford any of it.

The demand-function concept is actually pretty simple: Every time the price of something goes up, the demand for it falls. Double the price of coffee, some people buy tea instead.

If your budget is based on projections that only, say, 10 percent of your customers will give up coffee, and it turns out that 40 percent of them do, then you wind up laying people off or going bust. So anybody running a coffee business has to have the best possible information about the effect a price increase will have on sales.

But if the price of tea goes up too, at a different rate, and your advertising costs go down so you can do more promotion, and the unemployment rate goes up half a percent, and your major competitor offers a volume discount to retailers, and the federal government issues a new warning about caffeine and heart disease ... the picture can get pretty muddy.

That's where linear regression analysis comes in. A regression is a formula that takes historical data on every relevant factor and measures precisely how each one individually affects demand for a product, at any given time. The rate of change is called "elasticity."

Which brings us to the latest increase in Muni fares.

I got this report the other day from my old friend David Pilpel, who also knows the loneliness of late-night data addiction. The report was done by the American Public Transit Association in Washington, D.C.; it's called "Fare Elasticity and Its Application to Forecasting Transit Demand." It runs about 110 pages.

"You're going to love this," David told me. "It's better than sex."

And yeah, I've been drooling into the binding, and spending afternoons on the phone long distance with Dr. Larry Pham, who wrote the report. I've been lying awake at night, wondering why Muni still believes that it can raise cash fares by 17 percent and lose only 1.5 percent of its riders.

In regression-speak, Muni's official estimate — a 1.5 percent drop in unit sales on a 17.6 percent increase in price — amounts to an elasticity factor of -0.08. That, the study suggests, is absolutely nuts.

Dr. Pham has one of the sweetest little regression programs I've ever seen, and a set of computers to match. He ran at least 24 months of data from 52 transit systems around the country. He factored in everything from the price of gasoline and parking to the daily rainfall in each location. His conclusion: A lot of transit planners have their heads up their buses.

The average elasticity across his study was -0.4 — almost six times the estimate that San Francisco is using. By Pham's average figure, every time you raise bus fares 10 percent, you lose 4 percent of your riders. In some cases, the number goes as high as -0.8 — raise fares 10 percent, lose 8 percent of your riders. At that level, a fare increase is almost a wash, financially — and a very bad idea for traffic, pollution, etc.

Back in 1982, the PUC budget folks had no elasticity data. They just made a random guess at the effect of a fare increase, and it came back to haunt them. This time around, Assistant General Manager Michelle Witt told me the staff "used our past experience." Not a pretty thought.

Perhaps before they go off half-cocked on this fare increase, the commissioners ought to plug into Dr. Pham's computers, and see what a good elasticity study can do for their bottom lines. ●

Reprinted, with permission, from Tim Redmond, "Warning Signs", *San Francisco Bay Guardian*, Vol. 26, Number 40 (July 8, 1982): 12.

7.1 The Simple Linear Regression Model

So far, when we have studied a random variable, say Y, we've ruled out any effect of other variables. In this chapter, we study a model which accommodates the systematic effect on Y of one other variable, X—the ONLY systematic effect on Y. We think of X as known or under our control. This X is sometimes called the "explanatory" variable in view of its role as a known or controlled "effect" on Y. So Y is the variable in question and X plays the role of input information relevant to Y. Sometimes, Y is described as the "response" to the "factor" X. In itself, X might or might not be a random variable, but from the point of view of the model, X is not random because the model focuses on particular values of X which are known or in some sense controlled. In other words, it's a model for the *conditional* distribution of Y given X.

One word of caution at the beginning: The model says nothing about the real-world nature of the effect of X on Y. In particular, X may or may not be causally related to Y. We'll see a number of examples where there is no causal relationship.

Our model, the ***simple linear regression model***, assumes that X affects only the mean of Y. In all other respects, Y is assumed to be independent of X. In other words, our new model takes one careful step forward. It admits only a very limited effect of X on Y. This is necessary as a first step. After all, when you begin speaking of "effects" of one variable on another, you introduce the potential for overwhelming complications!

In fact, the effect of X is restricted even further: For the simple linear regression model, the mean of Y should be determined by a *linear* function of X. That's why the model is called "linear" regression. Other models are possible. The mean of Y might, for example, be determined from X by a quadratic or cubic polynomial or some other function.

Why is it "simple" linear regression? Because we allow only one X. By contrast, there are "multiple" regression models which allow a number of different explanatory variables accounting for a number of different systematic effects on the mean of Y. Multiple regression works just like simple regression from a theoretical point of view, but the mathematical complications have inspired us to stick to the case of simple linear regression.

And why is it called "regression"? This name was given to the model by its discoverer Francis Galton (1822–1911), known for his work in heredity. Galton, first cousin of Charles Darwin, was a medical doctor by training whose interests "ranged over psychology, anthropology, sociology, education and fingerprints" [Stigler]. His discovery of the regression model was a gradual revelation over 20 years, starting with hints in the book *Hereditary Genius* (1869, ten years after Darwin's *Origin of the Species*), a book "naive and flawed" [Stigler] which sought to show that genius "runs in families." From this humble beginning through numerous studies of extensive data on sweet peas and on human populations, studies culminating in his book *Natural Inheritance* (1889), Galton gradually saw his way through to a clear articulation of the regression model. His work in discovering one of the most powerful mathematical tools in statistics is all the more remarkable for the fact that he was a very indifferent mathematician, requiring help on relatively simple details from a mathematician friend, J. Hamilton Dickson, at St Peter's College, Cambridge. Galton chose the term "reversion," later "regression," from the effect which he observed in his sweet pea data that the progeny of very large or very small peas would not have the same average weight as the parent peas, but would "regress" to the average of all peas.

Here are some examples of real-world situations which might reasonably be modeled by simple linear regression:

	the variable Y is . . .	**the "known" quantity X is . . .**
1.	height of a bean plant	number of days since planting
2.	toxic chemical in plant	amount of toxic chemical in soil
3.	crop yield	amount of fertilizer used
4.	household expenditures	household income
5.	manufacturer's production costs	number of units produced
6.	a measure of on-job performance	score on a job-skills test
7.	percent of votes for a candidate	campaign expenditures
8.	income at age 45	years of schooling completed
9.	height of adult daughter	height of mother

There are four characteristics of the model which derive from our description of X as the only systematic factor affecting Y, affecting the mean of Y only, with that effect expressed through a linear function. First, for a fixed value of X, Y is approximately *normally distributed*. This normally distributed random variable is denoted $Y|X$ although it's often just written as Y if the context makes it clear. Second, for different X's these $Y|X$'s must be *independent*. If not, the effect of X on Y would go beyond its effect on the mean of Y, contrary to our description of the role of X.

Third, because X affects only the MEAN of Y, the variance must be the same from one value of X to another. This is expressed by saying the model is ***homoscedastic***. For a fixed X, the variance of Y is denoted by the symbol $\sigma^2_{Y|X}$. This symbol is read "the-variance-of-Y-given-X," or simply, "sigma-squared-sub-Y-given-X" (or some other variant). Because $\sigma^2_{Y|X}$ is the same for every value of X, it's often denoted simply by σ^2, as long as the context makes it clear we're speaking of this particular regression model for Y and not something else. Finally, the fourth characteristic of the model implicit in our description is that β must not be zero (why not?).

Here's a picture of the model:

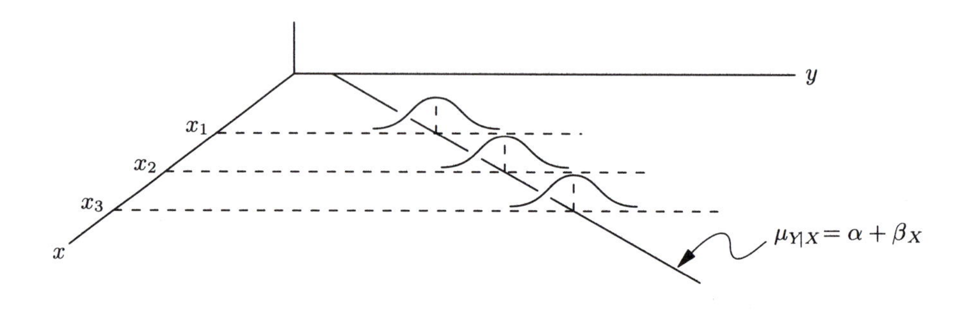

In the picture, you see that the means of the $Y|X$'s—the centers of the various normal curves—lie along a line, reflecting the assumption that the mean of $Y|X$ is affected by X through a *linear* function of X. The notation for this linear function is

$$\mu_{Y|X} = \alpha + \beta X.$$

The symbol $\mu_{Y|X}$ can be read "mu-sub-Y-given-X" or, more completely, "the-mean-of-Y-given-X." Note that the slope of this line is β and the y-intercept is α. With this notation, $Y|X$ can be expressed as

$$\begin{aligned} Y|X &= \alpha + \beta X + \epsilon \\ &= \mu_{Y|X} + \epsilon, \end{aligned}$$

where ϵ is a normally distributed random variable with variance σ^2. Now, to help you understand the model, we'll ask you to . . .

Try Your Hand

7.1.1 In this problem, you should ignore the regression model until you come to part (f). We're going to look again at what you know about normally distributed random variables and then show you what's meant by saying the simple linear regression model "takes one careful step forward" beyond what we've already done.

(a) Show that a normally distributed random variable Y has the form $Y = \mu + \epsilon$, where ϵ "looks like random error."

(b) What is the "systematic part" of a normally distributed random variable?

(c) What's "variable" for a normally distributed random variable?

(d) Suppose you know Y has the form $Y = \mu + \epsilon$, where $\epsilon \sim N(0, \sigma^2)$. Show that, even though you may not have known it, Y is indeed normally distributed.

(e) How does our Chapter 4 criterion for normality accommodate the systematic part of a normally distributed random variable Y?

(f) Now, why do we say the simple linear regression model "takes one careful step forward"?

7.1.2 For the nine examples listed in the text above, think about the character of the simple linear regression model:

(a) In examples one and three some obvious restrictions would be required on X to make the model relevant. What restrictions?

(b) In which examples might the roles of X and Y be reversed? In which examples would it be obviously absurd to reverse the roles of X and Y?

(c) In which examples might X reasonably be thought to CAUSE an effect on Y? If the "effect" is not causal, what kind of effect is there?

(d) In which examples would X probably be a random variable?

7.1.3 Let's think some more about the personality of the simple linear regression model.

(a) If it's really true that X is the only systematic factor affecting Y , then for a fixed value of X, $Y|X$ should be approximately normally distributed. Explain.

(b) In the simple linear regression model, $\epsilon \sim N(0, \sigma^2)$. Explain.

(c) Why must β not be zero?

7.1.4 For the simple linear regression model:

(a) What are the roles played by the variables X and Y?

(b) Identify the parameters for the model.

(c) Identify the assumptions which underlie the model and the consequences implicit in those assumptions.

7.1.5 For each of the nine examples listed in the text above, give the real-world interpretation of the parameters α and β.

7.1.6 Among the nine examples listed in the text, several would require restrictions on X because otherwise β almost certainly would not be constant [see Problems 7.1.2(a) and 7.1.5]. Why does it violate the assumptions of the model for β to be variable?

7.1.7 Give an example of a real-world situation which might be appropriately modeled by

(a) a multiple regression model,

(b) a simple linear regression model with $\beta < 0$.

7.1.8 Sketch a picture of a *quadratic* regression model similar to the picture given in the text above of the simple linear regression model.

7.1.9 One way to understand the assumptions of our model is to see what the model would look like if one assumption fails while all the others continue to hold. Sketch a picture like the one in the text for which all the assumptions hold

(a) except that $\sigma^2_{Y|X}$ is NOT the same for all values of X,

(b) except that β IS zero,

(c) except that the relationship between X and $\mu_{Y|X}$ is NOT linear.

7.1.10 Sketch four pictures of the simple linear regression model, one illustrating each of the following conditions . . .

(a) $\alpha = 0$;

(b) $\alpha = 1$;

(c) $\beta < 0$;

(d) $\beta = 0.5$.

7.1.11 Now that we've got the model, what must we do next to actually use the model?

7.2 The Least Squares Estimates for α and β

Now that we understand what the simple linear regression model is, let's see how we can estimate the parameters of the model. As usual in this text, we assume the data for our estimators has been properly generated by an appropriate random sampling experiment through which we have obtained n random observations on the pair (X, Y).

For example, suppose a research worker is studying a certain type of bamboo within the first three weeks of its growth cycle. Every second day beginning with the tenth day after planting the rhizomes (the "stem" which is put into the earth and which produces the roots and shoots of the bamboo), she measures three bamboo shoots randomly chosen from the hillside where they're growing. Continuing through the twentieth day, she measures 16 shoots. Unfortunately, on each of two days, one of the shoots was damaged and had to be discarded. Here's a record of her observations:

Day #	10	12	14	16	18	20
Height	9	16	34	61	87	113
(cm)	5	21	26	46	91	124
	6	19	—	53	—	110

We'll see below how this data gives the "point estimates"

$$\begin{aligned} a &= -112.7872 \quad \text{for the parameter } \alpha, \\ b &= 11.0319 \quad \text{for the parameter } \beta. \end{aligned}$$

A ***point estimate***, as you can see, is the value of an estimator calculated from a specific set of data. Taken by itself, a point estimate is meaningless because it gives no indication at all of the accuracy of the estimate or the certainty with which that accuracy is attained.

In the discussion which follows, we begin at the beginning by asking how to calculate point estimates for the parameters of the simple linear regression model. Once we know how to do that, we'll turn to a discussion of the "total context" of the numbers we've calculated. Then, once we have the total context, we'll be able to draw some meaningful real-world conclusions based on our model.

Before we turn to the actual calculation of these point estimates, you should first think a little about what the estimates mean. In the exercises which follow, you'll look at the values given above for a and b, the point estimates for α and β, and see how they relate to the simple linear regression model. To do this, you plot a "scatter diagram" for the observed data. The ***scatter diagram*** is simply the observed (X, Y) values plotted on (x, y) coordinates. Then, on the scatter diagram you'll plot the estimator for $\mu_{Y|X}$. That estimator is denoted by the symbol $\hat{Y}$. Finally, we'll ask you to interpret the relationship of $\hat{Y}$ to the abstract model.

Please . . .

Try Your Hand

7.2.1 First, let's explore the researcher's bamboo data a bit:

(a) What does the symbol (12, 21) refer to? Be as specific as possible.

(b) Assuming the data is recorded in the order of observation, how tall was the shoot observed just after the observation (16, 46)?

(c) What's the value of n here?

(d) Assuming the notation (X_i, Y_i) for the ith observation, what are the values of i? What's (X_8, Y_8)? And what's (X_3, Y_{16})?

(e) What is the average number of days after planting in this researcher's observations?

(f) What is the average height of all the observed bamboo?

(g) Note that the average of the daily average heights is not the average of all 16 observations. Why not?

7.2.2 For the bamboo data given in the text:

(a) Plot the scatter diagram.

(b) Give an equation for $\hat{Y}$ in terms of a and b (whose values are given in the text above).

(c) What's $\hat{Y}$ for 13 days? For six days?

(d) Plot $\hat{Y}$ on the scatter diagram.

(e) How does the scatter diagram relate to the abstract simple linear regression model?

(f) How does the line determined by $\hat{Y}$ in your scatter diagram relate to the simple linear regression model?

7.2.3 Let's think about point estimates:

(a) Explain the statement in the text: "A point estimate taken by itself is meaningless . . . ."

(b) With specific reference to β explain the following phrase which completes the quote in part (a): ". . . because it gives no indication at all of the accuracy of the estimate or the certainty with which that accuracy is attained."

(c) If the data is atypical, that means there's something wrong. What are some of the possibilities for what could be wrong?

(d) To make sense of a point estimate, we require something much more than just that one number. What more is required?

7.2.4 Let's look again at our researcher's scatter diagram:

(a) Look carefully at the way the scatter diagram relates to the estimated regression line which you drew in Problem 7.2.2(d). There appears to be a difficulty with the model. What is it? [Hint: See Problem 7.1.6.]

(b) A difficulty such as we've identified in part (a) may have a number of possible explanations. Suggest one possibility.

(c) How would you deal with the difficulty we found in part (a)?

(d) Restricting the model as suggested in part (c) would probably be very unsatisfactory for our researcher. Why?

The Principle of Least Squares

Let's look at the scatter diagram for our observed data in more detail. From that data, we'll determine the ***estimated regression line***, the line $\hat{Y} = a + bX$. That line estimates $\mu_{Y|X} = \alpha + \beta X$, the model's ***true regression line***. What criterion should determine the line $\hat{Y} = a + bX$? It should be the line which best "fits" the data in a technical sense which we explain below. You'll get the idea of "fit" if you look at the several lines in the following picture, NONE of which fit the data:

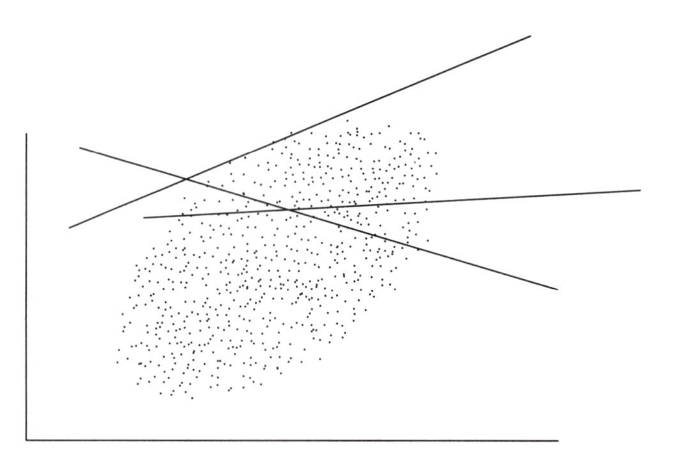

Now, if you try to draw a "good" line through this scatter diagram—one that really seems to fit the data—you'll probably draw a line through the point $(\overline{X}, \overline{Y})$ and that's perfectly correct. But still you may draw the wrong line! In situations where X is generated randomly along with Y, the scatter diagram often takes on a more or less elliptical shape. In that case, most people would take the major axis of the ellipse as the best fitting line, intuitively minimizing the *perpendicular* distance away from the line. They're choosing the line for which the total of those distances is as small as possible. Here's a picture, showing the line which minimizes the "perpendicular distances":

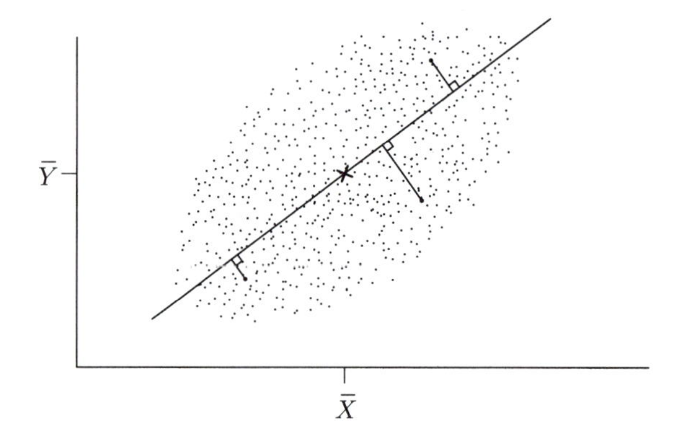

The line indicated in the picture above is NOT the estimated regression line; the regression line is slightly less steep. It's the line that passes through the point on the ellipse where there is a vertical tangent:

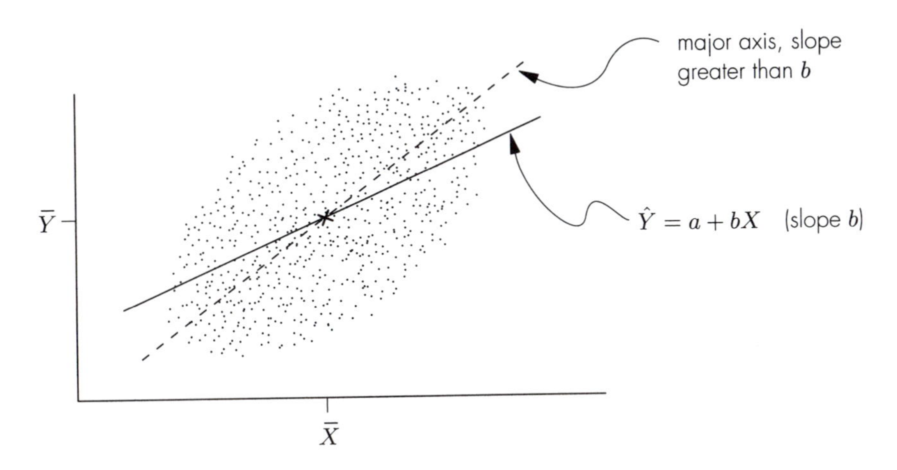

This line is determined by ***the principle of least squares***: It's the line for which the SQUARES of the VERTICAL distances from the line are minimized:

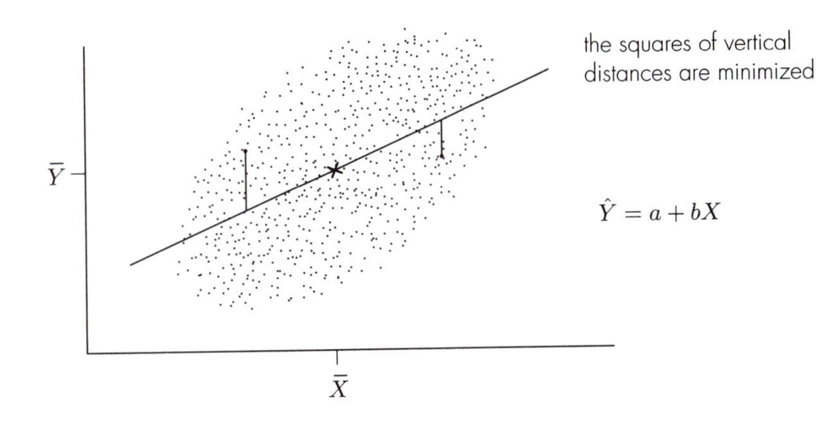

In the exercises, you'll see why it's the *squares* of the vertical distances that are minimized rather than the distances themselves and why it's the *vertical* distances rather than the perpendicular distances that are relevant. Because the line minimizes the squares of those distances, it's called the ***least squares estimate*** for the true regression line.

Before attempting to see how this line is actually determined—before deriving formulas for a and b—we should think a bit more about what we're trying to do. After you've done the following exercises, we'll see how to actually calculate. So please . . .

Try Your Hand

7.2.5 Here, we figure out some algebraic facts related to the least squares criterion. We're thinking about an observed point $(x,\ y)$ in the scatter diagram.

(a) Explain why the vertical distance of the point $(x,\ y)$ from the estimated regression line is $Y - \hat{Y}$, provided the point is above the line.

(b) If the point $(x,\ y)$ is below the line, $Y - \hat{Y}$ is NOT the vertical distance. Why not?

(c) How can you get rid of the possible negativity in the expression $Y - \hat{Y}$?

(d) What would be an algebraic expression for the "total" of the squared vertical distances away from the estimated regression line?

7.2.6 For the researcher's bamboo data, show that the point $(\overline{X},\ \overline{Y})$ lies on the estimated regression line.

7.2.7 In fitting the regression line to the data, why do you think we would want to minimize the vertical distances away from the line rather than the perpendicular distances?

7.2.8 The regression line is the key to the relationship between Y, the variable in question, and the information about Y captured in X. So it's the variability of Y about that line that's the key issue.

(a) Write out an algebraic formula for the total variability of the observed Y's about the regression line $\hat{Y} = a + bX$.

(b) The estimated regression line is determined by the principle of least squares. What does that principle say about the formula from part (a)?

Calculating the Least Squares Estimates of α and β

You just saw in Problem 7.2.8(b) how the principle of least squares finds the line which minimizes variability away from the line in the direction of Y for the observed data points. In other words, it minimizes the ***error sum of squares, SSE.***

$$\text{SSE} = \Sigma(y - \hat{y})^2 = \Sigma(y - a - bx)^2.$$

Note that everything in the formula for SSE is known except the values of a and b. After all, the X's and Y's which appear in this sum are just our observed data values. They are specific numbers. For the bamboo data, there are sixteen X values and sixteen Y values.

So, SSE is an algebraic expression involving two unknowns a and b. Using techniques of calculus, one can show there is one and only one choice for a and b which minimizes this expression as the principle of least squares requires. Apply that technique from calculus and you get the so-called ***normal equations***

$$\begin{aligned}\Sigma y &= na + b\Sigma x,\\ \Sigma xy &= a\Sigma x + b\Sigma x^2.\end{aligned}$$

Solving these equations gives formulas for a and b:

$$\begin{aligned}b &= \frac{n\,\Sigma xy - \Sigma x\,\Sigma y}{n\,\Sigma x^2 - (\Sigma x)^2},\\ a &= \overline{Y} - b\overline{X}.\end{aligned}$$

The equation for a requires knowing b. That's why we give the equation for b first. You should calculate b first, then use it to get a.

It's certainly not obvious that these are the "correct" estimators for α and β. The principle of least squares has a long and complex history. That story is one of the main themes in the first half of Stigler's *The History of Statistics*. After some 50 years of struggling with problems in geodesy and astronomy which really called for the principle of least squares, scientists finally found that principle clearly enunciated for the first time by the French mathematician Adrien Marie Legendre (1752–1833) in an appendix dated March 6, 1805, an appendix to a brief monograph on the determination of orbits of comets.

Working out the theoretical justification of the technique required another 20 or so years; work that was intimately bound up with the discovery by Gauss and Laplace of the Central Limit Theorem and with the discovery of the normal distribution as the appropriate model for random error. And that was not the end of it! Our use of least squares for the regression model was not discovered until around 1900. Since then, further theoretical development of the theory of probability distributions has found other, deeper relationships which justify the method. So if you think it's not immediately clear why the a and b given above are good estimates for α and β, CONGRATULATIONS, you couldn't be more right! It wasn't immediate; it took about 150 years.

The distance between two points, given by the Pythagorean Theorem, is the square root of a sum of squared differences. That's suggestive of the formula $\text{SSE} = \Sigma(Y - \hat{Y})^2$. In fact, if you take a sophisticated enough mathematical view of our data, regarding the values of X as determining one "point" in an n-dimensional space and Y another such "point," it turns out that the principle of least squares finds a "point" which minimizes the distance between the observed Y and the "direction of X." That gives us $\hat{Y}$, the component of Y in the direction of X. It estimates the systematic part of Y, the part determined by X. The component of Y in the direction *perpendicular* to X estimates the purely random part of Y, what we've denoted by σ.

By now you may be thinking, "Well again, not too clear!" Don't worry, we're just trying to give you a feel for the wonderful geometric ideas which back up the theory. Then maybe you'll get excited enough to pursue a career in mathematical statistics. If it all sounds a bit impractical, that's alright too. You can take up a career in applied statistics using all these powerful tools to answer real-world questions. Or—okay—maybe you'd rather do something else.

Well, let's return to the bamboo study once again. Our research worker, you'll recall, after looking at the scatter diagram for her data, discovered that her original data had been distorted when fertilizer was put on the bamboo halfway through her study (see Problem 7.2.4). Suppose she repeated the study with newly planted rhizomes, generating the following data:

X = day #	10	12	14	16	18	20
Y = height	7	22	43	62	70	84
(cm)	11	21	38	48	74	82
	6	25	33	50	—	68

To calculate a and b, rearrange the data into a table like the one below. All the quantities in the formulas for a and b are just column sums from this table.

X	Y	XY	X^2	Y^2
10	7	70	...	
10	11			
10				
12				
...	...		...	

Note that there were no repetitions in our bamboo data. With continuous data, repetitions are typically rare and so a frequency column becomes redundant. If a data observation should occur several times, you can just list it separately in the table as many times as it occurs.

Before going further, you need to get some practice calculating. So please . . .

Try Your Hand

7.2.9 Set up a table for our researcher's new data on the height of bamboo shoots with columns for X, Y, XY, X^2, and Y^2 like the table begun in the text just above. Then calculate a and b from the column sums.

7.2.10 H.G. Wilm wanted to predict the April to July water yield (in inches) in the Snake River watershed in Wyoming from water content of snow on April 1. Here's Wilm's data (after [Weisberg]) for the years 1919 to 1935:

X	Y	X	Y	X	Y
23.1	10.5	39.5	23.1	12.4	8.8
32.8	16.7	24.2	12.4	35.1	17.4
31.8	18.2	52.5	24.9	31.5	14.9
32.0	17.0	37.9	22.8	21.1	10.5
30.4	16.3	30.5	14.1	27.6	16.1
24.0	10.5	25.1	12.9		

$$\sum X = 511.5, \quad \sum Y = 267.1, \quad \sum XY = 8653.45,$$
$$\sum X^2 = 16628.65, \quad \sum Y^2 = 4549.43.$$

(a) Evidently X refers to what? And Y? Explain.

(b) Give the estimated regression line for this data.

(c) The equation in part (b) is suspicious—it predicts three-quarters of an inch of water yield when there was NO SNOW to yield that water! In other words, we know $\alpha = 0$, so the model is $Y|X = \beta X + \epsilon$. For this special case of the model, the least squares estimate for β is $b = \Sigma XY / \Sigma X^2$. Give the estimated regression line for Wilm's data with this more realistic version of the model.

(d) Plot a scatter diagram for the data with two regression lines, the ones you gave in parts (b) and (c).

7.2.11 Now, in the abstract, with no reference to specific data, but using the formulas for a and b:

(a) Show that the point $(\overline{X}, \overline{Y})$ is on the regression line.

(b) Show that $\Sigma(y - \hat{y}) = 0$.

(c) What's the point of part (b)?

(d) Derive the formulas for a and b by solving the normal equations.

7.2.12 In Problems 3.6.16 and 6.2.33, you studied Edwards and Eberhardt's data on 135 cottontail rabbits observed in a protected enclosure. They were studying various capture–recapture techniques for estimating the size of wildlife populations. Here's how one of the techniques works:

In Problem 6.2.33 you showed that the geometric model which had been proposed by Edwards and Eberhardt on the basis of several different analyses is not unreasonable. Assuming a geometric model with Y as the number of rabbits which will be recaptured X times, the *proportion* of rabbits which will be captured $X = x$ times is pq^x. So if the model is valid, the expected value of Y is Npq^x. We want N. Using logarithms,[1] we can isolate N this way: $\ln(Y) = \ln(Np) + x\ \ln(q)$. This is a linear equation in x and you can use your observed data to "fit" this equation and then estimate N. First, transform the observed data from (x, y) to $(x, \ln y)$:

X	1	2	3	4	6	7
Y	43	16	8	6	2	1
$\ln Y$	3.7612	2.7726	2.0794	1.7918	0.6931	0

When you look at the original data (Problem 3.6.16), you'll see we've omitted the "observation" for $X = 5$. Because no rabbit was caught five times, there was no such observation. Note that the 59 from the original data is also not part of your data because you do NOT know how many rabbits were never captured.

(a) We will use a regression model for $\ln(Y)$ with X as explanatory variable. What are α and β?

(b) Fit a least squares line to the transformed data.

[1] For an explanation of the natural logarithm, $\ln(x)$, see page 112 and Problem 3.8.9.

(c) Use the line in part (a) to estimate $\ln(Np)$. Then estimate $\exp[\ln(Np)] = Np$.

(d) Use the regression line in part (a) to estimate $\ln(q)$, then q, and, finally, p.

(e) Use parts (c) and (d) to estimate N. Compare your estimate with what in this study we know to be the true value: $N = 135$.

(f) Note that when $X = 0$, the model says $Y = Np$. In other words, Np is the number of rabbits never captured. So then $Nq = \Sigma_{x \neq 0} Y$. But you would know that number from the data; here, $Nq = 76$. So there are two more ways to estimate N. What are they?

(g) Suppose in part (e) you had estimated p by its maximum likelihood estimator (MLE) as you did in Problem 6.2.33(b). What happens and why?

(h) Critique the analysis of this problem.

7.3 Using the Simple Linear Regression Model

As we saw in the first section of this chapter, the parameter β for the simple linear regression model has a very concrete real-world meaning. It's the change in the average of Y for a unit increase in X (see Problem 7.1.5). For example, with our researcher's bamboo study, β is the daily increase in the mean height of all bamboo shoots of this type grown under similar conditions during the first three weeks of the growth cycle.

So the first real-world question we might think to ask is, "How fast does this bamboo grow in the first weeks?" ($\beta = ?$). Because we have only partial information in the form of sample data, we'll need to construct a confidence interval for β. That requires knowing the total context, the sampling distribution, of its estimator b. After all, taken out of context, the data by itself tells us nothing. That data could be—though PROBABLY it isn't—quite atypical of what's actually happening. This probability assertion is the basis for our confidence interval. To justify the assertion, we need a probability distribution—the sampling distribution for b. That probability distribution puts our specific data into the context of all theoretically possible data. Here it is:

the sampling distribution of b:

b is an approximately normally distributed, unbiased estimator for β

$$\mu_b = \beta.$$

The squared standard error for b, σ_b^2, can be estimated from the data (thus requiring the t-distribution) by

$$s_b^2 = \frac{s_{Y|X}^2}{\Sigma(X - \overline{X})^2}$$

$$\text{d.f.} = n - 2.$$

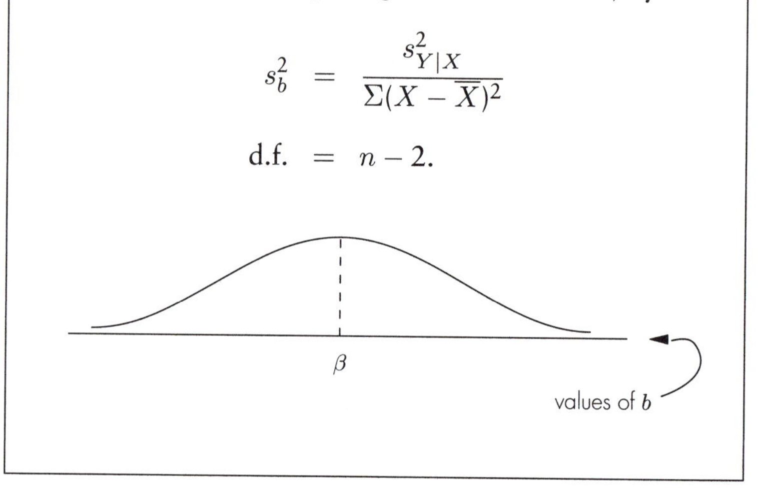

The standard error for b is σ_b; we estimate it by s_b. As you see in the box above, the numerator of s_b^2 is $s_{Y|X}^2$ (estimating $\sigma_{Y|X}^2$):

$$s_{Y|X}^2 = \frac{\text{SSE}}{n-2}, \quad \text{where SSE} = \Sigma(y - \hat{y})^2.$$

SSE, recall, is what's minimized when we find the regression line by the principle of least squares. It's called the error sum of squares. We'll give computing formulas below for the various parts of s_b^2.

Now you see why there are $n - 2$ degrees of freedom for b: There are TWO estimates in the standard error. In SSE, there's an estimate, a, for the parameter α, and b for the parameter β (a and b appear in $\hat{Y}$). This is exactly parallel to the situation with the standard error for $\overline{X}$. When σ is unknown, it's estimated from the data by s, so $\sigma/\sqrt{n} \approx s/\sqrt{n}$. Then, because one parameter was estimated, we had $n - 1$ degrees of freedom. Here's the principle:

In estimating a standard error, if some one other parameter is being estimated from the data, you have $n-1$ degrees of freedom; if two other parameters are being estimated from the data, you have $n-2$ degrees of freedom, and so on.

With a little algebra we get computing formulas for the pieces of the standard error of b

the computing formulas:

$$\begin{aligned} \text{SSE} &= \Sigma(y-\hat{y})^2 &= \Sigma y^2 - a\,\Sigma y - b\,\Sigma xy, \\ \Sigma(X-\overline{X})^2 &= \Sigma x^2 - n\overline{X}^2 &= \Sigma x^2 - (1/n)(\Sigma x)^2. \end{aligned}$$

Putting these pieces together computes s_b^2. With the data organized into an appropriate table, the calculation uses only column sums.

Now, how fast does our researcher's bamboo grow? Let's construct a confidence interval for β using the new data. First, estimate σ_b^2:

$$\begin{aligned} s_b^2 &= \frac{(1/15)(43,566 - 744a - 12,484b)}{3936 - (252)^2/17} \\ &= \frac{29.36744913}{200.4705882} = 0.146492557. \end{aligned}$$

We've departed from our usual convention of rounding to four decimal places to emphasize that you should NOT round in the course of a long calculation. Every time you round at an intermediate step, you compound the roundoff error. Learn to use your calculator in a way that minimizes the need to record a rounded result, reentering it later.

Taking the square root of s_b^2, we get a standard error of 0.3827. Thus, a 95% confidence interval estimate for β will have endpoints

$$b \pm t s_b,$$

$$7.2594 \pm 2.1315 \times 0.3827 \quad (\text{d.f.} = 15),$$

giving the interval (6.4436, 8.0752). So . . .

Question: How fast does this bamboo grow in the beginning?

Answer: With a 5% risk of error, based on our data we conclude that under similar conditions this type of bamboo will grow between 6.4436 and 8.0752 cm per day, on average, during the first three weeks of the growth cycle.

In addition to the question "What's the value of β based on our data?" we might also be asked for a test of significance or an hypothesis test for β. For example:

Question: Does our data seem to cast doubt on our previous contention that in the early stages of its growth cycle bamboo of this type gains about 9 cm per day?

Analysis:

$$\begin{aligned} p\text{-value} &= \mathrm{P}(b < 7.2594) = \mathrm{P}(t < -4.5477) \\ &< \mathrm{P}(t < -2.9467) \\ &= 0.005. \end{aligned}$$

This means our p-value is quite small and so the data does seem to challenge the hypothesis.

Answer: Based on our observations, contrary to our previous contention it would seem bamboo of this type grown under similar conditions gains less than 9 cm per day in average height during the first three weeks of its growth cycle.

Note that nothing new shows up here other than the specific formulas. The ideas and techniques of confidence intervals, tests of significance, and hypothesis tests are already firmly in your grasp!

Testing Hypotheses Concerning β

We've seen seven characteristics of the simple linear regression model any one of which, if it did not hold, would invalidate the model [see Problem 7.1.4(c)]. This problem of "diagnostics" is a topic in its own right, usually presented in a second course. Most of the diagnostic techniques are not within our ability to carry out in any rigorous way. The one exception is the requirement that β not be zero. The ***hypothesis test for linearity*** that is conventionally carried out is

$$\begin{aligned} \mathrm{H_o}&: \quad \beta = 0, \\ \mathrm{H_A}&: \quad \beta \neq 0. \end{aligned}$$

Carrying out this test is going to be routine for you because the logic is exactly the same as for any other two-tailed hypothesis test. For the research worker's second set of bamboo data, we'll reject H_o at the 5% significance level if our data standardizes to a value of t less than -2.1315 or greater than 2.1315. Of course, this really should be a test of significance (why?); we're just following the usual convention in treating it as an hypothesis test.

Note that other hypothesis tests for β may arise which have nothing to do with the validity of the model. The validity question for the model is a theoretical question. But certainly there will also be real-world questions about β requiring hypothesis tests. After all, β has a very concrete real-world meaning as the change in the mean of Y for a unit increase in X. So, for example you may be monitoring the height of bamboo shoots once every three days during the early weeks of the growth cycle with the intention of altering the watering schedule if it ever appears the daily rate of growth falls below 8 cm

$$\begin{aligned} H_o: &\quad \beta = 8 \qquad (\beta > 8 \text{ irrelevant}), \\ H_A: &\quad \beta < 8. \end{aligned}$$

Well it's past time for you to . . .

Try Your Hand

7.3.1 (a) Use our researcher's bamboo data to test the appropriateness of the linearity assumption for the simple linear regression model.

(b) In reference to the test in part (a), why did we say in the text, "Of course, this really should be a test of significance"?

(c) Give the decision rule for the situation described in the text just above where you are ". . . monitoring the height of bamboo shoots . . . with the intention of altering the watering schedule if it ever appears that the daily rate of growth falls below 8 cm." Assume[2] $\alpha = 5\%$.

(d) Would the research worker's second set of data given in the text cause you to alter the watering schedule?

(e) Compare the tests in parts (a) and (c). Why is the first a two-tailed test and the second not?

[2] This α is the significance level of the test; it has NOTHING to do with the α in $\mu_{Y|X} = \alpha + \beta X$!! And the β in the regression equation has nothing to do with the symbol β for the probability of type II error in a hypothesis test. This is all just a sorrowful accident of notation.

7.3.2 A mountain climber can determine altitude above sea level by using a barometer to measure atmospheric pressure—lower pressure meaning higher altitude. However, the barometers available in the mid-nineteenth century were very delicate and difficult to carry on expeditions into mountaineous terrain. The Scottish physicist James D. Forbes hoped to get around this difficulty by determining barometric pressure from the boiling point of water. Forbes made measurements at 17 locations in the Alps and in Scotland. Here is the data as published in his 1857 paper. See Weisberg for a very complete analysis of this data. Boiling point (BP) is measured in degrees Fahrenheit and pressure (Pr) in inches of mercury,

BP	Pr
194.5	20.79
194.3	20.79
197.9	22.40
198.4	22.67
199.4	23.25
199.9	23.35
200.9	23.89
201.1	23.99
201.4	24.02
201.3	24.01
203.6	25.14
204.6	26.57
209.5	28.49
208.6	27.76
210.7	29.04
211.9	29.88
212.2	30.06
3450.2	426.10

$\sum BP \times Pr = 86,755.435,$

$\sum (BP)^2 = 700,759.02,$

$\sum (Pr)^2 = 10,825.6366.$

(a) If Forbes wants to use a regression model, which variable is X and which Y? Explain.

(b) Without doing any calculations, should b be positive or negative?

(c) Give the estimated regression line for the model.

(d) Does this data support use of a linear model?

(e) How fast does the atmospheric pressure decrease as you climb?

The Coefficient of Determination

Here, we'll get a deeper insight into the test of the linearity assumption presented above, by introducing a measure of the strength of the linear relationship between the observed X's and Y's. We begin with an algebraic fact about the observed data which is not at all easy to prove[3] (you're welcome to try if you wish):

$$\begin{array}{ccccc} \Sigma(Y-\overline{Y})^2 & = & \Sigma(Y-\hat{Y})^2 & + & \Sigma(\hat{Y}-\overline{Y})^2, \\ \text{SST} & = & \text{SSE} & + & \text{SSR}. \end{array}$$

You've already encountered SSE, the error sum of squares. The notation SST is the ***total sum of squares*** and SSR is the ***regression sum of squares***. Note that SSR measures how far the estimated regression line is from the mean of all observed values of Y. Now, if X were not available, inferences about Y would all be based on $\overline{Y}$ with variability measured by SST. Once we bring X into consideration, we'll use the more informative $\hat{Y}$ and measure variability with SSE. You've already seen how this occurs: SSE is the numerator for the standard error of b. It shows up in other places as well, as we'll see later.

Note that SSR is the *reduction* in variability that becomes possible by introducing the explanatory variable X through the regression model. Without X, SST measures variability. With X, SSE measures variability. SSR is the discrepancy between the two. For this reason, SSR is sometimes called the "explained" variability—that much of the total variability accounted for by X. Now, if you divide by SST, you get the *proportion* of total variability explained by X, the so-called ***coefficient of determination***, denoted by r^2

$$r^2 = \frac{\text{SSR}}{\text{SST}}.$$

The choice of symbol—writing it as a square—would suggest that r^2 must be non-negative. Indeed, it is because it's made up of squared numbers. Furthermore, it's at most one, as you can see if you divide the

[3] At least one book claims this fact is obtained by "squaring and adding the terms" of the obvious identity: $(Y-\overline{Y}) = (Y-\hat{Y}) + (\hat{Y}-\overline{Y})$. But that's completely fallacious: Does $5 = 2+3$ imply $5^2 = 2^2 + 3^2$??!

original equation by SST:

$$1 = \frac{\text{SSE}}{\text{SST}} + \frac{\text{SSR}}{\text{SST}},$$
$$1 = \text{positive} + r^2.$$

Because it's a number between zero and one, we're justified in describing r^2 as a proportion. Or multiply by 100 and interpret it as a percentage. Then r^2 is the percentage of variability explained by X.

Now, the numerator of r^2 contains $\hat{Y} - \overline{Y}$ which simplifies algebraically to

$$\hat{Y} - \overline{Y} = a + bX - \overline{Y} = (\overline{Y} - b\overline{X}) + bX - \overline{Y} = b(X - \overline{X}).$$

So we get a useful identity:

$$r^2 = \frac{b^2 \, \Sigma(X - \overline{X})^2}{\Sigma(Y - \overline{Y})^2}.$$

For our researcher's new set of bamboo data, all the pieces of this expression have been calculated already except the denominator which is $\Sigma(Y - \overline{Y})^2 = \Sigma Y^2 - (1/n)(\Sigma Y)^2 = 11{,}005.0588$. And we get

$$r^2 = \frac{(7.2594)^2 \times 200.4706}{11{,}005.0588} = 0.9600.$$

Thus, about 96% of the variability in the height of these bamboo shoots is "explained" by elapsed time since planting. By introducing "number of days since planting," we significantly reduce the observed variability in the height of the bamboo shoots.

If you've entered your data into the statistical mode of your calculator, the variances for the two variables will be available at one or two key strokes. So you may prefer to calculate using the following expression for r^2

$$r^2 = b^2 s_X^2 / s_Y^2.$$

To get this from the previous formula, divide the numerator and denominator by $n - 1$. A similar analysis gives $s_b^2 = [(1 - r^2)/(n - 2)] s_Y^2 / s_X^2$, which is convenient if your calculator gives r (or r^2).

At first blush, one might expect the change in the average of the Y's for a unit increase in X to be determined by s_Y / s_X. But it's not; it's given by b instead. And $|b| \leq s_Y/s_X$. This inequality holds because $r^2 = b^2 s_X^2 / s_Y^2$ and $r^2 \leq 1$. So the change in the average of the Y's for a unit increase in X is actually less than s_Y/s_X. This accounts for the so-called "regression effect." It's the effect noticed by Galton, who saw it as a tendency for the values of Y to "regress" toward the mean, $\overline{Y}$. In one study, Galton was "explaining" the heights of adult children from the "midparent" height (essentially, the mean height of the two parents). For that data (as well as for his sweet pea data), $s_Y/s_X = 1$. He seems to have felt that a three-inch difference in midparent height should go with a three-inch difference in average height for the children. In fact, there was a difference of only two inches in the average children's heights.

You can see the "regression effect" in the picture below. Note how points on the regression line are always closer to $\overline{Y}$ than are the points on the line with slope s_Y/s_X.

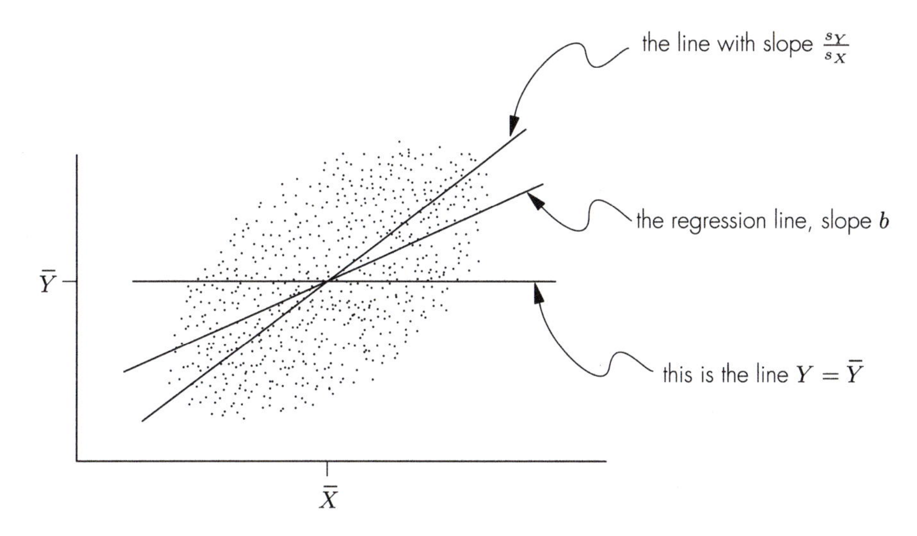

You can see that r^2 measures, in some sense, the correlation between the observed X's and Y's. The degree of correlation between two variables is very important, as you can imagine. Think about these examples: Is lung cancer correlated with smoking? Is high income correlated with education? Is IQ determined by nurture or nature? Is it correlated with parents' IQ or with factors in the early childhood environment? On and on! We'll not pursue this topic further. Instead, we refer you to the excellent elementary presentation—explaining potential abuses—given in Chapters 8 and 9 of [Freedman, Pisani, and Purves].

More often than not in actual practice, correlation is not expressed through r^2, but rather through the ***correlation coefficient***, $\boldsymbol{r}$, which can be calculated as the square root of r^2 with the sign of b. Here's the formula: $r = bs_X/s_Y$. This formula gives a deeper insight into the test of the linearity assumption. Not only are you testing against $\beta = 0$, but at the same time you're testing the parameter ρ (the Greek letter rho) for the model. The parameter ρ is estimated by r and is for the model what r is for the data—if there is such a parameter! The model doesn't always have a parameter ρ. It does only if X is randomly generated along with Y. In that case, ρ is a measure of the linear correlation between X and Y. And if X and Y are "jointly normally distributed," our test for β is simultaneously a test for ρ. So in some cases, the test for linearity can be thought of as testing for a significant degree of linear correlation between X and Y.

We don't study correlation in its own right. Freedman, Pisani, and Purves give an excellent overview of this topic. Look at it! They have a very interesting discussion of a well-known study by Skeels–Skodak from the 1930s of IQ for adopted children, addressing the nurture/nature question. This example also illustrates quite dramatically how r taken by itself without the associated means and standard deviations can be very misleading. Look at that discussion!

Note also the discussion by Freedman, Pisani, and Purves of the so-called "fallacy of ecological correlations," correlations that appear with rates or averages but vanish when the actual underlying numbers are considered. For example, based on U.S. Census Bureau data from 1970, they find a strong correlation ($r = 0.7$) between average age and average income for the men in nine geographical regions of the United States, but the correlation for the men as individuals is quite weak ($r = 0.4$). Well, who cares about the regions?! It's individuals we're interested in! By looking at r for rates or averages instead of r for the actual numbers, you could be seriously misled into accepting a fallacious correlation when, in fact, there is no correlation at all. The discussion of Freedman, Pisani, and Purves shows quite clearly how this fallacy arises. Look at it. Chapters 8 and 9 of their book. It's an easy read if you got THIS far in THIS book.

Now please . . .

Try Your Hand

7.3.3 We've seen that $r^2 \leq 1$. Show that equality holds if all the observed values of Y actually lie on the line $\hat{Y} = a + bX$.

7.3.4 Think of a test–retest situation. Let X and Y be the scores on the first and second tests, respectively. You might expect to have $\sigma_X = \sigma_Y$. Assume that's true. Explain why, as a group, persons who score far above average on the first test will, on average, score less on the second test. What about low scorers?

7.3.5 For each of the following pictures, guess the least squares line (the estimated regression line) and guess a value for the coefficient of determination. Verify your guess by computing a, b, and r^2 and then actually plotting the line.

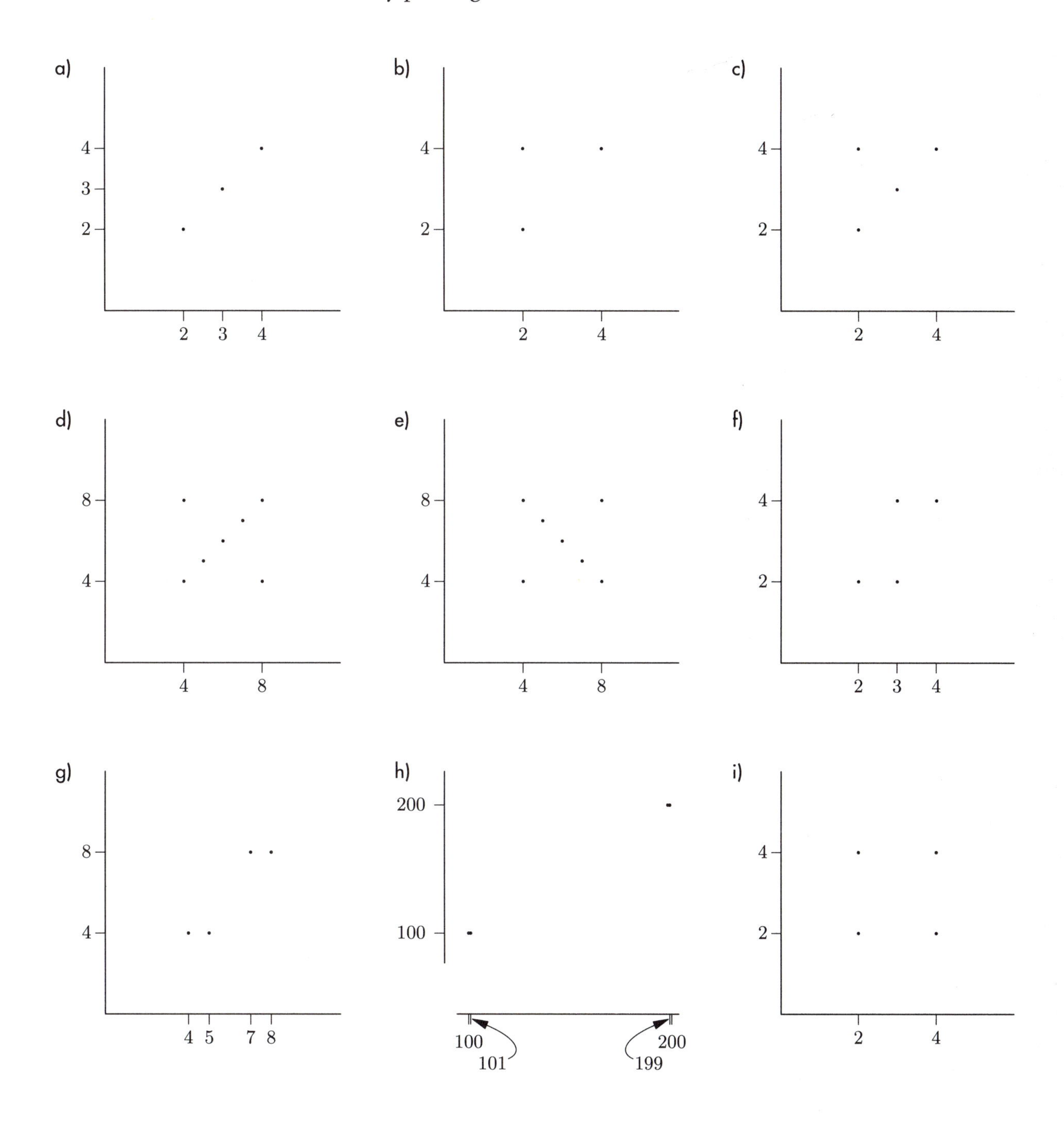

7.3.6 Let's look at some real data.

(a) Is it reasonable to think the boiling point of water would serve to guage altitude above sea level? Use Forbes' data from Problem 7.3.2.

(b) Why does the coefficient of determination not have much meaning in the Edwards–Eberhardt study (Problems 3.6.16 and 7.2.12)?

(c) Why have we not asked about Wilm's study in Problem 7.2.10?

7.3.7 Let's look at some of the ways the coefficient of determination (or the correlation coefficient) could be misleading as a measure of "strength of correlation between X and Y."

(a) Show that r^2 can be radically influenced by the number of observations, being relatively meaningless when n is small. Take an extreme case—suppose $n = 2$.

(b) What does part (a) say about the coefficient of determination?

(c) If you have a large number of observations and r^2 is quite close to one, you can conclude that X and Y seem correlated. Then it's a question of seeing which variable is actually causing the other. Comment.

(d) Plot a scatter diagram for the following data and compute r^2. Then for each X, replace the Y's observed at that level of X by their averages and plot that scatter diagram. Then compute r^2. Comment on the results.

X	1	2	3	4
Y	1	1	1	1
	2	2	2	3
	2	3	3	3
	4	4	4	4

X	1	2	3	4
Average of y's	2.25	2.5	2.5	2.75

(e) Plot the scatter diagram for the data in part (d) where instead of the Y's, you record for each X the "proportion above two." Calculate r^2. Comment.

(f) For the following data, sketch a scatter diagram and then calculate and interpret r^2:

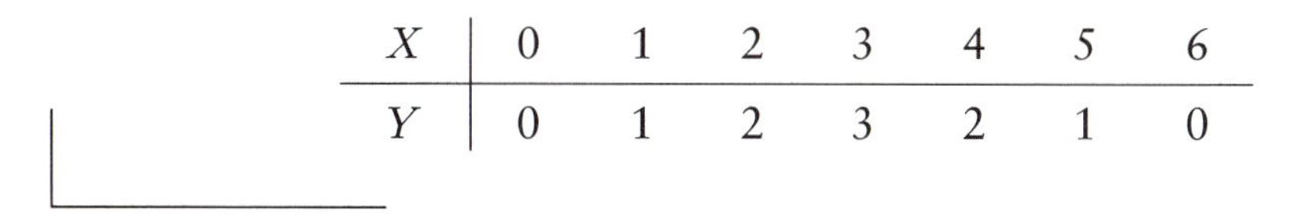

X	0	1	2	3	4	5	6
Y	0	1	2	3	2	1	0

Confidence Intervals to Predict $\mu_{Y|X}$ or the Average of a Few Y's for X_p, a Particular Value of X

Now we fix attention on one particular value of X, denoted by X_p. It should fall within the range of X's covered by our observed data, of course, but otherwise is unrestricted. So how can we use our model to answer questions about Y when X is fixed at X_p? For example, what's the average height 15 days after planting ($X_p = 15$) for all bamboo shoots of the type studied by our researcher and grown under similar conditions? Because your answer to this will be based on sample data, a confidence interval is required with endpoints

$$\hat{Y} \pm t \times \text{s.e.}$$

What's the appropriate standard error for this problem? Well, in fact there are two problems here: estimating the mean, $\mu_{Y|X}$, is one. The other is the "prediction" problem: estimating the average of a few, let's say m, values of Y. The second includes the problem of predicting one value of Y ($m = 1$). You'll see that the standard error in each case is analogous to the standard error for the corresponding nonregression prediction problems.

the squared standard error for predicting $\mu_{Y|X}$ at X_p:

$$s^2_{Y|X}\left[\frac{1}{n} + \frac{(X_p - \overline{X})^2}{\Sigma(X - \overline{X})^2}\right], \quad \text{d.f.} = n - 2.$$

Now, for the bamboo, what's the average height 15 days after planting? Here are the endpoints for a 95% confidence interval, using our researcher's new (valid) bamboo data:

$$\begin{aligned} \hat{Y} &\pm t \times \text{s.e.}, \\ a + 15b &\pm 2.1315 \times 1.3161, \\ 45.0458 &\pm 2.8052. \end{aligned}$$

So we can be about 95% sure that bamboo shoots of this type 15 days after planting will be, on average, between about 42 and 48 cm tall. Note that we calculated $s^2_{Y|X}$ and $\Sigma(X - \overline{X})^2$ earlier when we computed s_b. The value of t here is the same as we used earlier in the 95% confidence interval for β. After all, it's the same data and the same confidence coefficient and so we have the same degrees of freedom and the same column of the t table.

The standard error is calculated as the square root of

$$29.3674\left(\frac{1}{17} + \frac{(15 - 14.8235)^2}{200.4706}\right) = 1.7321.$$

Finally, the last question we'll address using the regression model is "What's the average of m values Y when X takes the value X_p?" For the bamboo shoots: "How high will these four shoots be 15 days after planting?" We use the estimator $\hat{Y}$, the same as before. But we're asking a harder question here—it's harder to predict the average of a few rather than to predict the mean of all. For the overall mean, you can get a few values wrong and still get the mean right if the errors cancel. But when you're looking at only a few, you have little leeway.

Because the question is harder, you can expect to see a larger standard error. The standard error for this problem is calculated from

the squared standard error for predicting a particular Y at X_p:

$$s^2_{Y|X}\left[\frac{1}{m} + \frac{1}{n} + \frac{(X_p - \overline{X})^2}{\Sigma(X - \overline{X})^2}\right], \quad \text{d.f.} = n - 2.$$

Note that the only difference between this standard error and the previous one is the addition of $1/m$ within the large parentheses. To understand this, suppose $m = 1$. That puts a "$+1$" in the square bracket. It has the effect of adding $s^2_{Y|X}$ to the squared standard error. That's not surprising because $s^2_{Y|X}$ is an estimate of $\sigma^2_{Y|X}$, which measures the variability of $Y|X$ about its mean. So the "$+s^2_{Y|X}$" measures how much harder it is to predict just one value of Y than it is to predict simply the mean itself. To see the logic of this analysis more precisely, look at the analogous nonregression model which we studied in Problem 5.5.6(g).

Well, now please . . .

Try Your Hand

7.3.8 Based on the new set of bamboo data (see page 302 and Problem 7.2.9):

(a) See this little shoot? How tall will it be 15 days after planting?

(b) How high will this field of bamboo be 15 days after planting?

(c) See these seven little shoots of bamboo coming up? They were planted just one week ago. How high do you think they'll be in eight more days?

7.3.9 Let's think about the formulas for the standard errors given in the text above:

(a) What's the difference between the standard error for predicting a particular Y as opposed to the average of all Y's?

(b) Why does $1/n$ appear in each of the standard errors?

(c) Why does it make sense that $(X_p - \overline{X})^2$ would be in the numerator?

(d) Why does it make sense that $\Sigma(X - \overline{X})^2$ would be in the denominator?

7.3.10 Here are some simple sets of data just for your practice:

(a)

X	3	3	5	7	7
Y	2	5	4	4	5

(b)

X	1	4	4	5
Y	8	3	6	2

(c)

X	8	5	2	2
Y	8	4	5	1

(d)

X	3	5	7	9
Y	2	6	10	14

For each set of data, (i) draw the scatter diagram and plot the regression line. Then calculate (guess first, if feasible): (ii) $s^2_{Y|X}$; (iii) s_b; (iv) the check for linearity; (v) a confidence interval for β; (vi) r^2; (vii) a confidence interval for $\mu_{Y|X}$ with $X_p = 3.5$; (viii) a confidence interval for Y_p with $X_p = 3.5$.

Be sure you can do all of this by completing the tables with appropriate columns and then calculating from the column sums. You can check your work by entering the data into the statistical mode of your calculator.

7.3.11 The following data is from a report of the U.S. Department of Agriculture entitled "Results of Fiber and Spinning Tests for Some Varieties of Upland Cotton Grown in the United States, Crop of 1944." We've significantly abbreviated the data for ease of analysis, see [Duncan], page 813, for a more complete description and analysis.

Here S is skein strength, "perhaps the most important single index of spinnning quality" for cotton, and L is fiber length. S is measured in pounds (to breaking point) and L in hundredths of an inch. See [Duncan] for a more exact description of these measurements.

S	91	114	99	87	100	103	91	110
L	76	79	65	74	74	92	68	77

(a) Does this data suggest that a linear model would be appropriate to study spinning quality on the basis of fiber length?

(b) How do the two approaches to part (a) relate? [Hint: Both S and L are generated together randomly and it's not unreasonable to think they are "jointly" normally distributed.]

(c) What was the change in average skein strength per hundredth of an inch in fiber length for the 1944 crop of cotton sampled here?

(d) What would be the spinning quality of a batch of cotton from the 1944 cotton crop which had a fiber length of about 0.7 inch?

7.4 Some Review Problems

Here's a mixed set of problems covering this and the previous two chapters. Give complete real-world answers to questions involving real-world situations. For hypothesis tests, give a formal statement of the hypotheses and the formal conclusion, "reject H_o" or "fail to reject H_o" as well as the real-world conclusion and a statement of the possible error you might be making with that conclusion. For tests of significance, calculate the *p*-value, identify it as SMALL or NOT SMALL, and then interpret that in real-world terms. It's very important that you be able to identify the TYPE of each problem. You would do well to go through the problems first just to identify the problem type. Before doing a regression problem, evaluate the model by interpreting r^2; you need not do an hypothesis test of the linearity assumption unless requested to.

All questions are based on the data described below. You will have to determine what part of the data is relevant to each question.

The Data

For budget planning purposes, Residential Management, Inc. wants to be able to estimate at the beginning of the quarter the quarterly maintenance supply costs for any one of their large apartment complexes. They manage several hundred complexes. A complex is considered large if it contains at least ten units. You are brought in as a consultant on this and a number of other questions. Residential Management, Inc. routinely allows a ten percent risk of error in its statistical analyses. Using an appropriate sampling process, you obtain the following information from the corporations's records:

A #of units in complex	B Quaterly maintenance supply costs ($100)	C # of residents on first day of quarter
38	17	110
62	19	125
51	18	140
72	23	150
32	22	165
64	20	170
91	25	190
87	24	200
115	27	210
102	29	215
714	224	1675

where $\sum A^2 = 57,672$ $\quad \sum AB = 16,808$

$\sum B^2 = 5,158$ $\quad \sum AC = 126,800$

$\sum C^2 = 292,375$ $\quad \sum BC = 38,700$

7.4.1 Without any reference to the number of units or residents, how would you answer Residential Management's original question? In other words, what would you estimate to be the "quarterly maintenance supply costs for any one of their large apartment complexes"?

7.4.2 The management is considering a tentative plan which it may put into operation. Among other considerations, they would like to know if this data seems to call into question the records from the Leasing Office of two years ago that there are at least 200 residents per complex on average in Residential Management's large apartment complexes.

7.4.3 For the management's tentative plan, it would also be helpful to know if this data casts doubt on today's validity for the records of two years ago, indicating that at least 90% of the large complexes have more than 130 residents. Does it?

7.4.4 Give an equation to estimate the average quarterly maintenance supply costs as a function of the number of residents in a complex on the first day of the quarter.

7.4.5 Do an hypothesis test to check if the linearity assumption of the model in Problem 7.4.4 seems to be satisfied.

7.4.6 The manager of a complex with 130 residents asks for an estimate of the probable maintenance supply costs for the coming quarter. What do you report?

7.4.7 Estimate the increase in maintenance costs for one of the complexes run by Residential Management, Inc. incurred with one new resident.

7.4.8 How many units are there in one of the large apartment complexes run by Residential Management, Inc., on average?

7.4.9 You are asked how many complexes have at least 200 residents on the first day of a quarter.

(a) Because our sample has only ten observations, we cannot use our normal distribution for the estimator. Why not?

(b) What minimum sample size would be required? [Hint: Use the present observations.]

(c) What sample size would be required to estimate the percentage of complexes with at least 200 residents to within ten percentage points with no more than a one percent risk of error?

(d) If you use the sample size of part (c), what's the smallest value of p for which the normal approximation would be valid?

7.4.10 What is the average number of residents in a complex with 50 units?

7.4.11 Management wants to make sure maintenance supply costs don't get out of hand—they should remain below $2500 per complex per month. Each month, they will examine the records from a small randomly chosen sample of several complexes. If that data ever suggests a problem, a careful examination of the records of all complexes will take place. What would you suggest as a criterion for determining when the more careful examination seems to be required? Suppose you have decided on a sample of size ten.

7.4.12 The complex across from the maintenance building has 93 units. How many residents do you suppose there are in that complex?

7.4.13 Based on your data, how much must Residential Management pay out in maintenance supply costs per unit in one quarter for a given complex?

7.4.14 At the beginning of the quarter, it was projected that this quarter's average maintenance supply costs would be below $2500 per complex on average. Does your data bear out this projection?

7.4.15 In this quarter, how many residents were there on average in those complexes which incurred maintenance supply costs of $2000?

7.4.16 A procedure is in place with a ten percent risk of error which alerts management to the need for a further investigation if the quarterly data ever suggests the maintenance supply costs have risen above about $13 per resident. Is a further investigation called for this quarter?

7.4.17 A journalist who wants to know how many units Residential Management operates in all of their complexes has managed to get hold of your data. What conclusion might she come to on the basis of this data?

7.4.18 Is this quarter's data consistent with the report released earlier that less than half of the complexes this quarter would incur maintenance supply costs in excess of $2500?

7.4.19 One of the managers claims no more than 20% of the complexes this quarter incurred maintenance supply costs in excess of $2500. Does the data bear her out?

7.4.20 Another one of the managers says, "Well, surely no more than 30% incurred costs in excess of $2500!" Does the data bear her out?

7.4.21 Let's settle this once and for all! How many of the complexes incurred maintenance supply costs in excess of $2500 this quarter?

7.4.22 How many residents would you expect on average this quarter in those of Residential Management, Inc.'s complexes which reported $2500 in quarterly maintenance supply costs?

7.4.23 Suppose, in fact, the maintenance supply costs this quarter for all Residential Management, Inc.'s complexes averaged $2387 with a standard deviation of $318. What's the probability of data like yours or worse for this variable? What assumption must you make here? How might it fail?

7.4.24 How likely is it that less than half of a sample of 25 complexes would report maintenance supply costs in excess of $2500 if, in fact, the true proportion for all complexes is 37%?

7.4.25 Of the complexes you sampled this quarter, how many had more than 150 residents?

Level

I

Answers for Try Your Hand Section

Chapter 1

1.1.1

(a) For the die, X is the number of dots on the uppermost face after rolling the die. For the coin, X is . . . ?

(b) Each of the six possible values of X are equally likely (the die is fair), so all the probabilities are the same. What are they?

(c) In the text, you find that "probability" means "theoretical relative frequency." Explain this in terms of the probability ($= 1/52$) of drawing the ace of spades. What's "theoretical" about it and what is the "relative frequency"?

1.1.2

(a) Here you want the probability that X is 1 or 2. Symbolically, you're asked for $P(X < 3)$. The condition $X < 3$ means $X = 1$ or $X = 2$. What is it?

(b) Here you want the probability of a six on the first die and a six on the second die. The probabilities do not add here. Make a guess!

(c) Yes?

1.1.3

Addition corresponds logically to "or" and multiplication to "and." This rule has important restrictions but is valid in our examples. Restate the examples in Problem 1.1.2 in these terms.

1.1.4

(a) Both terms refer to a repeatable "something you do." This rather vague expression is made more precise by . . . ? What's the difference between the two definitions?

(b) A random experiment is a kind of scientific experiment. What is it that makes it "random"?

1.1.5

(a) Give the probability distribution in a table. This requires knowing the probabilities for the other faces. Did you notice this information was missing? If so, congratulations! Let's assume the other faces are all equally likely. Now give the probability distribution for X.

(b) Yes?

(c) The verbal descriptions are the same: "The number of dots on the uppermost face." And the two random variables have the same values: the positive integers 1 to 6. The random variables differ only because they are talking about two different dice. Dice are physical objects. So the difference lives in the real world.

The difference is in the underlying random experiments. But exactly what about the experiments is different? After all, even the outcomes of the experiments have the same verbal description: the die sitting on the table top in some position. So, with specific reference to the definition of the term "random experiment," where's the difference? Think about the picture of a random variable as a bridge. Be as exact as possible. You can specify the exact phrase in the definition which accounts for the difference.

1.1.6

Y	$P(Y)$
1	0.26
2	0.26
3	0.22
4	0.26
	1.00

You should include the total of the $P(Y)$ column as a check on the condition $\sum P(Y) = 1$. If that column doesn t add to one, there s a mistake somewhere.

What's the meaning of ΣY for this example?

1.1.7

(a) This problem will be very useful to help in understanding random variables. Here's how you can organize the answer: Make a two-column table as shown below. In the first column, show the "doing" of the random experiment and in the second column give a verbal description of the rule which associates a number to each outcome:

	andom experiment	*andom variable*
1.1.2 a		
b		
c	[FILL IN THIS TABLE]	
1.1.5 a		
b		
1.1.6		

(b) and (c) If you didn't get these two parts of the problem, try again using the information above. Don't forget, you should complete part (a) through level II before trying parts (b) and (c).

1.1.8

(**a**) It's a probability distribution alright, a presentation of the values of X together with their probabilities, but it's not the probability distribution of a *random variable*. Why is X not a random variable? Think about this carefully before you read level II; it's an important point.

(**b**) This new X, is indeed, a random variable! Why? Note that X has a simple verbal description. What is it?

(**c**) Yes?

1.1.9

(**a**) For n rolls, what's the *fewest* number of dots possible? The *most* number of dots possible? After thinking about that you should be able to say how many dots would you expect altogether for a typical n rolls.

(**b**) Yes?

(**c**) Half the time you get two dollars and half the time three. So what do you get on average?

(**d**) Think this way: Nine times out of ten you get two dollars and once out of ten, three dollars. This is theoretical, of course. On a specific ten tosses, the results could be quite different. Theoretically, you get two dollars nine times and three dollars once. So what's your "on average" take?

(**e**) Before you can talk about a random variable you must first identify clearly the underlying random experiment AND ESPECIALLY THE OUTCOMES! Then you can verify that you have a random variable by giving the rule which associates a number to each outcome.

(**f**) Yes?

1.1.10

This X IS a random variable! X satisfies the definition of random variable exactly in every detail: It's a rule which associates a number to each of the possible outcomes of a random experiment. But then how do you account for the fact that the value one is certain? What's random about that?

1.1.11

Let Y be your gain/loss on one roll with this die. For example if you roll five dots, Y takes on the value one—you get five dollars for the five dots, but you paid out four dollars to play. Because Y is a rule which assigns a number to each of the possible outcomes of the random experiment, it's indeed a random variable.

(a) What's the random experiment? Verify each of the conditions in the definition of random experiment.

(b) What are the possible values of Y?

(c) Give an equation for Y in terms of X, where X is the number of dots on the uppermost face. This may look hard, but don't get discouraged! Think exactly how you would determine your gain/loss on any particular roll.

(d) Give the probability distribution for Y. This is easier than it looks! Which outcome of the experiment corresponds to which value of Y? What's the probability of that outcome?

1.1.12 Yes?

1.2.1 (a) Yes?

(b) Did you guess that $E(X) = 1$? The phrase "on average" is crucial. For one particular toss who knows what would happen?! The expected value only refers to what happens *on average* when you toss the two coins many times. Thus, for example, $P(X = 2) = P(HH) = P(\text{head AND head}) = 0.5 \times 0.5$, and so on. Now, fill in the table to compute $E(X)$.

X	$P(X)$	$XP(X)$
0	??	??
1	??	??
2	0.25	??
	1.00	1.00

so $E(X) = 1$.

(c) Here's how to guess: You should expect more dots on average than for a fair die, more than 3.5. Five is the most likely number of dots (50% chance) and there's only one value larger than five, but four smaller, so the expected value should be something short of five. So guess that $3.5 < E(X) < 5$. Now calculate $E(X)$.

(d) Answer: \$2.50. You got this heuristically in Problem 1.1.9. Now calculate it from the probability distribution for the appropriate random variable—the random variable X which takes the value two for the outcome "coin lying on table with heads up" and takes the value three for the "coin lying on table with tails up."

1.2.2 The key here is the phrase "in the long run." You have to think about ten plays or 100 plays. Then think about n plays, the "general case." How much will you pay out and how much are you likely to take in? Of course, to come out even, what you pay should equal what you expect to take in. So, what would you pay to play?

1.2.3 **(a)** The question is, what's $P(X = 1 \text{ or } 6)$? Use the "or" rule (see Problem 1.1.3). That rule doesn't always apply, but it's valid here. Later, we'll see the general rule.

(b) and (c) Yes?

1.2.4 For the value $X = 1$, the deviation from the mean is $X - \mu = -2.5$.

1.2.5 **(a)** Yes?

(b) For example, $X - \mu > 0$ means the deviation is positive. Now, a simple manipulation of this inequality will show you its heuristic meaning, its *practical* meaning.

(c) Here you don't know anything specific about X, so you'll have to rely on the abstract formula for the average deviation: $\Sigma(X - \mu)P(X)$. If you play with this algebraically, you'll discover it's just zero. First, distribute the $P(X)$ over $(X - \mu)$ by writing it as $XP(X) - \mu P(X)$ and go from there.

1.2.6 **(a)** Not true! Give an example of a loaded die that's MORE predictable than a fair die, keeping the expected number of dots per roll at 3.5. So you must "invent" appropriate probabilities. Show that your example is valid by computing the variance and then comparing it with the variance of a fair die. Make sure your die has an expected value of 3.5.

(b) You should give a graph for the die which you constructed in part (a) similar to the one below for the fair die. Be sure you see how your graph illustrates the point made in part (a).

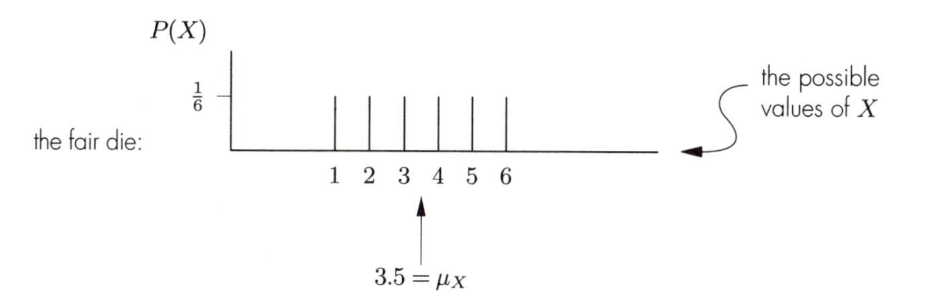

1.2.7 Note that 90% of the probability for this coin is concentrated at one, very close to the mean of 0.9. For the fair coin, both values, $X = 0$ and $X = 1$, are comparatively far from the mean of 0.5. Thus, the unfair coin is *less dispersed* about its mean than a fair coin.

(a) So, should the variance of our unfair coin to be larger or smaller than for a fair coin?

(b) and (c) Yes?

(d) Note that X takes the value zero with probability $1 - p$ and the value one with probability p. Set up the probability distribution table with these values. You'll be able to show that the mean is p and the variance is $p(1 - p)$.

1.2.8 **(a)** Yes?

(b) You can calculate the mean absolute deviation by just adding an appropriate column just as you did to calculate σ^2.

(c) Here are the answers: 1.5133, 1.6, and 1.1356. How?

1.2.9 Here's an example of a constant random variable: Every time you roll a die, I give you two dollars no matter what face of the die comes up. For this random variable, you get two dollars per roll, on average. *Exactly* two dollars per roll. The phrase "on average" is just a technicality because you always get the same amount.

Similarly, suppose the random variable X always takes on the constant value c, so that $X = c$ is true for all outcomes. Then we expect to get exactly c on every repetition of the random experiment. So we guess that $\mu_X = c$. Now, how much variability is there in the values of X? NONE. So the variance ought to be zero. Show analytically (use the formulas) that these values are correct: If $X = c$, then $\mu_X = c$ and $\sigma_X^2 = 0$.

1.2.10 This is a very instructive problem. Give it some careful thought before you look at the answer in level II. Let's think about the die described in the text for which $P(X = 1)$ and $P(X = 6)$ are both 0.3, whereas all the rest of the faces are equally likely. Consider these questions:

(a) The difference in the two dice is reflected in their variances. How?

(b) What does part (a) say about the expected number of dots for the loaded die compared with the fair die?

(c) What about the unfair coin which comes up heads 90% of the time? Which is more predictable, the fair coin or the unfair coin?

(d) Use the result in the level II answer to Problem 1.2.7(d) to calculate the variances for the coin which comes up heads 90% of the time and for the fair coin.

1.2.11 What does the variance measure? What does the standard deviation measure? Exactly what is the difference?

1.2.12 (a) Call the gain/loss random variable G. Then if X is the number of dots on the uppermost face,

$$G = \underset{\substack{\text{what you}\uparrow \\ \text{receive}}}{X} - \underset{\substack{\uparrow\text{what you} \\ \text{pay to play}}}{\mathrm{E}(X)}$$

Because you get one dollar per dot, your "expected receipts" is exactly $\mathrm{E}(X)$, the expected number of dots. If you pay that amount to play, your expected gain/loss will be zero. You come out even in the long run; no gain, no loss (Problem 1.2.2).

The predictability of your gain/loss is measured by how *variable* it is about this "break even" value of zero. If your gain/loss is concentrated close to zero, the game is highly predictable, otherwise not. Now complete part (a): For each of the three dice described in the problem, compute the variance of the gain/loss random variable. Then using these computed variances, compare the predictability of that gain/loss for the three dice.

(b) For the three dice, the house should charge respectively $4.00, $3.40, and $3.70. How do you derive these numbers?

(c) First, identify clearly the underlying random experiment (there are four things to verify) and then identify the random variable itself.

1.3.1 So $\hat{p} = 0.72$ (NOT 72, $\hat{p}$ is the proportion, not the percentage). Note that 72% of 150 is 108. Now write out the equation with the correct numbers substituted for the symbols.

1.3.2 (a) Let P be the amount you pay to play. Because you receive one dollar per dot, your "expected receipts" is $\mathrm{E}(X)$. Thus, $P = \mathrm{E}(X)$ and so $G = X - \mathrm{E}(X)$. Write this in the form $G = a + bX$. Note that in the general form for a linear equation, $Y = a + bX$, a is the constant term and b is the coefficient on X.

(b) Use part (a) above together with the formula: $\sigma_G^2 = b^2\sigma_X^2$.

(c) You must explain why it makes sense in this game that the variability of the gain/loss would be the same as the variability of the die.

(d) It's obvious that $\mu_G = -0.5$, why? Derive this formally using the fundamental equations for linearly related random variables. What's the variance?

1.3.3

(a) $\mu_Y = \Sigma Y \mathrm{P}(Y) = \Sigma Y \mathrm{P}(X)$ because, for the same outcome, the probabilities of Y and X are equal. Now, substitute $Y = a + bX$ into this equation and simplify.

(b) Again, just substitute $Y = a + bX$ into the formula.

1.3.4

(a) $WY = a + bSC$. So . . . ?

(b) About 12 inches. How? Use the fundamental equations for linearly related random variables.

(c) If $Y = a + bX$, we can get $Y = a$ if . . . ?

(d) The second. Why?

(e) Think about $SC = 0$.

1.3.5

(a) You may say you're not a gambler and wouldn't play such a game at all! But suppose you ARE willing to play such games. Still, you can't say what will happen because you don't know the probabilities for this die. Did you notice the problem would require more information before you could say anything meaningful?

Suppose the face with one dot comes up half the time and all the other faces are equally likely. Would you play this game? In the long run, you should expect about 2.5 dots on average. Why? What would be your long run gain/loss in this game?

(b) How risky compared with what? If there's no other obvious comparison, you might want to compare it with the game involving a fair die. But for the comparison to make sense, the games should have the same expected gain/loss. What would you have to pay in playing with the fair die to get the same net result assuming that you still receive three dollars per dot? And which game is riskier?

(c) You evaluate the risk by comparing what you get (that's X) with what you *expect* to get (that's $\mathrm{E}(X)$). So it's not just a question of half the probability being on one value. What else must you consider?

1.4.1

(a) You know a club was drawn. So forget the 52 cards; you need think only about the clubs.

(b) $P(A \text{ and } C) = 1/52$ because exactly one of the 52 cards is an ace of clubs. Apply one of the rules to $P(A \text{ and } C)$ to get . . . ?

(c) An "event" is a set of possible outcomes of some random experiment. What is the random experiment for the event A? For C? How many outcomes?

(d) "Ace" is independent of "club" on one draw from a deck. Explain.

(e) $P(A|B)P(B) = P(B|A)P(A)$ because both are equal to $P(A \text{ and } B)$. So . . . ?

(f) "Equivalent" means when one condition holds so does the other. So you have to do two things: First, show that when the simple product rule holds, then A and B are independent. Second, show the "converse": Show that when A and B are independent, then the simple product rule holds.

(g) We've already had an example like this. Were the events independent?

(h) Well, $P(A|B)P(B) = P(B|A)P(A)$. Why? How does that help us?

1.4.2

(a) Under what conditions does $P(A \text{ or } B) = P(A) + P(B)$? There is a very specific technical term for this condition. What is it? What about the "and" rule?

(b) You might interpret this probability as $P(X = 2 \text{ OR } 3 \text{ OR } 4 \text{ OR } 5)$ or you might interpret it as $P(X \geq 2 \text{ AND } X \leq 5)$. One way is "natural" in the sense of being easy and direct (using one of our simple rules). The other way is possible, but hard. Which is which?

1.4.3

Yes?

1.4.4

(a) Use $P(T) = P(T \text{ and } D) + P(T \text{ and } D^c)$. In other words, analyze "tests positive" into two mutually exclusive events "tests positive and has the disease" and "tests positive and does NOT have the disease."

(b) Show that $P(D|T) \approx 60\%$. Is this good or not?

(c) The predictive value declines as the disease becomes less common. Why?

(d) $P(D|T_1 \text{ and } T_2) \approx 74\%$. How? Independence means $P(T_1 \text{ and } T_2|D)$ is just: $P(T_1|D)P(T_2|D) = 0.95 \times 0.98$, similarly for $P(T_1 \text{ and } T_2|D^c)$.

(e) You must show either that $P(D|T) \neq P(D)$ or that $P(T|D) \neq P(T)$, whichever is easier. If one is true, so is the other.

1.4.5 With T as "fits the profile" and D as "should be denied boarding," what is $P(D|T)$?

1.4.6 (a) What is the predictive value of the test? In other words, how likely is it a defendant is lying when the test says she is? In symbols, with L as "actually lying" and T as "tests says lying," show that $P(L|T) \approx 68\%$. Interpret this.

(b) $\approx 95\%$.

(c) Yes?

1.4.7 The absolute value has the following meaning:

$$\begin{array}{lll} \text{case I} & |u| = u & \text{if } u \geq 0; \\ \text{case II} & |u| = -u & \text{if } u < 0. \end{array}$$

Now, apply that to $|X - \mu|$. Note that "X is within k standard deviations of μ" just means that X takes a value somewhere in the interval with endpoints $\mu \pm k\sigma$

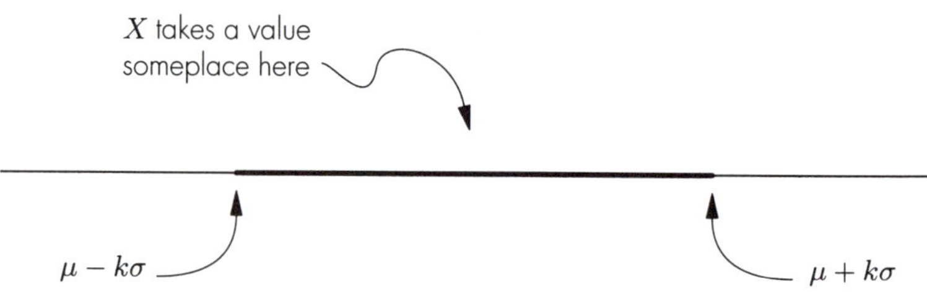

Think of μ as "my house" and σ as "one block." If Joe lives "within k blocks of my house," to find Joe you first go k blocks below my house, $\mu - k\sigma$. Then—all the while calling "Joe, Joe, Joe . . . "—walk toward my house and beyond until you've gone k blocks above my house, $\mu + k\sigma$. That stretch of road is the part of the road "within k blocks of my house." If Joe doesn't answer, either he's not at home or he doesn't want to talk to you. Or he's deaf. Or you're not calling loud enough. Because he lives somewhere in that stretch of road "within k blocks of my house."

1.4.8 (a) You can't answer this question exactly. You can only say there's at least a 75% chance to be within one standard deviation of the mean. Explain.

(b) Yes?

1.4.9 **(a) and (b)** Don't use Chebyshev here. Why not?

1.4.10 **(a)** 90% . This is still much stronger than what you would get from Chebyshev.

(b) 30% .

(c) Who knows what Chebyshev would say? But his theorem says the probability is at most about 83%. How? Why does this conflict with the 70% in part (b)?

1.4.11 **(a)** Give a "proof by contradiction." That is, suppose for all X it's true that $|X - \mu| < \sigma$. Show this leads to a contradiction and, therefore, it CANNOT be true.

(b) This does not contradict part (a) because it doesn't say "strictly."

(c) Parts (a) and (b) ask about "within one standard deviation of the mean." So in Chebyshev's Theorem, $k = 1$. So what?

1.4.12 We've seen a number of such examples. If you didn't get this, try again!

1.5.1 **(a)** The values of a constant random variable are all the same. There's only one value. What's *random* about such a variable? Suppose, for example, you toss a coin and the random variable always assigns the number 735, no matter which outcome you get. What's random about that? While you're thinking about this, what's the mean and variance of a constant random variable?

(b) ΣX is meaningless. Why? Each of the other expressions has a very specific meaning. What is it?

1.5.2 **(a)** There are four conditions to verify—what are they? Note that the outcomes could be described as "the tack lying on the table top in some position," but the problem suggests a simpler description of the outcomes.

(b) You can't do this! Why not?

(c) Your gain/loss random variable is $Y = 60X - 20$. How do you get this? Note that it gives the right result: \$40 if $X = 1$ and *minus* \$20 if $X = 0$. Here's the distribution of X:

X	$P(X)$	$XP(X)$	$(X-\mu)^2P(X)$
1	0.3	0.3	0.147
0	0.7	0	0.063
	1.0	0.3	0.210

Now, use this to compute the mean and standard deviation (it's $27.50) of your gain/loss.

(d) Tossing this thumbtack is, in an abstract sense, no different than tossing a coin which comes up heads 30% of the time. How do you use Problem 1.2.7(d)?

(e) Are you going to end up ahead in the long run? If not, would you play?

(f) Here your gain/loss is $Y = 10X - 5$ and your *expected* gain/loss is $10\mu_X - 5 = -2$, a loss of two dollars per toss, just as before. The two games are identical as far as your expected gain/loss is concerned. So, it's not a question of which game costs more on average per toss. Rather it's a question of which game is more exciting. In other words, which one offers the opportunity for big gains (and, consequently, big losses). A true gambler would prefer the game which is very exciting. A conservative or reluctant player would prefer the less risky game. Which game is more exciting/risky?

1.5.3

(a) The model describes barometric pressure as a __________ function of the boiling point of water. The parameters of the model are . . . ? Exactly what kind of object is the model in this case?

(b) About 206 degrees Fahrenheit. How?

(c) The degree measure of the boiling point would be more variable than the inches of mercury for barometric pressure. Why?

(d) For example, change *BP* from Fahrenheit to centigrade and show that now, in the new units, *Pr* and *BP* have virtually the same degree of variability. The relationship between Fahrenheit (F) and centigrade (C) is given by $F = 32 + 1.8C$.

1.5.4

(a) Look at the definition for the term "probability distribution"!

(b) We've given our distributions mostly in the form of a table. But two other modes of presentation have been mentioned—what are they?

1.5.5 (a) For each value of X, you must take the deviations from the mean, $X - \mu$, weight them by the appropriate probabilities, and then add that column. It should give you zero.

(b) Yes?

(c) The answer is 77%. How? Here's a hint: "One standard deviation from μ in either direction" means one σ to the left of μ (that's $\mu - \sigma$) and one σ to the right of μ ($\mu + \sigma$). So you need to find which values of X fall within that range: $\mu - \sigma < X < \mu + \sigma$. This is just an interval centered at μ extending one σ in either direction:

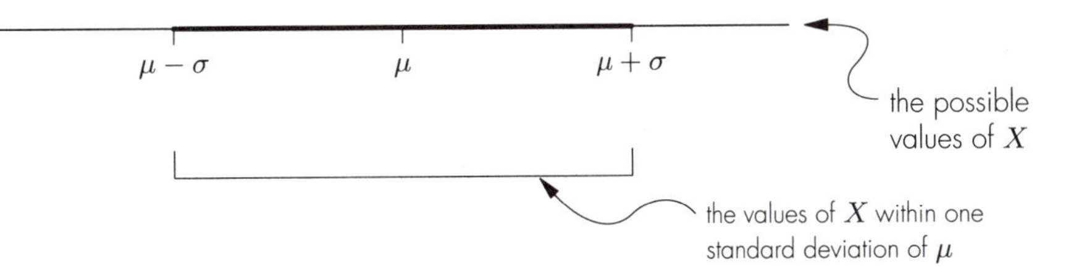

Now, first determine which values of X satisfy this condition and then answer the question: What proportion of the distribution is represented by these X's ?

(d) This problem will have to be done in a spirit of exploration, by trial and error. The statistical mode of your calculator will make the exploration go faster. To get a smaller variance, make Y less spread about its mean than X by putting less probability at the extreme values and more probability close to the mean. To avoid changing the mean as you change the probabilities, make the changes *symmetric* about the mean.

(e) This is like part (c). The answer is 88%. How?

(f) What is the real-world component of a random variable?

1.5.6 (a) You've got five things to verify. What are they?

(b) You cannot do this. Why not?

(c) You would want to know at least their "expected score." But that doesn't tell the whole story—what else is required?

(d) "Reliability" can only mean how predictable their game is. A player whose score on 100 throws is within one point of their expected score is much more reliable than a player who is only within ten points of their expected score. What must you know to make this comparison?

(e) You want to compute μ and σ for the score of an individual player. What information is required for that?

(f) Each player generates a different random variable. Explain.

(g) Your expected score is just shy of eight points with standard deviation a bit less than three points. How?

(h) You must concentrate the probability closer to the mean, and it should be a lower mean. What's meant by "improving" the measure of exactness?

(i) Your opponent. Why? Your score should fall between 3.6079 and 12.3121 for you to be within 1.5σ of your expected score. And your opponent?

(j) You are, by a narrow margin. Why? The range for you is (2.1572, 13.7628), and for your opponent, (2.0082, 7.9118).

1.5.7 First of all, what does the symbol $\Sigma_{X\neq 3}\mathrm{P}(X)$ mean? Make a guess!

1.5.8 (a) The repeatable "doing" is to roll the die. The outcomes? There is NO DIFFERENCE in the experiments for X and Y. Explain.

(b) Yes?

(c) Such a question could be quite vague. What kind of relationship are you looking for? Ordinarily, the first thing to look for is a simple equation relating the two random variables. There is such an equation for this X and Y—what is it?

(d) X is a linear function of $Y : X = a + bY$. Use the fundamental equations for linear functions of random variables.

1.5.9 (a) and (b) Here's how you could guess the answer: Suppose, for example, you concentrate half the probability on $X = 1$. That makes the mean relatively close to one. And the variance will be relatively large because the values 5 and 6 are far from the mean and together carry 20% of the probability. Of course, you could make up for that if you could put only a tiny amount of probability on those extreme values far from the mean. But you're required to make all other faces equally likely! Reasoning this way, you can see that for the variance to be small, you should concentrate the probability near the middle of the values of X.

(c) These are TWO DIFFERENT random variables. Go back through the definition and say *precisely* where in the definition the difference occurs.

1.5.10

(a) $\mu = 23.28$ and $\sigma = 0.8841$.

(b) $\mu = 0.986$ and $\sigma = 0.2608$.

(c) $\mu = 1.934$ and $\sigma = 0.1790$. What proportions of these distributions are within the required distances of the mean?

1.5.11

Here are the tables. Complete them to calculate the mean and variance:

(b)

X	P(X)
2	0.01
5	0.08
8	0.34
11	0.42
14	0.15

(c)

X	P(X)
17	0.22
18	0.17
19	0.14
20	0.17
21	0.12
22	0.18

1.5.12

(a) There's a simple and meaningful description of Y in terms of X. What is it? To see that, first calculate a couple of values of Y.

(b) Yes?

(c) The values of Y are described in part (a) and have the corresponding probabilities. Set up a table and compute μ and σ for Y.

(d) Because Y is the last digit of X and because Y has mean 5.78, the mean of X is 21,478.1578. Now, reasoning intuitively like this, what's the standard deviation of X? By the way, should you round this to 21,478.16 to avoid suggesting more decimal accuracy than the original numbers contained?

(e) Solve for X from the equation $Y = 100X - 2,147,810$. Then identify a and b for the equation $X = a + bY$, and use our "fundamental equations" to get the mean and variance of X.

Chapter **2**

2.1.1

As Tufte says, "To generate the thoroughly false impression of a substantial and continuous increase in spending, the chart deploys several visual and statistical tricks—all working in the same direction, to exaggerate the growth in the budget." You are not likely to identify the statistical "tricks," but see if you can spot any of the graphical tricks which contribute to the misimpression of a "substantial and continuous increase in spending." Note, in particular, the placement and use of labels and arrows. Also, take a careful look at the last three years and how they relate visually to the previous years—do you see the trick?

2.1.2

Oops! We forgot to record the control for gender:

Percentage Recovering

	Treatment		
Gender	I	II	
Female	20% (20/100)	24% (50/210)	23% (70/310)
Male	67% (40/60)	75% (15/20)	69% (55/80)
	38% (60/160)	28% (65/230)	32% (125/390)

Note that 38% recover under the first treatment as opposed to only 28% under the second. Without the control for gender, it's "obvious" that the first treatment is better. Obvious, but WRONG! Once gender is taken into consideration, it becomes obvious that *just the opposite conclusion* should be drawn. Both men and women do better with the SECOND treatment!

(a) Explain why controlling for gender leads to exactly the opposite conclusion for this data.

(b) State the basic principle involved in this example.

2.1.3

Oops! We forgot to record the control for on-job experience (measured by length of employment):

Average Annual Salaries ($1000)

	Female		Male	
	No. of employees	Average salary	No. of employees	Average salary
Less than 5 years on-job experience	10	100	40	125
More than 5 years on-job experience	40	175	10	200
	50	160	50	140

(a) Is there evidence of salary discrimination? Identify and explain the anomaly.

(b) Here's a more technical question to help you understand the calculation of weighted averages: Show how to obtain the 160 and 140, the total averages, as *weighted* averages.

2.1.4

(a) The authors of the NSM seem to feel that a mere promise of confidentially given by a total stranger would put to rest all concerns of a strictly closeted gay man even though the stranger knows that man's home address, place of work, social security number, and other very personal information. But it's a well-known fact that being "out" as a gay man can have disastrous consequences—loss of employment, eviction from housing, violent physical assults, estrangement from family, friends, church, and community, and so on. It is for that reason among others that public health officials are so adamant that HIV test status remain strictly confidential. Otherwise, the AIDS epidemic, which in the public mind in the United States is closely associated with the gay community, would be forced underground and totally out of control.

The authors of the NSM were evidently oblivious to the fact cited above. With this fact taken into account, how would you interpret the 2.3% and 1.1% results of the NSM?

(b) Eleven percent of U.S. males would be gay under this assumption, based on the results of the NSM. Show precisely how to obtain this.

(c) These percentages seem consistent with the interpretation given in level II of part (a), but not with the interpretation given by the authors of the NSM. Explain.

(d) Nonresponse is always a problem. But if the reasons for not responding are unrelated to the issues under study, there's hope that the responses you actually do obtain are not significantly biased. Is there any such hope here?

(e) In the NSM, a variation of Warner's technique would work like this: A respondent is given two questions such as "Are you gay?" and "Do you like ice cream?" Then, with the help of a spinner or some other random device, the respondent privately makes a random choice of which question to answer. Why does this work?

2.1.5

(a) Each time you toss the coin, record the number of heads. Then you'll have a simple random sample as required. "Verify" means match each condition of the definition exactly with the corresponding aspect of the example. Be sure to identify the value of n.

(**b**) You need to write down all possible ordered sets containing three values of X.

(**c**) There are 1.074 billion such samples! Analyze it this way: For a sample of size 30, you have 30 blanks to fill with either a zero or a one. There are two ways to fill each blank: two ways to fill the first blank AND two ways to fill the second blank AND two ways, and so on. So how many ways altogether? [Hint: Try to see the counting principle involved and check yourself against simpler cases. For example, how many samples of size three are there? Of size two?]

(**d**) 0.2401. How?

(**e**) There are several ways you could get a sample for which $\Sigma X = 2$:

1100 1010 1001 0110 0101 0011.

You either get the first sample OR the second OR the third and so on.

(**f**) For "half a head" per toss, you expect two heads on four tosses. So . . . ?

2.1.6

(**a**) There are two conditions in that definition. What are they?

(**b**) There are 36 possible samples. What are they?

(**c**) There are six possible values of X on each roll. That means there are well over sixty million such samples! How do you show that?

(**d**) 0.003. How? Which samples have $\Sigma X = 4$? What's the probability of each of those samples (use the "and" rule)? Finally, use the "or" rule to compute the probability that $\Sigma X = 4$.

(**e**) Yes?

2.1.7

(**a**) First specify the experiment, being clear about the possible outcomes. Ask yourself, "What is the 'doing' here?" Once the experiment is clear, go for the random variable. Give the rule which associates a number to the possible outcomes. Don't make the mistake of saying the *diameters* are the values of the random variable.

(b) You'll have to make an assumption about this manufacturing process. What assumption? This may not be obvious at first. Read the definition of "simple random sample" carefully.

(c) Statistical Process Control makes use of samples like this to monitor the production process. How might you detect something wrong in the process by means of this sample?

2.1.8

(a) First specify the experiment, being clear about the possible outcomes. This is an example where a physical (as opposed to numeric) description of the outcomes will be artificial—the outcomes really are numbers. What numbers? Once the experiment and its numeric outcomes are clear, specify the rule which "associates a number" to the possible outcomes.

(b) Yes?

(c) The difference is to be found in the underlying random experiments. Be very precise in identifying what exactly the difference is.

2.1.9

(a) Have you noticed that in setting up an abstract model for a real-world problem as we're doing here, it's not neccessary to describe what actually happens; the point is to have a consistent way of THINKING about what happens. It's possible to describe the experiment here so that "one silicon wafer" is the outcome, but a simpler description more relevant to the problem is possible, a description which makes it more like a coin toss. What is it?

(b) You can obtain the mean and variance of X in two distinct ways: Calculate from the probability distribution of X or use a known formula. Just as an exercise, do it both ways (making sure, of course, you get the same answer).

(c) You'll have to assume a lot is made up from silicon wafers taken off the production line in order (why?). And you'll have to make one other assumption about this production process. What is it? Justify this assumption.

(d) It's $0.2(\Sigma X)$. Explain.

(e) Be careful here. Work backward from the value of the random variable (which is described in the problem statement) to the outcomes to the "doing" which produces that outcome.

2.1.10

(a) What's the "doing" here? You can correctly describe the "doing" in more than one way, but make it look like the toss of a coin by having just two outcomes. Then, what's the random variable?

(**b**) As always, independence is the crucial thing and requires an assumption about this disease, an assumption we've not mentioned. What's the assumption? Verify that if the assumption is satisfied, the children of that neighborhood constitute a simple random sample.

(**c**) This is parallel to part (e) of the previous problem. Try to make your answer very specific to this real-world situation with as much detail as the problem statement allows.

(**d**) Under the assumption discussed in part (b), the incidence of this disease is modeled by the Poisson distribution which we'll study in Chapter 3. What practical conclusion might you be able to draw once you have the theoretical model in hand? Here, the phrase "incidence of the disease" means "how many children in the sample have contracted the disease." [Hint: You've probably encountered news media accounts of conclusions like the one we're thinking about. Apply common sense and imagination.]

(**e**) Reread the answer to part (c).

2.1.11

(**a**) It will be very helpful if you've got a clear verbal description of the population in mind. Try to formulate that description on the basis of the question. Then check yourself by referring to the original list of populations from the previous section. Be careful; it's easy to get caught by specifying the wrong population. For example, the first question does NOT refer to the population of "all voters in the upcoming election" nor even to "all eligible voters." What exactly is the population? For the last question, the population is NOT "all glaucoma patients."

(**b**) For the first population, you could ask for the average age or mean annual income, and so on. What about the others?

2.1.12

What percentage would "burn out too early" if "too early" means before 1200 hours?

2.1.13

(**a**) Deal a card, record its value, replace the card, shuffle many times, and deal again. Do this until you have five cards. Match this with the definition.

(**b**) First, what probability distribution are you sampling from? What's the random variable X? And the experiment? Here's a hint: The experiment is "select one member of the population at random." So, what is X? Then show that a simple random sample of size n from the numeric population yields "an ordered set of n values of X obtained from n independent repetitions of the random experiment for X." In other words, it yields a simple random sample from the distribution of X.

(c) There are two cases: sampling *with* or *without* replacement. In each case, you can give a formula for the probability of a sample. That formula involves the population size, N, and the sample size, n. But the formula is the same no matter which sample you're talking about, so any two samples have the same probability. What are those formulas? [Hint: sampling WITH replacement is easier, do it first!]

(d) You have to do two things: First, show that any two samples for a fair die have the same probability. Then give an example to show that this won't necessarily be true if the die is loaded. That means give a specific loading of the die and then show us two samples with different probabilities. Make it easy, take samples of size $n = 2$.

2.1.14

(a) There's no random number generator here! What is the random mechanism?

(b) The sample size is denoted by n. Now, concretely in the terms of this problem, give a verbal description of n. What's its value?

(c) Well?

2.1.15

(a) The first one is $2^4+2^2+2^1 = 16+4+2 = 22$. Here's how: Look at 10110. Counting from the RIGHT starting with zero, there's a 1 in the first, second, and fourth places (a zero in the zeroth place). Add those powers of two. The next binary number is 13 (how?) and then . . . ?

(b) 128, how? First, think about how many numbers you could generate by tossing the coin once. Twice?

(c) How many numbers did we need to generate?

(d) Be careful. This question does not say "simple random sample from a *population*." Go back and recall the definition of "simple random sampling from the distribution of a random variable." What's the random variable here?

(e) You'll have to toss the coin four times—why? You also need to decide which member of your population corresponds to which binary number. Suppose your tosses result in the sequences: THTT, TTTT, HTHT. What will your sample be?

(f) What's the purpose of sampling?

(g) To make this work, the coins have to be clearly distinguishable. Otherwise, any of the following four base-two numbers is generated by three heads and a tail:

0111, 1011, 1101, or 1110.

So toss a penny, a nickel, a dime, and a quarter. Or toss four quarters but from different years, or paint them different colors, or . . . ! As a random number generator, how is this different from tossing one coin four times as we did in part (e)?

2.1.16 Counting all occurrences of the word "the" throughout the entire text would be very tedious and time-consuming. Such a "census" of an entire population is usually not feasible. Even if you could do it, a complete census would result in a wrong count because it's difficult if not impossible to control human error in such an enormous project. The theory of random sampling, on the other hand, provides precise control over SAMPLING error (by contrast with human error) through the theory of random sampling. So you can control sampling error. Furthermore, when you're sampling instead of attempting an entire census, human error is a much more tractable problem. It's controlled through proper training of the relatively small number of people involved and by giving them small enough tasks and enough time so they can do the job with a high degree of accuracy.

So, in this problem, you should take a random sample of the words of the text and count occurrences of the word "the" in that sample. Here are some of the questions that will arise:

(a) What's the population here and how do you classify it?

(b) What exactly do you mean by "a word"? Are you going to count numbers such as 2403? What if they're simply written out numbers such as one or two? What about proper names? What about parts of a date? Addresses? What's the basic problem here? For any population, this problem will arise. Can you say briefly and concisely what it is?

(c) How do you index the text for selecting the sample? If you're just going to go through the text word by word, you might as well do a complete census. In fact, cluster sampling would be better here. Why?

2.1.17 Stratifying by income should work. You might put all those workers making more than 40,000 dollars per year in the top stratum, between 20,000 and 40,000 in the middle stratum and all those below 20,000 in the bottom stratum. This is just a suggestion. After all, determining a truly appropriate stratification is a statistical problem in its own right, a problem in the design of experiments. And that's a topic for your next statistics course.

(**a**) Can you suggest another possibility?

(**b**) Identify and classify this population.

2.1.18

(**a**) Yes, they're both examples of random sampling experiments. And a random sampling experiment is by definition a random experiment. What are the outcomes of a random sampling experiment?

(**b**) No, why not? How are they different?

(**c**) Throughout most of this text, a population will be either numeric or dichotomous. So give one answer to this question for each of the two types of population. Note that no matter what the design of your sampling plan, an outcome is just ONE RANDOM SAMPLE.

2.2.1

(**a**) Here's one possibility: 3, 4, 5, 12. Alter one of these four values to make the mean still larger without changing the median.

(**b**) The median is determined by the middle two data values alone; nothing else affects it. So, if you keep the number of observations the same, you can change all the other data values without affecting the median. What's the general condition for the mean to be larger than the median?

(**c**) If one or two (or a few) of the data values are quite "out of range," quite large, or quite small compared with all the rest, then the median will be a better measure of centrality than the mean. Give an everyday example where this would usually be true. Think about this carefully. Can you think of a real-world situation where you would typically expect to find a few especially large (or small) values which would distort the mean?

(**d**) Because each distinct value occurs only once, all the f's are just 1. Note that the symbol f belongs to a frequency distribution; the symbol rf to a *relative* frequency distribution.

2.2.2

(**a**) The median and mode (or modes, if there is more than one) will be easier to identify after you organize the data into a frequency or relative frequency distribution.

(**b**) All integers from two through eight were observed *except* four. Still, you should list four as a "possible" value. Otherwise, someone looking at your data presentation might suspect four was omitted by mistake. Remember, the point of a data presentation is to provide immediate answers to any questions that might arise. By omitting the value four, you would *raise* an irrelevant question without providing an answer. Of course, because the value four is missing, its frequency is . . . ?

(c) Yes?

(d) What exactly is different when you have sample data?

(e) Here are the answers: 12, 12, 60, 1, 5. Explain by giving the verbal descriptions.

2.2.3 Think about this carefully. Remember that a random variable is a rule which associates a number to each of the possible outcomes of some random experiment.

(a) What's the random experiment?

(b) What are the outcomes?

(c) What number is associated with each outcome?

2.2.4 This calculation will be done exactly like the corresponding calculation for a probability distribution. The difference is only that . . . ?

2.2.5 (a) The word "parameter" for observed data refers to a population. If we had sample data instead, the corresponding calculations would give the value of a ______. This would vary from one sample to the next. That's why it's not called a parameter. A parameter is a ______ number. Because for sample data the calculation gives a quantity which varies from sample to sample, it is a ______.

(b) Read the definition of the term "range" again.

(c) The calculation for $\hat{\sigma}^2$ is the same as for σ^2, but the notation in the formula will be different. What is the formula exactly?

2.2.6 Here's one possibility:

X	f
3	1
4	3
5	1
6	2
7	3
8	1

Now, compute the mean, variance, and standard deviation and identify the mode(s) and the median.

2.2.7 The formula for the variance given in the text $\sigma^2 = (1/N)\Sigma(X - \mu)^2 f$ assumes you have a frequency distribution for the data. For a relative frequency distribution, the formula would be "essentially" the same as for a probability distribution. What's that formula and why is it only "essentially" (not "exactly") the same as for a probability distribution?

2.2.8 The mean could fall virtually anywhere between the largest and smallest values. So the range has nothing to do with the mean.

(**a**) Show this by making up an example of two small data sets consisting of a few integers. Let your two data sets have the same size, the same data values, and the same range. Only the frequencies will be different. Then by choosing appropriate frequencies, you can have the mean of one data set be close to the smallest value in that data set, whereas the other mean is close to the largest value of its data set. So with the same range you have two very different means. [Hint: This problem will be useless if you don't at least TRY to do it without looking first at the level II answer.]

(**b**) Guess which of your two data sets has the larger variance and then verify your guess by calculating. [Hint: Draw a line graph for the data. We haven't told you how to do this. Guess how it ought to be done by analogy with the line graphs for random variables in Chapter 1.]

2.2.9 Yes?

2.2.10 (**a**) To make the standard deviation larger, you must make the data be more...?

(**b**) To make the mean large without changing the median, you should . . . ?

2.3.1 Here are the issues you need to address:

(**a**) What are the classes? Classes of what?

(**b**) Evidently the f means "frequency," but what does it mean to say, "The class 20 to 39 occurs 202 times"? How can a class "occur"?

(**c**) If you said X is the symbol for the data values, you were not careful enough. Note that these X values are not even possible test scores because they're fractional values. Test scores as we've given them here are integers.

We use X for grouped data in a way that's inconsistent with our earlier usage and we depend on context to resolve the ambiguity. This is sometimes whimsically referred to as "abuse of notation," but it's a very good way to make USE of notation. After all, you should keep the context clear at all times. If you're clear about the context, if you remember whether you're talking about grouped data or ungrouped data, you'll not be confused by the two inconsistent uses of the symbol X. So, for grouped data, the variable X refers to . . . ?

For parts (d) through (g), give a verbal description of:

(**d**) the product Xf is the number of . . . ?

(**e**) the column sum, Σf, is . . . ?

(**f**) the column sum, $\Sigma X f$, is . . . ?

(**g**) if we let $N = \Sigma f$, then $(1/N)\Sigma X f$ is . . . ?

(**h**) How would you use the grouped frequency distribution to calculate the variance, median, and mode?

2.3.2 Suppose the salary distribution is

salary (in thousands)	f
0–15	523
15–30	318
30–50	84
50–100	70
100–250	5
	1000

Class intervals include the left endpoint but not the right. It would be wrong to indicate the first class by 0–14 because then incomes *between* 14,000 and 15,000 are not accounted for.

Why would it be desirable to avoid equal class widths?

2.3.3

(a) The second class was observed 12 times. But what does that mean exactly?

b) Why is the second class, for example, not "15 – 29"? Because the data is not presented that way, presumably it's not appropriate. Under what conditions would it be appropriate?

(c) and (d) Yes?

(e) This is simple random sampling from the probability distribution of temperature readings. What's the random variable?

2.3.4

Note that, roughly speaking, the 137 scores in the original first class divide evenly between the two new classes. Now, if the relative frequencies are represented by area, how does the area of the old first class divide into the two new classes?

2.3.5

Suppose you were the quality control inspector. What would you think if your measurement of a rod was 0.999 cm? Remember that discarding good rods is expensive and a drag on the productivity of your plant. Furthermore, any measuring procedure is subject to error.

2.3.6

(1)

Class	f	X	Xf	$(X-\mu)^2 f$
0–10	10	5	50	1000
10–20	15	15	225	0
20–30	10	25	250	1000
	35		525	2000

$\mu = 15$,
$\sigma = 7.5593$.

0 10 20 30

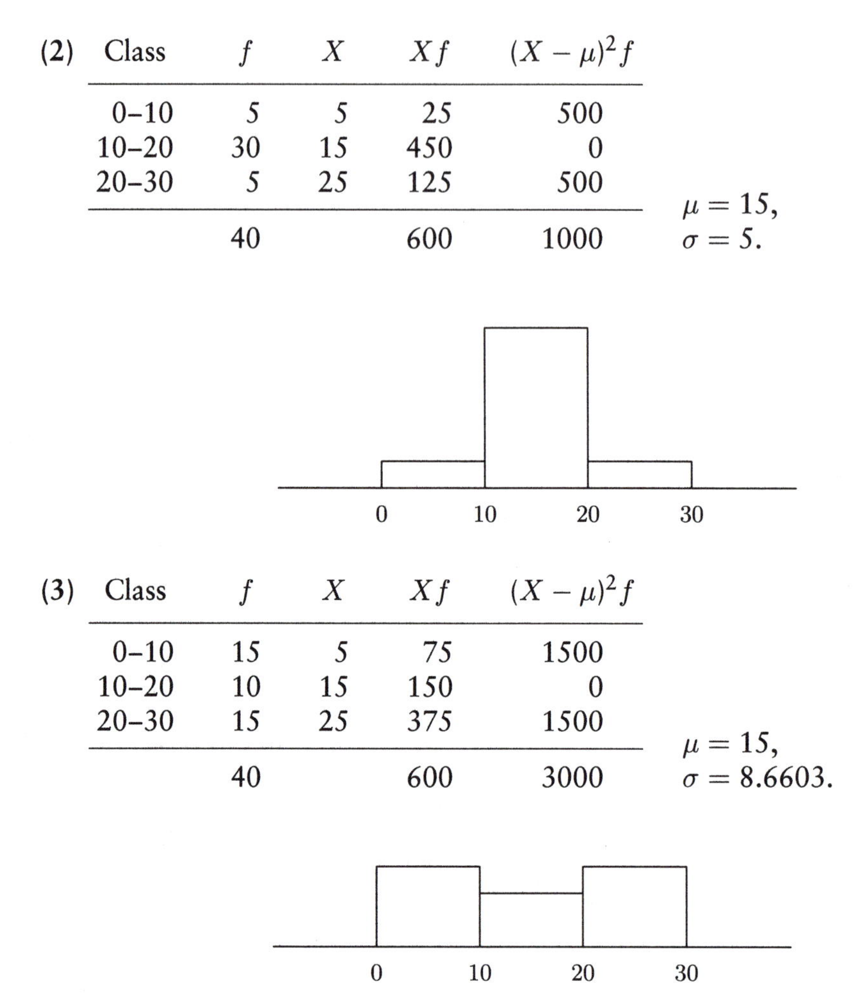

(2)

Class	f	X	Xf	$(X-\mu)^2 f$
0–10	5	5	25	500
10–20	30	15	450	0
20–30	5	25	125	500
	40		600	1000

$\mu = 15,$
$\sigma = 5.$

(3)

Class	f	X	Xf	$(X-\mu)^2 f$
0–10	15	5	75	1500
10–20	10	15	150	0
20–30	15	25	375	1500
	40		600	3000

$\mu = 15,$
$\sigma = 8.6603.$

Now, try (4) and (5) on your own. For each histogram, be sure to indicate the mean, the mean plus one standard deviation, and the mean minus one standard deviation, and shade in the area of the histogram which represents the percentage of the data within one standard deviation of the mean.

Chapter 3

3.1.1 "Variance zero" means $\sigma^2 = \Sigma(X-\mu)^2 \mathrm{P}(X) = 0$. Thus, each $(X-\mu)^2$ must be zero because a sum of positive numbers can add to zero only if each term is zero. In other words, to get a sum of zero you need negative as well as positive numbers so everything can cancel. Now, what can you say about the values of X?

3.2.1 (a) $\sigma^2 = \Sigma(X-\mu)^2 \mathrm{P}(X) = \Sigma[X^2 - 2\mu X + \mu^2]\mathrm{P}(X) = ??$

(b) Instead of a column for the $(X - \mu)^2\mathrm{P}(X)$'s you need a column for the $X^2\mathrm{P}(X)$'s. The EASY way to form this column is to multiply the X column entries by the $X\mathrm{P}(X)$ column entries.

(c) Clearly, the computing formula is quicker for computation, hence its name. But it's also more accurate in a certain sense. In what sense? Why the term "conceptual" for the conceptual formula?

(d) There are eight cases here: frequency distributions or relative frequency distributions for grouped or ungrouped data for population or sample data. Relative frequency distributions introduce nothing new, nor do grouped data, nor does the population/sample distinction. In all these cases, it's only the notation and the interpretation of the numbers that's new. For grouped data, the class mark estimates the actual observation and you calculate as if it really were the actual observation, so there's nothing new in calculating. For a relative frequency distribution, the computing formula is $\sigma^2 = \Sigma X^2(\mathrm{rf}) - \mu^2$. Here, we just replaced $\mathrm{P}(X)$, the *theoretical* relative frequency by rf, the *observed* relative frequency. Notice how it's only the interpretation (and so, the notation) that changes. For sample data the formula is $\sigma^2 = \Sigma X^2(\mathrm{rf}) - \overline{X}^2$.

That leaves observed data organized into a frequency distribution. In this case, the computing formula is a bit different. Derive the following formula for the variance of population data organinzed into a frequency distribution:

$$\sigma^2 = \frac{N\Sigma Xf - (\Sigma Xf)^2}{N^2}.$$

(e) Because X is a random variable, so is X^2. Why? So X^2 has an expected value, the average of the values weighted by the corresponding probabilities: $\mathrm{E}(X^2) = \Sigma X^2\mathrm{P}(X)$. This has another more interesting form. Do you see what it is?

3.2.2

(a) We had dice and coins and cards and so on. So, for instance, give an example of a uniformly distributed random variable associated with the roll of a die. Specifically, what's required to make this be a UNIFORMLY distributed random variable? What's the parameter?

(b) Give a second example of a uniformly distributed random variable.

(c) If you want to give an example of a uniformly distributed random variable involving the roll of a pair of dice, you run into a problem. What is it?

(d) Now, just for review, recall the probability distribution for the number of dots on the two uppermost faces for one roll of a pair of fair dice. First, guess the mean and then verify your guess by making up the distribution table and computing the mean. While you're at it, compute the standard deviation also. As you do this problem, keep in mind that it provides an example of a random variable that's NOT uniformly distributed.

(e) Verify that the 36 possible outcomes described in part (d) are equally likely. Where does independence come in? And where does the fairness of the dice come in?

(f) The 36 possible outcomes described in part (d) are equally likely. So why is this NOT a uniformly distributed random variable?

3.2.3 Nothing further is required! The mean should be somewhere around 19. What about the standard deviation? Now, set up a distribution table and compute the mean and standard deviation exactly.

3.2.4 (a) It's the VALUES of the random variable that must be equally likely, not the outcomes of the experiment. Give an example of a real-world situation involving equally likely outcomes modeled by a random variable which is NOT uniformly distributed.

(b) What's the key fact that makes an example such as you gave in part (a) possible? Think about the relation between the values and the outcomes.

3.2.5 Yes?

3.2.6 You require only one piece of information, a set of numbers. So . . . ?

3.2.7 The mean of a uniformly distributed random variable is just the average of its values because the probabilities are all the same. In other words, you don't need a weighted average because the weights (the probabilities) are all the same. Show that $\mu_X = (1/n)\Sigma X$ by using the formula $\mu_X = \Sigma X \mathrm{P}(X)$. [Hint: It's not too hard. Give it a try!]

3.3.1 (a) Because there are n objects, think of n positions:

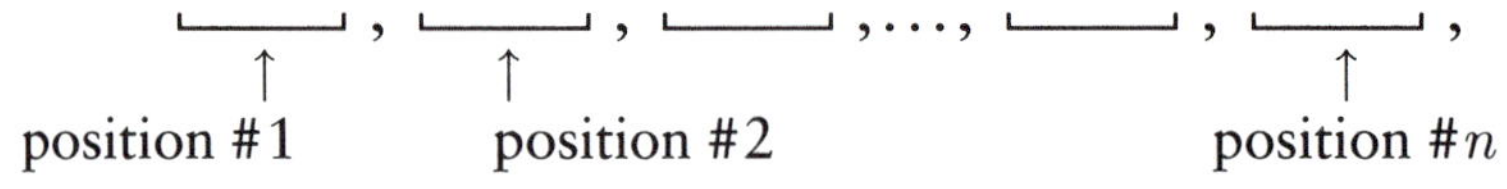

An "arrangement" of the n objects is obtained by choosing one object for the first position, another for the second, and so on. How many ways can you do that?

(**b**) By the fundamental principle of counting, the number of ways to choose x objects and then arrange them is the product of $C(n,x)$ and $x!$. Do it this way: Step 1, choose x from the n objects. There are $C(n,x)$ ways to do that. Step 2, arrange them. There are $x!$ ways to do that. Now write down a *different* expression for this count and you'll be able to solve for $C(n,x)$. This time think about it like this: We have x positions to be filled from a set of n objects. So there are n ways to fill the first position. And so . . . ?

(**c**) Learn to do these calculations without the factorial key of your calculator. Factorials get large very fast. So much so that after about 70! the calculator will give you an error message. Let's do one of these now, then you do the rest.

$$C(7,3) = \frac{7!}{3!4!} = \frac{7 \times 6 \times 5 \times 4!}{3!4!} \quad \text{(cancel the 4!'s)}$$

$$= \frac{7 \times 6 \times 5}{3 \times 2} = 35.$$

(**d**) Yes?

(**e**) A lot! 5040 ways. How?

(**f**) 30,856 ways, how?

(**g**) 720 ways, how?

(**h**) About 40 million. Use the factorial key on your calculator. For 80 students you probably get an error message on your calculator. Why?

(**i**) About two million ways. Do this in two steps, first choose six seats for your six students. Then seat them. Use the fundamental principle of counting.

(**j**) There are about 143,000 committees possible.

3.3.2

(**a**) Think about the population. See if you can suggest a further condition which would make the example into a true hypergeometric situation.

(**b**) In this example, why would we sample at all? The population has only 50 members. Why not just deal with the entire population? After all, sampling is used when the population is too large to deal with in its entirety. But this is NOT a large population.

3.3.3

(**a**) It's a *sampling* experiment. Just for review, verify that sampling without replacement is, in fact, a random experiment. Start with the outcomes and work backward to the "doing." What else must be verified?

(**b**) First think what is the smallest possible value. Then think about the largest possible value and go from there.

3.3.4

(**a**) The pool of 40 candidates. Is it a dichotomous population?

(**b**) Certainly the mayor does not choose the committee at random. However, from the point of view of gender (our only consideration) and assuming no gender bias, the choice should look random. So what's the underlying random experiment? Is it appropriate for the hypergeometric model?

(**c**) For any sampling experiment, an outcome is one sample. Real world?

(**d**) 658,008. How?

(**e**) A random variable is a rule which assigns a number to an outcome. Now, real world?

3.3.5

(**a**) If you said "the population of books," you missed it—what's wrong with that answer? What is the population? Note that it's supposed to be dichotomous. So, what are the objects which make up the population and what's the characteristic of interest? Be specific to the real-world detail of the problem.

(**b**) Work backward. First, what's an outcome for the experiment? What kind of object is it? Be very physical and real world in your description. Then work backward to find out what the "doing" must be.

(**c**) What's the smallest possible value? The largest?

(**d**) Two. Write that symbolically (without words).

(**e**) One of the characteristics of this model is not characteristic of our real-world example and so we shouldn't use the model. What characteristic is not satisfied in this example?

3.3.6 Don't be discouraged; it's a short list. Try to determine which are the really key words and underline them, as we did in the hypergeometric list.

3.3.7 (a) How many ways can we select $n = 3$ of these 50 components?

(b) For the numerator, use the fundamental principle of counting. You get the sample in two steps: first, select $X = 2$ components from the four which have the characteristic. Then select the rest of the sample.

(c) $\text{P}(X = 2) = 0.0141$, how?

(d) Yes?

3.3.8 (a) Wait a minute. We need more information. What's the representation of women among the 40 candidates? Let's suppose 15 are women. Now, what's the probability of no women on the committee? [8.07%]

(b) With only an eight percent chance of such a result, one might be suspicious of the mayor's selection procedure. Of course, you'd better be careful about what you have actually shown. What have you shown?

(c) We need to clarify the phrase "such a committee." It's natural to interpret it as "a committee with one woman" until you remember the point of the problem. The phrase "such a committee" really means "such a *bad* choice of committee"—a choice reflecting discrimination. The usual interpretation is "a choice of one woman or less" because "no women" would be an even worse choice. So the question becomes: What's $\text{P}(X \leq 1)$ if the choice was random? What is this probability and what does it mean in real-world terms? [0.3691]

3.3.9 (a) Are you clear about the notation used here? What's the difference between X and x? That is, what's the difference between the meaning of the uppercase and lowercase letters for a random variable?

(b) What's the denominator of $\text{P}(X = x)$?

(c) For the numerator, you need a product of two numbers. Write the formula for each of those two numbers.

(d) What's the formula for $\text{P}(X = x)$?

3.3.10 (a) Describe the experiment as completely as possible with the given information. That means incorporate $N = 10$, $R = 4$, and $n = 4$ into your description.

(b) First, guess what *proportion* of the sample you think ought to have the characteristic on average. As for the variance . . . ?

(c) and (d) Yes?

3.3.11

(a) There's about a 36% chance of at least one burnt out bulb among the three you choose. [Hint: $N = 30$, $R = 4$, $n = 3$.]

(b) Yes?

(c) 10%. Show this. Now, give the mean and variance for this situation and discuss the finite population correction factor. Compare this with the situation in part (a).

(d) Explain from the formula and intuitively why $P(X = 3)$ is zero.

X	$P(X)$	$XP(X)$	$X^2P(X)$
0	0.8069	0	0
1	0.1862	0.1862	0.1862
2	0.0069	0.0138	0.0276
3	0.0000	0	0
	1	0.2000	0.2138

$\mu = 0.2$,
$\sigma = 0.4169$.

(e) For 90%, $n = 4$. For 80%, $n = 7$. How?

3.3.12

(a) You should expect about one and a half hearts *on average* in a six card hand. Of course "one and a half hearts" doesn't make real-world sense, it's an average—a theoretical number. Just like the "average" family with 1.2 children. Now give a verbal description, as specific as the problem allows, of the random variable X including the possible values of X and a description of the underlying random experiment (What's the "repeatable doing"? What's an outcome?). Then justify our answer ("one and a half hearts") precisely by reference to the model.

(b) About three. Justify this precisely by reference to the model just as in the previous part.

(c) What's the appropriate measure of "predictability" for the values of a random variable?

(d) Note how tedious it would be to actually compute the probabilities and make up a probability distribution. There are seven values for each random variable and the formulas are a mess. Nevertheless, you can get a good qualitative idea of the two random variables without actually computing any probabilities. It's this qualitative information you can sketch in your graph. Get your sketch by thinking about the mean, the standard deviation, symmetry considerations, and so on. Think *qualitatively*. For example, values more than three standard deviations from the mean should—in a rough sketch—be shown with probability zero. Why?

(e) Remember, in the finite population correction factor you should think of the denominator as if it were just N.

3.3.13 Put the two effects together: analyze that part of the variance determined by the product of n and $\frac{(N-n)}{(N-1)}$:

$$n\left(\frac{N-n}{N-1}\right) = an^2 + bn.$$

Here, $a = -1/(N-1)$ and $b = N/(N-1)$. Thus, a and b are constants because N is a fixed number. In other words, the effect of n on the variance is through a quadratic function of n. Because the coefficient on n^2 is negative, the graph of this quadratic is a parabola opening down:

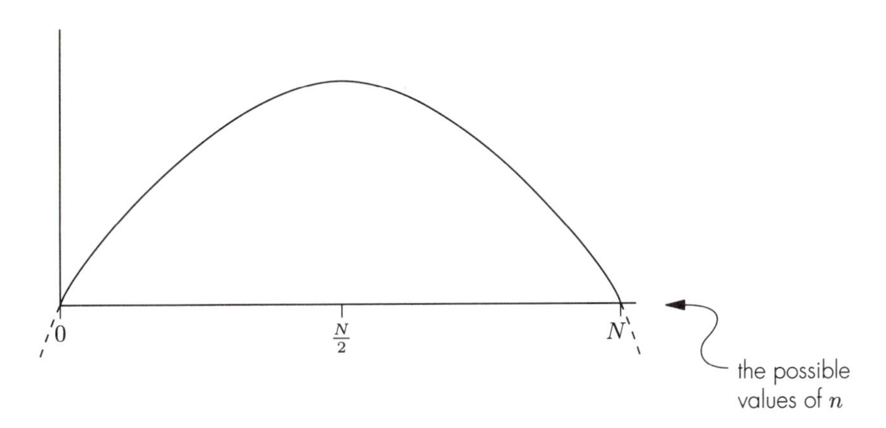

What does this tell you? Which effect dominates the variance? Note that the sample size is the only thing allowed to vary here. N, p, and q are fixed numbers.

3.4.1 (a) Yes?

(b) As with any sampling experiment, an outcome is a sample. But here you can describe the outcomes a little more precisely in terms of the characteristic of interest: Label each sample member as either "yes" or "no," depending on whether it does or does not have the characteristic of interest. Now we get a simple schematic form for any one sample. What is it?

(c) What's the smallest possible value for this random variable? That is, what's the smallest possible number of observations in the sample which could have the characteristic? Then, what's the largest possible value?

3.4.2

(a) This is the event that our outcome looks like

yes, yes, yes, no, no, no, no, no, no, no.

What's the probability of this event? Be sure to justify the probability rule you use here.

(b) The event in part (a) is only ONE way to have $X = 3$. Explain.

(c) Think of your potential sample as consisting of ten blanks to be filled with yes or no:

$$\_\_\_, \_\_\_, \_\_\_, \_\_\_, \_\_\_, \_\_\_, \_\_\_, \_\_\_, \_\_\_, \_\_\_.$$

For example,

$$\underline{yes}, \underline{no}, \underline{yes}, \underline{yes}, \underline{no}, \underline{no}, \underline{yes}, \underline{yes}, \underline{no}, \underline{no}.$$

How many ways can you choose three of these blanks to contain the three yes's for your sample? All the rest must contain no.

(d) Put the results of the previous two parts together. You'll use the "or" rule for mutually exclusive events. How?

(e) About four chances in 1000. How?

(f) No role at all. Explain.

3.4.3

(a) This is exactly like the argument of the previous problem, except that you have x yes's instead of three. See if you can imitate the argument to get the more abstract formula.

(b) No role. Explain.

3.4.4

Note that the condition is equivalent to $1/10 \geq n/N$.

3.4.5

(**a**) Obviously, because the problem is in this section, we're going to use the model of "sampling with replacement from a dichotomous population." But what justifies that? Isn't it clear that you would NOT do sampling with replacement? What would be the point of looking at the same catalog listing twice? Do you see the two possible responses to this question?

(**b**) Yes?

(**c**) Hmmmmmm! You cannot answer this question because the model requires a value for the parameter p. Suppose, in fact, one percent of the listings correspond to lost books. Then there's about a 32% chance for three or more lost books among our 200. How?

Here's a hint: $P(X \geq 3)$ will require 198 calculations! Work with the complementary event, it will only require three calculations:

$$P(X \geq 3) = 1 - P(X \leq 2).$$

This is a little messy to be sure, but TRY. Even if you don't get it, you'll benefit from trying. Note that X cannot be both zero and one at the same time—you can't say that a sample has no lost books AND at the same time has one lost book! This is a general fact—the values of a random variable represent mutually exclusive events. So we just add probabilities:

$$\begin{aligned} P(X \leq 2) &= P(X = 0 \text{ OR } X = 1 \text{ OR } X = 2) \\ &= P(X = 0) + P(X = 1) + P(X = 2). \end{aligned}$$

(**d**) If, in fact, one percent of the listed books are lost, in our sample of 200 we should expect to find about ?? lost books.

(**e**) $P(X = 2) = 0.0328$ and $P(X > 2) = 0.0033$. How?

(**f**) About three-tenths of a book. How?

(**g**) The parameters of the model are n and p. What values do they take?

3.5.1

(**a**) Something you do with two outcomes? Be specific in identifying S, F, p, and q.

(**b**) Make up a table for the probability distribution, using p as the probability for $X = 1$, and so on. Then calculate. You'll find that $\mu = p$ and that $\sigma^2 = pq$.

3.5.2 They had better be the same. Or at least approximately the same. Otherwise, one of the models is not valid. Are they the same?

3.6.1 (a) What's the "repeatable doing"?

(b) What are the outcomes? Can you give one instance?

(c) For the outcome you gave in part (b), what's the value of the geometric random variable? Describe it verbally.

(d) Give another possible outcome in addition to the one given in part (b) and give the value of X for that outcome together with a verbal description.

(e) How can you be sure the outcomes cannot be predicted in advance?

3.6.2 A Bernoulli trial is a random experiment with exactly two possible outcomes. So . . .

(a) What's the "repeatable doing"?

(b) What are the outcomes?

(c) How can you be sure the outcomes cannot be predicted in advance?

(d) Give a verbal description of the parameter p.

3.6.3 Is it true that the blood type of one person is totally unconnected with and unaffected by the blood type of another person?

3.6.4 (a) Note that the geometric random variable is a COUNT and, thus, takes on only positive integer values. What's the smallest possible value? The largest? Be careful, there's a trick here!!

(b) A parameter for a model is a *number* associated with the model. What are the parameters here? Give verbal descriptions.

3.6.5 (a) The answer is 0.0211—HOW? Here, "after 17 trials" might be ambiguous. Assume it means the person of the desired blood type was the seventeenth person tested. Also note that there's an eight percent probability for a person to have the desired blood type because eight percent of the population have that blood type. But how do you obtain this result? If you didn't get the answer, try again. It's not too hard.

(b) This is $P(X = 1)$ which is the probability of the outcome: "yes." You only had to do the trial once. So what's the probability that you get the desired blood type with any one given person?

(c) This is not possible. Why?

(d) The outcome is no, no, yes. The probability is 0.0677. How?

(e) The outcome is no, no, no, no, no, no, yes. What's the probability?

(f) The outcome is . . . ??

3.6.6 Well, what assumptions are required for the model?

3.6.7 Well, try to do it. You'll see it won't be possible. Why not? Think about the X column.

3.6.8 **(a)** A "statistic" is a random variable for which the underlying random experiment is a sampling experiment. Is that true here? Recall that the hypergeometric random variable was our first example of a statistic.

(b) Again, the distribution of a random variable is a "sampling distribution" provided the underlying random experiment is a sampling experiment, so here and in part (a) we're asking the same question: Is the underlying experiment for the geometric random variable a sampling experiment?

3.6.9 **(a)** 12.5, how?

For each of parts (b)–(k), here is the percentage chance for the event to occur. Show how you justify these probabilities using the formulas.

(b) 2.5%; **(c)** 71.37%; **(d)** 28.63%; **(e)** 8%; **(f)** Yes?; **(g)** 43.44%; **(h)** 84.64%; **(i)** 4.85%; **(j)** 100%; **(k)** 84.03%. Notice that part (j) is the same as $P(X \geq 1)$. Explain that.

3.6.10 Three standard deviations for us is about 36. The probability of being more than three standard deviations from the mean is 1.83%. This is much smaller than Chebyshev's one chance in nine ($\approx$ 11%). But we have much more information here than is assumed by Chebyshev, so of course we should be able to come to a stronger conclusion.

(a) Where did this 1.83% come from? Be sure you also verify that three standard deviations is 36.

(b) What information do we make use of beyond what's assumed by Chebyshev's Theorem?

3.6.11

(a) Group the values into classes. Ten classes would be more revealing, but, out of compassion for the poor student, let's do only five classes:

Class	$P(X \in \text{class})$
1–10	
11–20	
21–30	
31–40	
41–50	

Hint: Use the relationship

$$P(a \leq X \leq b) = P(X \leq b) - P(X \leq a - 1),$$

for example,

$$P(X \in \text{ class \#2}) = P(X \leq 20) - P(X \leq 10).$$

(b) Instead of a line graph as we have been giving for random variables, give a histogram.

3.6.12

(a) Look at the graph in Problem 3.6.11(b).

(b) The mean is 12.5. The graph in Problem 3.6.11(b) seems to suggest that about 50% of the distribution falls on either side of the mean. However, the value which cuts the distribution in half is the median, not the mean. Only if the distribution is symmetric about the mean (or median) will they be equal. Show that the mean splits the distribution by 63% below and 37% above.

(c) Parts (a) and (b) explain the "expected value" sense of the mean of a random variable—it's the "typical value." But in what sense?

3.6.13

The variance for this new situation is much smaller and so the model will give more accurate predictions. Explain all of this by

(a) telling what you would use for making predictions,

(b) computing the two variances and comparing them.

3.6.14

$$\begin{aligned} P(X \le x) &= p + qp + q^2p + \cdots + q^{x-1}p \\ &= p\left[1 + q + q^2 + \cdots + q^{x-1}\right]. \end{aligned}$$

Does the sum in the square brackets look familar?

3.6.15

(a) Write down the formulas for $P(X = x)$ and for $P(X = x + 1)$.

(b) In using the recursion formula, be sure you do not reenter the result of each calculation. You'll certainly round it off, losing significant accuracy in repeated calculations. With many calculations, this would be a serious and unnecessary source of error. You can store q in memory. This is convenient when q has so many digits that reentering it for each calculation would be a nuisance.

So, start with $P(X = 1) = p = 0.08$. Then, with $x = 1$ calculate $P(X = 2)$ by the recursion formula: multiply by $q = 0.92$ to get 0.0736. Now, record the rounded value in the table and continue your calculations to get the rest of the probabilities. Here, in this particular case, 0.0736 doesn't require rounding, but, in general, you will have to round. Now, please complete the rest of the problem.

(c) Yes?

(d) For the geometric random variable, the probabilities decrease as X increases. See the comment at the end of the Problem 3.6.11(b), level II. For example, if $p = 0.08$, $P(X = x)$ is less than 4%. How?

3.6.16

For example, $P(X = 5) = pq^5 = 0.0238$. Because there were 135 rabbits, if the model is valid you'd expect 3.2 (2.38% of 135) rabbits to be caught five times. Finish the model and compare with what was observed. Note that you'll truncate the model at $X = 7$ because the trapping was repeated only seven times. The model itself assumes an infinite number of repetitions. [Hint: Use the recursion formula.]

3.7.1

(a) Be careful here, there's a trick! The "doing" is NOT just to do the Bernoulli trial. It's to do the Bernoulli trial once and then . . . ?

(b) If you don't see this, look again at part (a).

(c) A typical outcome looks like

$$S,\ F,\ S,\ F,\ F,\ F,\ S,\ S,\ S,\ S,\ F,\ F,\ S,\ F,\ F,\ F,\ S,\ F,\ F,\ F.$$

Verbal desecription?

(d) The unpredictability is in the Bernoulli trial. Explain.

(e) There are two parameters for the binomial experiment, n and p, where n is the number of repetitions of the Bernoulli trial and p is the probability for success on one trial. What does this mean for the coin?

(f) An outcome is a string of n S's and F's . For example, if $n = 20$, an outcome could look like

$$S,\ F,\ S,\ F,\ F,\ F,\ S,\ S,\ S,\ S,\ F,\ F,\ S,\ F,\ F,\ F,\ S,\ F,\ F,\ F.$$

Here $k = 8$. When you think about the possible outcomes, think about having a string of n blanks to fill with S's and F's. So how many ways can you have k successes? For $n = 20$ and $k = 8$, it's about 126,000. Explain.

(g) Be systematic. List all the possible outcomes which have no successes, then the ones which have exactly one success, and so on.

3.7.2

(a) You must somehow code the coins so you can tell them apart, otherwise it's hopeless. If you number them, you can model the toss of twenty coins this way: Let the Bernoulli trial be "observe one coin." For example, the "third repetition" would be "observe the third coin." Clearly, the repetitions are independent: Whether one coin comes up heads is not affected in any way by whether some other coin comes up heads. So we have $n = 20$ independent repetitions of a Bernoulli trial as required. But still, there's a potential problem with the model. Do you see what it is?

(b) With "1500 coin tosses," we're tossing the same coin 1500 times. Now somebody might say we're producing the "same" electronic component, but that "same" refers to design not to the physical objects. In fact, we get 1500 DIFFERENT physical objects. So the better analogy is "toss 1500 coins" like part (a). This analogy is better specifically because it highlights an important consideration that might otherwise be overlooked. What is that consideration?

(c) In both "coin" examples, independence is automatic. Is it reasonable for the electronic components model?

3.7.3

(a) What's the Bernoulli trial? Describe it carefully as "something you do which is repeatable, with TWO clearly specified outcomes which cannot be predicted in advance."

(b) What is "success" on the Bernoulli trial?

(c) How many repetitions of the trial are there?

(d) What else must you verify?

3.7.4 For each part of this problem you must, IN THE REAL-WORLD TERMS OF THE PROBLEM:

(i) Identify the Bernoulli trial, a random experiment with exactly two possible outcomes. What is S?

(ii) Verify that you're repeating the SAME trial.

(iii) Verify that the repetitions of the trial are independent.

(iv) Identify the parameters. Are their values known or unknown?

(v) Specify how the outcome corresponds to a string of S's and F's.

If any parts of this model verification look questionable, say so.

3.7.5 What is the smallest possible value? The largest?

3.7.6 (a) The outcome is a sequence of *independent* events connected by the word "and":

$$S \text{ AND } F \text{ AND } S \text{ AND } F \text{ AND } F \text{ AND } F \text{ AND } \ldots$$

and so its probability is obtained by . . . ?

(b) This is the same as part (a). Explain.

(c) Note that A and B are mutually exclusive events. It's not possible to have "success" on the first trial and at the same time have a "failure" on the same first trial. So P(A or B) = ??

(d) About 126,000. It's how many ways you can pick eight positions for the S's among the $n = 20$ trials. Aren't you glad we didn't try to write them all out? Exactly how many such outcomes are there?

(e) Show that $P(X = 8) \approx 11\%$.

(f) Suppose you round p^8q^{12} to four places. What answer would you get?

3.7.7 There's a 30% chance for a success (think "heads") on each trial. There are 20 trials. So how many successes would you expect? If a coin comes up heads 30% of the time, how many heads would you expect on 20 tosses? It's the same of course.

3.7.8

(a) X = "the total number of successes on all the repetitions." Each X_k is the number of successes on ONE repetition; so X_k takes on only two values, zero or one. X is related to the X_k's in a simple way. Try to discover this for yourself before you turn to level II. [Hint: The X_k's relate to the outcomes this way:

$$F,\ S,\ S,\ F,\ F,\ F,\ S,\ S,\ S,\ S,\ F,\ F,\ S,\ F,\ F,\ F,\ S,\ F,\ F,\ F$$

$$0,\ 1,\ 1,\ 0,\ 0,\ 0,\ 1,\ 1,\ 1,\ 1,\ 0,\ 0,\ 1,\ 0,\ 0,\ 0,\ 1,\ 0,\ 0,\ 0$$

↑
This is the value of X_7, for example.]

(b) Because $X = \Sigma X_k$, the expected value of X is just the sum of the expected values of the X_k's. What is that? To get the variance, use the fact that the variance of a sum is the sum of the variances IF THE EVENTS IN QUESTION ARE INDEPENDENT. So what are the mean and variance of X?

3.7.9

(a) Work by analogy with Problem 3.7.6.

(b) Look at the quotient:

$$\frac{P(X = x + 1)}{P(X = x)}.$$

Write this out using the formula from part (a) and then simplify algebraically to obtain the recursion formula.

3.7.10

(a) Look at the definitions for various types of sampling in Chapter 2 and then think how the X_k's of Problem 3.7.8 are generated.

(b) First identify the random experiment: Work backward—What is an outcome, then what is the "doing"? You're going to obtain an outcome by repeating a Bernoulli trial. What is the Bernoulli trial (something you do with TWO possible outcomes)? What is "success"? How many times do you repeat it? Finally, identify the random variable.

(c) Section 3.4 is "sampling with replacement from a dichotomous population." We said we would see later why that model doesn't have a special name of its own. Now we know. Why doesn't it?

(d) Part (c) of this problem explains why we'd use the binomial model for sampling with replacement, but what justifies that model when we're sampling withOUT replacement?

3.7.11 (a) A player's skill is modeled by the Bernoulli trial: "attempt a basket," where S is "sinks a basket." Note that we give exactly the same verbal description for each of our two players. But we have two different Bernoulli trials depending on who is attempting the baskets. Two different "doers" means two different "doings" even though the "doings" are described with the same words. So we have two different "doings," therefore two different random experiments (two different Bernoulli trials). Now, by precise reference to the appropriate formula, show that Juan's game is more predictable than Shu Wen's even though she's more likely to be successful.

(b) We don't know! Reinterpret. This question only makes sense if we interpret it to mean "on average." It's like asking, "How many children in a San Francisco family?" You would anticipate an answer like "On average, 1.2." (Don't worry about the family with only two-tenths of a child! An average is a theoretical number.)

Now, with this reinterpretation show that Shu Wen should expect to make about five attempts before sinking a basket. Note the phrase "should expect"; it implies you're giving an average (an "expected value").

(c) There's about an 11% chance. Show how to get that. Note that Juan makes $n = 10$ repetitions of his Bernoulli trial.

(d) How do you measure the predictability of a player's game?

(e) Take the extreme cases: Suppose p is zero or one, values as far from one-half as you can get.

3.7.12 There's less than a five percent chance of such a result if the choice was random. Explain. That's a pretty small probability to the "mind of justice."

3.7.13 (a) If you don't round at intermediate steps (and you shouldn't), if you do the entire calculation with the calculator, you'll get

$$\mathrm{P}(X = 0) = q^5 = 0.03125$$

and

$$\mathrm{P}(X = 1) = [(5 - 0)/1] \times (1 \times 0.03125) = 0.15625$$

and so on. You shouldn't round at intermediate stages because further calculations with rounded numbers will magnify the roundoff error. Of course, when you record the probabilities in the distribution table, you should round them to four places (that's our convention).

(b) and (c) Yes?

3.8.1 Here are the answers. Show how you get them.

(a) 0.0780.

(b) 0.3796.

3.8.2 (a) 0.4040.

(b) 0.0300.

(c) Recursion may look harder, but when you see what's involved you'll see it's really much easier. After all, the recursion formula just says, "Multiply by λ and divide by the next value." As you calculate the various probabilities, accumulate the sum in the memory of your calculator. That way you don't have to write down intermediate values and waste time later adding them all together. We could make use of part (a), but let's ignore that and do this from scratch. Use recursion to calculate $P(X \leq 5)$ and then get the required probability as $1 - P(X \leq 5) = P(X > 5)$.

(d) The recursion formula gives the probability of the next value from the probability of the current value, where x is the "current value" and $x + 1$ the "next value." Now, give a verbal description of the formula.

3.8.3 (a) The Poisson distribution cannot be presented completely in a table. Why not?

(b) But from a practical point of view, this is no serious restriction. Explain.

3.8.4 What kind of explanation or reason could you give for the equation $\sigma^2 = \lambda$? How could you understand the equations of the Poisson model?

3.8.5 Which condition of the Poisson model would be violated if the disease were contagious?

3.8.6 There are certain mathematical conditions which are required to derive the Poisson model from the binomial. The rules of thumb are attempts at real-world formulations of those mathematical conditions. Now, if some observed data seems to fit the Poisson model, does that mean the underlying situation satisfies the three rules of thumb?

3.8.7 (a) What does it mean to say there were 109 corps-years with $B = 0$?

(b) Use the appropriate technique from Chapter 2.

(c) λ is the expected value of B, the expected number of fatalities in any one year. Based on these observations, it should be 0.61. Why?

(d) The phrase "empirical probability distribution" means you calculate *observed* relative frequencies rather than *theoretical* relative frequencies. Observed relative frequencies are sometimes called "empirical probabilities." The word "empirical," after all, means "based on observation." Before you do the computations, guess what the mean and $\hat{\sigma}^2$ should be.

(e) Assuming B has a Poisson distribution (theory!), compute the theoretical relative frequencies (the probabilities). There is one minor difficulty: B has an infinite number of possible values, but your table cannot be infinite. So, take B only through the values actually observed. Here, the approximations for the mean and variance will be less than their true values. Why?

(f) What comparison should you make?

3.8.8

(a) Note that we can't answer the question as asked because it asks for the value of a variable quantity. We must reinterpret the question as asking for the "EXPECTED value" of that quantity. Call the variable quantity X. So we interpret the question as asking for the mean of X. After some thought, you'll see that X, verbally described, is "the number of years over a ten-year period in which YES" where YES means "a Prussian army corpsman would have seen more than one of his comrades killed as a result of a horsekick."

(b) Let $X =$ the number of years the corpsman must observe before encountering a year in which YES.

(c) You're looking at a random sample of 15 corps-years chosen from the 200 observed by Bortkiewicz. Obviously, you're choosing without replacement. Because it's not given, we must assume the five years were chosen randomly; otherwise, we have no technique for answering the question. So we make the assumption remembering that we should check up on it before accepting the answer. Finally, the question is $P(X > 4) = ?$

3.8.9

The answer is 1.3195. How? Note that—you can check this with your calculator—just as $\log(10) = 1$, $\ln(e) = 1$, so . . . ?

3.8.10

(a) Well, what information picks out a particular Poisson random variable from the whole class? For $B(300, 0.01)$, which Poisson random variable should you use? What's the general rule?

(b) Think of the n repetitions of the Bernoulli trial as taking place over time—an hour, say. Any convenient period of time is acceptable here because we're just trying to show that it's *possible* to think of the binomial situation as a Poisson situation. Then divide the time into n equal subintervals of time. If $n = 60$, for example, then each subinterval is one minute long.

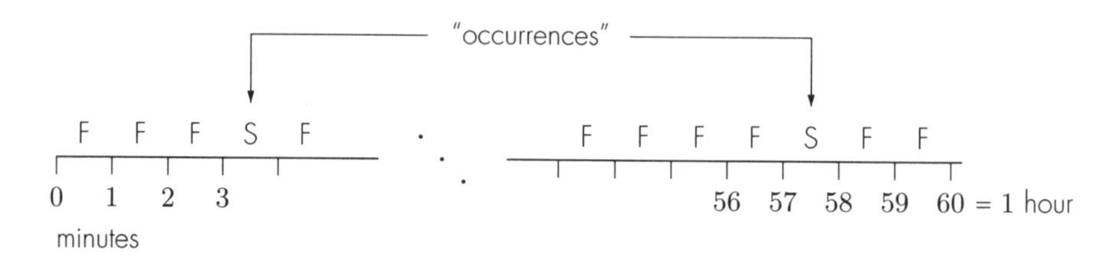

The binomial counts "successes" and the Poisson counts "occurrences of some event," so, let the "event" be "observe a success in this one minute interval."

Now show that the three rules for the Poisson distribution will be satisfied:

- Simultaneous occurrences should be impossible,
- Any two occurrences should be independent,
- The expected number of occurrences in any interval should be proportional to the size of the inteval (length, area, volume, depending on the type of interval).

(c) 0.2244 and 0.2240. How?

(d) Write down the formulas for the two means and the two variances.

(e) Well, for the Poisson approximation, set $\lambda = np$. You want to say that $\lambda \approx npq$. What does that tell you?

3.9.1

(a)–(c) Yes?

(d) Be careful in specifying the smallest possible value of X; it's not zero or one.

(e) Remember, the repetitions of the Bernoulli trial are independent.

3.9.2

For each of these, think of $X = x$ blanks in which you must put k S's and the rest F's . Don't forget that the last blank MUST contain an S because you stop exactly when you have observed that kth success. So you have no choice for the last blank. Here are the answers, show how you get them:

(a) 36; **(b)** 330; **(c)** 8; **(d)** 1; **(e)** look for the pattern in the previous parts.

3.9.3 Hint: Do part (c) first because it has the smallest number of different outcomes without being trivial. Part (d) is trivial; there's only one outcome. Write out all the ways you can have $X = 9$. Then compute $P(X = 9)$.

3.9.4 This means "Don't set $k = 8$; let it be unspecified." So, go back to the previous problem and, wherever eight refers to the number of successes you seek to obtain, replace it by k.

3.9.5 Here, "correct result" means that if you set $k = 1$ in the formulas for the negative binomial model, you should get the formulas for the geometric random variable. Do you?

3.9.6 (a) 75; (b) 0.0002; (c) 0.0007; [Hint: Use recursion.]; (d) Oh, you know! (e) 12.5; (f) 25.

3.9.7 In fact, the probability is much less than one in 1000. Why?

3.9.8 This is a difficult problem until you see it breaks down into two separate problems, each of which by itself is straightforward.

(a) Let $X =$ the number of bolts you must receive to get three with more than one defect. So, X should be negative binomial. What is the Bernoulli trial? What is "success" on the trial? What is p?

(b) Let $Y =$ the number of defects in one bolt. What is the distribution for Y?

(c) The question is $P(X \geq 5|p = ?)$. What's the answer?

(d) There are two models here. What assumptions are required in each case?

3.9.9 Suppose, for example, you have three traps.

3.10.1–3.10.14 Yes?

Chapter **4**

4.1.1 Think about all the possible values X can assume. How many are there? Suppose just to simplify things all the values were equally likely. What would the probability of each value have to be?

4.1.2 This is important! If all probabilities are zero, it looks like the model is totally trivial and useless. In fact, it's not. Think about the dart board mentioned in the solution to the previous problem. You'll also get a hint from the definition of the continuous uniformly distributed random variable. Look at that definition again. After thinking about all this, you'll see that

> although any specific value is virtually impossible (zero probability), it *is* possible to have X fall within . . . ??

4.1.3 **(a)** What does that shaded area represent? Write down an explicit expression for this using the symbols X, 0, and 100.

(b) So far, we have only defined the median for observed data, but obviously for a random variable, it should be defined in a similar way: that value of the random variable which cuts the distribution in half. So here there is a 50% chance to fall below 30 and a 50% chance to fall above 30 since 30 is the median. What is the shaded area?

(c) Use Chebyshev's Theorem because you really do not have any other information about this random variable.

4.1.4 First, an intuitive analysis: Because you know nothing at all about your sample, making use of the principle of "indifference"—averaging "across your ignorance," as in Section 1.4—you should assume your sample is evenly distributed across the distribution from which it's chosen. In other words, assume the sample cuts the distribution into $n+1$ sections of equal probability each.

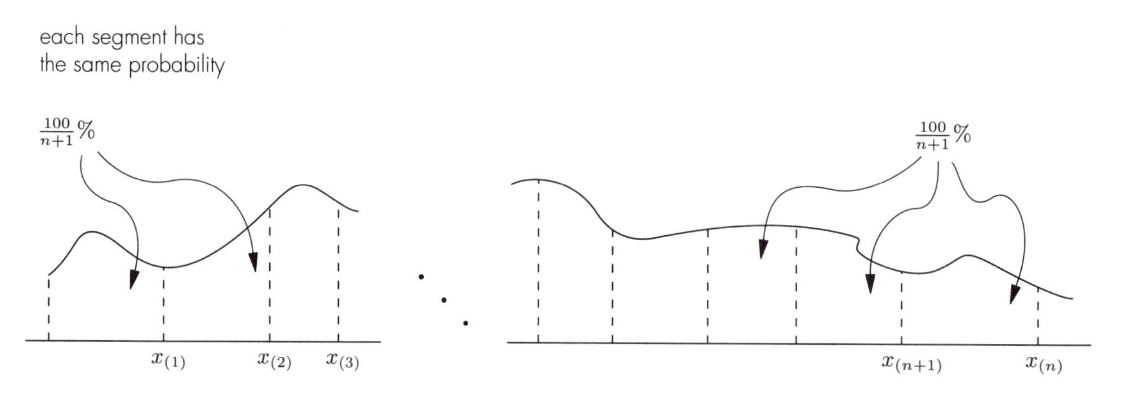

Note that there are $n+1$ segments. Also note that the members of the sample are NOT evenly spaced in the distribution. It's the probabilities that are equal. Now, how do you get the formula given in the problem statement? It's possible to get this result just by counting, looking at just your one sample. How?

4.1.5 (a) Use the fact that X is bound to take a value somewhere between two and three and, therefore,

$$\text{ONE} = P(2 \leq X \leq 2.25) + P(2.25 \leq X \leq 2.5) + P(2.5 \leq X \leq 2.75) + P(2.75 \leq X \leq 3).$$

From this together with the definition of the uniform distribution, you can compute $P(2.25 \leq X \leq 2.5)$.

(b) Now draw a graphical representation of this probability.

(c) Explain how in part (a) we can get away with just adding the four probabilities. What addition rule are we using?

4.1.6 Hey, don't be so impatient! A little calm exploration will show you the answer. Take a deep breath, relax, and—now, try looking at just ANY CURVE and see why it wouldn't be appropriate for a uniform distribution. Then you'll see what's required for the density function of a uniform distribution. Don't forget that any two subintervals of the same length must have the same probability. Try the following curve just for starters

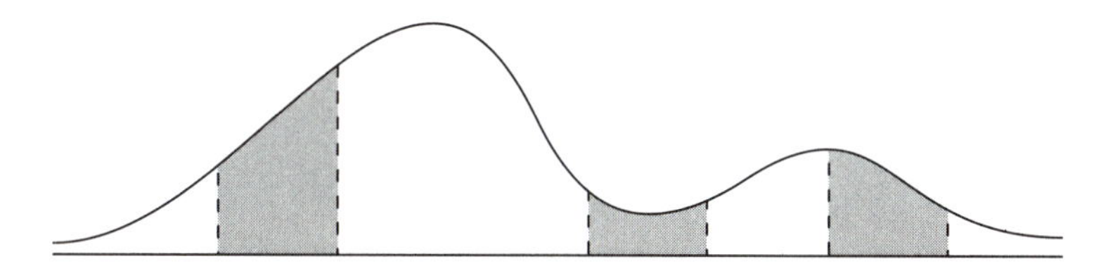

and note that even though they have the same length, the three subintervals we have marked do not have the same probability. The associated shaded areas are NOT the same.

Now do a few more sketches and see if your exploration doesn't show you what's required to make all subintervals of the same length have the same probability.

4.1.7 X is certain to fall between a and b after all, and so $P(X \in [a, b]) = 1$. Look at the corresponding picture and you'll see what c must be:

4.1.8 (a) A uniform distribution is completely symmetric. So the mean and median should be equal. Look at the graph of the density function (Problem 4.1.6) and you'll see that the median, and therefore the mean, is . . . ?

(b) Look at the picture and figure out what the area should be. After all, for a continuous distribution probability is given by area under the curve.

(c) When we say two things are proportional to each other, we mean

$$\text{“thing one”} = (\text{a constant}) \times \text{“thing two.”}$$

Now, $P(X \in [c, d]) = (d - c)/(b - a) = K(d - c)$, where $K = 1/(b - a)$. Because a and b are fixed, K is indeed a constant. Now what's the VERBAL description of $P(X \in [c, d])$?

4.1.9 (a) All you have to go on here is the probability formula in the box preceding this "Try Your Hand" exercise set. For that, you need to interpret the event $X \leq x$ as $X \in [x_1, x_2]$. So, what is x_1 and what is x_2?

(b) Yes?

(c) Note that $|X - \mu| \leq (b - a)/2$.

4.1.10 (a) The answer is 0.25—HOW?? Here's the picture of $f(x)$. Where is this 0.25 in the picture? Show that the answer of 0.25 is obvious from the picture and then show how to get this answer from the appropriate formula.

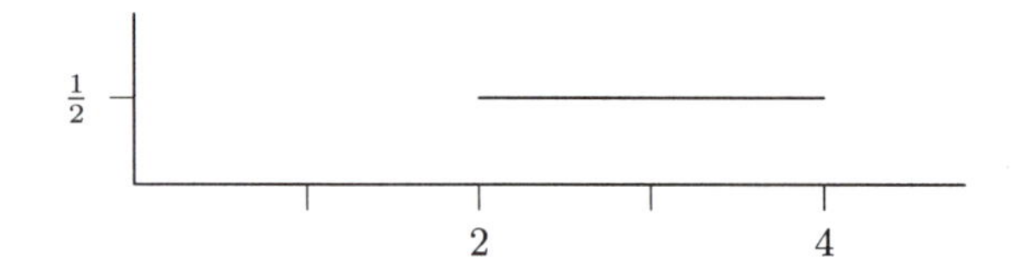

(b) The answer is 0.625. How? And the picture?

(c) Yes? Draw the picture.

(d) Yes?

(e) About a 87% chance.

4.2.1

(a) Let X be the number of failures during the period of useful life. X should be Poisson. Verify first that X really is a random variable. What's the "doing"? Is it repeatable? What does an outcome look like? Are they unpredictable? Then describe X as a rule assigning a number to each outcome. Finally, show that the period of useful life (the random failure period) satisfies the three rules of thumb for the Poisson distribution.

(b) This is just $P(X = 0)$, where X is the number of failures from time zero to time t. It's customary to express this in terms of λ, the expected number of failures during a unit of time. For example, if t is measured in hours, λ is the average number of failures per hour. Now, if failure is really due to random causes, the expected number of failures over a period of t units of time will just be $t\lambda$. For example, for two hours, you would expect 2λ failures; for half an hour, $\frac{1}{2}\lambda$ failures. This should look familiar; it's the proportionality assumption of the Poisson model. So, $P(T > t) = P(X = 0) = ??$

4.2.2

(a) What is the underlying random experiment? Is it really a random experiment? And so on.

(b) What is a continuous random variable?

(c) It's just $1 - e^{-t\lambda}$. How? Recall that the cumulative probability distribution for T is just $P(T \leq t)$. Here's a hint:

> Look at the complementary probability: $P(T \geq t)$. What is this in terms of the variable X from Problem 4.2.1(a)? Then use Problem 4.2.1(b).

4.2.3

(a) About 2.1739 minutes. How?

(b) About 39%. You'll have to assume that any five-minute period for this telephone switchboard looks just like any other five-minute period. With that assumption, the number of calls received in a quarter of an hour is a random variable with what distribution?

(c) About one percent. Because the question is in terms of minutes, model this with t in minutes. Use the exponential distribution with λ as the average number of calls in one unit of time (one minute). Note that the question requires you to evaluate $P(T > 10)$.

4.2.4

(a) About one chance in three. How? Define the random variable T be "time to burn-out" for this component. What is the distribution of T? [Hint: Measure time in thousands of hours.]

(b) Five. You can do this two ways: (1) in terms of the distribution for the number of burn-outs in 10,000 hours or (2) in terms of the distribution of lifetime (time to burn out) for the system where you have, say, M components.

(c) About 45%. For that to happen, both components must operate for at least 1000 hours. What's the probability of that? You'll have to make an assumption about how these components operate within the system. Otherwise, there's not enough information for you to answer the question. What's the assumption?

4.2.5

(a) Recall that $\mathrm{P}(A|B)$ is just $\mathrm{P}(A \text{ and } B)/\mathrm{P}(B)$. Apply that to $\mathrm{P}(T > t+s|T > t)$, making use of the cumulative probability formula.

(b) Take specific values for t and s and then say (verbally) what the equation tells you about T and these values t and s. For example, start at noon and let $t = 1$ and $s = 1$ (so $t + s$ is 2:00 P.M.), measuring in hours.

4.3.1

(a) You'll have to use the "fundamental fact" in the box from the text. What does it tell you about Z?

(b) $Z \sim N(?, ?)$. Explain.

(c) Here's the picture. Use the fact that a normally distributed random variable is completely symmetric about its mean.

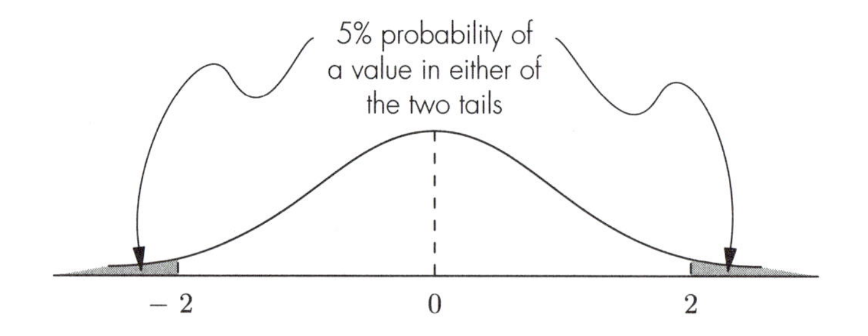

Where did we get this picture? What is $\mathrm{P}(Z < -2)$?

(d) $\mathrm{P}(Z > 1) = 16\%$, $\mathrm{P}(0 < Z < 1) = 34\%$, $\mathrm{P}(-1 < Z < 2) = 81.5\%$, $\mathrm{P}(1 < Z < 2) = 13.5\%$. How?

4.3.2 All of these are obvious from the appropriate picture. Do this carefully before you look at the answers.

4.3.3 Suppose the unknown number you're trying to estimate is 2. The estimates might be

$$2.18,\ 1.09,\ 1.37,\ 2.26,\ 2.08,\ \ldots.$$

What are the errors here? What is it about the errors that should make the mean of the errors zero? You'll have to make a certain assumption; otherwise, a mean error of zero would NOT be reasonable.

4.3.4 (**a**) Be sure you label μ and $\mu+\sigma$ taking into consideration the fundamental fact that about 68% of the probability should fall within one standard deviation of the mean.

(**b**) You don't have enough information to locate zero exactly, but you can do it approximately. Think about how many standard deviations below the mean you would have to go to find zero.

4.3.5 (**a**) This is like the previous problem.

(**b**) How many standard deviations below the mean is zero?

4.3.6 (**a**) Impossible, why? Redraw the curve.

(**b**) Impossible, why? Redraw the curve.

(**c**)–(**f**) Do you get the idea? What about these last four pictures?

4.3.7 What are the values of the parameters when you specialize to Z?

4.3.8 (**a**) The usual questions: What's the underlying experiment? Is it really a random experiment? What are the outcomes? Then, once you have the random experiment clear, what's the rule which assigns numbers to the outcomes?

(**b**) Suppose the measurements are called M; the question is "What's the value of μ_M?" You can guess on intuitive grounds what it ought to be and then you can prove it analytically based on what you've understood about measurement error.

(**c**) The measurement error is exactly as variable as the measurements themselves. Why?

(**d**) Think physically! What did Simpson say? For the measurements, there are two sources of variability. What about the errors?

4.3.9

(**a**) First, what's the "situation giving rise to numbers"? Then, what are the "many independent random factors" which should determine the difference between any two values? Why do we say "approximately" normally distributed; isn't the criterion fulfilled exactly?

(**b**) What can you DO to get a randomly determined number from the population?

4.3.10

(**a**) You can answer this in complete detail like the answer in Problem 4.3.9(a). But briefly, in intuitive terms, specification error should be normally distributed because it really is just a form of . . . ??

(**b**) Under what circumstances would specification error not look like random error?

4.3.11

(**a**) Apply our criterion for normality: Look at the difference in two tests scores. There are, indeed, many independent random factors which would account for why the score of one student might be different from that of another student. Name some.

(**b**) It looks as if a small group of students did better than the rest of the class. How could that have happened? There's more than one possible explanation of course. How does our criterion fail in such a case?

4.3.12

(**a**) The "relationship" is through an equation. What is it?

(**b**) First, in what sense is T a random variable? Then, why should it be normally distributed? Think about the population of all students who take the test that year.

(**c**) E is just a certain kind of . . . ??

(**d**) By part (a), T is the sum of two normally distributed random variables. The general rule: The sum of any two normally distributed random variables is normally distributed. Why is that true? In other words, show that if X and Y are both normally distributed, then $X+Y$ is. Use our "rule of thumb" criterion.

4.3.13

(**a**) You have to use our rule of thumb, of course, because you have no other criterion for normality. So, assume X satisfies the conditions of the rule of thumb and show that any linear function of X also satisfies those conditions. In other words, show that if $Y = a + bX$, then Y will satisfy the conditions if X does. For this, you must look at the difference between any two values of Y. You can write such a difference as $Y_1 - Y_2$.

(b) Take a specific example like $Y = 2X$, or $Y = X + 1$, where X is binomial. Then think about the possible values of a binomial random variable.

(c) First recall X; it's the Bernoulli random variable with parameter p. It takes only the values zero and one, taking the value one with probability p. Now, first, what are the values of Y? Then give the probability distribution for Y. You can compute the mean and variance of Y two ways: (1) from the distribution and (2) from the linear relationship with X. Remember, the mean of X is p and its variance pq.

(d) Yes?

(e) Use our criterion.

4.3.14

(a) $X = \mu + \sigma Z$. That's the answer. And the slope is σ which is, indeed, positive. How do you show that X really is $\mu + \sigma Z$? First, assume for the moment that $X = a + bZ$. You can discover that $a = \mu$ and $b = \sigma$ by writing down the equations for the mean and variance of X. You will get two equations in two unknowns which you can solve for a and b. Once that's done, you can say that IF X is a linear function of Z, then $a = \mu$ and $b = \sigma$. Next you should prove that $X = a + bZ$ for these values of a and b. That will complete part (a). Now you carry all this out in detail.

(b) Recall that $Z \sim N(0, 1)$. Now write $Z = c + dX$. What are the constants c and d? Note that d, the slope, is positive. Why?

4.3.15

(a) The random experiment is identical to the random experiment for the measurement error, E. You described that in Problem 4.3.8. In other words, the measurement and the error both arise from the same real-world situation. The only difference is in the rule which assigns numbers to outcomes. For M, what is that rule? Contrast it with the rule for E, the error.

(b) Use the fact that E is a linear function of M.

(c) Compare M with E. Sketch the curves for M and E.

(d) In both cases, the random experiments are identical. For E and M, the experiment is "make a measurement"; for X and Y, it's "roll the die." Because in each case the experiments are the same, of course the outcomes are also the same. For E and M, an outcome is . . . ? For X and Y? So what's the difference between E and M? Between X and Y? How is the situation for E, M similar to the situation for X, Y? Finally, identify some differences between the two cases.

4.3.16 For $X \sim N(2, 25), \sigma = 5$. Just using the fundamental fact about the probabilities of a normal distribution (draw a picture!), you can see that

$$P(X < -3) \approx 16\%, \qquad P(X < 2) = 50\%, \qquad P(X < 7) \approx 84\%.$$

So, in this way, you can make approximate guesses before calculating the exact probabilites from the Z table. For example, the answer to part (a) should be a bit larger than 84%.

(a) $P(X < 7.4) = 0.8599$. $X < 7.4$ if and only if

$$Z < \frac{7.4 - 2}{5} = 1.08$$

and $P(Z < 1.08) = 0.8599$ from the Z table. The corresponding pictures are given below. You should always draw such pictures. ALWAYS! We, on the other hand, will from now on usually omit the pictures, but only in the interests of space. This book is already too expensive!

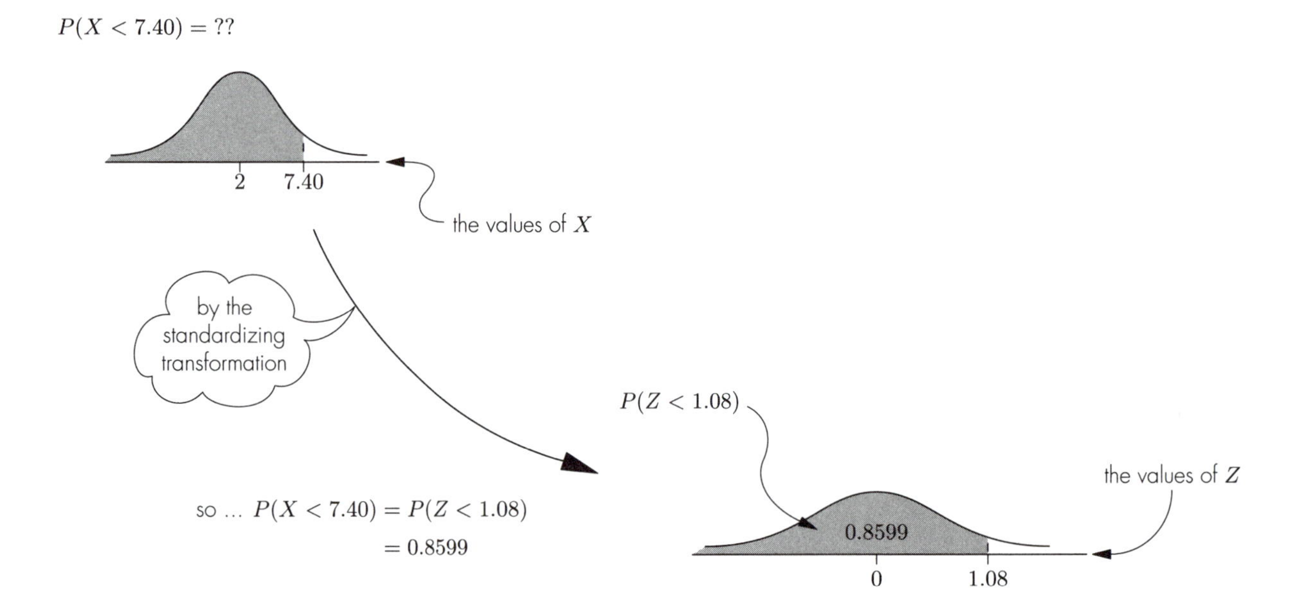

(b) This one requires no calculation at all. Why?

(c) $P(X > 1) = 0.5793$. Here's how to guess: This X value is only slightly below the mean; so the probability to be above that should be slightly more than 50%. Now get this precisely from the Z table.

(d) 0.7257. Here $X = -1$ is less than a full standard deviation below the mean. So the probability should be larger than 50% but less than 84%. Now do it precisely from the table.

(e) 0.2743. How?

(f) This probability is necessarily less than 50%. Bear this in mind as you do your calculations.

(g) The answer is 0.4392. How? Here's the picture:

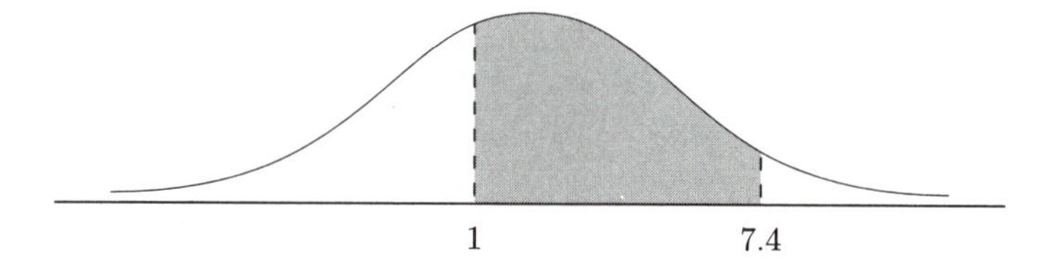

4.3.17 The answers for all of these can be read off the picture, making use of the standardizing transformation and the Z table. However, you may have to go through several steps to put the picture in the form of the Z table picture.

4.3.18 (a) Looking in the body of the Z table—probabilities are in the BODY of the table—you discover that $Z = 1.64$ puts a probability of 0.0505 in the RIGHT tail. And so, by symmetry, the corresponding negative value, $Z = -1.64$, puts that same probability in the left tail. Or take $Z = -1.65$, because what you're looking for is exactly halfway between 0.9495 and 0.9505. In fact, it's usual to interpolate, taking $Z = 1.645$ because you're exactly halfway between the two values from the table. Show: $X = 0.6091$.

(b) $X = 1.6864$, how?

(c) $X = 5.77$, how?

(d) $X = 4.23$, how?

(e) $X = -0.6580$, how?

(f) $X = -1.9450$, you should have guessed X to be just slightly bigger than -2.

(g) $X = 13.54$, how?

(h) This is not possible, do you see why?

(i) $X = -30$ standardizes to $Z = 1.2$. So, $X = -27.525$. Explain.

(j) $X = -31.05$, how?

4.3.19

(a) If you divide the picture for Z into ten equal probabilities, each probability will be ten percent. So you're looking for the 10th, 20th, 30th, and so on, percentiles of Z. The picture in the text shows how to obtain the tenth percentile for Z. It's -1.28, because $P(Z < -1.28) = 10\%$. Now go ahead and obtain the others. Note that you're only looking for nine (not ten) values of Z. Why?

(b) Draw the bell-shaped curve for Z and label it with the probabilities and the values of Z determined in part (a). Why are those values (the percentiles) not evenly spaced?

(c) The text says you will plot the observations in the sample against the corresponding percentiles of Z. The "corresponding percentiles of Z" are given in part (a). So the normal probability plot is a set of ordered pairs of the form (z, x), where z is a percentile of Z, and x is the "corresponding" observation in our sample. The word "corresponding" means the smallest percentile of Z should be matched with the smallest observation in the sample. List these (z, x) points.

(d) First, suppose the sample is EXACTLY like the distribution $N(\mu, \sigma^2)$. That means we have the "ideal" sample. So the points (z, x) in our probability plot lie on the graph of a linear function. In other words, they lie on a line. To see this, remember that our percentiles for Z were obtained by standardizing the "ideal observations"—here by standardizing our sample, the x's. So the z's are just the standardized x's. Because the standardizing transformation is a LINEAR transformation, the points (z, x) lie on the line which is the graph of that function. What if the sample is only "more or less" like the distribution of X?

(e) Plot the points from part (b).

(f) Yes?

(g) For us, $n = 9$, and the first value of Z was the tenth percentile, from $1/(n+1) = 0.10$. And then . . . ?

4.3.20

Because there are nine observations, you'll use the same percentiles of Z determined in part (a) of the previous problem. Make a normal probability plot by plotting the following points:

$$(-1.28, 65),\quad (-0.84, 80),\quad (-0.52, 90),\quad (-0.25, 110),\quad (0, 137),$$
$$(0.25, 145),\quad (0.52, 170),\quad (0.84, 200),\quad (1.28, 270).$$

What does your plot tell you? [Hint: Increment the vertical scale by 15's.]

4.3.21 (a) This is given in the text. TRY TO RECALL IT!

(b) This is not mentioned in the text, try to guess.

(c) Yes?

4.3.22 Approximately 0.0475. How?

4.3.23 (a) $P(\widetilde{X} > 7.5)$. Explain.

(b) $P(6.5 < \widetilde{X})$. Why $<$ instead of $\leq$?

(c) $P(2.5 < \widetilde{X} < 8.5)$. Explain.

4.3.24 (a) $np = n/2 \geq 5$ implies that n is . . . ?

(b) $n \geq 20$—how? (c) 100—how? (d) 2500—how?

(e) $n \geq 20$ (not $n \geq 7$); think about this before you look at the answer.

(f) $p \geq 0.4167$—how? (g) 0.05—how? (h)–(j) do these on your own.

4.3.25 (a) 0.0963. Why is it inappropriate to use the normal approximation here?

(b) $P(\widetilde{X} > ?|\mu = ?, \sigma = ?)$, this probability is about $2\frac{1}{2}\%$.

(c) 0.8062. If you got 0.8078, you used the normal approximation. What's wrong with that?

(d) There's no hope to calculate this binomial probability directly because the coefficient $C(200, 24)$ would be ridiculous to attempt. However, the normal approximation is appropriate here because $np = 24$ is larger than five. The normal approximation gives 0.0876. How? Look carefully at the continuity correction.

4.3.26

(a) Because you have a large population, the "with–without replacement" distinction won't much matter. So use the simpler model, sampling WITH replacement (binomial). You want $P(X \geq 5|n = 12, p =?)$. The answer is 0.3032, but HOW? First, determine p. That involves a bit of work! Think clearly about the characteristic you're looking for. Then the probability of getting that characteristic for ONE selection of a machine part is p:

$$p = P\,[\text{one machine part has the characteristic}].$$

(b) Let X = the number of parts having the desired characteristic. State clearly what the characteristic is. It involves the probability distribution for "lifetime" of these parts. Now calculate the probability of having that characteristic. This information about X is required before you can solve the original problem. The original question is then of the form

$$P\,[\text{condition on } X \mid \text{information about } X] = ??.$$

(c) This is similar to the previous parts.

4.4.1

(a) This value of χ^2 is found in the column headed 0.975. What does that mean?

(b) The answer is 7.5%. How? Draw the picture!

(c) The answer is one-half of a percent. You have 20 values of Z and you're looking for the probability that $Z^2 + Z^2 + Z^2 + \cdots + Z^2 > 40$. So . . . ?

(d) Where did the 20 Z's come from?

(e) You can't answer this; why not?

4.4.2

(a) This is straight from the table!

(b) Draw the picture.

(c) $Z^2 + Z^2 + Z^2 + Z^2 + Z^2 + Z^2$ is χ^2 with d = ?. Assuming what?

4.4.3

(a) Think about the definition of this random variable.

(b) If the degrees of freedom is seven, there are seven squared Z's; if ten, ten squared Z's, and so on. What does that say about the *expected value* for your sum of squared Z's? Now, forget intuition. Do this precisely. Hint: See Problem 3.2.1(e).

(c) Draw the picture for a particular χ^2 distribution, for say $d = 4$. What does it mean to have a value in the far right tail?

(d) Yes?

(e) Hint: See Problem 3.2.1(e).

4.5.1 (a) $\mu = 1.5769\,\text{cm}$, $\sigma = 0.5325\,\text{cm}$. How?

(b) Assume the diameters normally distributed. Is that justified?

(c) You could report that no screw is more than 50 cm long, but that's not meaningful. Think about the "average number of children in a San Francisco family." You could certainly say that no family has more than 100 children, but that's not meaningful. If you know the average is 1.6 children with a standard deviation of 1.3, you know by Chebyshev's Theorem that it's very unlikely you'll ever find a family with more than six children (three standard deviations above the average). Now for the problem at hand, you need not appeal to Chebyshev, you can reason in a more precise way.

4.5.2 Assume an exponential distribution for "waiting time," T. Think of this as a Poisson situation in which you are observing a period of time for the "occurrence" of a "free checkout clerk." You're given $P(T > 10) = 0.045$. That gives an equation you can solve to find $\lambda = 0.3101$. Then?

4.5.3 (a) The cutoff for passing is $\mu - 1.2\sigma$. The probability of a score above that can be interpreted as the proportion of students who pass.

(b) $P(X = 10|n = 100, p = ?) = ??$ Of course, the problem should be modeled as "sampling WITHOUT replacement," but because N is large ($N \geq 60$) and n is less than ten percent of the population ($N \geq 10n$), the sampling WITH replacement model (a special case of the binomial) is a reasonable approximation.

4.5.4 $P(X > 0|n = 25, p = ?) = ??$

4.5.5 (a) Use a normal distribution for weight of stones, W. Then, for example,

$$P(\text{size} = 2) = P(-2 < Z < -1) = 0.1359.$$

(b) For each size, you want the expected number in a sample of $n = 12$ taken without replacement.

(c) P(size 4 or 5) = P(size 4) + P(size 5). Explain. So . . . ?

(d) Yes?

(e) $\mathrm{P}(X \geq 24 | n = ?, p = 0.0228) \geq 0.9$.

4.5.6 Yes?

4.5.7 Zero, 0.052.

4.5.8 (a) Because $n = 23$, the sample median will be the twelfth in succession once the sample is ordered. So? What's $x_{(6)}$?

(b) If $M < x_{(h)}$, how many observations in the sample are less than M? That number is the value of Y. What if $M > x_{(k)}$?

(c) The probability of "success" for Y is just the probability for one observation to be below the median. Thus, Y is $B(n, 1/2)$. So you can use a normal approximation for Y with mean $n/2$ and variance $n/4$. This approximation is valid, provided $np = n/2 \geq 5$, in other words, provided $n \geq 10$. Now, we analyzed the confidence coefficient in the problem statement [just above part (a)], and in part (b), we expressed it in terms of Y. The two probabilities in the square bracket add to ten percent. Because the endpoints play symmetric roles, divide this 10% equally into two parts and determine values of h and k to satisfy

$$\mathrm{P}(Y < h) = 5\% \quad \text{and} \quad \mathrm{P}(Y \geq k) = 5\%.$$

(d) You should report to the Boss: "Boss, we can be about 90% sure the median fill of cups from the drink machine in the employee lounge is somewhere between 6.3 and 6.5 ounces." Evaluate the formulas from part (b) to find the values of h and k. Then identify the values of the corresponding order statistics. Those are the required endpoints.

(e) Yes?

(f) The 90% probability cannot refer to the median, saying something like "90% of the time the median is between 6.3 and 6.5, the rest of the time not." Why not? What's wrong with this interpretation? What does the confidence coefficient refer to?

(g) "Boss, we can be about 95% sure the median fill of cups from the drink machine in the employee lounge is somewhere between 6.3 and 6.7 ounces." To get this new report, look at the argument in part (b) and see how to modify it. You'll find that z_0 should be 1.96. Explain. Use this z_0 in the formulas from part (b).

(**h**) You want a z_0 for which $P(Z < z_0) = 99.5\%$, putting 0.5% in the right tail. Because 0.995 is exactly halfway between two values in the table, we split the difference. Thus, $z_0 = 2.575$. How do you obtain the endpoints?

(**i**) Yes?

Chapter **5**

5.1.1

(**a**) So far, we've discussed two types of population. What are the two types? A population in this problem must be of what type?

(**b**) Take the case of sampling from a population first, and take a specific example—suppose you want to know the mean score on a test you've taken. What information would be required?

Then take the case of sampling from the distribution of a random variable—suppose you're trying to estimate the mean length of rods from a manufacturing process. Here the manufacturing process is the random experiment, a rod is an outcome, and the length of the rod is the number assigned to the outcome.

(**c**) You're asserting that the unknown true mean is the same as the sample mean. The assumption must be that . . . ?

(**d**) When you say the true mean is *approximately* the same as the sample mean, what assumption must you be making? How is this different from part (c)?

(**e**) Just how typical of the original distribution is your random sample?

(**f**) A random sample might be typical; it might be atypical. How do you find out whether it's typical or not? Think of sampling from a population first, then think about sampling from the distribution of a random variable.

(**g**) What does this value of $\overline{X}$ tell you?

5.1.2

(**a**) It's a population of people of course, but we think only of the *age* of each person—that's a number. Still, be careful—there's a trick here. Is this a numeric or a dichotomous population?

(b) Well? If you can't seem to focus on this question (it's easy, after all), you should think about a simple example—just like you did in Problem 5.1.1(b). Two pieces of information would be required to compute p—what are they?

(c) You certainly don't mean to assert that 18% is the exact percentage of the population over 50 years of age; no sample is exactly like the population. If the sample proportion is 0.18, you should say that *approximately* 18% of the population is over 50. Comment.

(d) If the sample you obtained happens to be typical of the population as a whole, then you would be justified in concluding that $p \approx 0.18$, but . . .

5.1.3

(a) Write the second three as $\bar{3}$ to distinguish it from the first three; so the population is

$$0, \ 1, \ 2, \ 3, \ \bar{3}, \ 4, \ 5.$$

There are seven objects in this population and so there are $C(7,2)$ ways to choose two of them for a sample of size two (without replacement). Write out all the possible samples of size two by writing first all the samples which contain 0 (there are six), and then all the ones containing 1, and so on.

(b) This is easy! For example, the mean of the sample $\{1, 3\}$ is just two.

(c) $\overline{X}$ starts out with samples and then . . . (does what?)? What role do the samples play for $\overline{X}$?

(d) With simple random sampling from a population, each sample has the same chance to be chosen. For example, what's the probability that $\overline{X}$ takes the value 1.5?

(e) The mean of the sample means, $\mu_{\overline{X}}$, is 2.5715 and the variance, $\sigma^2_{\overline{X}}$, is 1.0544—how?

(f) The mean of the sample means is 2.5715. You've seen a number like that someplace else, it's just . . . ?

(g) This is subtle! It's a question of the variability in the population "averaging out" in the samples. Will the sample means be more or less variable than the numbers making up the population? Compare them.

the sample means

$\overline{X}$	f
0.5	1
1.0	1
1.5	3
2.0	3
2.5	4
3.0	3
3.5	3
4.0	2
4.5	1
	21

the population

X	f
0	1
1	1
2	1
3	2
4	1
5	1
	7

(h) A sample is "typical" if its mean is close to the population mean, close to 2.5714. There are four samples which give a mean of 2.5—they seem to be fairly typical. Of course, the term "typical" is not precise. Let's say you'd be willing to include any sample whose mean was within one point of the true population mean. By that criterion of "typical," there are 13 typical samples. What are they?

(i) Well, of course, you will compute $\overline{X}$ for that sample. What would that value of $\overline{X}$ tell you?

5.1.4

(a) $p = 2/7$. But, of course, you're supposed to pretend the population as a whole is not accessible—you don't know that p is $2/7$. So you choose a random sample and compute a value of $\hat{p}$. What does that value of $\hat{p}$ tell you about the unknown p?

(b) Obviously it's a numeric population because . . . ? (Careful . . . !!)

(c) $\hat{p}$ looks at a random sample and . . . ?

(d) There are only three possible values. What are they?

(e) Guess that *on average* the sample proportions should be $2/7$. Why? Explain on intuitive grounds and then show it by making up a probability distribution for $\hat{p}$ and using it to compute the mean. Compute the variance as well, just for review.

(f) This question is identical to part (h) of the previous problem. And you're talking about exactly the same set of possible samples. So why do we ask the same question twice? Because it's not the same QUESTION! How is it different? Think about this carefully before you look at level II.

(**g**) It could only say one thing: Because $\hat{p} = 1$ for that sample, it would have to suggest that almost all the numbers in the population are even and positive! But only about 30% of the population have that characteristic! What's wrong??

(**h**) 1/21. For simple random sampling, with or without replacement from a population, all samples have the same probability of being drawn. This is one of the 21 possible samples and it has one chance in 21 of being drawn—exactly the same as any other sample! So what good is random sampling if atypical samples are just as likely as typical samples?

5.1.5 Look again at the discussion in the text. See why we said the conclusion "$\mu \approx 2.37$" is not justified. Then you say, "So μ must be far from 2.37." What's the defect in that argument?

5.1.6 (**a**) Is it a number? A set? An interval of numbers? No, it's none of these. Think about the estimators we've discussed (the sample mean, the sample proportion). To evaluate an estimator, you obtain a sample and compute a number, for example, $\overline{X} = 2.37$ or $\hat{p} = 0.18$. This number is one of the possible values of the estimator. So an estimator is a . . . ?

(**b**) For any random variable, we would always want . . . ?

5.1.7 (**a**) $\hat{p}$ is the proportion of a sample having a certain characteristic. To compute it, you first must COUNT how many in the sample have that characteristic. And then . . . ? So $\hat{p}$ is $a + bX$, where a, b, and X are . . . ? Now use this linear relationship to obtain the formulas.

(**b**) You cannot give a line graph for $\hat{p}$; there are too many possible values. What are the possible values? What kind of sketch should you give? As you think about this, don't forget that $\hat{p}$ is "essentially" like a binomial random variable.

(**c**) Simple random sampling from a Bernoulli distribution is just another name for the binomial experiment. By definition such a sample is "n values of the random variable," in this case n zeros and ones, "obtained from n independent repetitions of the underlying random experiment." The experiment being repeated is a Bernoulli trial. But n independent repetitions of a Bernoulli trial is the *binomial* experiment.

Now let X be the sum of the zeros and ones in the sample. Note that X can be described as the "number of observed ones"

$$0,\ 0,\ 0,\ 1,\ 0,\ 1,\ 1,\ 1,\ 1,\ 0,\ 0,\ 1,\ 0.$$

These add to six, which is exactly the "number of ones." So X is the number of ones in the sample. How does X relate to $\hat{p}$?

5.1.8

(a) We've led you up to this question, but we've not answered it. See if you can guess what we mean by "the entire context" or "the whole picture" for an estimator. Here's some help . . .

For a question about a parameter like p, you're trying to find out what more information is available beyond just one value computed from one sample. That "more information" is the "entire context" of the estimator, "the total picture" of the estimator. [Hint: Think about the previous problem.]

(b) There is a very specific answer to this question; you can see it in the solution to the previous question.

5.1.9

(a) The "total picture" of the estimator is its probability distribution. So where in the picture for the probability distribution of $\hat{p}$ does our 18% fall? There's a trick to this question; see if you can avoid getting caught!

(b) Because the distribution of $\hat{p}$ is centered on p ($\mu_{\hat{p}} = p$) and its value is unknown, the $\hat{p}$ value of 18% could be anywhere. So the first picture is certainly possible. What about the other pictures? The third one is highly unlikely, but remotely possible. Why?

5.1.10

(a) p is a proportion, but proportions CAN be zero. But here, p cannot be zero. Why not? This is not hard, but you have to be alert. Take another look at the "situation"!

(b) Go back to Problem 5.1.7(b) where you first saw that $\hat{p}$ is approximately normally distributed. But first, try to recall on your own: What is it about $\hat{p}$ that would justify a normal approximation, why should the population and sample be large? In fact, there's one situation where the population need not be large, but even then the sample would have to be large.

5.2.1

(a) Think about why it would be important to have $\mu_{\hat{p}} = p$ or to have $\mu_{\overline{X}} = \mu$. You saw this in some of the earlier problems! [Hint: What's the basic question we're trying to answer in this chapter?]

(b) The standard error is a measure of the variability of the values of $\hat{p}$ from one sample to the next. What does the standard error of $\hat{p}$ tell you about this variability? Think about how this relates to the "typicality" of a sample. To answer these questions, you should *look very carefully at the standard error formula.*

(c) The smallest possible value of $\hat{p}$ is zero, the next is 0.1, and so on. Why does this make any reasonable approximation impossible with $n = 10$? What happens if $n = 100$?

(d) The answer is $n = 625$. How? [Hint: You want $\sigma^2_{\hat{p}} \leq (0.02)^2$].

(e) The answer is four. Why? Draw the picture.

5.2.2

(a) About half of one percent. Draw the picture! Use a normal distribution with center $\mu_{\hat{p}} = ??$ and standard deviation $\sigma_{\hat{p}} = ??$

(b) The phrase "variability in response" is ambiguous here. It's not clear whether it refers to the *number* of yes's, or to the *proportion* of yes's. For a larger sample the number of yes's should be more variable, but the proportion less variable. Explain this intuitively and by reference to the appropriate measure of variability.

(c) What is the real-world question being studied in this chapter? How is this problem different? If the question is unrealistic, why do we ask it?

(d) A bit less than a 35% chance. This, as stated, is a binomial problem, but you can convert it to a question about $\hat{p}$.

(e) This is deceptively simple—be careful!

(f) A bit more than 10%. You can do this problem two ways: in terms of the proportion of your sample who say "yes" or in terms of the proportion who say "no." The question is asked in terms of those who say "no." Do the problem both ways!

(g) About a four-percent chance.

(h) Yes?

5.2.3

(a) Are all these ten-year-olds the same height? Of course not. This information can only mean that the *average* height is 129 cm. Apart from just common sense, we're also told that $\sigma = 17$ which again means not all heights are the same. If they were, the standard deviation would have to be zero! Show that when all values are the same, then $\sigma = 0$.

(b) About 89%—how? This is stated as a binomial problem; it refers to *how many* in the sample have the characteristic of interest. You can convert to a $\hat{p}$

problem and use the normal approximation. Why is the normal approximation justified? You're asked about $\hat{p} = 15/24$. Now look at the "total context" of this value and calculate $P(\hat{p} < 15/24)$. Note that $p = 1/2$. Show why that's true.

(c) About 11%—how? To get this, you need to know that $p = 0.1736$. How do you get that value of p?

(d) You had to assume a normal distribution for the heights of all the ten-year-olds in our geographic region. How exactly is that assumption used in each of parts (b) and (c)? Would it seem reasonable? Besides the assumption of normality, you made another important assumption in each of parts (b) and (c). What was it?

5.2.4 There are two parts to the definition of a confidence interval. Show that both parts are satisfied by the answer given in the box.

5.2.5 **(a)** A probability necessarily refers to something that's variable. Try to think of the total situation and ask yourself what's variable.

(b) Ask yourself what's wrong in saying that p is sometimes in the interval (0.15, 0.21) and sometimes not.

(c) A probability is a theoretical relative frequency and so you should be able to interpret the 95% by saying something like "On average, 95 times out of 100 . . . ." Remember what it is that varies here.

5.2.6 **(a)** This should be pretty obvious!

(b) To have a "high degree of confidence," you would have to choose a confidence coefficient which is quite large. This means you'll be more sure of your conclusion. But clearly, to be more sure of your conclusion, you would require more . . . ? And that costs money! Explain.

5.2.7 **(a)** The answer in the box does not provide an exact value for p; it contains two sources of uncertainty. What are they?

(b) There's no way to say which values in the "range of possible values for p" are more likely and which are less likely. So if you have to choose one value, which one would you go for? Go ahead—guess!

(c) There are two "aspects" to a confidence interval. Each is a source of possible error. What are they and how do you control the error?

5.2.8

(**a**) Your interval consists of a range of possible values together with a probability, the confidence coefficient. Think about the confidence coefficient. What information will be required to determine that?

(**b**) The conclusion, "Based on this sample, we believe $p \approx 18\%$," is unacceptable because it's incomplete. What's missing to make it valid?

5.2.9

(**a**) The quantity δ should be $1.96\sigma_{\hat{p}}$. To see this, standardize L and R and work backward from the Z table. The Z picture is

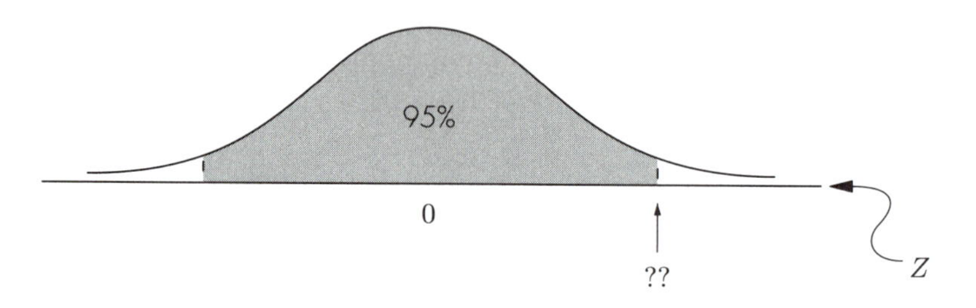

(**b**) If you could evaluate the standard error for $\hat{p}$, you would have no need of a confidence interval estimate for p. Why? The answer to this is obvious if you think about the formula for $\sigma_{\hat{p}}$.

(**c**) Look at Problem 5.2.1(b). The standard error measures the accuracy of $\hat{p}$ as an estimator. A high degree of accuracy means little variability in the estimator from one sample to another and that means a *small* standard error.

You want $\sigma_{\hat{p}}$ to be small; the "worst case" would be the value of p which makes it large. Because $\sigma_{\hat{p}} = \sqrt{pq/n}$ and n is fixed, the standard error will be maximum when pq is maximum. Think of pq as $p(1-p) = p - p^2$. For p very small, this is close to zero; for p very large, close to one; again $p - p^2$ will be close to zero. Now, guess what value of p would make pq take on its maximum possible value. Using that p and assuming $n = 1000$, show that the standard error would be 0.0158.

Analytically, you can maximize $pq = p - p^2$ by looking at the graph. Sketch a graph of the parabola $p - p^2$ and see what value of p gives the maximum value for pq.

5.2.10

What is "the question it addresses"? Why would you expect a simple answer to that question? What is the "complexity" of the confidence interval?

5.2.11

(**a**) YOU WILL NEVER KNOW! It could be that none of them do; it could be that they ALL do! Still, you can say something—what?

(**b**) This problem is very unrealistic! Why?

5.2.12

(**a**) We can be about 80% sure that the unknown value of p is at least 40.39% and not more than 45.61%. How did we obtain this?

(**b**) We can be about 90% sure that the unknown value of p is at least 67.12% and not more than 74.88%. How did we obtain this?

(**c**) We can be about 72% sure that the unknown value of p is at least 41.77% and not more than 48.23%. How?

(**d**) We can be about 92% sure that the unknown value of p is at least 29.17% and not more than 56.83%.

5.2.13

When you look at the normal distribution, what is its most visually obvious characteristic? What does that say about the samples?

5.2.14

(**a**) (0.1428, 0.2243); (**b**) (0.1269, 0.2331); (**c**) (0.1392, 0.2208).

(**d**) Compare the centers and the interval widths. Which interval has the greatest maximum error of the estimate and why? Smallest? Recall that the maximum error of the estimate is half the width of the interval. The maximum error of the estimate is also called the "error tolerance" because it's the maximum error you will have to "tolerate." Which is better, a large error tolerance or a small error tolerance?

(**e**) That "exact" formula was derived using an *approximate* model! So, in fact, the endpoints are still only approximate. What model?

5.2.15

(**a**), (**b**), **and** (**c**) No hints this time!

(**d**) Compare the error tolerances (maximum error of the estimate); which is largest and why? Which interval has the greatest uncertainty about the actual value of p? Which interval has the greatest uncertainty to actually contain the unknown p?

5.2.16

The answers are: CONSERVATIVE, (0, 0.1780); LESS CONSERVATIVE, (0.0268, 0.1332); EXACT ENDPOINT, (0.0411, 0.1500). Compare the centers of the less conservative and exact intervals. What are some other relevant points of comparison? In particular, why exactly would you expect these intervals to be "substantially different," as we said in the problem statement?

5.2.17

(**a**) What unstated assumption must you make to solve this problem? In fact, there are two different possibilities for modeling the problem to account for this assumption. What are they? [Hint: How did you get your data, the 318 printheads?]

(**b**) The "probability a printhead . . . will fail . . . " is just the proportion p of all your printheads which have this characteristic. Just as the probability of drawing a club from a deck of 52 playing cards is the *proportion* of clubs in that deck. Remember that probabilities are theoretical relative frequencies (proportions).

The question seems to ask for a number, but the available information is not adequate to give a simple numeric answer. A confidence interval is required. First, you must select $1 - \alpha$ because it was not given. If you choose 99%, the interval is (0.0144, 0.0736). How? Then . . . ?

5.2.18

(**a**) You might wonder if a random sample from the telephone directory can provide information about stuttering for ALL residents of Tallahassee. Suppose it does. But still, the respondents were not asked if they stutter, so the question we've asked you cannot be answered on the basis of this information. Give the best answer possible. The answer is NOT 21.31%.

(**b**) You don't know how many people in Tallahassee stutter. What can you say?

(**c**) You still don't know. What can you say? This time, for practice, let $\alpha = 0.1$. The interval is (0.0227, 0.0484). How?

5.2.19

(**a**) What does the confidence coefficient measure? And the maximum error of the estimate?

(**b**) The maximum error of the estimate will vary depending on which of the three procedures you use to construct the interval. Give a numeric value for the maximum error of the estimate assuming you use:

(i) the worst case estimate of the standard error,

(ii) the less conservative estimate of the standard error.

(**c**) Think how we choose that approximation.

5.2.20

The specificity of such a test is the probability an uninfected person would not test positive, $P(T^c|D^c)$. Of course that should be a high probability. This problem shows how difficult it is to get a reasonably high specificity even on repeated tests when the condition you're testing for is rare. This is a serious problem for employers wanting to do "workplace drug testing" (think about that!). And it is particularly a problem for blood banks testing their blood supplies when infected persons have already been discouraged from donating blood. So, what is the specificity here? Note that you have 17,053 persons tested, of whom 17 tested positive.

5.2.21 (a) Before you can solve this equation for n you will need some value for pq. You might try to use the approximation $\hat{p}\hat{q}$, but that's not possible—YOU HAVE NO SAMPLE! After all, you're trying to decide on an appropriate sample size so clearly you have not yet obtained a sample. If you have no sample, you have no value for $\hat{p}\hat{q}$. You'll have to approximate pq somehow; what approximation will you choose?

(b) The answer is $n = 505$. [Hint: You know for sure that n (it's $2401pq$) is greater than or equal to what?]

5.2.22 (a) $n = 66,307$.

(b) $n = 30,625$. For $1 - \alpha = 0.92$, $z = 1.75$.

(c) $n = 2626$.

(d) $n = 5000$. If you said $n = 265$, think again. Think about the conditions required for the normal approximation to be valid.

5.3.1 (a) This is very straightforward if you think clearly about X_k as the kth number in the sample. But there are two ways to think of your sample: It's drawn either (1) from the distribution of a random variable X or (2) from a numeric population. The second is a special case of the first, but to understand X_k clearly it helps to take the two cases separately. So take them separately and show in each case that the mean and variance of X_k are μ and σ^2.

(b) Here there's no need for two cases because you've already taken a clear look at X_k. For $\overline{X}$ to be "unbiased" means it is, on average, equal to the parameter it's estimating. In other words, you must show that . . . ??

(c) Yes?

5.3.2 (a) Any X_k is the value of X obtained on the kth repetition of the random experiment. So, abstractly as an unknown variable, each X_k is just a version of the normally distributed X. Why does that make $\overline{X}$ normally distributed?

(b) Using our criterion for normality, show that the difference between two values of $\overline{X}$ is determined by "many independent random factors." To avoid abstraction, take the specific case with $n = 2$. When you see how the argument works in that case, you'll see how to make it work for any n, large or small. So look at $\overline{X}_1 - \overline{X}_2$, assuming each sample consists of two numbers, and account for this difference as determined by many independent random factors.

Now go back through this argument and note that it would also work as an alternative argument for part (a). Why?

5.3.3

The argument must be different from the argument in Problem 5.3.2 because now the underlying situation may not be normally distributed. Suppose $\overline{X}_1$ and $\overline{X}_2$ are two different values of $\overline{X}$. If you write each of these sample means out as a sum and think about the difference of the two sums, you'll see why that difference is "accounted for by many independent random factors."

5.3.4

(a) The total of the sample, ΣX_k, is a linear function—it's just a sum—of the X_k's. If the distribution you're sampling from is normal, the X_k's are normal and so their sum is normal [Problem 4.3.13(e)]. Then your argument is complete.

But the distribution you're sampling from may not be normal, so you must argue differently. Here's one simple way to proceed: Show that ΣX_k is a linear function of the *sample mean*. In other words, show it has the form: $a + b\overline{X}$ for some constants a and b. Then . . . ??

(b) Random error should be

THE RESULT OF MANY SMALL INFLUENCES FROM MANY DIFFERENT SOURCES.

If we could exercise control over some or all of these influences and their sources, the error would not be random. (We exercise control over something only if it is to some extent systematic.) Match this description of random error—"the result of many small influences from many different sources"—to the statement you are to verify. Be precise: Do an item by item matching of this description with the statement in the problem.

(c) Put parts (a) and (b) together!

5.3.5

What could you mean by the "true" measurement of an object apart from any actual physical measurement you might make? After all, any two physical measurements will differ slightly (measurement error). So what is the "true" measure of the object? This question could get us into deep philosophical problems. Instead, please just accept the idea that

the "true" measurement of an object is the average
or mean of all possible such measurements.

Philosophy comes in when you realize it's impossible to make "all possible" measurements or even to *make sense* out of that phrase. You can't evaluate the "true" measurement by averaging up "all possible" measurements! That's why we have confidence intervals to *estimate* the "true" measurement.

Now, show that repeated measurements of the same object should be approximately normally distributed. Call the measurements M and show that they are a linear function of some normally distributed variable. Think about this carefully before you look at the level II answer; it's not very hard. What is the "variable" part of M?

5.3.6 The formula was $2.37 \pm 1.645\sigma_{\overline{X}}$. So how did we get the specific values?

5.3.7 (a) Use the relationship $s^2 = [n/(n-1)]\sigma^2$.

(b) Do this in two steps. First, the computing formula for $\Sigma(X - \overline{X})^2$:

$$\Sigma(X - \overline{X})^2 = \Sigma X^2 f - (1/n)(\Sigma X f)^2;$$

then, to get s^2, multiply by $1/(n-1)$ in the form $n/n(n-1)$.

(c) 1.5526. How?

(d) What is the underlying random experiment? Is it, indeed, a random experiment? Identify the outcomes clearly. What is the "assignment of numbers to the outcomes," the rule which defines the random variable?

(e) $\hat{\sigma}^2$ is biased as an estimator of σ^2. Its expected value is not σ^2; rather it's $[(n-1)/n]\sigma^2$. Use this fact to show that s^2 is an unbiased estimator of σ^2. Note that the question was about s and σ, but it's the unbiasedness of s^2 which is critical (in fact, as an estimator of σ, s is biased, not unbiased).

5.3.8 (a) This is easy! Take another look at the standard error formula for p.

(b) There are three different ways to handle this. Briefly, what are they?

(c) In the case of both proportions and means, the standard error raises a problem because it contains an unknown parameter. In the case of proportions, the unknown proportion itself appears in the standard error formula. In the case of means, the standard deviation of the distribution you're sampling from appears in the standard error formula. But if the mean of that distribution is unknown, the standard deviation is probably also unknown. This is the difficulty we have with the standard error: How, in each case, do the resolutions of this difficulty "mimic" each other?

5.3.9

Here are the answers: in level II we show you how to calculate them. But try on your own first!

(a) 6.2217 and 6.5783;

(b) 0.3035 and 0.3365;

(c) 121.46 and 123.56;

(d) 1.6815 and 1.7185;

(e) 0.1964 and 0.2636.

5.3.10

(a) First, of course, you must decide what degree of certainty you want to allow. You might take this issue back to the manager and decide in consultation with her or you might simply determine it on your own. Let's suppose you decide on 85% certainty; then your answer will be:

> We can be about 85% sure that the children who would use the planned daycare center are somewhere between two years eight months and three years old, on average.

How do you get this answer?

(b) You can't say "how many"; you can only estimate the *percentage* of children having the characteristic in question. What estimate would you give?

(c) Think about the data you're using.

5.3.11

When you enter the data into your calculator in the statistical mode, you'll discover $n = 76$, $\overline{X} = 7.2449$, and $s = 0.0090$. From this data you can obtain the following answers:

(a) (7.2384, 7.2513); but this answer is not complete! Be sure you complete it.

(b) (7.2372, 7.2525); (c) (7.2358, 7.2540); (d) Yes?

(e) Did you note that a *proportion* (or percentage) is required here? But 0.1184 is not the right answer either. Try again!

(f) Yes?

(g) The data is certainly not a sample from a population; there's no population anywhere in sight. Presumably, you observed the first 76 cups and recorded the amount of fill in each cup. So now think again, how could the resulting data be considered a random sample?

(h) First identify the underlying random experiment, then the random variable.

(i) Are np and nq both at least five?

5.3.12 When you choose your confidence coefficient, you "choose" the following picture. In the picture, we're assuming your confidence coefficient to be 82%:

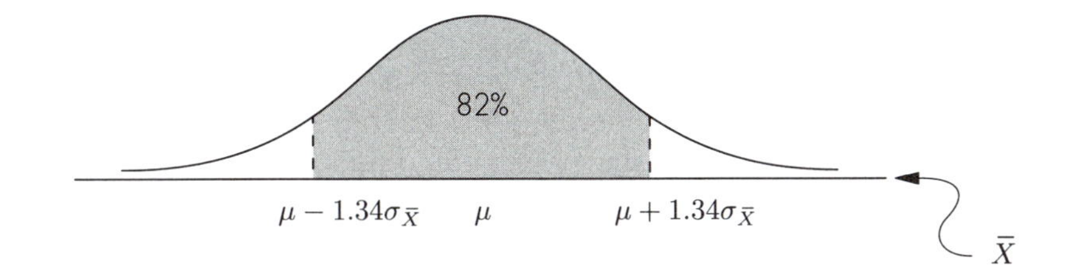

When you choose this 82% confidence coefficient, this picture is your way of saying:

> Any sample which gives a sample mean within 1.34 standard errors of μ will be thought of as "typical."

In other words, the picture determines what you mean by "a sample whose mean is close to the true mean." Those will be considered the "typical" samples. They're the samples giving a mean close to the center of the picture.

But when you look at any particular observed sample, you'll never know whether it's typical or not.

(a) Explain why not.

(b) Explain why this theoretical notion of a "typical sample" is useful even though you'll never know whether a particular sample is typical or not.

5.3.13 (a) You require the "total context" of that one sample. What is the "total context"? Make your answer general; don't talk specifically about a mean or a proportion, talk about "an unknown parameter" and "the estimator of the parameter." Be as complete as possible in giving your answer. For example, what is the connection between the "total context" and the parameter being estimated?

(**b**) It won't differ at all. If the estimator for the parameter is unbiased and normally distributed, for a 90% confidence interval you'll have the following picture:

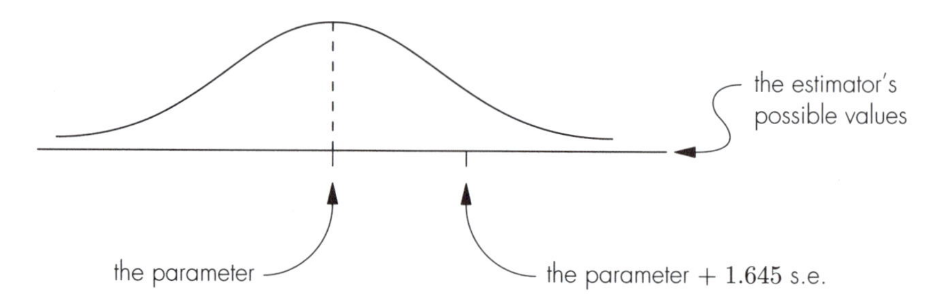

Suppose your data gives a value for the estimator of 3.12. Give the endpoints for the confidence interval. You won't be able to give exact numbers, but you can give the *form* of the endpoints. Look carefully at the picture and use the terms and symbols in the picture.

5.3.14 (**a**) (3.1926, 4.3674); (**b**) (11.8741, 12.9260).

5.3.15 (**a**) You shouldn't use s as an estimate here—why not (There are two reasons!)? "With a five percent chance of being wrong, the mean of this population is at least 16.4342, but not more than 17.5658."

(**b**) You cannot answer this question until you choose a confidence coefficient. Because nothing has been specified, you get to choose any one you like. Let's choose 95%.

(**c**) To get a more accurate estimate, keeping the confidence coefficient at $1 - \alpha = 0.95$ will require more information. You should supplement the present sample. What sample size is required? How many more must you select from the population?

(**d**) 17 is no longer the correct center because the larger sample will give a new, more accurate mean! Suppose your new sample mean is 16.7833; what is the new interval estimate?

(**e**) $n = 156$.

5.3.16 That tool is Student's t-distribution, but what kind of *theoretical object* is it? It's not a number. Not a set. It's a __________ __________ (two words!).

5.3.17

(**a**) First, identify the underlying random experiment being clear about the outcomes. Then specify the rule by which s^2 assigns a number to that outcome.

(**b**) Gosset's analysis requires that s^2 be intimately related to the chi-squared distribution. That will only be true if you're sampling from a normal distribution. In the rest of this discussion, we'll attempt to make all this seem plausible, at least.

Note that s^2 does seem to be related to the chi-squared distribution, provided the sample comes from a normal distribution. Recall that $(n-1)s^2 = \Sigma(X-\overline{X})^2$. Divide this by the constant σ^2. Now, if you replace $\overline{X}$ by μ, you get something similar:

$$\frac{(n-1)s^2}{\sigma^2} \quad \text{is similar to} \quad \frac{\Sigma(X-\mu)^2}{\sigma^2} = \frac{1}{\sigma^2}\Sigma(X-\mu)^2.$$

Distribute the $1/\sigma^2$ into the sum, getting $\Sigma\big[(X-\mu)/\sigma\big]^2$. How is this related to the chi-squared random variable? Why does the sample have to come from a normal distribution?

5.3.18

(**a**) Like Z, Student's t is "bell shaped" and centered on zero. The formulas tell us it's centered on zero because $\mu_t = 0$. That's how t and Z are alike. They are different because the variance for Z is one, whereas for t the variance is *larger than one*; this is to be expected because Student's t is a model for a less precise situation. Why is the situation less precise?

(**b**) The "modified" standardizing transformation is

$$\frac{\overline{X}-\mu}{s/\sqrt{n}} = t;$$

what's been "modified'?

(**c**) Student's t-distribution arises in place of Z from uncertainty about the value of σ. As the sample size increases, that uncertainty is reduced—why?

(**d**) These two situations are discussed in Problems 5.3.2 and 5.3.3. But before you look them up, see if you can't remember what they are! And also see if you can remember how to justify the normality in each case using our criterion for when to expect a normal distribution (page 148).

5.3.19

(**a**) (1.1059, 1.2941).

(b) Guess whether this interval is wider or shorter than the one in the previous part. Give an intuitive justification for your guess.

(c) (1.1503, 1.2497); this is the "correct" answer, but, in fact, a more accurate answer would be the interval (1.1490, 1.2510). Where does this more accurate answer come from? Why is it more accurate and why is the less accurate answer considered correct?

(d) (1.0889, 1.3111). Here, again, the population must be normally distributed, but why?

(e) (1.1728, 1.2272). No assumptions are required—why?

(f) (2.7727, 4.0273). Assumptions?

(g) (0.1827, 0.2373).

(h) (0.1729, 0.2471).

(i) Same as (e)—why? And what is the difference? Why does it not change the answer?

5.3.20

(a) In other words, when the population is normally distributed, what justifies using Z if σ is known and using t otherwise?

(b) The justification for the large sample case is more complicated than the small sample case. See if you can analyze the cases. There are four cases starting with the distinction between "σ known" and "σ unknown"—then each of these splits into two subcases.

5.3.21

(a) First, if the population is not normally distributed, what would you have to do? How does that introduce less accuracy into the final answer?

(b) In any case, for small samples, we must assume the population to be normally distributed. Then there are two cases to consider, but in both cases, our models are exact. Explain this.

5.3.22

(a) (0.3819, 0.4581); **(b)** (0.1252, 0.2748); **(c)** (22.4161, 23.5839); **(d)** (1.0910, 1.4490); **(e)** (84.2472, 89.7528); **(f)** Ouch! You need a large sample here—why? **(g)** (0.3446, 0.6354); **(h)** (6.5427, 7.8573); **(i)** (43.9785, 44.0215).

5.4.1

(a) Recall that $s^2 = 1.0828$. First, get the endpoints for a 95% confidence interval for σ^2. Then, taking square roots, you obtain the endpoints for our interval.

(b) The estimator s^2 is a constant times χ^2. What is that constant? So s^2 has "essentially" a χ^2 distribution, not a normal distribution.

(c) Even though the average amount of contaminant is twice as high for the first supplier compared to the second, you should prefer the first supplier. Why?

(d) You must show that $(n-1)s^2/\sigma^2$ is of the form $a + bs^2$.

(e) The χ^2 table is set up in terms of left-tail areas. Draw the picture for determining L.

5.4.2

(a) (0.5708, 2.0573); (b) (0.7676, 1.5919).

(c) We can be about 95% sure that the standard deviation of fill after resetting the fill mechanism is between 0.0073 and 0.0101 ounces.

5.5.1

(a) Any kind of interval estimate for μ_{MD} is a probability statement involving inequalities such as $a < \mu_{MD}$ or $\mu_{MD} < b$. So . . . ?

(b) With a 5% risk of error, your candy apples have an average maximum diameter of at most 3.9 cm. How do you obtain this conclusion? Because we want to find U so that $95\% = \text{P}(\mu_{MD} < U)$, you need z_0 so that $95\% = \text{P}(Z < z_0)$.

(c) With a 5% risk of error, your candy apples have a median maximum diameter of at most 3.9 cm. Note that the analysis for the median and mean give the same result. How do we obtain this solution? Study the analysis for the two-sided confidence interval for a median given in Problem 4.5.8. What's required here is considerably easier!

(d) First, the box certainly has to be a bit bigger than the apple to accommodate the tissue paper and cushioning. But that's easily resolved; just determine how much to add to the dimension of the box for that purpose. Beyond that, there's a more serious problem. We've really used the wrong approach to answering this problem. Do you see why?

(e) We can be 90% sure that about 9.8% of your candy apples have a maximum diameter below 3.7 cm. How?

5.5.2

(a) We can be 99% sure that a steel cart of pennies will weigh on average less than 725 pounds. Explain. A gram is 0.035 ounces.

(b) With a one percent risk of being wrong, a typical bag of pennies should weigh between 3.0967 and 3.1189 kg. But in fact, this is not an appropriate answer to the question. Why not?

5.5.3

For an upper confidence interval, the endpoint is $\overline{X}+1.645s/\sqrt{n}$. The picture of the normal distribution for $\overline{X}$ will have 95% of the area located in the RIGHT of the distribution. Explain. How does the probability statement about $\overline{X}$ translate into a statement about the interval?

5.5.4

(a) Naturally, we'll use $\mu = 6.3$ to predict with. But there's a zero chance that I'll get exactly 6.3 ounces! That's not a reasonable answer. You should choose a confidence coefficient, let's say 95%, and give a prediction interval. So, "we can be 95% sure I'll get somewhere between 5.8 and 6.8 ounces." Explain.

(b) Verify that the conclusion in part (a) satisfies the definition of a "prediction interval" for the question.

(c) What key information have we used here which ordinarily would not be available?

5.5.5

We can be about 95% sure to obtain between 6 and 6.6 ounces each, on average.

5.5.6

(a) What's the "doing" here?

(b) In the cases accessible to us, it was normally distributed and allowed us to discover that the endpoints of the confidence interval should be $\overline{X} \pm z$s.e. What was the model?

(c) The model is a sum of two random variables. In Chapter 6 we'll officially see that the mean of a sum is the sum of the means and that if the two random variables are independent, the variance of the sum is the sum of the variances.

(d) Draw the picture for $X - \overline{X}$:

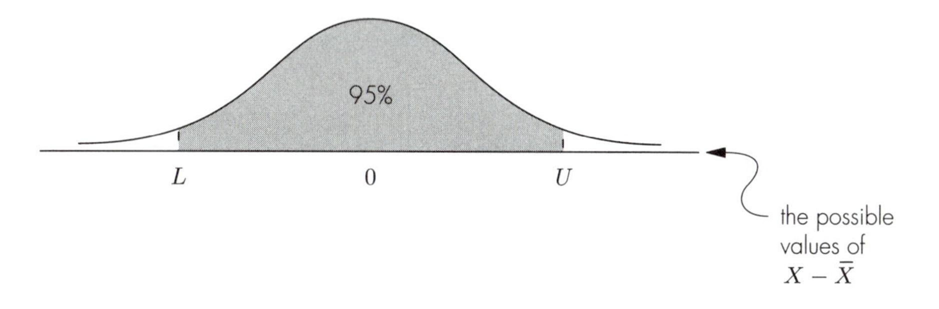

(e) Representing "an observation" by a value of X, a prediction interval is defined as "a range of possible values" for X, "together with the probability that that range of values actually does contain" that value of X. Does an interval with endpoints $\overline{X} \pm z$s.e. satisfy this definition?

(f) I can be about 95% sure my drink will contain somewhere between 5.96 and 7.24 ounces. Explain. What assumptions are you making here which might or might not be satisfied?

(g) Note that the standard error of A is $\sigma/\sqrt{m}$.

(h) Yes?

(i) Think back to the discussion of the model in part (c).

(j) It involves that future observation. Explain.

5.5.7 The box should be 4.6 cm on each side. There's a five percent chance that an apple would be so atypical of our sample that it would be too tight in its box.

5.5.8 (a) We can be about 99% sure this cart will weigh no more than 725 pounds. Explain.

(b) We should re-count any bag which weighs less than 3.0962 or more than 3.1194 kg. We've used a 99% prediction interval here, but the 99% does NOT mean that "the re-count should be unnecessary for about one in a hundred bags"—why not?

(c) The endpoints of our prediction interval incorporate a value of Z. That's only justified if the model you're using is normally distributed. So what's the model here and what was it in Problem 5.5.2? Why does one require X to be normally distributed and the other not?

5.5.9

(a) You did this problem earlier! See if you can remember how it goes.

(b) There are h segments below $x_{(h)}$, so there should be h segments above $x_{(k)}$. So . . . ?

(c) From part (a), $1 - \alpha = [(n - h + 1) - h]/(n + 1)$. So . . . ?

(d) The interval has endpoints $x_{(45)}$ and $x_{(56)}$. How?

5.5.10

(a) For the example in the text, we didn't actually calculate the tolerance interval. We'll see later how to do that. But we told you it turned out to be the interval $(6.4, +\infty)$. So we know it's very likely (90%) that the interval $(6.4, +\infty)$ contains 93% of all values of X. If that's true, here's the picture for X:

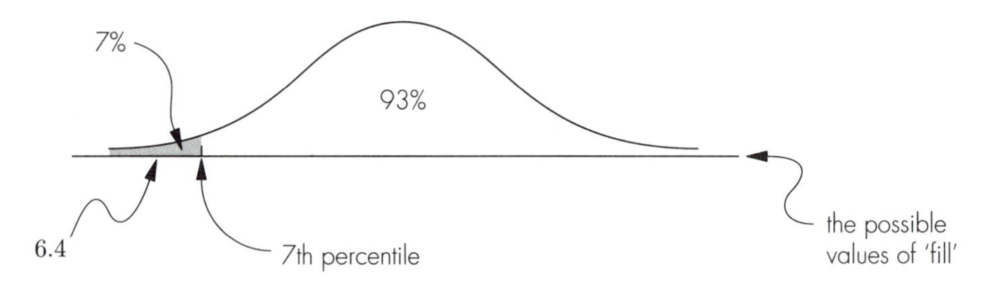

The 100pth percentile (see the discussion of the normal probability plot in Chapter 4) is that value of X which puts 100p percent of the values of X to its left. (If p is a proportion, 100p is the corresponding percentage.) For example, the median is the 50th percentile. In the picture, the seventh percentile puts seven percent of the values of X to its left. Now, explain how all this answers the question.

(b) This formula's derivation is parallel to the argument in Problem 4.5.8(c). But now, instead of the median, you have either the 100pth percentile of X or the 100qth percentile of X. To see the argument, DRAW THE PICTURE OF THE DISTRIBUTION OF X. Let Y = # of observations in the sample which are less than the appropriate percentile of X.

(c) It's the distribution of X which need NOT be normally distributed. All we need to know about the probability distribution of X is that there's a negligible chance for it to take the specific value p or Q. That's true, for example, if X is continuous. That's how we've drawn the picture. So where does the z in the formula for part (b) come from. Obviously there's a normal distribution somewhere!

(d) The "one-half" percentile is the median.

(e) Suppose the endpoints are a and b, respectively. Draw a picture of X showing the 100pth percentile and showing a and b. Recall part (a) of this problem.

5.5.11 Yes?

5.5.12 First decide how certain you need to be. Let's say 95%.

(a) We can be about 95% sure that U.S. pennies weigh between 3.0993 and 3.1162 g, on average.

(b) This penny should weigh between 3.02 and 3.19 g. But there's a five percent chance that either Youden's data or our penny, one or the other, is sufficiently atypical that the true weight is outside this range.

(c) We can be about 95% sure that 90% of all U.S. pennies weigh less than 3.17 g.

(d) We can be 95% sure that this penny would weigh between 3.01 and 3.19 g.

(e) The information utilized in the two techniques is NOT the same. Part (d) uses more information from the sample. Part (b) uses extraneous information. Explain.

(f) n must be at least $(2 - \alpha)/\alpha$. Explain.

(g) Think about part (f).

Chapter **6**

6.1.1 There is not one necessarily correct answer to each part of this question. The choice of procedure depends not so much on the hypothesis as on the intention of the person who poses the question or the situation in which the hypothesis arises. It's a question of what exactly is being asked about the hypothesis. See what you think for each one and compare with our discussion in level II.

6.1.2 With a test of significance, if your data seems inconsistent with the hypothesis, you get a conclusion. The hypothesis seems false. On the other hand, if the data seems consistent with the hypothesis, you get no conclusion. Why?

6.1.3 (a) Be careful. If you said 25, you went astray! In fact, to answer this question, you need more information. That five percent "error rate" is five percent of what? Recall our description of this five percent error rate: "We determine in advance an acceptable probability, say five percent, of erroneously rejecting a shipment which, in fact, does meet specifications."

(b) Just make it smaller! Like one percent! Or better yet, one tenth of a percent. After all, for an hypothesis test, you choose the probability of error. You set it in advance. So why not make it ZERO?! That, obviously, would be the best. Why not? [Hint: Think about how you control sampling error in a parallel situation—in a confidence interval problem.]

6.1.4

(a) In the text, we listed nine pairs of contrasting characteristics for these two procedures. See if you can identify them all. For example, a test of significance provides a numeric measure of consistency, whereas an hypothesis test is a decision procedure. Present the contrasts in a table with two columns, one for each procedure.

(b) Recall our two examples: For the test of significance, the question was, "Is it believable that this data came from a normally distributed population?" For the hypothesis test, the example was monitoring incoming shipments of chain links.

6.2.1

(a) Now $z = 1.6$ giving a p-value of . . . ?

(b) Here $z = 2.4$, giving a p-value of 0.0082. The coin does not seem fair. What do you think the probability of heads is?

(c) The standard error is 0.049. With confidence intervals, the standard error poses a problem (The value of p is unknown). That problem does NOT arise here because you're assuming the hypothesis to be true. The hypothesis *gives you a value of* p. We assume $p = 0.4$; so $(\text{s.e.})^2 = (0.4 \times 0.6)/100$.

Thus, $z = 3.67$, giving a p-value of zero. This observation seems very unlikely if the probability of heads is really 40%. Would you conclude that the coin is fair?

(d) Here the kids' observation is *below* what's expected if the hypothesis is true. So the p-value is the probability of any data as far or further *below* what is expected. What's the p-value? And what's the conclusion?

6.2.2

The data supports the negation of the hypothesis if the p- value is small. Furthermore, the data will suggest a "direction" for the negated hypothesis. Explain that with reference to Problem 6.2.1(c).

6.2.3 (**a**) The "data seems extreme" means: The data is very atypical of what you should expect if the hypothesis is true. There are two possible explanations for the discrepancy. What are they? [Hint: Look for the answer in the text!]

(**b**) Given a small p-value, either explanation in part (a) is possible. But one is unlikely. Which one?

(**c**) Think about the probability models we use for our estimators.

(**d**) If you said something like "There's a 7% chance of data like theirs or worse if the coin is fair," you're not being as detailed as the situation allows!

(**e**) Neither of these statements is correct. Why not?

6.2.4 "Consistent with" means "not small p-value." In the first case ($p = 0.5$), the p-value is 0.2743. In the second ($p = 0.56$), the p-value is almost the same. For $p = 0.53$, $z = 0$ and the p-value is as large as a p-value can ever be, namely, one-half. But here a p-value calculation is really unnecessary: With a probability of 0.53 for heads, 53 heads out of a hundred tosses is EXACTLY WHAT YOU EXPECT. So, OF COURSE, that observation is consistent with the hypothesis. What does all this say about tests of significance?

6.2.5 (**a**) With a standard error of 0.3521, $z = -2.5$ and the p-value is 0.0062, about half of a percent. That's small by almost any standard. Thus, it seems the mean is smaller than 70.18. Do you see how to get the p-value? It's . . .

$$P(\overline{X} < 69.3 | n = 50, \sigma^2 = 6.2).$$

How much smaller than 70.18 is the mean?

(**b**) Unlike the previous problem, you have to estimate the standard error because σ^2 is unknown. That estimate is 0.7874, giving $t = -1.12$. So the p-value is larger than ten percent, not small by any standard. This says the data is consistent with a mean of 70.18. Your "good reason" stands unchallenged. Note that you have not *proven* that the mean is 70.18. After all, your data is consistent with many other hypotheses as well. Show that the data is also consistent with means of 68.42, 68.7, and 69.3.

(**c**) Part (a) yields a more certain conclusion. Why?

6.2.6 Are they going to play with the same coin every day or a different coin each time?

6.2.7 In the text above, we said " . . . the standard errors can be estimated if necessary from the samples, replacing p's and q's by $\hat{p}$'s and $\hat{q}$'s and replacing σ's by s's. Of course, if the true values are known, this is not necessary." Under what circumstances would they be known so that estimating the standard error would be unnecessary?

6.2.8 **(a)** We can be 90% sure second graders in the inner city are 3 to 6.5 cm shorter on average than their peers in the suburbs. Explain.

(b) Our two samples lead us to believe that second graders in the inner city are shorter on average than their peers in the suburbs. In other words, the observed difference between the two samples is "statistically significant" (the p-value is small), so we don't accept that it's due just to sampling error—that's possible, but highly unlikely.

You can get the same bare conclusion from the confidence interval of part (a) because it does not contain the value zero and so precludes (with a 10% risk of error) the possibility that $\mu_1 - \mu_2 = 0$. But the logic of the analysis is very different. The p-value carries specific and direct meaning in comparing our hypothesis with the observed data. It measures the consistency of the hypothesis with the data. That insight is not available from the confidence interval. Throughout this course, we follow the convention that a question of this type calls for a p-value (Question type: "Does the data seem to challenge the hypothesis?'). What's the p-value here?

6.2.9 Here, you'll have to estimate the standard error from the pooled sample proportion. See the discussion in the text just after the table of parameters and their estimators. This pooled proportion is 53/200 and it gives a standard error of 0.0624. How? Do you still think the two proportions are the same?

6.2.10 Clearly, the mean of W is seven because W always takes that value and only that value. Show how $\mu_W = 7$ comes from the "appropriate formula." Then, because W is constant, it's variance should be zero (there's no variability at all). But that does NOT follow from the formula $\sigma^2_{X+Y} = \sigma^2_X + \sigma^2_Y$ because no matter how the die might be loaded, the variances for X and Y (which are equal) are NOT zero. Explain why this formula for the variances is not applicable here.

6.2.11 **(a) and (b)** Let E_1 be the first estimator ($\hat{p}_1$ or X_1) and E_2 the *negative* of the second. So, in each case, your estimator is $E_1 + E_2$. Then the mean is just the sum of the means. Each E_i is approximately normally distributed (why?) and so . . . ? Finally, if the samples are chosen independently, the variance is the sum of the variances. All of this follows from the equations "for the sum of two random variables" given in the text. Write this out in detail for each of parts (a) and (b) separately.

6.2.12 (a) There's only one chance in 10,000 that such a difference is due just to chance. Explain.

(b) With a 5% risk of error, we conclude that the pass rate for blacks is below that of whites by somewhere between 38.3 to 12.44 percentage points.

6.2.13 Statistical significance should be measured by a p-value.

6.2.14 (a) The little kid seems much too sharp to make this mistake. What exactly is the mistake?

(b) There are two possibilities. Maybe they've already done a test of significance as in the dialogue and concluded bias in favor of heads. Then, the question is, "How biased?" How should they test the coin?

Or maybe the little kid thinks, "Well, I'd like to play this game. If it's just a little biased in favor of heads, so what? That's of no significance." This means *practical* significance! First, the little kid should decide how much bias there would have to be to be significant in a practical sense. Then, they should do a test of significance for *more than that much bias*. For example, she may feel that one or two heads extra for the big kid in 100 tosses is not a reason to walk out of the game. That is, she thinks a probability for heads of one or two percent more than 50% is of no practical significance. What hypothesis should they test?

(c) A 95% confidence interval for p would be go from 0.4833 up to 0.6767. This actually includes $p = 0.5$. What would the little kid do?

On the other hand, if we do a test of significance to allow for one or two percentage points bias above $p = 0.5$, the standard error is 0.0499. So $Z = 1$ for a p-value of about 16%. What would the little kid do?

(d) Yes?

6.2.15 (a) What parameter are we concerned with here?

(b) State the hypothesis which you seek to challenge in terms of the parameter. That is, formulate an appropriate *statistical* hypothesis.

(c) Calculate the p-value.

(d) Interpret the p-value in real-world terms.

(e) Is the p-value small or not?

(**f**) Give a real-world answer to the question. Be careful that you answer the question asked, not some other related question!

6.2.16

Follow the format of the preceding question. You should think of the cancer as the occurrence of a rare event. That is, the incidence of cancer is usually modeled by a Poisson random variable. How will you deal with the fact that the residents of the southern part of your city are not a random sample?

6.2.17

(**a**) Clearly, the word "number" here is intended to be relative to the total number of passengers; that is, you are actually dealing with proportions.

The phrase "any pattern of *significant* difference" suggests that a small difference in the two proportions is of no practical significance. Thus, before the hypothesis can be properly formulated, some further clarification is required. Suppose it's decided that a difference of no more than two percent is of no practical significance. Now follow through with the format of the previous two questions.

(**b**) The model is only valid for the large sample case. The criterion for proportions is that np and nq must be at least five. But you don't know the values p_1 and p_2, so . . . ?

(**c**) Gather more data! Suppose over many flights you observed 63 requests for vegetarian meals out of 741 passengers going from New York to San Francisco and 11 such requests among 1236 passengers going from New York to Chicago. Do you think there's a significant difference in the number of such requests?

6.2.18

The only probability you obtain from a test of significance is the p-value. It does NOT provide the probability that the hypothesis is true! Explain.

6.2.19

When the p-value is small, there are two possible explanations. The test of significance goes for the second of the two: We accept the hypothesis as false. Here, that suggests "real bias" in the coin. What's the other possibility?

6.2.20

(**a**) This is the fallacy of repeated sampling which we mentioned in the text. Explain it for this particular example.

(**b**) In part (a), evidently you were NOT doing a test of significance. What were you doing? So, what's this question got to do with tests of significance?

6.2.21

(**a**) With a negative binomial, the p-value is 0.037. With a binomial model, it's 0.073. What's your conclusion about the coin?

(**b**) This example is quite infamous among statisticians; it's very disturbing in its implications. It shows that a p-value calculation is not independent of *what's in the mind* of the person gathering the data. What was she thinking of when she stopped tossing? Did she stop because she had obtained one head or because she had completed 12 tosses? Note that the observable physical process of gathering the data is the same in both cases. It is only what's in the mind of the coin tosser that's different. Why is that disturbing?

6.2.22

(**a**) The p-value would be $P(\hat{p} < 7/90) = P(Z < -4.12) = 0$, with a standard error of 0.0468. However, this is NOT what the Court reported. Explain.

(**b**) When $p = 0.5$, the binomial distribution is symmetric about the mean and so the mean is the most probable value, but yet when n is large, it has a SMALL probability of occurring. Explain.

6.2.23

(**a**) Translate the real-world question into the standard form for a test of significance: Does this data seem to challenge the claim that people are just as likely to die in the three months preceding their birthdays as not? Express the question in terms of an appropriate population parameter. Then calculate the p-value. Is the p-value small or not small? What does the p-value mean in real-world terms? Is the claim believable?

(**b**) In Problem 6.1.4(a), you compared and contrasted tests of significance with hypothesis tests. Carry out that comparison with reference to this particular situation [see Problem 6.1.4(b)]. Omit comparison #3 (testing for randomness) which we have not yet discussed.

6.2.24

The p-value is $P(X = 0) = q^{100} = 0.1800$. So . . . ?

6.2.25

When the face with two dots comes up more than half the time as it did in our observations, it seems obvious the die is not fair. Still you might like to double check with a χ^2 test. You're testing the hypothesis that the die is fair, so each E_i is . . . ? What's the p-value?

6.2.26

(**a**) Bortkiewicz observed 200 corps-years. Theoretically, if the Poisson model is appropriate (if our hypothesis is true), 108.68 of those corps-years should have been accident-free. How? You do the rest of the $E_i's$.

(**b**) $(O_0 - E_0)^2/E_0 = (109 - 108.68)^2/108.68$. And $\chi^2 = 0.7197$. How?

(**c**) There's a trick here! There are five cells, but there are only three degrees of freedom, TWO less than the number of "cells." Because we had to *estimate* λ from the data, we lose one more degree of freedom. This is parallel to what happens with Student's t-distribution. There, you're estimating an unknown σ^2 from the data by using s^2, and the degrees of freedom is $n-1$ instead of n. We will see other cases of this principle in Chapter 7. The principle is that you lose one degree of freedom for each unknown parameter which must be estimated from the data.

(d) $\chi^2 = 0.7197$ was calculated from Bortkiewicz' observed data. That data was assumed to be a sample of 200 from the Poisson distribution with $\lambda = 0.61$. What does the p-value of 95% tell us?

6.2.27 $\chi^2 = 0.0128 + 0.0865 + 0.1538 + 0.2571 = 0.5103$. So . . . ?

6.2.28 **(a)** $E_i = n_i p_i$; for the first cell, it's $587 p_i$. But the p_i's for the three precincts are unknown, although you know they're equal (You assume the hypothesis true for a test of significance). That means you can pool your three samples and estimate the unknown common proportion by the observed $\hat{p} = 676/1553 = 0.4353$. So $E_1 = 587 \times 0.4353 = 255.5132$. You do the other E_i's.

(b) The first term is $(253 - E_1)^2/E_1 =$?? And then?

(c) There are three cells, so the degrees of freedom starts with two. But because you're estimating one parameter, you lose one more degree of freedom. This brings you down to ONE degree of freedom. So . . . ?

6.2.29 **(a)** Let M = male and R = recovers. Then assuming independence, $P(M \text{ and } R) = P(M) \times P(R) = 0.5 \times 0.84 = 0.42$. So, IF THE HYPOTHESIS OF INDEPENDENCE IS TRUE, 42% of our sample of 194 should be recovered males. That means $E_1 = 81.48$. Now, you make a table for the E_i's.

(b) There are four cells here, a cell being one entry in the table. For the first cell, we calculate $(O_1 - E_1)^2/E_1 = (127 - 81.48)^2/81.48 = 25.4304$. Now, you do the other three. So for our data, $\chi^2 = 48.3447$. How?

(c) With three degrees of freedom, our observed value of χ^2 is far off the table. The largest value in the table is 12.838. What does that suggest?

(d) Yes?

(e) If the 84% recovery rate were not known for the population, you would have to estimate it from the data. Similarly, if you were doubtful about the proportions of females versus males, you would have to estimate that also from the data. How would such estimates affect the test?

(f) Now $E_1 = 194 \times 0.7423 \times 0.8711 = 125.4433$. It's convenient to put the observed and expected numbers together into one table, with the expected numbers in parentheses.

	Recovered	Died within 5 years
Male	127 (125.4433)	17 (18.5567)
Female	42 (43.5567)	8 (6.4433)

Now you calculate χ^2 and interpret the results.

(g) This is NOT the same as the questions in the previous parts! There, we asked if *recovery* is independent of gender. Here, we're asking if *incidence* of this cancer is independent of gender. If so, the population of persons who fall victim should split about 50/50 between men and women. How would you test this against our data?

6.2.30 It's virtually impossible that we would observe such a result just by chance if passing were truly independent of race.

6.2.31 Think generally; think about any test of significance. Suppose the p-value is small; suppose it's not small. What conclusions can you draw?

6.2.32 You might do a χ^2 goodness of fit test. We cannot retrieve Kunz' exact numbers; clearly he rounded his percentages. But the following numbers give his reported percentages when rounded (beginning with the 8% figure): 60, 344, 231, and 112. That is, it appears that 60 of the 747 persons sampled by Kunz died in the three months prior to their birthday, 344 in the three months after the birthday, and so on. Now, you carry out this test and interpret the results.

6.2.33 (a) χ^2 is 6.9801. What's the p-value and what's your conclusion?

(b) First, set up the model (as in Problem 3.6.16), then compare with the data. One-place decimal accuracy is adequate. MLE for p is 0.5352. How?

6.3.1 (a) The decision of whether to stop production is based on . . . ? Why might that lead to error?

(b) How did you control error for confidence interval problems? How is that similar to the situation of part (a)?

(c) You'll analyze the sample through the estimator $\hat{p}$. That estimator is normally distributed. So . . . ?

6.3.2

(a) The evidence comes from a random sample, of course, summarized in this case into the sample mean. So, if the sample mean is less than 3.15, we have evidence that μ is less than 3.15. Comment.

(b) Would a mean diameter of 3.199999 mm be too small?

(c) The mean diameter might be right on target even though a significant percentage of the parts were unusable. How could that happen?

(d) It may be that any part with a diameter less than 3.15 mm is entirely useless, but otherwise the part is acceptable. Then you might want to monitor a proportion. What proportion?

6.3.3

(a) For any hypothesis test, the "evidence" is . . . ?

(b) The uncertainty of the diameters is measured by . . . ?

(c) Less than two percent. How?

(d) Now it's $P(Z < -1) = 0.1587$. What's the point of this and part (c)?

(e) Use σ^2, not σ. You don't want to work with square roots! What distribution will we use? What are the hypotheses?

(f) Be specific to the real-world context!

(g) No—not necessarily! Even if $s^2 > 0.000625$, we may not have evidence for H_A. Why not?

6.3.4

The words are "logical" and "practical"—which is which? Which characterizes the null hypothesis and which the alternative hypothesis? Explain the characterization in each case. What's "alternative" about the alternative hypothesis?

6.3.5

(a) See Problem 6.3.2(d). You want to monitor p, the proportion of parts with a diameter less than 3.15 mm. Why would it be wrong to try monitoring the mean diameter with $H_A : \mu < 3.15$? What are the hypotheses?

(b) You're looking for evidence that p is too small. What's p? What's "too small"? What are the hypotheses?

(c) Be careful! Is this a right- or left-tailed test?

(d) There are several questions to be answered here:

(i) The "one-third" condition guarantees the "one percent" requirement. Explain. [Hint: How many chains would vary from the mean length by more than 1 cm if $\sigma_L < 1/3$ (with L = length of a chain)?]

(ii) In monitoring incoming shipments, you will be looking at the standard deviation for the length of links, not the length of chains. You will reject a shipment if $\sigma_X^2 > 0.0012\,\text{cm}^2$. Explain.

(iii) What should your hypotheses be?

(iv) What assumption are you making in the analysis above?

(e) The contract is with a supplier of *links*, not chains. So . . . ?

6.3.6

(a) We're attempting to flag the need for an exceptional or "alternative" course of action. That's the focus of the monitoring procedure. Thus, H_A becomes the focus of the hypothesis test. But this practical focus on the alternative hypothesis is also reflected in the very *structure* of the test. What is it about the test itself that puts a practical focus on the alternative hypothesis? There is a very specific answer to this. If you can't think what it is, find it in the text!

(b) First, what *is* the significance level? It's a number, but that's vague. What does it measure?

(c) The power of the test is the probability that the test *succeeds* in what it attempts to do. It's the probability you're led by the data to act on H_A when indeed H_A is true. It's a probability, but it is not a single number. It takes on various values depending on . . . ?

(d) and (e) Yes?

(f) Our action in each period of the monitoring process ("act on H_o," "act on H_A") is real world. Where else does action take place?! And it's based on the sample data for that period. Sample data is observed in the real world! But here's the unrealistic part: Any discussion of the "state of the world" is purely theoretical. After all, we're using a statistical procedure only because the "state of the world" is not known. So the table is partly realistic and partly theoretical. To the extent that it's theoretical, the table cannot help us *avoid* error ("avoiding error" happens in the real world). So what's the table for?

(g) Be realistic!

(h) Why are small values of p irrelevant from the point of view of the test itself? And beyond the test?

(i) "Act on H_A" is the more conclusive decision. Why?

(j) What is the alternative hypothesis supposed to be, anyway?

6.3.7 (a) There are two possible ways to determine the direction—what are they?

(b) Think of the trade-off between risk and effectiveness. Take an extreme case to see what happens. Suppose the disease is very serious and some kind of treatment is urgent. Further, suppose the old procedure is hardly effective at all—for example, suppose there's NO old procedure—and suppose the new procedure poses little or no risk. Now which of the two errors given in this problem is the more serious one?

(c) The alternative hypothesis here should be determined by the appropriate choice of type I error.

6.3.8 The probability statement is vague. What are the specific instances we're talking about? Five out of 100 of what? To make this clear, suppose we do this quality control check once a week. Then out of 100 weeks we can expect to stop production unnecessarily about five times. True or false? This is like Problem 6.1.3.

6.3.9 (a)–(c) Follow the pattern of the text. You must choose a significance level—suppose you choose 5%. In your answer, be careful you're faithful to the details of the problem!

(d) This is like part (c) because it's a question of returning a shipment of chain links when it doesn't meet specifications. But you'll have to change some of what was said in part (c). What must be changed?

6.3.10 (a) Formally, you just have "formal terminology." In the real world, you take a certain action. Contrast these.

(b) Follow the pattern of part (a) for the two possible errors.

(c) Take the quality control example where you're seeking evidence that p is more than one percent.

(d) There's a catch here. Do you see what it is? What role does the alternative hypothesis play?

(e) Yes?

6.3.11 It does unless you're also checking . . . ??

6.3.12 $H_o: \mu = 310 \qquad (\mu < 310 \text{ irrelevant}),$
$H_A: \mu > 310.$

Cutting through all the real-world complexity, in essence: You're prepared to remain with the present supplier unless there's evidence the new supplier's part has a mean life greater than 310 hours. What's the parameter? The estimator, conclusions, errors? The rejection region?

6.3.13 $H_o: p = 0.25 \qquad (p > 0.25 \text{ irrelevant}),$
$H_A: p < 0.25.$

Give the parameter, the estimator, the conclusions, and errors in real-world terms.

6.3.14 $H_o: p = 0.25 \qquad (p < 0.25 \text{ irrelevant}),$
$H_A: p > 0.25.$

Here the direction of the test is determined by specifying type I error. It's specified in the problem statement that you're willing to allow "no more than a 5% risk of launching this series if fewer than 25% of the registered voters will see the spots." To have this be type I error, you must set up a *right*-tailed test. The parameter and estimator are the same as in Problem 6.3.13. What are the hypotheses, the conclusions and errors?

6.3.15 $H_o: \mu_1 - \mu_2 = 15 \qquad (\mu_1 - \mu_2 < 15 \text{ irrelevant}),$
$H_A: \mu_1 - \mu_2 > 15.$

You're going to act on the basis of the test if $\mu_1 - \mu_2 > 15$ where μ_1 and μ_2 are the average SAT scores of students at your school and at Bad U, respectively. In this problem, does the word "significant" refer to statistical significance or practical significance? Finish the problem.

6.3.16 $H_o: \mu_1 - \mu_2 = 15 \qquad (\mu_1 - \mu_2 > 15 \text{ irrelevant}),$
$H_A: \mu_1 - \mu_2 < 15.$

Why is this a left-tailed test? Finish the problem.

6.3.17 Here, $H_A: \mu_1 - \mu_2 > 15$. Do you see why? Can you finish the problem from here?

6.3.18 Let p be the proportion of returns that reveal attempts at tax evasion. Then (why *right* tailed?):

$$H_o: p = 0.1 \qquad (p < 0.1 \text{ irrelevant}),$$

$$H_A: p > 0.1.$$

6.3.19 $H_o : p = 0.2$ ($p < 0.2$ irrelevant),
$H_A : p > 0.2$.

6.3.20 Quality control should probably be interpreted as "continue in production UNLESS there is evidence of a weakening of quality control." Thus,

$$H_o : \sigma^2 = 0.3 \qquad (\sigma^2 < 0.3 \text{ irrelevant}),$$

$$H_A : \sigma^2 > 0.3.$$

6.3.21 **(a)** If $\alpha = 5\%$, the rejection region is $\{\hat{p}|\hat{p} > \hat{p}_c = 0.0333\}$. What's the decision rule? What if $\alpha = 10\%$?

(b) At the 5% significance level, $\hat{p}_c = 0.7187$. If $\alpha = 10\%$?

(c) Because you don't know the variance for the length of the chain links, you'll have to give the decision rule in terms of the test statistic. State that decision rule at the 5% and 10% significance levels. Then, as a separate problem, suppose you DO know the variance for the length of links. Suppose it's about 0.1 cm. Give the decision rule.

(d) Recall that $s^2 = [\sigma^2/(n-1)]\chi^2$. Because you're assuming the null hypothesis, set $\sigma^2 = 0.0012$.

6.3.22 **(a)–(d)** You need only recall the decision rules from the previous problem.

6.3.23 **(a)** The sample mean standardizes to $t = -1.2136$. This does NOT reject H_o, but surely there's a problem with this shipment. Explain.

(b) Too many chain links will be too short. Explain.

c6.3.12 The test statistic is t. You should express the rejection region in terms of this test statistic. Why? With a 1% significance level, you compare your observed value of the test statistic with $t = 2.624$. Explain why part (e) is a bit surprising. You are assuming the lifetimes of the machine parts in question are normally distributed. That is NOT reasonable. Why? How could you avoid this assumption?

c6.3.13 The rejection region is $\{\hat{p}|\hat{p} < 0.1875\}$ at the 5% significance level. Identify the decision rule in real-world terms and identify the test statistic. Note that in part (a) the data obviously fails to reject H_o, no need looking at the rejection region—why is this obvious? What about parts (b)–(g)?

c6.3.14 For $\alpha = 5\%$, $\hat{p}_c = 0.3125$. What is the rejection region? Make sure you can state the decision rule in real-world terms.

c6.3.15 The standard error is 31.4885. Even if you had small samples, you would NOT use the t-distribution here. Why not?

c6.3.16 The critical value is -25.3052 at a 10% significance level.

c6.3.17 The rejection region here is determined by 55.3053 ($\alpha = 0.1$).

c6.3.18 The standard error is 0.0086 and $\hat{p}_c$ is 0.1201.

c6.3.19 The standard error is 0.0115; $\hat{p}_c$ is 0.2147.

c6.3.20 At a 5% significance level with six degrees of freedom, the critical value of χ^2 is 12.592. Give the decision rule in terms of the value of s, the observed standard deviation. Then determine your conclusion for each part.

6.3.24 Problem 6.3.20 is a true case of *monitoring*. The others are all one-time decisions. What justifies treating a one-time decision as an hypothesis test instead of a test of significance?

6.3.25 (a) We'll leave this to you.

(b) The p-value decision rule is redundant for a one-time decision. Explain why and then answer the original question.

6.3.26 (a)–(d) For each of these parts, you'll have to recalculate the standard error, because you have a different p each time. Before calculating, guess whether the power will be greater or less than 50%.

(e) Here, the normal approximation is no longer valid (why?), but you can easily determine what the power should be.

6.3.27 (a) Recall the definition of "power."

(b) For the power to be 75%, Z must be -0.675 (roughly halfway between -0.67 and -0.68). That means

$$\sqrt{700}\,\frac{0.0162 - p}{\sqrt{p(1-p)}} = -0.675.$$

Squaring and simplifying, you get this quadratic equation in p

$$0.183708 - 22.68p + 700p^2 = 0.455625p(1 - p)$$

which gives

$$700.455625p^2 - 23.135625p + 0.183708.$$

Using the quadratic equation, we get $p = 0.016514697 \pm 0.003235109$. This gives $p = 0.0133$ or 0.0197. But to get power above 50%, we know p must be greater than 0.0162, the critical $\hat{p}$. So, to get power of at least 75%, the unknown p would have to be at least 0.0197. Interpret this in real-world terms.

6.3.28 (a)–(c) Yes?

(d) The χ^2 table is too incomplete to actually calculate the power. But we can see that for $\sigma^2 = 0.0065$, the power is a bit above 95%. How?

6.3.29 We'll leave this to you. (Sorry!)

6.3.30 (a) $n = 996$. How? Draw the picture! The rejection region for the test as we've been doing it was determined by a sample size of n=700. Now it will be different. Here the 60% power at $p = 0.015$ gives an expression for the critical value, $\hat{p}_c$, of the rejection region. This expression has the unknown value of n in it. You'll get a second expression for $\hat{p}_c$ from the picture centered at $p = 0.01$. Use these to determine the sample size n. Be careful about the standard errors!

In fact, such a large sample is seldom practical in a quality control context. But with smaller samples, the normal approximation for $\hat{p}$ will often be invalid. What should you do then?

(b) If more than 14 out of the observed 996 parts is defective, stop production and take corrective action. How do you get this?

6.4.1 Yes?

6.4.2 We can be 95% sure the baggage in today's flight will weigh between 4213 and 4566 pounds total. But this is not the usual confidence interval problem. Explain.

6.4.3{6.4.9 Yes?

6.4.10 (a) From the data, the average number of hits per region was 0.927083333. You should obtain this from the statistical mode of your calculator; doing it by hand will take about five times as long! Now, what's P($X > 2$)?

(b) In fact, the answer would be $1/\lambda = 1.0787$, using the exponential model for "time between two occurrences." What's wrong with this? What's the unit of time? What information would be required to answer the question?

(c) Yes, because it would seem the bombs are falling more or less randomly. In other words, the data does not seem to call into question the "randomness" assumption. Explain.

6.4.11

(a) 0.0764.

(b) 90% (0.0596, 0.1501). Real-world conclusion?

(c) 90% (0.5598, 1.6755).

(d) 90% (0.0140, 0.0307).

(e) Yes?

6.4.12

Because all three of these situations are apparently one-time decisions, a test of significance would seem appropriate. And the conclusion would be very straightforward. First, you obtain an appropriate sample of air. Then for parts (a) and (b), either you do obtain evidence (small p-value) or you do not obtain evidence (not small p-value) to support your position. For part (c), if the data is statistically significant (small p-value), see what the data suggests—"clean air" or "problem with air pollution"—and report that in the newspaper.

However, it does make sense to treat such problems as hypothesis tests. That allows you to think clearly about the possible errors and, in particular, to decide on an appropriate sample size to guarantee good power for the test.

Still, these three problems together illustrate well how very self-serving an hypothesis test could be. The Chamber of Commerce and the environmentalist are only concerned with data favorable to their position. That would require a one-tailed test. However, if the results are going to be made public or retained for records, there is the possibility of misinterpretation if a one-tailed test is used. A disinterested party would look for statistically significant data, whatever it might suggest. From that point of view, a two-tailed test is required.

On the other hand, the analysis of error is confused by a two-tailed test because you're involved with two distinct and very different practical errors depending on whether you conclude air pollution is well under control or out of control. Furthermore, you might prefer to assign different probabilities to those two errors depending on the perceived seriousness of the errors. Note that it's not just a question of whether there's too much toxin in the air. It's a question of what you're going to DO about it. That's why we keep saying that an hypothesis test is an *action-oriented* decision procedure.

Possibly the best solution is to think in terms of two separate one-tailed tests with appropriately chosen error probabilities and with good power.

6.4.13 This problem illustrates how in the real world an hypothesis test is just one factor in a complex decision process. If you analyze the action appropriate to H_A and think about type I error, the situation will become much clearer. For instance, the FDA might reason as follows:

(a) In view of the seriousness of the present mortality rate, we will allow the drug on the market unless we find good reason to think the drug less effective than present treatment. What are the hypotheses?

(b) In view of the side effects . . . ? Now what are the hypotheses?

6.4.14 (a) Yes?

(b) If there's nothing special going on, then the number of deaths should be uniformly distributed over the year. Verify that the p-value is greater than 10%. How do you interpret that?

(c) This is similar to Problem 6.2.23.

(d) Yes?

6.4.15 (a) This can only be picture number 5. Why? What are the others?

Chapter **7**

7.1.1 (a) Let ϵ be $Y - \mu$, where μ is the mean of Y. Now, why does ϵ have the required form and why does ϵ look like a random error?

(b) First, real-world: Think about any real-world situation giving rise to a normally distributed random variable. For example, a measurement process. Or something like "fill" for a drink machine. In all such cases, there is a fixed systematic factor determining the values of Y: For a measurement process, it's the object being measured. For the drink machine, it's the setting on the fill mechanism. That systematic factor determines the values of Y. But not exactly; there's some variability. That's the nature of real-world processes! Now forget the real-world: Abstractly, what's the "systematic part" of a normally distributed random variable Y?

(c) The only thing variable about a normally distributed random variable is the random fluctuations about its mean. And that just looks like random error. Explain.

(d) We have two ways to show a random variable normally distributed. What are they? Which one is more natural here? Show how each of these two approaches will give that Y is normally distributed.

(e) The normality criterion eliminates the systematic part. How?

(f) For the regression model, the "one systematic effect" on Y is allowed to be *variable* instead of fixed. Allowing a VARIABLE effect on Y could lead to very wild complexities! To avoid that kind of chaos, our model restricts the effect even further. How? [two further restrictions]

7.1.2

(a) Think how the average height or average crop yield would change as you increase X. That average is $\mu_{Y|X}$ and it should be a linear function of X, but the relationship won't really be linear. To see this, try sketching a graph of Y as a function of X. Conjecture what it should look like. How high is a shoot one day after planting, for example?

Once you see why the mean of Y won't be a linear function of X, what sort of restriction on X would correct the problem to make the model at least approximately valid?

(b) Y is supposed to be the variable of interest about which we have a question. And X is supposed to be known or controlled by us in some sense. It should play the role of "input information" relevant to Y. So . . . ??

(c) In the first example, time does not CAUSE the height of the plant. The height is caused by the growth process (a very complex system of causes) and that process, like any process, takes place in time. The model makes use of the fact that the height of the plant is *correlated* with time. And it's a positive correlation: The height increases with increasing time (so in the model, $\beta > 0$). In the second example, by contrast, the toxic chemical in the plant seems to be directly caused by the chemical in the soil. Now, go through the rest of the nine examples and think about the nature of the effect of X on the mean of Y.

(d) From the point of view of the model, X is given. It's not random. In other words, the model is conditional on X. So when we come to use the model for answering real-world questions, those questions must give a value of X and then ask about Y. However, in generating data to specify the model—to estimate the parameters of the model—we may want to have the values of X generated randomly along with Y. For which of our nine examples would this probably be true?

7.1.3

(a) Recall the Chapter 5 criterion (page 148) for when to expect a normal distribution.

(b) ϵ is a linear function of $Y|X$. So it takes the form $\epsilon = a + b(Y|X)$ Make this explicit by saying what a and b are. Then use this linear function to show that ϵ is normally distributed with mean zero and standard deviation σ^2.

(c) What role does X play in the model and what happens to X if $\beta = 0$?

7.1.4

(a) Which variable is in question? What's the role of the other variable?

(b) We've identified three parameters. What are they? Remember the definition of the term "parameter"—it's a simple *number* associated with the model.

(c) There are three assumptions and there are four characteristics of the model implicit in those assumptions.

7.1.5

The parameter α is the y-intercept. It's the value of $\mu_{Y|X}$ when X is zero. If $X = 0$ is outside of the meaningful range of X, then this parameter is not meaningful in itself (i.e., in the real world).

The parameter β is the "slope" of the line on which the Y's are centered. The slope of a line is the change in Y for a unit increase in X:

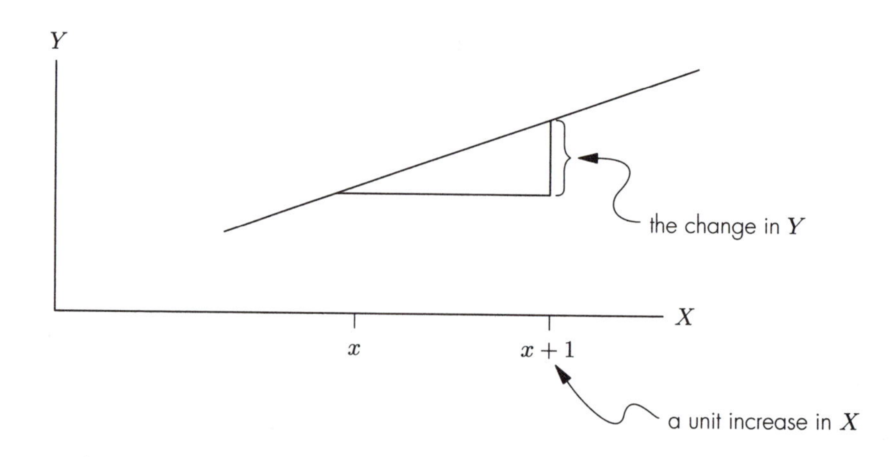

7.1.6 Yes?

7.1.7 (a) "Multiple" regression means there is more than one explanatory variable X. [Hint: Modify one of the nine examples listed at the beginning of this section.]

(b) You need an example of a pair (X, Y) where the expected value of Y would *decrease* as X increases. There are many, many instances of this. Give one.

7.1.8 A quadratic regression model would have $\mu_{Y|X}$ given as a quadratic function of X instead of a linear function of X. [Hint: The graph of a quadratic function is a parabola.]

7.1.9 (a) $\sigma^2_{Y|X}$ measures the spread of Y about its mean. So sketch a picture for which one of the values of X has a Y distribution which is more (or less) spread than at the other values of X.

(b) $\beta = 0$ means the slope of the line is zero, the line is parallel to the x-axis.

(c) Yes?

7.1.10 Think carefully about the given conditions. What exactly do they mean? [Hint: Reconsider Problem 7.1.5.]

7.1.11 Suppose you wanted to predict how high a bean plant will be ten days after planting or to determine how much toxic chemical will be found in a plant for a given level of the toxic chemical in the soil; what would be necessary before you could use the simple linear regression model

$$Y = \alpha + \beta X + \epsilon, \qquad \epsilon \sim N(0, \sigma^2)?$$

7.2.1 (a) (12, 21) is (X, Y) where X is . . . ? And Y is . . . ? Which observation is this—first, second?

(b) The observation is (16, 53) and so . . . ?

(c) n is the number of observations. How many were there?

(d) There are 16 observations, so $i = 1, 2, \ldots, 16$. So, (X_8, Y_8) is the eighth observation—what is it? And—there's a trick here, don't get caught!—what is (X_3, Y_{16})?

(e) $\overline{X} = 14.8750$. The question is asking for $\overline{X} = 1/n\Sigma X$. But there's a trick. You can't just add the six values of X across the top of the table and divide by six. Why not?

(f) $\Sigma Y = 821$.

(g) To get the "average of the daily average heights," you average each day's measurements and then average those averages. That's NOT the same as the "average daily height" which is the average of all 16 observations. Why not?

7.2.2

(a) Draw (x, y) coordinates and plot the observed (X, Y) values.

(b) $\hat{Y}$ is the esimator for $\mu_{Y|X}$. And so $\hat{Y} =$??

(c) The point estimate for the average height 13 days after planting is 30.6675 cm. How? What about six days after planting? There's a trick; what is it?

(d) This is a line. The easiest way to plot a line is to find two points on the line. Here, it's a good idea to evaluate $\hat{Y}$ at $X = 12$ and $X = 18$ and plot those two points. These are good points because they're near the extremes of the observed X's. Choosing such X's minimizes the loss of accuracy in drawing.

(e) When we speak of "the model," we're thinking of the picture in three dimensions with the normal curves for Y marching along the regression line, the line which gives $\mu_{Y|X}$. Now think about a particular X; for example, $X = 12$. In the model, what is Y for $X = 12$? Then, in general, how do the observations relate to the $Y|X$ of the model?

(f) In the model, what does $\hat{Y}$ estimate?

7.2.3

(a) A point estimate is a simple number calculated from partial information in the form of random data. If the data is atypical of the entire situation from which it is generated, that one number by itself, taken as an estimate of the true value of the parameter, could be very misleading.

Explain this in terms of our researcher's bamboo data by taking the parameter β and explaining the previous paragraph as it pertains to β. Be very specific to the real-world meaning of β (see Problem 7.1.5).

(b) The point estimate for β is $b = 11.0319$. How can we use that number and still indicate "the accuracy of the estimate" and "the certainty with which that accuracy is attained'?

(c) All kinds of things could go wrong in the data-collection process, but when we gave the data, we specifically said, "As usual in this text, we assume the data for our estimators has been *properly* generated by an appropriate random sampling experiment . . . ." So what could go wrong?

(d) We require the "total context" of the point estimate—the "total context" of that one number. What is that "total context"?

7.2.4 (a) Between the fourteenth and sixteenth days after planting, something seems to have changed. Do you see the pattern in the scatter diagram? If the model is valid, no such pattern should occur.

(b) Try to think of some practical situation which would cause the rate of growth for the bamboo to suddenly increase. Remember that β is the increase in the average height of the shoots corresponding to one more day of growth—the "daily rate of growth."

(c) Either get new data or restrict the model somehow. How?

(d) Think about how the researcher might actually want to use the model.

7.2.5 (a) You need an expression for the vertical distance in a picture like this:

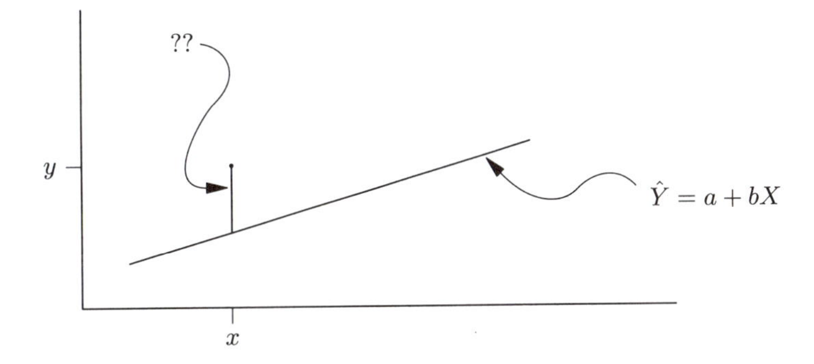

The point at the top of this "vertical distance" is just our observed (x, y) in the scatter diagram and the distance is just the difference in the y-values of this point and the bottom point on the line. What are the coordinates of the point at the bottom?

(b) Draw a picture like the one above, but with the observed point below the line.

(c) This is exactly like the problem we ran into in Chapter 1 when we wanted to construct a measure of "spread about the mean" for a random variable. What did we do in that case to avoid a possibly negative "distance" from the mean?

(d) Yes?

7.2.6 You sketched the graph of the bamboo data in Problem 7.2.2. Go back to that picture and plot the point $(\overline{X}, \overline{Y})$ and you'll see that it lies on the line. Then show this algebraically by showing that the point $(\overline{X}, \overline{Y})$ satisfies the equation for the estimated regression line. [Hint: The data is summarized below the table where it was presented. You need not do a lot of tedious calculations.]

7.2.7 Think about the role X and Y play in the model.

7.2.8 **(a)** Remember, you're only interested in the *vertical* variability of the observed (x, y)'s away from the line (see Problem 7.2.7). What formula would capture this "vertical" variability in the Y's ?

(b) The principle of least squares says find the line for which $\Sigma(y - \hat{y})^2$ is as small as possible. Explain the meaning of this.

7.2.9 Yes?

7.2.10 **(a)** What role do X and Y play in the model?

(b), (c), and (d) Yes?

7.2.11 **(a)** Show that the point $(\overline{X}, \overline{Y})$ satisfies the equation for the regression line. In other words, show

$$a + b\overline{X} = \overline{Y}.$$

This is easy! Try it before you look at the level II answer. Look at the formula for a and you'll see what to do.

(b) Here's how to get it started:

$$\Sigma(y - \hat{y}) = \Sigma[y - (a + bx)] = \Sigma\, y - na - b\,\Sigma\, x.$$

Now, you can make use of the fact that $\Sigma X = n\overline{X}$ and $\Sigma Y = n\overline{Y}$.

(c) You might think to measure the variability of the Y's about the estimated regression line by calculating the average deviation. But that won't work. Why not?

(**d**) The equation for a is immediate from the first of the normal equations. To get the formula for b, multiply the first equation by Σx and the second equation by n and then subtract. This will eliminate a from the equations.

7.2.12

(**a**) The model says $\ln(Y) = \ln(Np) + x\ln(q)$, so . . . ?

(**b**) Use the statistical mode of your calculator to get a and b.

(**c**) $\ln(Np)a = 4.0886$. $Np \approx$??

(**d**) $\ln(q) \approx b$ and so . . . ??

(**e**) $N = Np/p$, so $N \approx$??

(**f**) $N = (Np + Nq)$ and so . . . ? Or $Nq = \Sigma_{x \neq 0} Y$ and so . . . ?

(**g**) Recall: Assuming a geometric model for the data, the MLE for p is $1/\overline{Y} = 76/142$. This gives a poor estimate for N. What is it? Why poor?

(**h**) The assumptions of the regression model have not been addressed at all. For example . . . ?

7.3.1

(**a**) The critical values of t for the rejection region at the five percent significance level FOR A TWO-TAILED TEST are $t_c = \pm 2.1315$. Now standardize the observed value of b, assuming the null hypothesis.

(**b**) Yes?

(**c**) Don't attempt to give the decision rule in terms of b. Why not? What should it be?

(**d**) Yes?

(**e**) One of these tests is theoretical, the other real-world. Which is which? So . . . ?

7.3.2

(**a**) What are the roles of X and Y in the model?

(**b**) As altitude increases, pressure will decrease. That would seem to suggest that b is negative. But that's NOT right! What's wrong with this argument?

(**c**) $b = 0.5222$, and so $\hat{Y} =$??

(d) Yes, the p-value is zero. Explain. [Hint: $s_b = 0.0101$.]

(e) The question cannot be answered as asked! We have no information relating altitude to pressure. However, we can say, "On average, as we increase altitude, we should observe a decrease of between 0.4925 and 0.5520 inches in the reading of mercury in a barometer for one degree Fahrenheit increase in the boiling point of water. There is a one percent risk of error in this conclusion."

7.3.3 Note that when $Y = a + bX$, $Y - \overline{Y} = b(X - \overline{X})$.

7.3.4 This is the regression effect again. Note that $b < 1$.

$$\begin{aligned} b &= r s_Y / s_X \\ &= r \qquad \text{because } s_X = s_Y \\ &< 1. \end{aligned}$$

We say r is *strictly* less than one because equality holds only if there's an exact linear relationship and, as a practical matter, that would not happen. Now, for the second test, "how far above average" is captured by $\hat{Y} - \overline{Y}$. You need to show that $\hat{Y} - \overline{Y} < X - \overline{X}$. Sketch some possible scatter diagrams to see how this works.

7.3.5 Yes?

7.3.6 (a) Altitude above sea level is measured by a barometer. Think about the variability in barometric reading for Forbes' data. What proportion of that variability is explained by the boiling point of water? [$\approx 99.5\%$]

(b) You're not "explaining" Y in terms of X. Clarify.

(c) The coefficient of determination does, indeed, make sense for Wilm's data because X and Y were randomly generated together. But we don't know how to compute it. Why not?

7.3.7 (a) If $n = 2$, then necessarily $r^2 = 1$. Explain. Look at the scatter diagram.

(b) Thinking of r^2 (or r) as measuring the strength of correlation between X and Y, you could be misled by a small sample. Explain.

(c) It's true that for large n, the coefficient of determination is a reasonable measure of strength of correlation. But even when there's a very strong correlation between two variables, it's a fundamental fallacy to think there must be a cause–effect relationship between them. Explain. Think of some of the examples given at the beginning of the chapter. Or think about this example: X = wife's age, Y = husband's age.

(d) For the original observations, without calculating we can see that r^2 is close to zero. But when we replace the Y's by their averages, we get $r^2 = 0.9$, very close to one. Explain what this tells us.

(e)

X	1	2	3	4
Proportions	0.25	0.5	0.5	0.75

(f) Here, $r^2 = 0$. So that means there's no relationship between X and Y. Comment.

7.3.8

(a) You must give a confidence interval. Because no confidence coefficient was specified, why not choose 95% because you already know, with 15 degrees of freedom, $t = 2.1315$ for that case.

(b) Yes?

(c) Yes?

7.3.9

(a) Look at the squared standard error. Distribute the factor $s^2_{Y|X}$ over the sum. Explain the difference.

(b) When n is large, you have more information, and so . . . ?

(c) When you want to predict something about Y for a very atypical X, you should expect your answer to be less accurate. Why? So . . . ?

(d) $\Sigma(X - \overline{X})^2$ is a measure of the variability of the X's. Because it's in the denominator, when it's large the standard error will be small. That means we WANT $\Sigma(X - \overline{X})^2$ to be large. Large? Don't we always want variability to be SMALL? Explain.

7.3.10

Yes?

7.3.11

(a) You have two ways to answer this question, the quick way is to calculate r^2. It is a bit harder to carry out the hypothesis test that β not be zero.

(b) If the variables are jointly normally distributed, the hypothesis test to establish that $\beta \neq 0$ is equivalent to testing that ρ not be zero. So . . . ?

(c) By part (a), the simple linear regression model with only fiber length as regressor looks weak for answering this question. Still, if you must give an answer based on that model, how would you do it? Let $\alpha = 95\%$.

(d) The question must be made more specific. Because spinning quality is measured by skein strength, interpret the question as asking for $\mu_{S|7}$.

7.4.1–7.4.25 See level II.

Level

II

Answers for Try Your Hand Section

Chapter 1

1.1.1 (a) H for heads or T for tails, representing the face of the coin which shows after one toss.

(b) For the die all the P(X)'s are 1/6. For the coin the probabilities are each 1/2.

(c) If the deck is well shuffled, then *theoretically* each card has an equal chance to be drawn—that's what you mean by well shuffled! There are 52 cards, one of which is the ace of spades, and so the relative frequency with which you SHOULD (theoretically!) draw that card is one out of 52. That means there's approximately a two-percent chance to draw the ace of spades: $1/52 \approx 0.0192$.

1.1.2 (a) In this situation, you just add probabilities:

$$\begin{aligned} P(X < 3) &= P(X = 1) + P(X = 2) \\ &= 1/6 + 1/6 = 1/3. \end{aligned}$$

(b) You just multiply probabilities:

$$\begin{aligned} \text{P(six on each die)} &= \text{P(six on the first die)} \times \text{P(six on the second die)} \\ &= 1/6 \times 1/6 = 1/36. \end{aligned}$$

(c) $(1/2) \times (1/2) = 1/4$.

1.1.3 (a) Roll a one OR a two on one roll of the die.

(b) Roll a six on the first die AND a six on the second die.

(c) Toss a head on the first coin AND a head on the second coin.

1.1.4 (a) The "something you do" is made precise by clearly specifying the outcomes. Thus far, both definitions are the same: "Something you do which is repeatable, with clearly specified outcomes." The difference is that for a *random* experiment, the outcomes should be unpredictable. No such restriction is made for a scientific experiment. So a random experiment is a scientific experiment, but not the reverse. For many scientific experiments, there is a theory which predicts the outcome exactly. If it isn't what's predicted, you'll suspect something wrong in the theory or possibly in the performance of the experiment.

(b) A scientific experiment is a random experiment if the outcomes cannot be predicted in advance.

1.1.5

(a)

X	P(X)
1	0.1
2	0.5
3	0.1
4	0.1
5	0.1
6	0.1
	1.0

(b)

X	P(X)
1	0.1
2	0.4
3	0.1
4	0.1
5	0.2
6	0.1
	1.0

(c) The "doing" is different! Although the outcomes have the same verbal description, they're generated by two different random mechanisms, the two dice. Because a random experiment is "something you do," it's the "doing" that's different, depending on which die you roll. Note how we've pinpointed the difference between these two random variables in the *exact phrase* ("something you do") in the definition where the difference occurs.

1.1.6

ΣY, the total of the values of Y, has no meaning at all because it ignores completely the fact that these values occur with differing probabilities. That's why we don't put in the total of the Y column. Similarly for the die, the number $\Sigma X = 21$ has no meaning if the faces don't occur with the same probabilities.

1.1.7

(a)

	Random Experiment	Random Variable
1.1.2	(a) Roll the die	# dots on the uppermost face
	(b) Roll the two dice	# dots on the two uppermost faces
	(c) Toss the two coins	no r.v. given for this problem, but # heads is one possibility
1.1.5	(a), (b) roll the (unfair) die	# dots on the uppermost face
1.1.6	Draw one card from the shuffled deck of 50 cards	1 for spades; 2 for clubs; 3 for hearts; 4 for diamonds

(b) For Problem 1.1.2 parts, (a), (b), and (c) and for Problem 1.1.5, an outcome is the die or two dice or the coin in a particular position on some table top after the roll or toss. For Problem 1.1.6, you could describe the outcome as "me standing there holding a card in my hand." Or more simply, an outcome can be described as one of the 50 cards, whichever one is drawn. This second description or "modeling" of the outcomes recognizes that "my position" is probably irrelevant to the actual game.

Note, in each case, that the outcome is a real-world object or situation. On the other hand, the values of the random variable are abstract, they're the abstract part of the model. The *variable* is a bridge from the experiment and its outcomes in the real world to numbers in the world of theory. That's the picture in the text. The random variable is shown as a *bridge* between the real world and theory.

(c) There are four conditions in the definition for a random experiment. In part (a), we described the "doing" for each example. That's condition one. Clearly, in each case the "doing" is repeatable (condition two). In part (b), we "clearly specified" the outcomes (condition three). Finally, in each case the outcomes can't be predicted in advance (condition four).

To guarantee unpredictability for the draw of a card from the deck of 50, we were careful to include the shuffling of the deck in the description of the "doing." The idea of "shuffling" or "mixing thoroughly" is one of the most basic intuitive formulations of "randomizing." In 1970, at the height of the Vietnam War, a lottery was conducted to decide who would be drafted into the U.S. army. This was intended to correct perceived inequities in the existing draft procedures. Capsules representing the 366 possible birthdays (some people are born on the extra day in leap years) were put into a large device for mixing and then someone drew capsules "at random." All eligible men born on the first birthday chosen, September 14, were to be the first inducted into the armed forces in 1970, all born on the second birthday chosen were to be the next group inducted, and so on. However, after the drawing, there were complaints of bias. Subsequently, a careful study of the lottery results suggested the capsules were put into the mixing device by month and were not thoroughly mixed. In other words, the draw was not completely random. (See the article "The 1970 Draft Lottery," *Science,* January 22, 1971.) This example will give you some idea of the practical difficulties that arise in attempts to make the idea of randomness precise and to implement it in a concrete instance. It's not a simple matter!

1.1.8

(a) X is not numeric valued, its values are the letters H and T.

(b) When you flip the coin, an outcome is the coin sitting on the table top in some position. To that outcome (the "sitting coin"), X assigns the number of heads visible on the top face, none or one. This X is a random variable because it is a rule which assigns a number to each of the possible outcomes of a random experiment.

In fact, because X cares only about whether there is a head or a tail showing, we can simplify the description of the outcomes. We can group all the positions with heads uppermost together, calling it H for "heads." Then all other outcomes will be grouped together and called T for "tails." In this way, we have exactly two possible outcomes. This is a perfectly acceptable way of modeling this random experiment. And although it makes the outcomes seem somewhat less physical and less real world, it makes the probabilities clearer. If the coin is fair, clearly, H and T are equally likely.

(c)

X	P(X)
0	0.5
1	0.5
	1.0

1.1.9

(a) On one roll of a die, we get anywhere from one to six dots; on 100 rolls, anywhere from 100 to 600 dots; on n rolls, anywhere from n to $6n$ dots.

Because each of the faces are equally likely, you should expect 3.5 dots on average for one roll of the die—the average of the integers one through six. This does not mean we think there is a face with 3.5 dots! If the average family in our city has 1.8 children, it does not mean we think some family has eight-tenths of a child! You think of this expected 3.5 dots as the "middle" of the integers from one to six. Now, if on one roll of the die we expect 3.5 dots, obviously on 100 rolls, we should expect 350 dots. And so for n rolls, you expect $3.5n$ dots altogether, on average.

(b) On the "average" toss, we expect half of a head. That means on 100 tosses, we expect 50 heads. This is an "on average" figure, of course. For a specific 100 tosses, you would be surprised to get *exactly* 50 heads, but you would expect something close to that.

(c) Split the difference. Expect \$2.50 on average. For example, the "typical" result of ten tosses on a fair coin would be

$$H, H, H, H, H, T, T, T, T, T$$

for which you would collect \$25 for an average of \$2.50 per roll. Of course, this "typical" result would occur rarely, but it reflects what ought to happen on average.

(d) Theoretically, on ten tosses you receive \$21 and so the average is \$2.10. On 100 tosses, you receive \$210, for the same average.

(e) It's the same experiment we've been discussing all along. The "doing" is "toss the coin once (in such a way that it stays on the table top)." Obviously, that's repeatable. An "outcome" is the "sitting coin," the coin sitting on the table top in some position. That says we have a scientific experiment. Then, because we can't predict in advance what position the coin will take, this is a *random* experiment. Thus, we've verified the four conditions for having a random experiment. The "take" is a random variable because it is a rule which assigns a number to each possible outcome: "two" if the "sitting coin" has heads visible, "three" if the tail is visible.

(f) The smaller payoff—\$2.00—comes 90% of the time instead of 50% of the time. Because 40% (90% – 50%) of the time you're getting less money, you should expect less on average.

1.1.10 It's not whether you can predict the VALUES in advance! The definition requires that the OUTCOMES be unpredictable. Can you predict in advance how the coin will land on the table top when you toss it, where it will land and whether it will come up heads or tails? Of course not. So the experiment is indeed a random experiment.

Now, any rule which associates a number to the possible outcomes of the experiment is by definition a random variable. So our X is a random variable. Notice that . . .

THE RANDOMNESS IS NOT IN THE VARIABLE, IT'S IN THE EXPERIMENT!!

This makes perfect sense. A random variable is supposed to model uncertainty in real-world situations, so the randomness should be in the real-world part of the model. The *experiment* is the real-world part of the model. The *values*, on the other hand, are numbers; they're abstract, not real world. Because you're modeling uncertainty in the real world, it makes sense that the randomness would be in the experiment, not in the values.

Here's the very simple probability distribution for X:

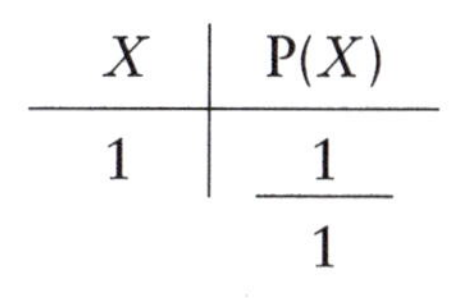

X	P(X)
1	$\frac{1}{1}$

1.1.11 (a) The experiment is to roll the die. We've already verified the four conditions of the definition in our solution for Problem 1.1.7(c).

(b) The values of Y are $-3, -2, -1, 0, 1, 2$. Make sure you see why. When you roll one dot, what's your net gain/loss?

(c) $Y = X - 4$.

(d) If you roll three dots, $X = 3$ and $Y = -1$ (you get three dollars for the three dots you rolled, but you paid out four dollars). Thus,

when X is	1	2	3	4	5	6
Y is	−3	−2	−1	0	1	2

Now, because the value −3 corresponds to a roll of one dot and one dot shows up with probability 10%, we get $P(Y = -3) = 0.1$.

The probability distribution is

Y	$P(Y)$
−3	0.1
−2	0.4
−1	0.1
0	0.1
1	0.2
2	0.1
	1.0

Note that the probabilities for Y are exactly the same as the corresponding probabilities for X. That should be true, after all the corresponding values of the two random variables are determined by exactly the same outcome. It's the outcomes which determine the probabilities. They're ultimately determined by what happens in the real world—by the uncertainty of the situation being modeled. That "situation" is in the real world.

1.1.12

(a) A chance mechanism is just a random experiment.

(b) For a random experiment, the abstract model is the probability distribution for some appropriate random variable associated with the experiment.

1.2.1

(a)

X	$P(X)$	$XP(X)$
1	1/6	1/6
2	1/6	2/6
3	1/6	3/6
4	1/6	4/6
5	1/6	5/6
6	1/6	6/6
	1	21/6

so $E(X) = 3.5$.

(b) $P(X = 1) = P(HT \text{ or } TH) = P(HT) + P(TH) = 0.25 + 0.25$ and so

X	$P(X)$	$XP(X)$	
0	0.25	0.00	
1	0.50	0.50	
2	0.25	0.50	
	1.00	1.00	$E(X) = 1.$

(c)

X	$P(X)$	$XP(X)$	
1	0.1	0.1	
2	0.1	0.2	
3	0.1	0.3	
4	0.1	0.4	
5	0.5	2.5	
6	0.1	0.6	
	1.0	4.1	$E(X) = 4.1.$

(d)

X	$P(X)$	$XP(X)$	
2	0.5	1.0	
3	0.5	1.5	
	1.0	2.5	$E(X) = 2.5.$

1.2.2 On one roll of a die, you expect $E(X)$ dots. Because you get one dollar per dot, you expect to take in $E(X)$ dollars per roll. To break even, you should pay that many dollars to play, $E(X)$ dollars per roll. Thus, you should pay

(a) \$3.50 per roll;

(b) \$1 per toss;

(c) \$4.10 per roll.

1.2.3 **(a)** "Or" means add, not always, but it's valid here. So

$$P(X = 1 \text{ or } 6) = P(X = 1) + P(X = 6) = 1/6 + 1/6 = 1/3.$$

(b) For the loaded die given in the text,

$$P(X = 1 \text{ or } 6) = P(X = 1) + P(X = 6) = 0.3 + 0.3 = 0.6.$$

For the loaded die, the probability of this event is almost twice as big as for the fair die, 60% as compared with about 33%.

(c) The probability distribution for the loaded die is more spread about its mean than the distribution for the fair die.

1.2.4

(a) The average deviation from the mean is ZERO! The deviations are -2.5, -1.5, -0.5, 0.5, 1.5, 2.5. Weight them by $1/6$ and then add.

(b) The systematic way to do a calculation like this is by extending the table to include a column for the weighted deviations from the mean. Then the sum of that column will be the required average. It's zero:

X	$P(X)$	$XP(X)$	$(X-\mu)P(X)$
1	0.3	0.3	−0.75
2	0.1	0.2	−0.15
3	0.1	0.3	−0.05
4	0.1	0.4	0.05
5	0.1	0.5	0.15
6	0.3	1.8	0.75
	1.0	3.5	0.00

1.2.5

(a) For a particular value of X, the deviation, $X-\mu$, tells us HOW FAR that value of X is from μ. So it's capturing the idea of "spread" or "dispersion" from the mean, μ.

(b) The deviation is positive exactly when X is larger than μ and is negative when X is smaller than μ. Here's why: $X-\mu>0$ is the same as $X>\mu$. Just add μ to both sides of the first inequality. Doing that preserves the inequality.

(c) Watch:

$$\begin{aligned}
\Sigma(X-\mu)P(X) &= \Sigma\left[XP(X)-\mu P(X)\right] \\
&= \Sigma XP(X) - \Sigma\mu P(X) \\
&= \mu - \mu\Sigma P(X) \quad \text{but } \Sigma P(X)=1! \\
&= \mu - \mu \\
&= \text{ZERO.}
\end{aligned}$$

1.2.6

(a) Here's one possibility . . .

X	P(X)	XP(X)	$(X-\mu)^2$P(X)
1	0.1	0.1	0.625
2	0.1	0.2	0.225
3	0.3	0.9	0.075
4	0.3	1.2	0.075
5	0.1	0.5	0.225
6	0.1	0.6	0.625
	1.0	3.5	1.850

$\mu = 3.5,$
$\sigma^2 = 1.85$

Do you get the basic idea? The expected value must be 3.5, and we concentrate the probability CLOSE TO 3.5. This way the die is closer to what's expected, on average. To verify this analytically, just observe that the variance for this die, 1.85, is smaller than the variance of a fair die, 2.9167.

(b) Here's the graph:

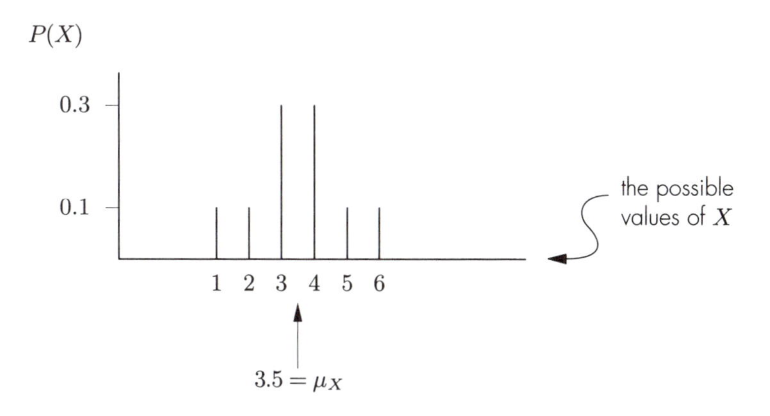

The probability is concentrated closer to the expected 3.5 than for the fair die. Because values close to what's expected are more probable, this die is MORE PREDICTABLE than the fair die.

1.2.7

(a) Less dispersed about the mean implies a *smaller* variance. Here's the calculation:

X	P(X)	XP(X)	$(X-\mu)^2$P(X)
0	0.1	0	$0.9^2 \times 0.1$
1	0.9	0.9	$0.1^2 \times 0.9$
	1.0	0.9	0.09

So $\sigma^2 = 0.09$, but for the fair coin $\sigma^2 = 0.25$ (verify).

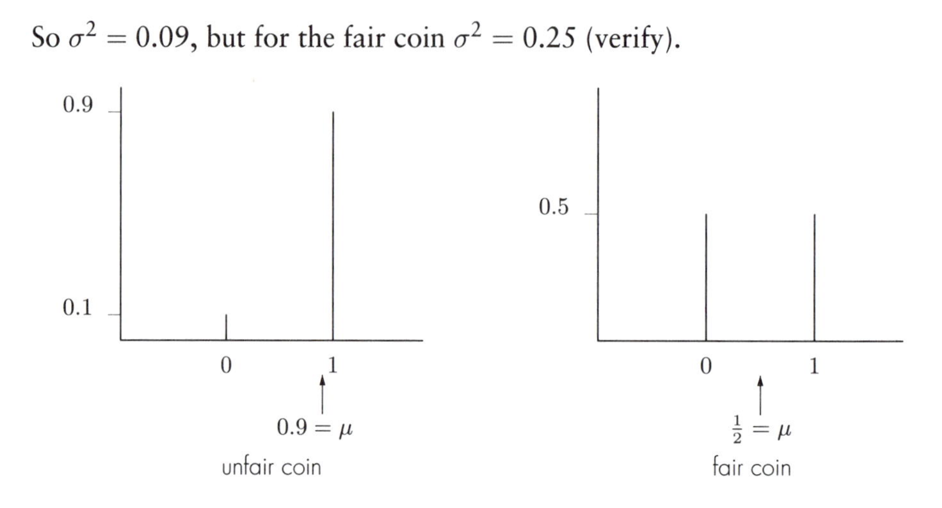

(c) The standard deviation is just the square root of the variance. For the fair coin, it's 0.5, and for the unfair coin, it's 0.3.

(d)

X	$P(X)$	$XP(X)$	$(X-\mu)^2P(X)$
0	$1-p$	0	$p^2(1-p)$
1	p	p	$(1-p)^2p$
	1	p	$p(1-p)$

$\mu = p$,
$\sigma^2 = p(1-p)$.

To get $p(1-p)$, you'll have to simplify the sum $p^2(1-p)+(1-p)^2p$. Factor out $(1-p)$.

1.2.8

(a)

X	$P(X)$	$XP(X)$	$(X-\mu)^2P(X)$
1	0.1	0.1	0.361
2	0.5	1.0	0.405
3	0.1	0.3	0.001
4	0.1	0.4	0.121
5	0.1	0.5	0.441
6	0.1	0.6	0.961
	1.0	2.9	2.290

so $\mu = 2.9$ and $\sigma^2 = 2.29$.

X	$P(X)$	$XP(X)$	$(X-\mu)^2P(X)$
1	0.1	0.1	0.484
2	0.4	0.8	0.576
3	0.1	0.3	0.004
4	0.1	0.4	0.064
5	0.2	1.0	0.648
6	0.1	0.6	0.784
	1.0	3.2	2.560

$\mu = 3.2$ and $\sigma^2 = 2.56$.

Y	$P(Y)$	$YP(Y)$	$(Y-\mu)^2P(Y)$
1	0.26	0.26	0.5695
2	0.26	0.52	0.0599
3	0.22	0.66	0.0595
4	0.26	1.04	0.6007
	1.00	2.48	1.2896

$\mu = 2.48$ and $\sigma^2 = 1.2896$.

(b) Here's the table for first of these random variables:

X	$P(X)$	$XP(X)$	$\lvert X-\mu\rvert P(X)$
1	0.1	0.1	0.19
2	0.5	1.0	0.45
3	0.1	0.3	0.01
4	0.1	0.4	0.11
5	0.1	0.5	0.21
6	0.1	0.6	0.31
	1.0	2.9	1.28

For the second and third random variables, the MAD is just 1.44 and 1.0192, respectively.

(c) The standard deviation is just the square root of the variance.

1.2.9

$$\begin{aligned}
\mu_X &= \Sigma XP(X) = \Sigma cP(X) && \text{since } X = c \\
&= c\Sigma P(X) && \text{since } c \text{ is constant, it can come out in front of the sum} \\
&= c && \text{since } \Sigma P(X) = 1
\end{aligned}$$

and

$$\sigma_X^2 = \Sigma(X-\mu)^2P(X) = \Sigma(c-c)^2P(X) = 0.$$

1.2.10

(**a**) The variance is larger for the loaded die than for the fair die, 4.25 versus 2.9167. Here's how to understand what that means: For the loaded die, two values far from the mean (1 and 6) carry 60% of the probability. Their *weighted* squared deviations are quite large compared with the fair die. The other squared deviations, the small ones, carry less weight. So for the loaded die, the small squared deviations carry less weight and the large squared deviations carry more weight. Because the variance is "the average of the squared deviations from the mean," it will be larger for this loaded die.

(**b**) For both dice, the expected number of dots is the same, but for the loaded die you're much further from that value, on average. So for the loaded die, the expected number of dots is less certain. This is seen in its larger variance.

In general, the variance of a random variable is a measure of the ACCURACY of the expected value for purposes of predicting what will happen in the long run with many repetitions of the random experiment.

(**c**) With heads coming up 90% of the time, the unfair coin is obviously more predictable; you're quite certain of getting heads. The variance tells you exactly that. Of course, this is such a simple situation you really don't need to consider the variance. Still, it's reassuring that the calculations work! For the unfair coin, $\sigma^2 = 0.09$ compared with 0.25 for the fair coin. With a much smaller variance, the unfair coin is more predictable, as we guessed.

(**d**) By Problem 1.2.7(d), $\sigma^2 = p(1 - p)$ and so the answers are 0.9×0.1 and 0.5×0.5.

1.2.11

The variance and the standard deviation measure the same thing. They measure "dispersion about the mean." They're not two *different* parameters! The difference is only one of convenience. The variance is more convenient for computational purposes because it doesn't involve the awkward square root. The standard deviation is more convenient for reporting your final conclusions because it returns you to the original units. The variance for the die is in units of "squared dots," the units for the standard deviation is "dots"—a more realistic unit!

1.2.12

(**a**) Here $E(X) = 3.5$; for two dots, $G = 2 - 3.5 = -1.5$.

G	P(G)	GP(G)	$(G-\mu)^2$P(G)
−2.5	1/6	−5/12	1.0417
−1.5	1/6	−3/12	0.3750
−0.5	1/6	−1/12	0.0417
0.5	1/6	1/12	0.0417
1.5	1/6	3/12	0.3750
2.5	1/6	5/12	1.0417
	1	0	2.9168

G	P(G)	GP(G)	$(G-\mu)^2$P(G)
−1.9	0.1	−0.19	0.361
−0.9	0.5	−0.45	0.405
0.1	0.1	0.01	0.001
1.1	0.1	0.11	0.121
2.1	0.1	0.21	0.441
3.1	0.1	0.31	0.961
	1.0	0	2.290

G	P(G)	GP(G)	$(G-\mu)^2$P(G)
−2.2	0.1	−0.22	0.484
−1.2	0.4	−0.48	0.576
−0.2	0.1	−0.02	0.004
0.8	0.1	0.08	0.064
1.8	0.2	0.36	0.648
2.8	0.1	0.28	0.784
	1.0	0	2.560

The gain/loss would be MOST PREDICTABLE for the second die because it has the smallest variance. You might have guessed that because the second die has fully half its probability concentrated on the value 2, not too far from the mean of 2.9.

The gain/loss is LEAST PREDICTABLE for the game with the fair die. Again, you might have guessed this because more than 30% of the probability for that die is on the two extreme values 1 and 6, far from the mean of 3.5. For each of the other two dice the probability of the two most extreme values—the values 1 and 6—is only 20%.

(b) For the house to make a \$0.50 profit, the player has to have a \$0.50 loss. That means the player should pay \$0.50 more than the break-even charge,

E(X). For the fair die, for example, you should pay \$4, \$0.50 more than the break-even charge of \$3.50.

(**c**) In each case, the underlying random experiment is to roll the die once (the "doing"). An outcome is the die sitting on the table top in some position (the "clearly specified outcomes"). Obviously, the "doing" is repeatable and the outcomes cannot be predicted in advance. That verifies the four conditions of the definition of "random experiment."

The gain/loss is a random variable because it is a rule assigning a number to each outcome. The rule is: one dollar for each of the dots on the uppermost face minus what you pay to roll the die once.

The verbal description of the random experiment and of the gain/loss random variable is EXACTLY THE SAME in each of the three cases. The only difference is in the dice—which of the three dice you're playing with. These critical real-world differences are seen at the abstract level when you write out a specific "abstract model" in the form of a probability distribution for the gain/loss. In the distribution tables, you see different gain/loss values, different probabilities, and different parameter values (different means and variances).

1.3.1 Here's the equation: $0.72 = (1/150) \times 108$. If you had not already observed that X is 108, you could have solved for X in this way:

$$\hat{p} = 1/nX \quad \text{so } X = n\hat{p},$$

from which you conclude that $X = 150 \times 0.72 = 108$.

1.3.2 (**a**) The coefficient on X is 1, so that's b. The constant term is $-\mathrm{E}(X)$ and so that's a. The equation is $G = -\mathrm{E}(X) + 1X$.

(**b**) Because $b = 1$, $\sigma_G^2 = \sigma_X^2$.

(**c**) You get exactly one dollar per dot, so your receipts look just like the number of dots. But that's the whole story as far as the *variability* of the game is concerned. The variability of the gain/loss won't be affected by what you pay to play because that's the same for every play; it's constant.

(**d**) The house gain is your loss, so, of course, μ_G represents a loss of \$0.50. Formally, $G = X - P$, where $P = [\mathrm{E}(X) + 0.5]$. So

$$\mu_G = \mu_X - P = \mathrm{E}(X) - \mathrm{E}(X) - 0.5 = -0.5$$

because μ_X is just another symbol for E(X). Finally, $\sigma_G^2 = \sigma_X^2$, because $b = 1$.

1.3.3

(a)
$$\begin{aligned} \mu_Y &= \Sigma(a + bX)\mathrm{P}(X) && \text{since } Y = a + bX \\ &= \Sigma a\mathrm{P}(X) + \Sigma bX\mathrm{P}(X) \\ &= a\Sigma\mathrm{P}(X) + b\Sigma X\mathrm{P}(X) \\ &= a + b\mu_X && \text{since } \Sigma P(X) = 1, \Sigma X\mathrm{P}(X) = \mu_X \end{aligned}$$

(b)
$$\begin{aligned} \sigma_Y^2 &= \Sigma(Y - \mu_Y)^2\mathrm{P}(Y) \\ &= \Sigma(Y - \mu_Y)^2\mathrm{P}(X) && \text{since probabilities of } Y \text{ and } X \text{ for the same outcome are equal} \\ &= \Sigma\big[(a + bX) - (a + b\mu_X)\big]\mathrm{P}(X) \\ &= \Sigma[bX - b\mu_X]^2\mathrm{P}(X) \\ &= \Sigma b^2[X - \mu_X]^2\mathrm{P}(X) \\ &= b^2\Sigma(X - \mu_X)^2\mathrm{P}(X) \\ &= b^2\sigma_X^2. \end{aligned}$$

1.3.4

(a) $a = 0.7254$, $b = 0.4981$.

(b) According to our model, the average April to July water yield over the ten years in question was $0.7254 + 0.4981(22.3) = 11.8330$. This uses the fundamental equation: $\mu_Y = a + b\mu_X$.

(c) $Y = a$ exactly when $X = 0$. So, $WC = 0.7254$ when $SC = 0$. In other words, a is the April to July water yield when there is NO SNOW to yield any water! So a should be zero. When you notice this fact, you realize the model is not correct. See part (d).

(d) We need to compare σ_{WY}^2 (or σ_{WY}) for the two models. In each case it's obtained from $b^2\sigma_{SC}^2$. We don't know σ_{SC}^2, of course, but we don't need to. It's the same for both models. σ_{WY}^2 is larger for the model which has the larger b. That's the second model.

(e) In the situation of this problem we know as a matter of physical fact that $a = 0$ (see part (c)). The model in part (d) does have $a = 0$.

1.3.5

(a) You'll lose \$0.50 per roll, on average. You expect to take in \$7.50. In other words, \$3.00 times the expected number of dots, E(X). But you pay \$8.00 to play. If you really like to gamble, you might be willing to run the risk of losing

$0.50 per play with the hope of "beating the odds" and ending up ahead at some point. On the other hand, if you're a conservative gambler, you might NOT be willing to play. Suppose you're only in this casino to please a potential client and you're really not willing to risk much money.

But this analysis is too simplistic. The long run is not completely described by the "cost per roll" of $0.50. There's also a question of predictability. Maybe your cumulative gain/loss is always very close to this expected $0.50 loss. For the dedicated gambler, that would be a boring game because there's little chance to beat the odds. On the other hand, if the gain/loss is highly variable—sometimes you're $80 ahead (cumulative gain), sometimes $75 behind (cumulative loss)—then the game will be very exciting and you do have a chance to "beat the odds." This variability is a crucial consideration. A conservative player, by contrast, might be willing to lose $0.50 per roll provided she can be fairly certain she'll never be too far from that. Her concern is that the game not be too exciting. Note that there's no correct answer to whether one should or shouldn't play the game. It depends on your individual motivations.

(**b**) You expect 3.5 dots on average per roll on the fair die, so you expect to take in $3 \times 3.5 =$ $10.50 per roll. That says you should pay $11.00 to roll once. Then you have an expected loss of $0.50 per roll, just as you did with the loaded die.

The standard deviation for the loaded die is about 1.8 dots compared with 1.7 for the fair die. So our game is slightly more risky than the same game played with a fair die. But is this correct? The risk of the game is not in the dots, after all. It's in the gain/loss. Yes, but for both games, the gain/loss is determined by the number of dots THROUGH A LINEAR FUNCTION WITH THE SAME b ($b = 3$). So we can assess risk through the variability of just the dots.

Still, instead of the standard deviation for the number of dots let's look at the standard deviation of the gain/loss. It's about $5.41 for our game compared with $5.13 for the game with the fair die. Same result: This game is slightly more risky than the game with a fair die. Note that each of the two standard deviations is determined by $3\sigma_X$ because $b = 3$ for both games. It's only σ_X which differs from one game to the other. So the variability in the gain/loss is determined by the variability in the number of dots. If b had differed, the variability of dots would NOT have sufficed to assess risk for the two games.

(**c**) One value carries half the probability alright, but how far is that value from the mean? "One dot" is NOT close to the expected 2.5 dots. So you have a large deviation carrying half the probability. Consequently, this die is MORE variable than a fair die. By contrast, the die with half the probability on "two dots" is LESS variable than the fair die [see Problem 1.2.12(a)].

1.4.1 (a) There are 13 clubs and one of them is an ace. So there's one chance in 13 to draw an ace.

(b) Using Rule 3:

$$1/52 = \mathrm{P}(A \text{ and } C) = \mathrm{P}(A|C)\mathrm{P}(C) = \mathrm{P}(A|C)13/52.$$

Solve this equation for $\mathrm{P}(A|C)$:

$$\mathrm{P}(A|C) = 1/52 \times 52/13 = 1/13.$$

(c) How many outcomes there are depends on how you "model" the outcomes. The "doing" here is "draw a card from a deck of 52." An outcome can be described as "you standing there holding a card in your hand." Then every position you might be standing in will give a different outcome, so there are an infinite number of outcomes for the experiment and the events A and C as well each consist of an infinite number of outcomes.

More simply, bearing in mind what's of interest in a card game, describe an outcome as "one card." If you're trying to model a card game, this is clearly better—it abstracts away from the irrelevant consideration of "your position." This model gives 52 possible outcomes altogether for the experiment. Now, because there are four aces in the deck, the event A is a set of four of the possible outcomes and the event C, a set of thirteen.

(d) In part (b) of this problem, $\mathrm{P}(A|C) = 1/13$ which is the same as $\mathrm{P}(A) = 1/13$. So A and C are independent. If you're going to guess the top card is an ace, would your chance to guess right be improved by knowing the suit? No! In each case you have a one in 13 chance to be right. Information about the suit tells you nothing about the face value.

(e) Suppose A is independent of B. Then $\mathrm{P}(A) = P(A|B)$. Multiplying by $\mathrm{P}(B)$, you get $\mathrm{P}(A)\mathrm{P}(B) = \mathrm{P}(A|B)\mathrm{P}(B) = \mathrm{P}(B|A)\mathrm{P}(A)$. Now cancelling $\mathrm{P}(A)$ from both sides, we get $\mathrm{P}(B) = \mathrm{P}(B|A)$. So B is independent of A.

(f) First, suppose the simple product rule holds. Then

$$\begin{aligned} \mathrm{P}(A|B)\mathrm{P}(A) &= \mathrm{P}(A \text{ and } B) && \text{by Rule 3} \\ &= \mathrm{P}(A)\mathrm{P}(B) && \text{by the simple product rule.} \end{aligned}$$

Cancel $\mathrm{P}(A)$ from each side and you get independence.

Now, supose A and B are independent, so $\mathrm{P}(A|B) = \mathrm{P}(A)$. Multiply each side by $\mathrm{P}(B)$ and you get $\mathrm{P}(A|B)\mathrm{P}(B) = \mathrm{P}(A)\mathrm{P}(B)$. By Rule 3 the left hand side is $\mathrm{P}(A \text{ and } B)$, and so the simple product rule holds.

(g) From the text, $P(H_1|H_2) = P(H_2|H_1) = 12/51 \neq P(H_1) = 13/52$. So . . . ?

(h) $P(A|B)P(B) = P(A \text{ and } B) = P(B|A)P(A)$. Now, if $P(A|B) = P(B|A)$, cancel from both sides and get $P(A) = P(B)$. The converse is similar.

1.4.2

(a) The simple "or" rule holds if the events are *mutually exclusive* because then $P(A \text{ and } B) = 0$. The simple "and" rule holds if the events are *independent*. That's Problem 1.4.1 (f).

(b)

$$
\begin{aligned}
P(X = 2 \text{ OR } 3 \text{ OR } 4 \text{ OR } 5) &= P(X = 2) + P(X = 3) + P(X = 4) + P(X = 5) \\
&= 1/6 + 1/6 + 1/6 + 1/6 \\
&= 2/3.
\end{aligned}
$$

This is easy and natural. This calculation is always possible because the values of a random variable are MUTUALLY EXCLUSIVE.

On the other hand, $P(X \geq 2 \text{ AND } X \leq 5)$ is not so obvious. You can't easily use the "and" rule because these are DEPENDENT events. This probability is

$$P(X \geq 2 \text{ AND } X \leq 5) = P(X \geq 2|X \leq 5) \times P(X \leq 5).$$

But recalling that $(A \text{ and } B)^c = A^c \text{ or } B^c$, you can do it as follows:

$$
\begin{aligned}
P(X \geq 2 \text{ AND } X \leq 5) &= 1 - \left[P(X < 2) + P(X > 5)\right] \\
&= 1 - [1/6 + 1/6] \\
&= 2/3.
\end{aligned}
$$

Here, of course, we've used the obvious fact that $X < 2$ and $X > 5$ are mutually exclusive events.

1.4.3

(a) Since there are 48 non-eight's, there's a 48/52, or about a 92% chance that you do not draw an eight.

(b) Let E be the event you draw an eight. Then

$$P(E^c) = 1 - P(E) = 1 - 4/52 = 48/52.$$

1.4.4 (a)

$$\text{So } P(T) = P(T|D)P(D) + P(T|D^c)P(D^c)$$
$$= 0.99 \times 0.03 + 0.02 \times 0.97 = 0.0491.$$

(b) By Bayes' Theorem, $P(D|T) = P(T|D)P(D)/P(T) = 0.99 \times 0.03/0.0491 = 0.6049$. You've little more than a 60% chance to have the disease when the test says you do! This, of course, is NOT GOOD.

(c) Here $P(T) = 0.02291$ and so $P(D|T) = 0.1296$. This is even worse! Now there's only a 13% chance to have the disease when the test says you do. Screening tests of this kind are very common, not only in medicine, but in education, psychology, business (remember drug tests in the workplace?) . . . on and on.

Various methods are used to get around the difficulty illustrated in this problem for such screening devices. First of all, the sensitivity and specificity of the test should be as high as possible. Then it's also commonly required that a positive test result be confirmed by another, independent test. That can substantially improve the predictive value of the test. It's important that the second test be *independent* of the first. In particular, it should not be a repeat of the same test. If some extraneous condition caused a false positve on the first test, it's highly likely that a repeat of that test will have the same problem, giving a second false positive.

(d)

$$P(D|T_1 \text{ and } T_2) = \frac{0.95 \times 0.98 \times 0.003}{0.95 \times 0.98 \times 0.003 + 0.05 \times 0.02 \times 0.997}$$
$$= 0.7369.$$

This is a significant improvement over the 13% for $P(D|T_1)$ you obtained in part (c). But still, suppose this is a drug test: There are only three chances in four that you really are a drug user given a positive test result on TWO such independent tests! You may object that more than three in 1000 persons (our assumption) are drug users. But really, are you so sure? Would there be so many drug users on the job after management has already instituted penalties and notified everyone of a testing program?

(e) From part (b), $P(D|T) = 0.6049 \neq P(D) = 0.03$ [or in part (c), $\neq 0.003$]. Alternatively, it's just as easy to show $P(T|D) \neq P(T)$. How?

1.4.5 $20/500{,}000 = 0.00004$ (do NOT round this to zero!). Then $P(D|T)P(T) =$

$P(T|D)P(D) = 0.9 \times 0.00004$, but

$$\begin{aligned} P(T) &= P(T|D)P(D) + P(T|D^c)P(D^c) \\ &= 0.000036 + 0.0005(1 - 0.00004). \end{aligned}$$

Thus, $P(D|T) = 0.000036/P(T) = 0.000036/0.00053598 = 0.0672$, by Bayes' Theorem. So less than seven percent of those who fit the profile should actually be denied boarding. The court ruled against Lopez' motion while (in the words of the court) "candidly recognizing the disquieting possibilities suggested by the techniques of the anti-hijacking program."

1.4.6

(a) Bayes' Theorem states

$$P(L|T) = \frac{0.88 \times 0.25}{0.88 \times 0.25 + 0.14 \times 0.75} = 0.6769.$$

About 68% of those identified by the test as lying actually are.

(b) Now about 94.96% of those identified by the test as lying actually are.

(c) The predictive value of a screening test can decrease sharply as the condition being screened becomes less common [see Problem 1.4.4(c)].

1.4.7

Case I: $X - \mu \geq 0$ and Chebyshev's condition says $|X - \mu| = X - \mu \leq k\sigma$. Now add μ to both sides (you can add anything to both sides of an inequality without changing it). So, $X \leq \mu + k\sigma$ which says X is below the right endpoint of the interval $(\mu - k\sigma, \mu + k\sigma)$.

Case II: $X - \mu < 0$ and we get

$$\begin{aligned} |X - \mu| = -(X - \mu) &\leq k\sigma \\ X - \mu &\geq -k\sigma \quad \text{multiplying by } -1 \text{ reverses the inequality.} \end{aligned}$$

Or, in other words, $X \geq \mu - k\sigma$, and we see that X is above the left endpoint of the interval $(\mu - k\sigma, \mu + k\sigma)$. So X takes a value somewhere in that interval. Thus "X is within k standard deviations of μ," as you were to show.

1.4.8

(a) According to Chebyshev's Theorem $P(|X - \mu| \leq 2\sigma) \geq 1 - 1/2^2 = 3/4$.

(b) There's at most a 44.44% (from $1/1.5^2$) chance to be more than one and a half standard deviations away from the mean.

1.4.9 (a) Chebyshev's Theorem states that this probability is at least 75%, but, in fact, it's 100%. If you know anything at all about your random variable, you can usually do better than Chebyshev's Theorem. Here you actually know the entire probability distribution of X. So which values fall within $(\mu - \sigma, \mu + \sigma)$? The endpoints are $3.5 \pm 2\sigma$ with $\sigma = 1.7078$. So which values fall within $(0.0843, 6.9156)$? Answer: ALL values of X are within two standard deviations of the mean. So there's a 100% chance to be within two standard deviations of the mean.

(b) Now you're talking about the range $(0.9383, 6.0617)$ which again encompasses ALL the values of X. Because none of the values fall outside this range, the answer is ZERO. There's a zero chance—it's impossible—to ever roll the die and obtain a value more than one and a half standard deviations from the mean.

1.4.10 (a) The range is (-0.1265, 5.9265) which includes all values other than $X = 6$. So $P(X \neq 6) = 1 - P(X = 6) = 1 - 0.1 = 0.9$.

(b) The range is (1.2354, 4.5646) which includes $X = 2, 3, 4$. So we need $P(X = 1, 5 \text{ or } 6)$.

(c) $(1/1.1)^2 = 0.8264$. This does not conflict with part (b). In part (b), we utilized more information about X and so we were able to give a precise answer to the question. Chebyshev's Theorem just gives an "upper bound" on the probability in a case where you know nothing at all about the random variable. It does not provide an exact answer.

1.4.11 (a) $|X - \mu| < \sigma$ implies $(X - \mu)^2 < \sigma^2$. If that's true for all the values of X, then

$$\sigma^2 = \Sigma(X - \mu)^2 P(X) < \Sigma\sigma^2 P(X) = \sigma^2 \Sigma P(X) = \sigma^2,$$

because $\Sigma P(X) = 1$. This says $\sigma^2 < \sigma^2$! Impossible!

(b) For X, σ and μ are both $1/2$, so $\mu - \sigma = 0$ and $\mu + \sigma = 1$. Are the two values of X "within" the range zero to one? Yes, although not strictly. To express this another way, X satisfies $|X - \mu| \leq \sigma$ because, in fact, *equality* holds: $|0 - \mu| = \sigma$ and $|1 - \mu| = \sigma$. Be sure you see this! Substitute $\mu = 1/2$ and $\sigma = 1/2$ to make sure we're telling you the truth.

(c) When $k = 1$ in Chebyshev's Theorem, you learn nothing because it tells you the probability is less than or equal to one. You already knew that!

1.4.12 Toss an unfair coin. Roll an unfair die; throw a dart at a dart board; and so on. Of course, in each case, you can't say the outcomes aren't equally likely until you've clearly specified what the outcomes are. Maybe there's a way of specifying the outcomes so they ARE equally likely. Think about that!

1.5.1

(**a**) The randomness is in the outcomes not in the values of the variable (see Problem 1.1.10). It's the outcomes that must be unpredictable. For a constant random variable, the value (there's only one) is predictable in advance, without changing the fact that the *outcomes* are not. For the given example, $\mu = 735$ and $\sigma^2 = 0$ (see Problem 1.2.9).

(**b**) Because ΣX does not take into account the probabilities with which the X's occur, it is meaningless. $\Sigma \text{P}(X) = 1$, $\Sigma X\text{P}(X) = \mu = \text{E}(X)$. Finally, for any random variable X, the average of the deviations of X from its mean, $\Sigma(X - \mu)\text{P}(X)$, is ZERO [see Problem 1.2.5(c)]. Of course, if you took the average of the *squared* deviations of X from its mean, you would have σ_X^2, the variance of X.

1.5.2

(**a**) Condition 1, the "doing": toss the thumbtack over the table in such a way that it stays on the table. That's clearly repeatable (condition 2). You can describe the outcomes very simply as just U and D (condition 3). Clearly, these outcomes cannot be predicted in advance (condition 4). And so yes, we do indeed have a random experiment.

(**b**) This X is not a random variable—it takes on *letters* as values. A random variable must have numeric values.

(**c**) Obtain the equation for the gain/loss, $Y = a + bX$, by solving for a and b in the equations $40 = a + b(1)$ and $-20 = a + b(0)$. Then the expected gain/loss is $\mu_Y = 60\mu_X - 20 = -2$. So you lose two dollars per toss, on average. Because $\sigma_Y^2 = 60^2 \times 0.21 = 756$, the *standard deviation* is \$27.50.

(**d**) With $p = 0.3$, the mean is 0.3 and the variance is 0.21. By Problem 1.2.7(d), the mean is p and the variance is $p(1 - p)$.

(**e**) You lose two dollars per play on average. A true gambler might be willing to pay that much per play. The real question is whether it's possible to get way ahead of the game. If so, you might be able to walk away from the game with a killing. See part (f) for a possible analysis.

(**f**) The variance of gain/loss for the new game is $10^2 \times 0.21 = 21$ giving a standard deviation of \$4.58 compared with \$27.50 for the original game. The first game seems significantly more exciting! Personally, I would prefer the second game (it's less risky). In fact, I'd rather see a movie.

1.5.3

(a) Pr is given as a LINEAR function of BP. $a = -81.0637$ (don't omit the negative sign!) and $b = 0.5229$. What kind of object? Answer: The model is AN EQUATION relating two random variables.

(b) Use the fundamental equation $\mu_Y = a + b\mu_X$:

$$26.7 = 0.5229\mu_{BP} - 81.0637 \text{ and so } \mu_{BP} = (26.7 + 81.0637)/0.5229.$$

(c) Use the fundamental equation $\sigma_Y^2 = b^2\sigma_X^2$. To get the standard deviation for barometric pressure, you multiply the standard deviation of BP by a number SMALLER THAN ONE ($b = 0.5229$). So, given the units we're using [see part (d)], Pr is less variable than BP.

(d) Let BPC be the boiling point of water in centigrade. Then

$$\begin{aligned} Pr = a + b(32 + 1.8BPC) &= a + 32b + 1.8bBPC \\ &= a' + b'BPC. \end{aligned}$$

After changing units, $b' = 1.8b$ and

$$\sigma_{Pr} = (1.8)(0.5229)\sigma_{BPC} = 0.9412\sigma_{BPC}. \text{ So, } \sigma_{Pr} \approx \sigma_{BPC}.$$

1.5.4

(a) A probability distribution for a random variable X is a "presentation of the possible values of X together with their probabilities." The two essential ingredients are (1) the POSSIBLE VALUES of X and (2) the PROBABILITIES of those values.

(b) A probability distribution can be given by a table, a graph, or through a set of equations. We've seen tabular and graphical presentations in this chapter. In Chapter 3, we'll see distributions presented by equations. Chapter 4 relies almost exclusively on graphical presentations.

1.5.5

(a)

X	$P(X)$	$XP(X)$	$(X-\mu)P(X)$
7	0.05	0.35	−0.3105
11	0.42	4.62	−0.9282
14	0.35	4.90	0.2765
17	0.11	1.87	0.4169
21	0.07	1.47	0.5453
	1.00	13.21	0

(b)

X	P(X)	XP(X)	$(X-\mu)^2$P(X)
7	0.05	0.35	1.9282
11	0.42	4.62	2.0513
14	0.35	4.90	0.2184
17	0.11	1.87	1.5801
21	0.07	1.47	4.2479
	1.00	13.21	10.0259

$\mu = 13.21,$
$\sigma^2 = 10.0259.$

(c) By part (b), $\sigma = 3.1664$ so the interval $(\mu - \sigma, \mu + \sigma)$ is (10.0436, 16.3764). $X = 11, 14$ are the only values in this interval. So, 77% of the distribution falls within one standard deviation of the mean.

(d) Here's one possibility:

Y	P(Y)	YP(Y)	$(Y-\mu)^2$P(Y)
7	0.01	0.07	0.3684
11	0.42	4.62	1.7997
14	0.45	6.30	0.3892
17	0.11	1.87	1.6989
21	0.01	0.21	0.6288
	1	13.07	4.8851

Now $\sigma^2 < 5$, whereas before it was more than ten. Note what happened—we put more probability on the value 14, close to the mean, making the variance smaller. How much probability to concentrate at 14 was determined by trial and error.

(e) $1.5\sigma = 4.7496$. So the interval is (8.4604, 17.9596):

8.4604 13.21 17.9596

This includes the values 11, 14, 17, representing 88% of the distribution.

(f) The random experiment—the "doing" and its outcomes—is the real-world component of a random variable. The random variable is a BRIDGE from the real world to the world of theory:

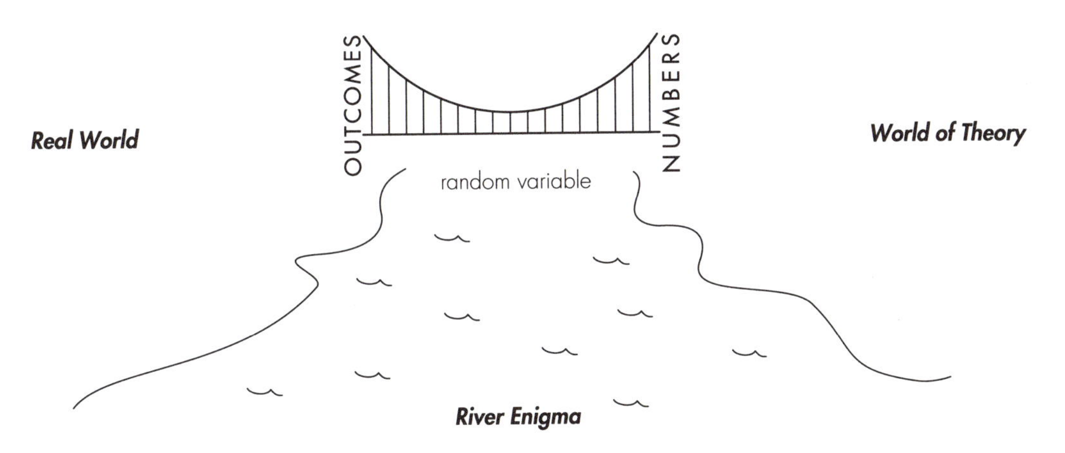

In this problem, X was defined by its probability distribution with no reference to any underlying random experiment. We don't know what that experiment might have been. The real-world component is totally lost from view. So the random variable is "abstract."

1.5.6

(a) The "doing" (1) is to throw the dart once at the dart board. That's clearly repeatable (2). An outcome (3) is the dart having come to rest someplace, hopefully lodged in the dart board. Clearly, the outcomes cannot be predicted in advance (4). So we have a random experiment. "Score" looks at the location of the dart (outcome) after one throw and assigns a number. This is a rule (5) associating a number to each outcome, so we have a random variable.

(b) You don't know the probabilities for the various possible scores.

(c) You would want to know how predictable their game is.

(d) You need the standard deviation for the two players' scores.

(e) You need their probabilities, the probabilities for each of the four possible scores. But they vary from one player to another; consequently, we don't have just one random variable here.

(f) Each player determines a different random variable modeling that player's game. The PLAYER is the random mechanism here. Different players give different mechanisms (different "doings" and so different random experiments). The abstract description in part (a) is exactly the same for each player but because the players are different, the realizations ("making real") of the game are different.

(g)

X	P(X)	XP(X)	$(X-\mu)^2 P(X)$
10	0.65	6.50	2.7050
5	0.23	1.15	2.0152
3	0.11	0.33	2.7062
−2	0.01	−0.02	0.9920
	1.00	7.96	8.4184

$\mu = 7.96$,
$\sigma = 2.9014$.

(h) You measure exactness of the score by σ. "Improvement" means a smaller σ. Here's one possibility:

X	P(X)	XP(X)	$(X-\mu)^2$P(X)
10	0.05	0.50	1.2701
5	0.83	4.15	0.0013
3	0.11	0.33	0.4226
−2	0.01	−0.02	0.4844
	1.00	4.96	2.1784

$\mu = 4.96$,
$\sigma = 1.4759$.

(i) You must have a score of 5 or 10 to be within 1.5σ of μ. There's an 88% chance of this. Your opponent is within 1.5σ of her expected score 94% of the time because her score is within 1.5σ of μ if it falls between 2.7462 and 7.1739. That means a score of 3 or 5.

(j) You're within 2σ of your expected score 99% of the time and your opponent 94% of the time. For your opponent, the two conditions (to be within 1.5σ or 2σ of the expected score) mean the same thing, a score of 3 or 5.

1.5.7 $\Sigma_{X\neq 3}$P(X) means add the P(X)'s for all values of X *except* $X = 3$. It's the probability for X to take any value other than three. Using the first probability rule P$(X \neq 3) = 1 -$ P$(X = 3) = 1 - 0.07 = 0.93$.

1.5.8 (a) For both X and Y, the "doing" is to roll the die. An outcome is the die in some position on the table after one roll. So X and Y are two different random variables for the SAME experiment. The difference is not in the experiment but rather in the "rule" (the rule which assigns numbers to outcomes). One rule looks at the top face, the other at the hidden face.

(b)

Y	$P(Y)$	$YP(Y)$	$(Y-\mu)^2P(Y)$
1	0.1	0.1	0.529
2	0.1	0.2	0.169
3	0.5	1.5	0.045
4	0.1	0.4	0.049
5	0.1	0.5	0.289
6	0.1	0.6	0.729
	1.0	3.3	1.810

$\mu_Y = 3.3,$
$\sigma_y^2 = 1.81.$

(c) $X + Y = 7$.

(d) $H = 7 - Y$, with $a = 7, b = -1$, and so

$$\mu_X = 7 - \mu_Y = 7 - 3.3 = 3.7$$

$$\sigma_X^2 = (-1)^2\sigma_Y^2 = 1.81.$$

1.5.9 (a) and (b) Think of changing the probabilities from those of the fair die. By concentrating half the probability on $X = 3$, you pull the mean down from 3.5. By concentrating it on $X = 4$, you pull the mean above 3.5. These two changes are exactly symmetric and so will yield the same net effect on the variance. Similarly, concentrating half the probability on 1 or 6 will yield the same net effect on the variance. In fact,

IF . . .		THEN . . .		
	$P(3) = 0.5,$		$\mu = 3.3,\ \sigma^2 = 1.81$	smallest
	$P(4) = 0.5,$		$\mu = 3.7,\ \sigma^2 = 1.81$	
	$P(1) = 0.5,$		$\mu = 2.5,\ \sigma^2 = 3.25$	largest
	$P(6) = 0.5,$		$\mu = 5.5,\ \sigma^2 = 3.25$	
	$P(2) = 0.5,$		$\mu = 2.9,\ \sigma^2 = 2.29$	middle
	$P(5) = 0.5,$		$\mu = 4.1,\ \sigma^2 = 2.29$	

(c) The "doing" of the experiment is different. In each case, a different face of the die carries half the probability and so you're talking about TWO DIFFERENT DICE. The random variable has the same verbal description in each case, but that description refers to two different random mechanisms. This is similar to part (f) of Problem 1.5.6.

1.5.10 (a) $\mu \pm \sigma$ gives the interval (22.3959, 24.1641) containing the values 23 and 24 which represent 71% of the distribution.

$\mu \pm 1.5\sigma$ gives (21.9539, 24.6061) containing the values 22, 23, and 24 and representing 84% of the distribution.

$\mu \pm 2\sigma$ gives (21.5118, 25.0482) containing 22, 23, 24, and 25 and representing ALL of the distribution, 100%.

$\mu \pm 2.8\sigma$ gives an even larger interval and so, of course, it contains all of the distribution too. Just to check your calculation, the interval is (20.8046, 25.7554).

(b) Within one standard deviation of the mean you'll find the values 0.8 and 1.1, representing 76% of the distribution. The interval is (0.7252, 1.2468). Within 1.5 standard deviations, you'll find exactly the same values of X, so it's still 76% of the distribution. Within two standard deviations, you'll find all but the first value comprising 99% of the distribution. Finally, 2.8σ below the mean, 0.2558, still excludes the first value and so we still have only 99% of the distribution.

(c) Within σ of μ is 60% of the distribution. All the other ranges contain 100% of the distribution!

1.5.11

(a) Because $Y = 10X$, solving for X, you get $X = 0.1Y$. So X is a linear function of Y. That is, $X = a + bY$, with $a = 0$ and $b = 0.1$, and so

(b)

$$\mu_X = 0.1\mu_Y = 0.986 \qquad \text{since } \mu_Y = 9.86,$$
$$\sigma_X^2 = 0.1^2\sigma_Y^2 = 0.0680 \quad \text{since } \sigma_Y^2 = 6.80.$$

(c)

$$\mu_X = 0.1\mu_Y = 1.934 \qquad \text{since } \mu_Y = 19.34,$$
$$\sigma_X^2 = 0.1^2\sigma_Y^2 = 0.0320 \quad \text{since } \sigma_Y^2 = 3.20.$$

1.5.12

(a) Y is just the last digit of X.

(b) $a = -2,147,810$ and $b = 100$.

(c)

Y	$P(Y)$	$YP(Y)$	$(Y-\mu)^2P(Y)$
4	0.17	0.68	0.5386
5	0.22	1.10	0.1338
6	0.37	2.22	0.0179
7	0.14	0.98	0.2084
8	0.10	0.80	0.4928
	1.00	5.78	1.3915

$\mu_Y = 5.78,$
$\sigma_Y = 1.1797.$

(d) $\mu_X = 21,478.1578$ and $\sigma_X = 0.0118$. You should NOT round to conform with the accuracy of the original numbers. A mean is a theoretical number and does not represent a "possible value." For example, suppose a city has 1.4 children on average per family. It would be very misleading to round this number to 1, even though it's true that no family has four-tenths of a child!

(e)

$$X = (1/100)[Y + 2,147,810]$$
$$= 0.01Y + 21,478.10 \qquad a = 21,478.10, b = 0.01.$$

So

$$\mu_X = 0.01\mu_Y + 21,478.10$$
$$= 21,478.1578.$$

Similarly, the variance of X is $(0.01)^2$ times the variance of Y, giving $\sigma_X^2 = 0.00013916$, from which the standard deviation of X is 0.0118.

Chapter **2**

2.1.1 Here is Tufte's analysis:

> Despite the appearance created by the hyperactive design, the state budget actually did not increase during the last nine years shown. To generate the thoroughly false impression of a substantial and continuous increase in spending, the chart deploys several visual and statistical tricks—all working in the same direction, to exaggerate the growth in the budget. These graphical gimmicks:

These three parallelepipeds have been placed on an optical plane *in front* of the other eight, creating the image that the newer budgets tower over the older ones.

This cluster of type emphasizes and stretches out the low value for 1966–1967, encouraging the impression that recent years have shot up from a small, stable base. Horizontal arrows provide similar emphasis.

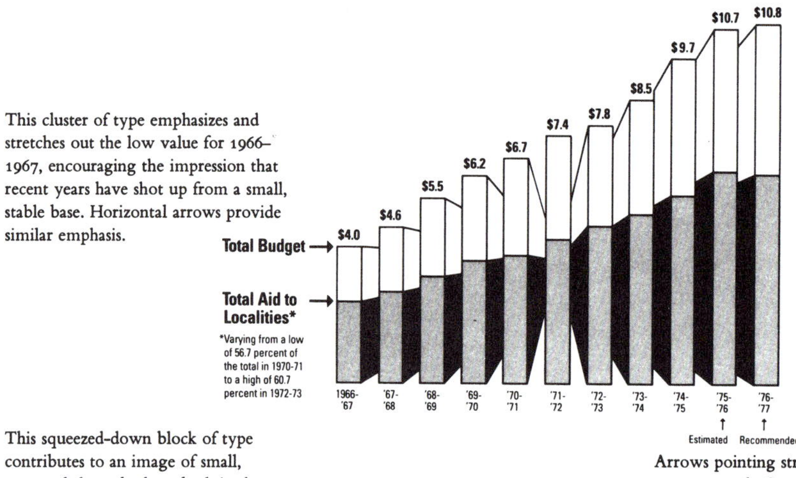

This squeezed-down block of type contributes to an image of small, squeezed-down budgets back in the good old days.

Arrows pointing straight up emphasize recent growth. Compare with horizontal arrows at left.

Leaving behind the distortion in the chartjunk heap at the left yields a calmer view:

Reprinted, with permission, from Edward Tufte, *The Visual Display of Quantitative Information*, (Cheshire, Connecticut: Graphics Press, 1982), AII30.

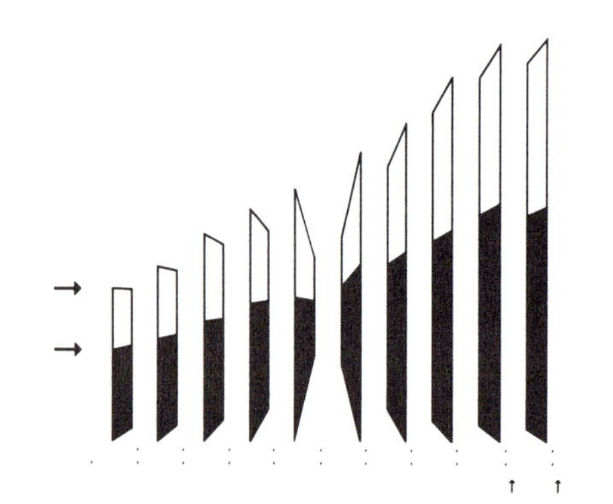

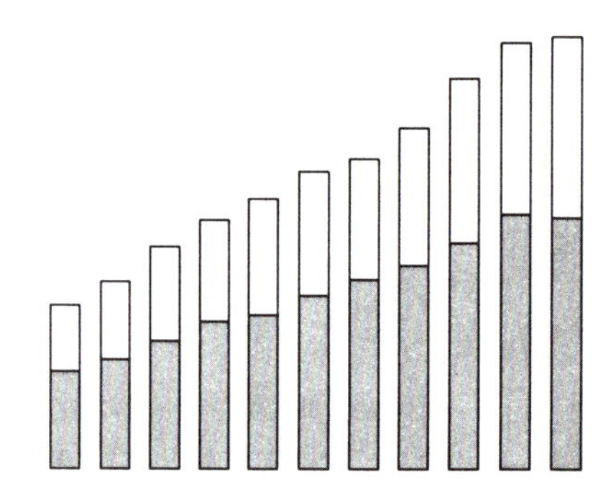

Two statistical lapses also bias the chart. First, during the years shown, the state's population increased by 1.7 million people, or 10 percent. Part of the budget growth simply paralleled population growth. Second, the period was a time of substantial inflation; those goods and services that cost state and local governments \$1.00 to purchase in 1967 cost \$2.03 in 1977. By not deflating, the graphic mixes up changes in the value of money with changes in the budget.

Application of arithmetic makes it possible to take population and inflation into account. Computing expenditures in *constant (real) dollars per capita* reveals a quite different—and far more accurate—picture:

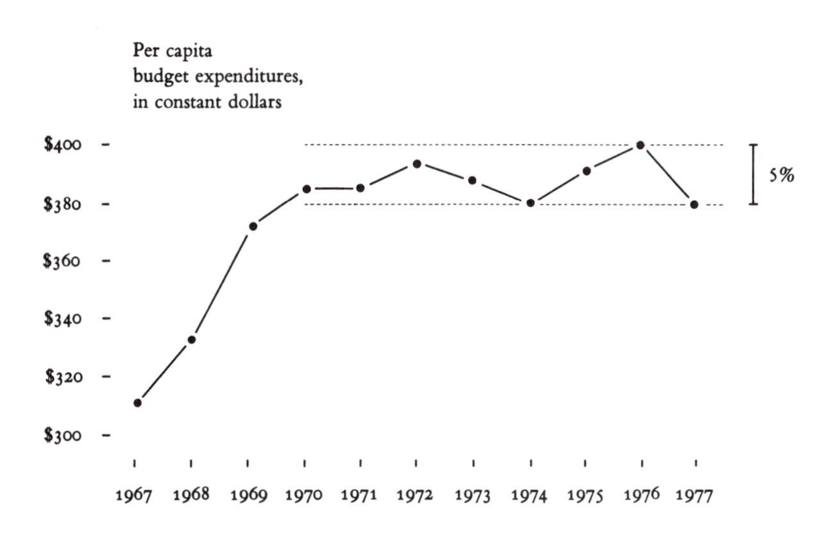

Thus, in terms of real spending per capita, the state budget increased by about 20 percent from 1967 to 1970 and remained relatively constant from 1970 through 1976. And the 1977 budget represents a substantial *decline* in expenditures. That is the real news story of these data, and it was completely missed by the Graph of the Magical Parallelepipeds. Of course no small set of numbers is going to capture the complexities of a large budget—but, at any rate, why tell lies?

The principle:

> In time-series displays of money, deflated and standardized units of monetary measurement are nearly always better than nominal units.

2.1.2

(**a**) Women fare much worse than men with this disease. Less than 23% of the women recover, whereas nearly 69% of the men recover. Taking both treatments into account, only 70 out of 310 women recover, whereas 55 out of 80 men recover.

(**b**) You must control for a variable if the effect you're studying is not constant for that variable. Otherwise, such a variable is confounded with the effect under study. As originally stated, the variable "gender" is confounded with "effectiveness of treatment" as measured by the recovery rate. This happens because the overall recovery rate is not the same—not constant—for different values (male/female) of the variable "gender."

To express this in other terms: The group "persons" is not homogeneous for the variable "recovery rate." The two genders within the group "persons" respond differently. So, to avoid confounding "recovery rate" (the variable under study) with the extraneous variable "gender," you must control for the variable "gender." That means you break the data down according to gender as we've done in the table from level I.

2.1.3

This is another example of Simpson's paradox.

(**a**) Disregarding the control for on-job experience, it looked as if women made more on average than men. But it's clear that men are making more than women for both categories of experience: for "less than five years" versus "more than five years" on-job experience. In the problem as originally stated, the effect we want to study—"salary level for men versus women"—was confounded with the variable "experience on job." In other words, the employees as a group are not homogeneous with regard to "salary level by gender," the picture looks very different depending on on-job experience.

(**b**) The total salary (in \$1000) for the ten women with less than five years on the job is 10×100(\$1,000,000). The salary figure "100" is "weighted" by ten. So the total salary for all the women is $10 \times 100 + 40 \times 175$. To get the average, divide by 50 (see below). This kind of "average" is called the arithmetic mean: Take the total of all the observations and divide by how many you have.

Here's another way to obtain this same *weighted* average: 20% of the women have less than five years of on-job experience and 80% more, so the weights for those two categories are 0.2 and 0.8:

$$100 \times 0.2 + 175 \times 0.8 = 160.$$

Compare these two calculations. It's just a matter of factoring out the denominator. The first two equations below are two versions of the weighted average, the second is just the usual "add and divide" average. Of course, they're the same, both yield the arithmetic mean:

$$\left.\begin{array}{l} 100 \times 0.2 + 175 \times 0.8 = 160 \\ 100 \times 10/50 + 175 \times 40/50 = 160 \end{array}\right\} \text{weighted average,}$$

$$1/50[100 \times 10 + 175 \times 40] = 160 \quad \text{usual average}$$

and for the men, the weighted average is

$$125 \times 0.8 + 200 \times 0.2 = 140.$$

Note how the weights make the nonhomogeneity clear. Eighty percent of the men have less than five years on-job experience, whereas only 20% of the women fall into that category. Part of the effect under study is gender, but "gender" is confounded with "experience on job" because "on-job experience" is not constant for "salary by gender."

2.1.4

(**a**) The results should be interpreted as referring to OPENLY gay men, open to the extent of being willing to acknowledge their homosexuality to a stranger who knows their home address, work address, social security number, and so on.

(**b**) Let M =U.S. male, G = gay, R = willing to respond and gay. Then by the NSM, $R = 0.011M$ and by the assumption of this problem $R = 0.1G$. Putting these together, $0.1G = 0.011M$ and so $G = (0.011/0.1)M = 0.11M$.

Note that the assumption we're making here—that we know the percentage of gay men who are OPENLY gay—is exactly what is not known. It's the missing information that makes impossible the conclusion drawn by the authors of the NSM about the proportion of the general population who are gay. Furthermore, this information will be impossible to obtain. Would you do a survey that asks, "Are you a gay man who doesn't admit to being gay?" Maybe you could get more truthful answers if you say, "Don't worry, I promise not to tell."

(**c**) Young gays are much more likely to be open than their older counterparts, given the increasing acceptance of homosexuality in the United States over the past ten years. So if they refer to OPEN gay men, these percentages from the NSM make perfect sense. They show that young gay men are more likely than their older counterparts to be openly gay. On the other hand, if they are taken to refer to all gay men, we would have to believe that homosexuality is ON THE RISE! Indeed, some people would make that argument in view of what they see as the "general decline in morality." They seem to believe that homosexuality is so enticing an option that everyone would be homosexual if only they had the chance—if only the authority of government and church didn't impose sanctions with heavy and burdensome penalties to keep everybody in line. Personally, I find that hard to believe. I think a lot of people would be HETEROsexual even if they didn't have to be.

(**d**) It's obvious, and admitted by the authors of the NSM, that this very high nonresponse rate is largely due to the sensitive and personal nature of the questions being asked. Experts have become concerned about increasing nonresponse rates for surveys of all kinds. For the ongoing "Current Population Survey," a federal labor-force survey sampling over 60,000 households a month, the nonresponse rate rose from 1.8% in 1968 to 2.4% in 1980.[1] This rise was considered cause for concern. For the Household Health Interview Survey, that rate rose from 1.2% to 2.2% over the same period.[2] So you can see that nonresponse rates of more than five percent are quite problematic. Here we have a nonresponse rate of 30%! In reference to survey sampling in general, Stopher and Meyburg.[3] say, " . . . when 30% of a sample fails to respond—particularly when the failure is a reflection of a lack of interest in or opposition to the survey objectives, or objection to being surveyed—a considerable bias will be introduced."

The authors of the NSM say, "Seventy percent is a respectable response rate for a survey of sexual and health behaviors, given the highly sensitive nature of the questions." They then go on to cite comparable rates of other such studies to conclude that their response rate is just as good as "others in the field." But this doesn't say their study is good, what it really says is that the other studies are just as meaningless as theirs. That's not to say all these studies are completely useless. They may have a limited use in the absence of any better information, provided they're understood IN CONTEXT and provided one does not lose sight of how very weak the conclusions actually are.

But the fact that neither you nor anybody else can do better than a 70% response rate doesn't make a 70% response rate good (or even "respectable"). And certainly it does not make your results valid for 100% of the population. To use the results of such a questionable study to make very official and authoritative sounding pronouncements—unqualified pronouncements—that will have a significant effect on public policy and negatively impact the lives of numerous individuals is irresponsible in the extreme.

(**e**) Only the respondent knows which question he's answering and so (we hope) he answers truthfully. However, the probability for the spinner to choose our question is determined in advance (it shouldn't be 50%, then the technique

[1] See Madow, William G., et al. (eds.), *Incomplete Data in Sample Surveys*, Academic Press, New York, 1983, Vol. 1, page 22.

[2] Ibid.

[3] See Stopher, Peter R., and Meyburg, Arnim H., *Survey Sampling and Multivariate Analysis for Social Scientists and Engineers*, DC Heath, Lexington, MA, p. 112.

"blows up"). With that information—even though we don't know any individual's response—we can deduce the proportion of respondents who are gay (assuming we get truthful answers). Probability theory can be a powerful tool!

The authors of the NSM did not use any such techniques to guard their respondents' anonymity.[4] See the *International Statistical Review*, volume 44, number 2 (1976) for a series of articles discussing extensions and refinements of Warner's technique.

2.1.5

(a) There are two conditions to verify. Each toss yields a zero or a one, so on three tosses we get three such values ordered according to the order of the toss. For example $(0, 1, 1)$, means we tossed a tail followed by two heads. So the three tosses generate "an ordered set of $n = 3$ values of X" and the first condition of the definition is satisfied. For the second condition: These values of X are obtained from "independent repetitions" because one toss does not in any way affect the outcome of the next toss.

(b) Here are the possible samples:

$$(0,0,0), \quad (0,0,1), \quad (0,1,0), \quad (1,0,0),$$
$$(1,1,0), \quad (1,0,1), \quad (0,1,1), \quad (1,1,1).$$

(c) "AND" means "multiply." We'll discuss such counting principles in more detail in the next chapter. This multiplication rule is one of the most basic. We think of the sample as 30 blanks

____ ____ ____ ____ . . . . . . ____ ____

which can be filled with zeros and ones:

$\underline{\ 0\ }\ \ \underline{\ 0\ }\ \ \underline{\ 1\ }\ \ \underline{\ 0\ }\ \ \ldots\ \ \ldots\ \ \underline{\ 1\ }\ \ \underline{\ 1\ }$ (30 blanks)

Each blank can be filled in two ways, so there are $2 \times 2 \times 2 \times \cdots \times 2 = 2^{30}$ ways to fill the 30 blanks. In other words, there are 2^{30} samples. Use your calculator: That's over a billion samples!

You can check this counting procedure with the simpler cases. Using this principle, there should be 2^3 samples of size three. And there are—you saw those eight samples in part (b) of this problem. There should be 2^2 samples of size two:

$$(0,0), \quad (0,1), \quad (1,0), \quad (1,1).$$

[4] I talked by phone with one of the authors, Koray Tanfer. He was reluctant to discuss the issue of anonymity. When asked if randomized response techniques had been used, he said, "One question, one question! It's all in the article." Evidently the answer is no because "the article" makes no mention of any credible effort to assure anonymity.

(d) $\Sigma X = 4$ means you got four heads, heads on every toss. So, $P(\Sigma X = 4) =$ P(head AND head AND head AND head) $= (0.7)^4$. The simple "and" rule for probability works here because the tosses are independent (see Problem 1.4.2).

(e) Using the simple "and" rule, each of the six samples with two heads has probability $(0.3)^2 \times (0.7)^2 = 0.0441$. Then use the simple "or" rule:

$$\begin{aligned} P(\Sigma X = 2) &= P(1100 \text{ OR } 1010 \text{ OR } 1001 \text{ OR } 0110 \text{ OR } 0101 \text{ OR } 0011) \\ &= 0.0441 + 0.0441 + 0.0441 + 0.0441 + 0.0441 + 0.0441 \\ &= 6(0.0441) = 0.2646. \end{aligned}$$

The simple "or" rule works because you can't get two of these samples at the same time. That is, any two samples are mutually exclusive.

(f) This is the same as part (e). The average is $1/2 = (1/4)\Sigma X$. Multiplying by four we get $2 = \Sigma X$.

2.1.6

(a) Ten rolls of this die constitute $n = 10$ independent repetitions of the random experiment for X. If we record for each roll the number of dots on the top face, we'll have an ordered set of $n = 10$ values of X, obtained from those n repetitions. This verifies the definition as required. Note the two conditions that must be verified: (1) You must have an ordered set of n values of X and (2) those values must come from n independent repetitions of the random experiment for X.

(b)

$$\begin{array}{cccccc} (1,1), & (1,2), & (1,3), & (1,4), & (1,5), & (1,6), \\ (2,1), & (2,2), & (2,3), & (2,4), & (2,5), & (2,6), \\ (3,1), & (3,2), & (3,3), & (3,4), & (3,5), & (3,6), \\ (4,1), & (4,2), & (4,3), & (4,4), & (4,5), & (4,6), \\ (5,1), & (5,2), & (5,3), & (5,4), & (5,5), & (5,6), \\ (6,1), & (6,2), & (6,3), & (6,4), & (6,5), & (6,6). \end{array}$$

(c) Think of a potential sample as ten blanks to fill in, each of which has six possibilities. So, it's $6^{10} = 60,466,176$.

(d) $\Sigma X = 4$ happens with three samples: (1,1,2) OR (1,2,1) OR (2,1,1) . Each of the values in these samples has probability 0.1 . So each one of the three samples has probability $0.1 \times 0.1 \times 0.1 = (0.1)^3 = 0.001$, using the "and" rule. Adding that to itself three times—the "or" rule—gives a probability of 0.003 for the event $\Sigma X = 4$.

(e) If the average is four thirds, then $4/3 = 1/3 \Sigma X$ and so, multiplying by three, the total number of dots is four: $4 = \Sigma X$. This is the same as part (d).

2.1.7

(a) The "doing" for the experiment can be described as "produce one machine part." Of course, that's repeatable. Otherwise, you're out of business. So, produce one part: An outcome will be that one particular part. This outcome is unpredictable in the sense that no two of these parts will be exactly alike. There will be variations in weight, size, strength, and so on. So the production process is a random experiment with a machine part as outcome. Then, the random variable "specification error" is a rule which associates the error in diameter (a number) to that machine part (an outcome). It tells how far off the diameter actually is from the specified 2.5 cm.

(b) Producing the next five machine parts is $n = 5$ repetitions of the random experiment (almost condition two, what about independence?), giving five specification errors, "an ordered set of $n = 5$ values of the random variable" (condition one). So, if the repetitions are independent, we have a simple random sample of size $n = 5$. But independence is not obvious. To have independence, you must assume that errors in the diameters are due to purely random factors and not to something wrong in the process. This will be true if the process is "in statistical control." Statistical Process Control is an important part of industial quality control.

(c) Suppose the average diameter for these five machine parts is significantly above 2.5 cm. That's evidence that the manufacturing process has moved "out of control." The specification error has become unacceptably large. Of course, the words "significant" and "unacceptable" must be defined more precisely. We'll return to this later in Chapter 6.

2.1.8

(a) The "doing" for the random experiment is to make the measurement one time. Clearly, that's repeatable. An outcome is "one measurement" (a number). Now, that measurement involves a certain error. That's the value of the random variable. To one measurement (one outcome), the random variable associates the error. So, "measurement error" is indeed a random variable.

If you're uncomfortable because you feel the measurement error will never be known, you're RIGHT. The whole point is to have a *theoretical* model for the error. Once that model is in place, it provides a very powerful analytic tool to help us understand and control measurement error. The abstract model for measurement error and specification error is called the normal distribution. We'll study that model in Chapter 4.

(**b**) Assuming the measurements are truly independent of each other, the next five measurements give us an ordered set of $n = 5$ measurement errors ("an ordered set of n values of the random variable") "obtained from $n = 5$ independent repetitions of the random experiment."

(**c**) In one case—specification error—the experiment produces *many objects*, the outcomes of the experiment. In the other case—measurement error—there's *one object* which is repeatedly measured. The outcomes are the many measurements. In the first case, the outcomes are many different real-world objects. In the second case, the outcomes are many different numbers associated with ONE real-world object.

2.1.9

(**a**) The "doing" for the experiment can be thought of as "produce one silicon wafer and examine it" (clearly repeatable) for which we can think of the outcome as that one silicon wafer labeled either "good" or "defective." Because we can't predict defects in advance, this is indeed a random experiment. Now, the random variable X will count the number (zero or one) of defectives. Note that X is a rule which associates a number to each outcome. If the silicon wafer is good, $X = 0$, and if defective, $X = 1$.

(**b**) The formula for the mean and variance of X is in Problem 1.2.7(d). After all, our X is just like counting the number of heads (= defect) for one toss of a coin which comes up heads three percent of the time. So the mean of X is 0.03 and the variance is $0.0291 (= 0.03 \times 0.97)$. Or

X	$P(X)$	$XP(X)$	$(X - \mu)^2 P(X)$
0	0.97	0	0.0009
1	0.03	0.03	0.0282
	1	0.03	0.0291

(**c**) We must assume "defective wafer" to be independent from wafer to wafer, a reasonable assumption. If it doesn't hold, there must be something systematically wrong in the production process. With this assumption, a "lot" comes from $n = 500$ independent repetitions of the random experiment for X (condition two). Further, each silicon wafer generates a value of X depending on whether it's defective or not. So, the lot generates an ordered set of $n = 500$ values of X (condition one). Because both conditions of the definition are satisfied, a lot does indeed generate a simple random sample from the distribution of X.

(**d**) The *proportion* defective per lot is $(1/n)\Sigma X$ where $n = 500$. To get the *percent* defective we must multiply by 100: $100/500 = 0.2$. The random variable ΣX is important in its own right. For independent repetitions of a random experiment like tossing a coin, ΣX is the binomial random variable which we'll study in detail in Chapter 3.

(e) This is just the sampling experiment described in part (c). The "doing" is to produce and examine, one after the other (in order), 500 silicon wafers. Obviously that's repeatable. The experiment produces a "lot" as outcome. This outcome is unpredictable with regard to quality. To that lot (outcome), the random variable assigns a number, the percent defective. Thus, "percent defective per lot" is a random variable, a rule which associates a number to each of the possible outcomes of a random experiment.

2.1.10

(a) You could describe the "doing" as "examine a child for this disease" with two possible outcomes: The child is either "positive" or "negative" for having the disease. Or you could describe the "doing" as "let the child live its life," again with two possible outcomes: "yes" if the child contracts the disease, or "no" if the child does not contract the disease. Maybe you found another way to describe this experiment. Then finally, let X count "incidence of the disease." In other words, $X = 1$ if the child has the disease and $X = 0$ otherwise.

(b) If incidence of this disease is to be independent from child to child, it better not be a contagious disease. A problem could also arise if the disease is genetically determined, for then you would have dependency within families (in some circumstances this dependency is weak enough to be ignored). Although, in fact, independence can sometimes be assumed for contagious diseases as well if everyone, for example, has been more or less equally exposed to the disease so that "incidence" is not dependent on exposure. This question of modeling epidemics of contagious diseases is quite complex and requires the informed judgment of experts.

Now, suppose there are, say, 150 children in the neighborhood and suppose they've all been examined, their names listed (ordered) and labeled 1 or 0 according to whether they do or do not contract the disease. Then those 150 children constitute an ordered set of $n = 150$ values of X. If the independence assumption holds, those values of X have been obtained from $n = 150$ independent repetitions of the random experiment for X. Because we've verified the two conditions of the definition, we do indeed have a simple random sample from the distribution of X.

(c) The "doing" of the experiment can be thought of as "examine 150 children for this disease and make a list labeling a child 1 if it has the disease and 0 otherwise." This is repeatable by examining another group of 150 children. An outcome would be a list of names labeled with zeros and ones. These outcomes are unpredictable because before examining the children there's no way to predict incidence of the disease. That shows we have a legitimate random experiment. Then, "percent who contract . . . " is a rule which assigns a number to any one of those lists (outcomes), and so it's a random variable. Finally, in our particular neighborhood, the percentage of children who contract the disease is one of the possible values of this random variable.

(d) With the help of the Poisson model, you might show that the number of children who've contracted the disease in that neighborhood is much higher than should be expected. You would then look for some extraordinary cause—for example, environmental toxins—to account for the high incidence of the disease among children within that neighborhood.

(e) The "incidence of the disease" in that neighborhood is the number of children who've contracted the disease. The symbol for "incidence" is ΣX.

2.1.11

(a) The first population is dichotomous (characteristic of interest: "this registered voter will vote for our candidate in the upcoming election"). The second and third populations are numeric; to compute an average, you must have numbers. The fourth population consists of numbers (SAT scores), but it's NOT a numeric population. It's a dichotomous population. You're only interested in distinguishing those scores "above 1000" (the characteristic of interest) from the rest. The fifth population is numeric and the sixth population is dichotomous (characteristic of interest: "will respond to treatment," for a patient exhibiting this clinical symptom of glaucoma). Note that the last population is not "all glaucoma patients." It's only those who exhibit the particular symptom of glaucoma that we're interested in; there will be glaucoma patients who happen not to have that symptom.

(b) For the second population, what proportion of a given day's output is defective? Do you see the pattern? Means require numeric populations, proportions require dichotomous populations. We'll let you finish by giving questions for each of the other four populations.

2.1.12

To say the average life is 1200 hours means the "typical" component will burn about 1200 hours. As a practical matter, most components will burn out a bit sooner or a bit later. You'd be very surprised if a component burned *exactly* 1200 hours (not a second longer or shorter)! Because many components—maybe roughly half—will burn less than 1200 hours, that could hardly be a criterion for "too early." If don't believe me, look at the problem: "for one supplier, 5% of the components burn out too early and for the other 15%." Clearly, more than five or even 15% would burn out before 1200 hours. Of course, someone with expert judgment has to tell us what's meant by "a bit less than 1200 hours" or "too far short of 1200 hours." She may say that "too early" means the component burns less than 1050 hours.

2.1.13

(a) Because you shuffle the deck after each draw, when you get ready to draw a card (a sample element) any of the 52 cards has "an equal chance to be drawn next." That's all the definition requires.

(b) An outcome of the random experiment is "one member of the population." Because it's a numeric population, that's a number. Well, let X be that same number. Now, when we repeat the selection process with replacement, we're just repeating this same experiment. And each repetition is *independent* of the previous one because we sample WITH REPLACEMENT. After we've done this n times, we have "an ordered set." The order is the order of selection. And it's a "set of n values of X," the possible values of X are just the numbers in this numeric population. And those n values were "obtained from n independent repetitions of the random experiment for X."

(c) For sampling with replacement, each member of the population has one chance in N to be drawn. That means a probability of $1/N$. Because you're sampling with replacement, each draw is identical to the previous one. So the draws are independent and a sample of size n has probabilty $(1/N)^n$. For sampling withOUT replacement, after the kth draw the next member of the sample has a probability of $1/(N-k)$ to be drawn. So a sample of size n has a probability given by the product of all these fractions.

(d) For a fair die, each value of X has probability $1/6$. Thus, a sample of size n has probability $(1/6)^n$. Now, think of a die for which the face with two dots comes up half the time, with all other faces equally likely. Then the sample (1, 1) has probability 0.01, but the sample (2, 2) has probability 0.25.

2.1.14

(a) The random mechanism is the shuffling of the deck.

(b) Verbal description: n is the number of cards in a hand (in a sample). You're choosing five cards, so $n = 5$.

(c) If you sample with replacement, you will end up with only one card. Whoever heard of a five-card hand with only one card?

2.1.15

(a)

$$2^3+2^2+2^0 = 8+4+1 = 13; \quad 2^4+2^0 = 17; \quad 2^2+2^1+2^0 = 4+2+1 = 7.$$

(b) One toss generates two possible numbers, zero and one. Two tosses generate four numbers: 0, 1, 2, 3 (in binary, 00, 01, 10, 11). There are two possibilities for each toss, and so with two tosses you have 2×2 possibilities, with three tosses $2 \times 2 \times 2$ possibilities, and so on. With seven tosses, there are $2^7 = 128$ possibilities.

(c) We were to generate random numbers between 0 and 125. That's 126 numbers. With fewer than seven tosses, we would not have enough numbers. Six tosses give only $2^6 = 64$ possibilities. But seven tosses generate 128 random numbers, 0 through 127, two more than we require. If we generate a number out of the range 0–125 (126, for example) that number is aborted and we simply generate another number.

(d) The random variable is the number of heads on ONE toss of this perfectly balanced coin. So this is similar to Problem 2.1.5(a). The only difference is that here $n = 7$, there $n = 3$.

(e) Because the population size is 15, you need to generate at least 15 numbers. Three tosses generates only eight numbers, four tosses generates 16 numbers, zero through 15. So you'll have the possibility of one number which is out of range. Now, index the population so that it's clear which random number selects which population member. Here's one possibility:

$$21_1,\ 22_2,\ 23_3,\ 24_4, \ldots, 35_{15}.$$

This means if our toss generates 0110 (which evaluates to "six") then we select the number 26, because it's indexed by the subscript "six."

Of course, you could have started the indexing with zero, in which case the index "six" selects the number 27. The way we've indexed the population above, if we toss four tails in a row generating the number 0000, we would abort that series of tosses because we have no population member indexed by zero. Operational decisions such as how to index the population have to be made in advance as part of the sampling procedure. Those conventions will go into some kind of "procedures manual" so that everyone involved is able to follow exactly and consistently the same procedure. Our decision to count "heads" instead of "tails" is another such operational decision which would have to go into a procedures manual.

The three tosses suggested in level I give the following binary numbers: 0100, 0000, and 1010. Because the unusable number 0000 was generated, you should abort that series of tosses. That leaves you with only two numbers. To get one more random number, you need another series of four tosses. Suppose it turns out to be TTHT giving the number 0010. Then your sequence of three random numbers is 0100, 1010, and 0010 yielding the sample

$$\{24, 30, 22\} \quad (\text{sample size } n = 3).$$

With the index starting at zero, the sample is $\{25, 31, 23\}$. We're using { }'s instead of ( )'s just because it's conventional to think of samples from a population as unordered.

(f) Nobody in her right mind would ever sample from a population of only 16 members. The only reason for sampling is to say something about a population which in some sense is inaccessible as a whole.

(g) Abstractly, it's not different at all. The real-world "doing" is different, but everything else is the same. Recall that a random number generator is a random experiment which gives numbers as outcomes where all the numbers generated are equally likely to occur. Both of our procedures give the same outcomes (base-two numbers from zero to 15), all equally likely to occur (assuming the coins are fair, of course). So they're the same random number generators. This shows how different real-world situations may be modeled by the same abstract model.

2.1.16

(a) The words of the text are the population, it's a dichotomous population. Every word falls into one of two categories, either it is or it is not a instance of the word "the."

(b) The problem is to identify the population clearly and unambiguously before you begin to discuss sampling.

(c) An obvious case for cluster sampling: Just take a simple random sample of page numbers (note that the pages are automatically "indexed") and count occurrences of the word "the" on each of the pages selected.

Simple random sampling is possible, of course: Introduce an index like (17, 12, 6), meaning "take the sixth word on line 12 of page 17." So the index has the form (page #, line #, word #). Using this index you could obtain a simple random sample of the population of "all words" of the text. But a sample of any reasonable size, say 500 words, would be very tedious to obtain and prone to human error—a lot of it! And there's no theory for "human error," so we have no way to control it. By contrast, there is a theory for random sampling which allows us to control "sampling error." We'll see that later.

2.1.17

(a) Possibly stratifying by educational level would be appropriate.

(b) The term "computer skills" has not been made precise enough to identify the population. If "computer skill" is measured by some test, the test scores might be the population, a numeric population. Or possibly the population is to be regarded as dichotomous—either a worker does or does not possess some minimum criterion for "computer skill."

2.1.18

(a) An outcome of a random sampling experiment is a *random sample* of some fixed sample size n. If you repeat the experiment, you will get a different sample, also of size n. You could possibly get the same sample, but it is so unlikely as to be virtually impossible.

(b) Cluster sampling and stratified random sampling are both more complex than simple random sampling. Simple random sampling is the simplest form of random sampling (hence, its name). Both cluster sampling and stratified random sampling make use of simple random sampling as part of their design. How? Here's how: cluster sampling uses simple random sampling to choose the clusters in the sample. For example, in Problem 2.1.16, you choose a simple random sample of pages and then your cluster sample consists of all the words on those pages.

In stratified random sampling, you choose a simple random sample from each of the strata. If you've stratified the population by gender, you choose a simple random sample of males and another of females. These two simple random samples together constitute your stratified sample. So a stratified random sample consists of a set of simple random samples.

(c) The sample mean is a random variable. It's a rule which associates a number (the sample mean) to each outcome (each sample) of the random sampling experiment. For sampling from a dichotomous population, the sample proportion provides an example of a random variable, the proportion of the sample having the characteristic of interest. It associates a number (the sample proportion) to each outcome (each sample) of the random sampling experiment.

Note that in each case the random variable varies from sample to sample. Neither the population mean nor the population proportion vary at all because there is only one population. That's why the population mean for a numeric population and the population proportion for a dichotomous population are called *parameters*. Each is a fixed number associated with the population.

[Hint: For discussions like this you should always bare in mind a specific simple example of what's being discussed. So, think of some specific examples of numeric and dichotomous populations and compare what you read above with those examples. That's how you learn concepts. That's what we did for you in the part (b). We gave specific, simple examples.]

2.2.1

(a) Here's a possibility: 3, 4, 5, 75.

(b) The mean will be larger than the median if a few data values are very, very large, causing the mean to be large without affecting the median.

(c) You probably thought of test scores. If, as sometimes happens, one or two students already have significant knowledge of the material or for some other reason do much better than the class as a whole, their scores will "ruin the curve"—they pull the average way up. But their scores would have no affect at all on the median. Income figures are another case in point. Usually one or two lucky slobs make enormous salaries while all the rest of us get along on ordinary incomes. Suppose in a small company the boss makes \$220,000 a year while each of her nine underlings makes only \$20,000 a year. The median salary is \$20,000 and is a reasonable measure of "average" salary for that company. But the mean salary is \$40,000 per year, twice as much as anybody makes other than the boss.

(d) One of N and n is "4," but they cannot both be applicable. If the data is population data, n is not applicable and $N = 4$. If it's sample data, $n = 4$ and N is not applicable.

Then $\Sigma f = 4$, $\Sigma Xf = 87$, $\Sigma \text{rf} = 1, \Sigma X(\text{rf}) = 21.75$. Here are the verbal descriptions: Σf is the "number of observations," ΣXf is the "total of all observations." The symbol Σrf has a simple verbal description, it's "the number one." The sum of all the relative frequencies must necessarily add to 1. That sum serves as a double check on your accuracy, if it's not 1, then you've made a mistake. Finally, $\Sigma X(\text{rf})$ is the "arithmetic mean of all the observations."

2.2.2

(a) There's no "middle" value because there are an even number of observations. So the median is the average of the middle two, the average of the 6th and 7th values: $(5 + 5)/2 = 5$. The data is bimodal with modes 2 and 5.

(b) and (c)

X	f	Xf	
2	3	6	
3	1	3	
4	0	0	
5	3	15	
6	1	6	
7	2	14	
8	2	16	
	12	60	$\mu = 5$

X	rf	X(rf)	
2	0.2500	0.5000	
3	0.0833	0.2499	
4	0	0	
5	0.2500	1.2500	
6	0.0833	0.4998	
7	0.1667	1.1669	
8	0.1667	1.3336	
	1	5.0002	$\mu = 5.0002$

Note the round-off error. The second and fifth values in the X(rf) column should be 0.2500 and 0.5000, respectively. They will be if you keep all the accuracy of your calculator instead of rounding and then calculating again with your rounded number. In fact, that's what you should do.

(d) The only difference is in the notation. Instead of μ use $\overline{X}$.

(e) N is not applicable here because we have sample data. Then, n is the "sample size," the "number of observations." If you've done everything correctly, $\Sigma f = n$, because it too is the "number of observations." ΣXf is the "total of all the data," Σrf must necessarily be "one" and finally, ΣX(rf) is the "arithmetic mean of all the data."

2.2.3

(**a**) A statistic is computed from a random sample. That sample comes from some random sampling experiment. The *sampling experiment* is the random experiment requested.

(**b**) An outcome of a sampling experiment is one sample.

(**c**) To each sample—to each outcome—is associated the calculated value of the statistic.

2.2.4

There's no difference between this calculation and the calculation for a probability distribution. The difference is only in our understanding of what the numbers refer to—probabilities are THEORETICAL relative frequencies, here we have OBSERVED relative frequencies:

X	rf	X(rf)	$(X-\mu)^2$(rf)
2	0.0625	0.1250	1.1963
3	0.1250	0.3750	1.4238
4	0.0625	0.2500	0.3525
5	0.1250	0.6250	0.2363
7	0.2500	1.7500	0.0977
8	0.1875	1.5000	0.4951
9	0.1250	1.1250	0.8613
10	0.0625	0.6250	0.8213
	1	6.3750	5.4843

$\mu = 6.3750$,
$\sigma^2 = 5.4843$,
$\sigma = 2.3419$.

Note that the fourth column sum is already the variance—there's no need to divide that column sum by anything because division by 16 has already been done.

2.2.5

(**a**) For sample data, you calculate the value of a statistic. It varies from sample to sample. By contrast, a parameter is a fixed number. A statistic is a random variable (see Problem 2.2.3) which corresponds to a population parameter and serves to estimate that parameter.

(b) The range is supposed to be a single number, a parameter or a statistic (see part(a)). Here you are giving an interval of numbers. Note that the technical term "range" is more exact than its dictionary meaning.

(c) $\hat{\sigma}^2 = (1/n)\Sigma(X - \overline{X})^2 f$.

2.2.6

The two modes are four and seven. The median is the sixth value which by coincidence takes the value 6. We'll assume in our notation that we have population data

X	f	Xf	$(X-\mu)^2 f$	
3	1	3	6.4796	
4	3	12	7.1657	
5	1	5	0.2976	
6	2	12	0.4131	
7	3	21	6.3467	
8	1	8	6.0246	
	11	61	26.7273	$\mu = 5.5455$, $\sigma^2 = 2.4298$.

2.2.7

Distribute $1/N$ over the sum:

$$\sigma^2 = \frac{1}{N}\left[\Sigma(X-\mu)^2 f\right] = \Sigma(X-\mu)^2 f/N.$$

This is the formula to use for a *relative* frequency distribution of observed data. Now, a probability, $P(X)$, is just a theoretical relative frequency (as opposed to an "observed" relative frequency). Interpreting f/N as a theoretical relative frequency, we get

$$\sigma^2 = \Sigma(X-\mu)^2 f/N = \Sigma(X-\mu)^2 P(X),$$

the formula for the variance of a random variable.

2.2.8

(a) For both of the following data sets, there are only two distinct data values, 1 and 10. And in both cases, $N = 6$ and the range is nine. So they have the same size, same values, and same range. But they have very different means:

X	f	Xf	
1	5	5	
10	1	10	
	6	15	$\mu = 2.5$

X	f	Xf	
1	1	1	
10	5	50	
	6	51	$\mu = 8.5$

So the range has nothing to do with the mean. Note that you MUST use the symbol μ here because the problem used the symbol N, which implies this is population data.

(**b**) For the two data sets we've given above—because they're symmetric to each other—the line graphs are mirror images of one another. From that, you guess that the two variances are the same. The variance ignores *direction* of spread and addresses only the *distance* (squared) of data values from the mean. Your two data sets may not have been symmetric like this, and so it might not have been so easy to guess the relative sizes of the variances.

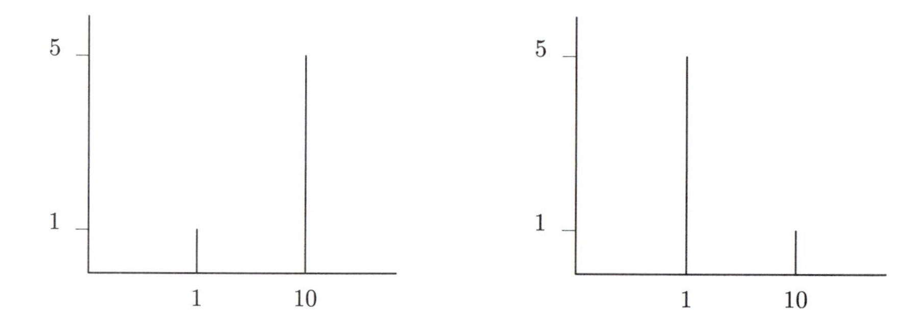

What's the difference between the line graph for observed data as we've given it here and the line graph for a random variable? The difference is only in what's recorded on the vertical axis. Here we've recorded the observed frequency of occurrence. For a random variable, the vertical axis records the probabilities, the theoretical relative frequency with which the values OUGHT (in theory) to be observed. Note the contrast between "observed" and "theoretical."

Now let's calculate the variances for our data sets above. Note, just as we guessed, they're the same:

X	f	Xf	$(X-\mu)^2 f$
1	5	5	11.25
10	1	10	56.25
	6	15	67.50

$\mu = 2.5$,
$\sigma^2 = 11.25$.

X	f	Xf	$(X-\mu)^2 f$
1	1	1	56.25
10	5	50	11.25
	6	51	67.50

$\mu = 8.5$,
$\sigma^2 = 11.25$.

2.2.9

X	f	Xf	$(X-\mu)^2 f$
1.8	3	5.4	1.6521
2.4	5	12.0	0.1010
2.6	6	15.6	0.0201
2.8	2	5.6	0.1330
3.1	1	3.1	0.3113
3.3	2	6.6	1.1488
	19	48.3	3.3663

$\mu = 2.5421$,
$\sigma^2 = 0.1772$,
$\sigma = 0.4209$,
mode $= 2.6$,
median $= 2.6$,
range $= 1.5$.

2.2.10

(a) . . . spread about the mean. That's what the standard deviation measures. Let the frequencies be 11, 1, 8, 2. Note what we've done. We have eight fewer observations of the value $X = 4$, which is located right at the mean ($\overline{X} = 4.0455$), and we spread those eight observations AWAY from the mean. That makes the standard deviation larger. Furthermore, we did it SYMMETRICALLY about the mean so as not to change the mean. Verify that this works.

(b) . . . have one very large value. By putting a lot of data at $X = 22$, the median becomes 22. Then put in one very large value to pull the mean up. Verify that the following choices work:

X	20	21	22	23	1000
f	1	6	20	3	2

2.3.1

(a) These are classes of test scores. For example, the second class is the class of all those scores which fall between 20 and 39 inclusive. There are a lot of practical considerations in grouping the data into classes. These problems are not resolved by abstract rules, they're resolved by common sense. Usually several attempts are necessary before you find a good solution. If there are too many classes, for example, you defeat the purpose of grouping the data. A grouped frequency distribution with 40 or 50 classes can't be readily understood. It's just as clumsy as the ungrouped distribution.

On the other hand, if there are too few classes, you lose too much information. The purpose of grouping the data is to wipe out irrelevant and confusing detail, what's usually called "noise." But an appropriate balance is required. To take an extreme example, suppose you have only ONE class as in the table below. This is too few classes. Now, everything about the test scores is lost except the fact that there were 1000 students:

Class	f	X	Xf	$(X-\mu)^2 f$
0–100	1000	50	50000	0

(b) In the second class, there were 202 test scores, 202 students got scores between 20 and 39 inclusive. So a class "occurs" by having a data value fall within that class. Or we could say the class is "observed" 202 times.

(c) The symbol X refers to the class mark, the midpoint of the class. It's the average of the upper and lower class limits. For the second class, the class mark is $(20 + 39)/2 = 29.5$. The class mark does not represent one of the actual, observed data values. No student made a test score of 29.5. The class mark is a substitute—an approximation—for the actual data values. An important criterion for choosing the grouping of the data is that the class mark should be a reasonable approximation to the actual data values within that class. If, for example, all the grades in one class are above the class mark, you should try another grouping to avoid this problem. Note that the class mark probably won't be a reasonable approximation to the data values in a class containing only two or three values. That gives a clue to the inappropriateness of that grouping. In other words, the classes should not be "spare," containing too few data observations. The very fact that we're given this grouped frequency distribution implies that the average of the 202 grades in the second class is about 29.5. If we use the value 29.5 in place of each of those 202 scores, any calculations we make will be good approximations on average.

(d) The product Xf is NOT the number of points earned by the students who got a score of X (as it would be for ungrouped data). For the second class, X is 29.5, but no student made a score of 29.5. There are 202 scores in that class. If each of those scores is approximated by the class mark, then

> Xf is an *approximation* to the number of points earned by the 202 students in that class.

In other words, these 202 students earned about 5959 points altogether on the test. This assumes, of course, that the grouping is reasonable for that class [see part (c)].

(e) Because f is the number of scores within a class, Σf is just the total number of data points. Here $\Sigma f = 1000$, the number of students taking the test. Note that this is NOT an approximation.

(f) ΣXf is an approximation to the total number of points earned on the test by all one thousand students.

(g) Because Σf is not an approximation, N is the *exact* number of data observations. Here, $N = 1000$. And from part (f), we see that $(1/N)\Sigma Xf$ is an *approximation* to the mean of all the test scores. Here it's 46.8864. In the table, to compute the last column, we've used this value for μ, even though it's only an approximation. So the heading of the last column has two approximations in it: X to approximate the various test scores and μ to approximate the true mean.

(h) You cannot *calculate* the variance, median, or mode because you do not know the actual data values. However, the variance can be *approximated*. With X approximating the various test scores, clearly we can estimate the variance by the sum of the last column divided by the sum of the second column, giving the approximation $\sigma^2 \approx 508.5129$.

Because you don't see the actual test scores, there's no way to even approximate the mode. It's usual to identify the "modal class," the class with the greatest number of observations. Here it's the third class. But that doesn't tell you the true mode. The true mode could even be in a class other than the modal class. Furthermore, the actual data could be bimodal or trimodal even though there's only one modal class. In other words, for grouped data we don't attempt to approximate the actual mode; we cite the modal class instead.

The median can be approximated by identifying the class containing the median and then assuming all the values of that class evenly spread across that class. But the computation is somewhat awkward—we omit it.

2.3.2

If all classes are the same width, either you'll have many classes with one or no data or you will have too few classes, losing too much information. For example, if every class is 15 units wide, then the last class will become ten classes, ten classes for only five salaries. If you make the classes, say, 100 units wide, then virtually all the data falls into one class.

2.3.3

(a) Twelve of the temperature readings fell between 15° C and 30° C. Exactly what those readings were, we cannot know. We have followed the usual convention for interpreting the endpoints; the class includes the left endpoint and excludes the right endpoint.

(b) Temperature readings are usually considered to be continuous. That means a reading of 29.5° C should be possible, or 29.87, and so on. To allow for such fractional readings, we follow the convention described in part (a). If your readings did not include fractional degrees, then the data would have been given without overlapping endpoints.

(c)

f	the number of temperature readings in the given class,
Σf	the total number of temperature readings altogether,
ΣXf	an ESTIMATE of the sum of all the temperature readings,
$(1/\Sigma f)[\Sigma Xf]$	an ESTIMATE of the mean of all the temperature readings.

Note that any symbol involving X is only an estimate because X itself is just an estimate for the actual readings which are unknown.

(d)

temperature (centigrade)	f	X	Xf	$(X - X)^2 f$
0–15	6	7.5	45	1990.5612
15–30	12	22.5	270	123.9796
30–45	8	37.5	300	1111.2245
45–60	2	52.5	105	1434.9489
	28		720	4660.7143

$\overline{X} \approx 25.7143$,
$\hat{\sigma} \approx 12.9017$.

Be sure you do NOT put in irrelevant column sums. In particular, ΣX is not meaningful. Also note that we've used the notation for sample data. We're not told if the data was sample or population data, so we must choose one interpretation. But be consistent, don't mix the notation for populations and samples in the same presentation. The mode (or modes) cannot be estimated from the grouped distribution. But we can identify the "modal class;" it's the second class. We did not ask you to estimate the median because we have not explained the details of that approximation, but it's approximated by 24.375 (in case you want to try on your own!).

2.3.4 It should divide evenly, of course. So the new histogram looks exactly like the old one, you just add a line to divide the first rectangle equally into two (so the total area remains the same):

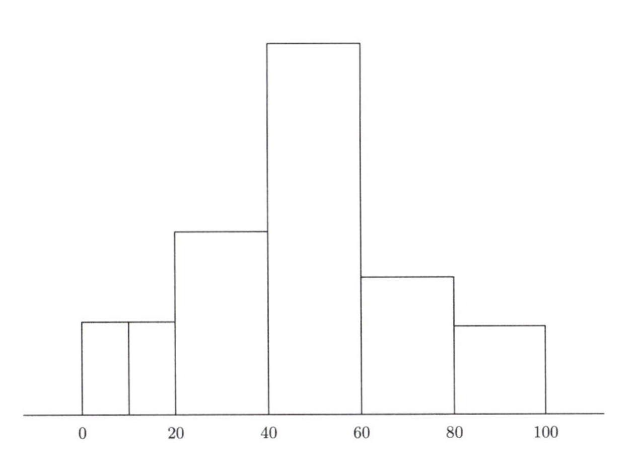

2.3.5 Inspectors were passing rods which were just under the lower limit of 1 cm thinking such measurements might be wrong and in any case would not be significant. Thus, in the histogram, we find no observed diameters of 0.999 cm and too many of 1.0000 cm. The inspectors thought to accommodate an "insignificant" error. They were, as Deming says, "unaware of the trouble that an undersized diameter would cause later on." When this fault in the inspection procedure was corrected, it was discovered that an unacceptably large proportion of the rods were undersized because of a faulty machine setting. That wrong machine setting was easily corrected once recognized. Note how the histogram led to the discovery of an important problem which otherwise would have gone undetected.

2.3.6

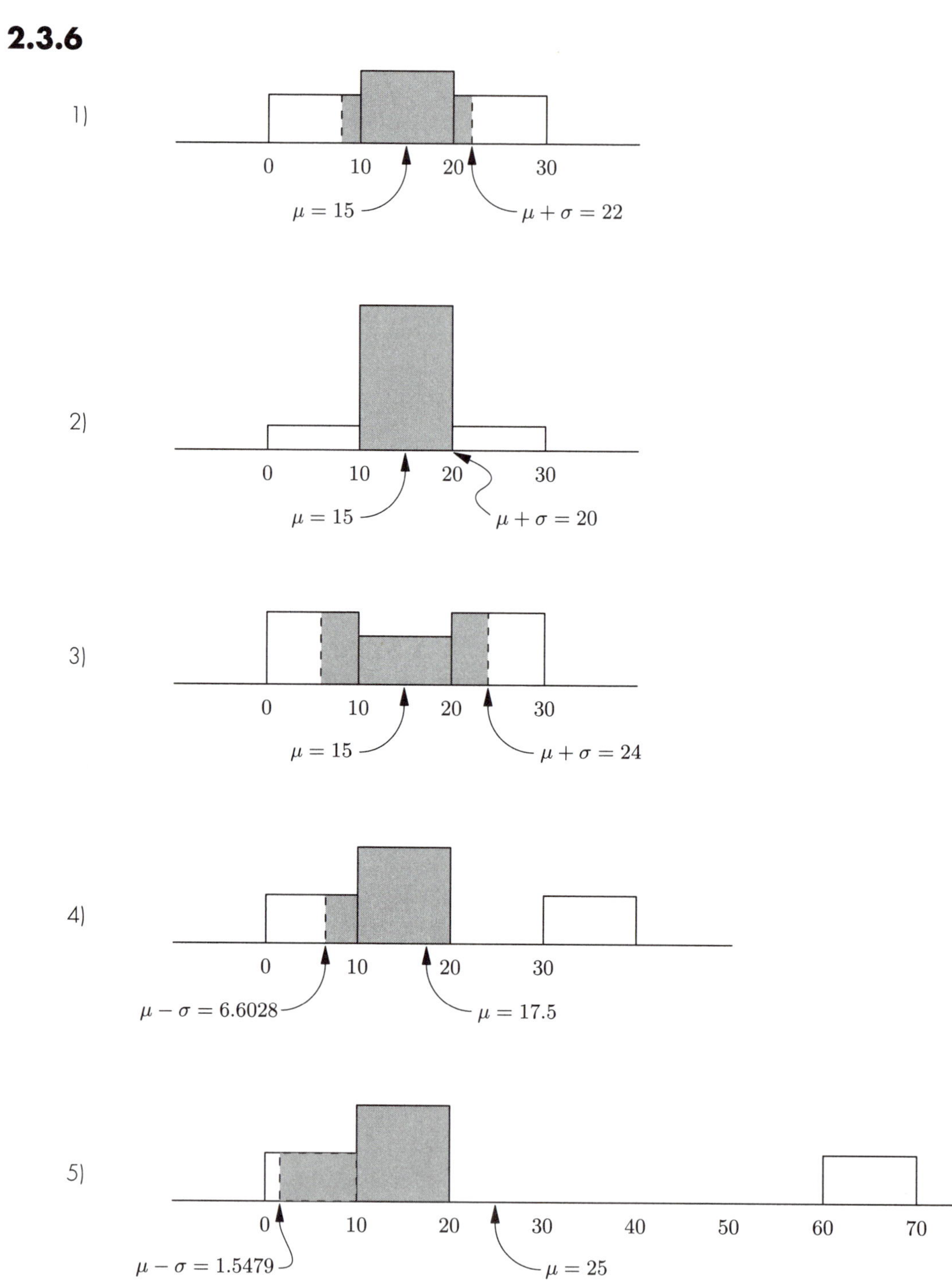
1)
0
10
20
30
$\mu = 15$
$\mu + \sigma = 22$
2)
0
10
20
30
$\mu = 15$
$\mu + \sigma = 20$
3)
0
10
20
30
$\mu = 15$
$\mu + \sigma = 24$
4)
0
10
20
30
$\mu - \sigma = 6.6028$
$\mu = 17.5$
5)
0
10
20
30
40
50
60
70
$\mu - \sigma = 1.5479$
$\mu = 25$

Chapter 3

3.1.1

If the square of a number is zero, the number itself is zero. So for each X, $X - \mu = 0$, which means $X = \mu$. But then all the values of X are the same. This is the class of *constant* random variables, random variables for which every outcome of the random experiment is assigned the same number.

Don't forget, it's the *outcomes* of the underlying experiment which must be "not predictable." Just because the values can be predicted in advance—as they can in this case—doesn't mean it isn't a random variable. The "randomness" of a random variable is in the outcomes of the experiment, not in the values of the variable.

3.2.1

(a)

$$\begin{aligned}\sigma^2 = \Sigma(X-\mu)^2\mathrm{P}(X) &= \Sigma\left[X^2 - 2\mu X + \mu^2\right]\mathrm{P}(X)\\ &= \Sigma X^2\mathrm{P}(X) - 2\mu\Sigma X\mathrm{P}(X) + \mu^2\Sigma\mathrm{P}(X)\\ &= \Sigma X^2\mathrm{P}(X) - 2\mu(\mu) + \mu^2(1)\\ &= \Sigma X^2\mathrm{P}(X) - 2\mu^2 + \mu^2\\ &= \Sigma X^2\mathrm{P}(X) - \mu^2.\end{aligned}$$

(b)

X	$\mathrm{P}(X)$	$X\mathrm{P}(X)$	$X^2\mathrm{P}(X)$
1	0.25	0.25	0.25
2	0.25	0.50	1.00
3	0.25	0.75	2.25
4	0.25	1.00	4.00
	1.00	2.50	7.50

$\mu = 2.5$,
$\sigma^2 = 1.25$.

Here $\sigma^2 = 7.5 - 2.5^2 = 1.25$.

(c) The computing formula often involves less serious round-off error. The conceptual formula contains μ, often rounded. But calculation with a rounded number compounds the round-off error.

The conceptual formula explains what the variance measures. Because the formula explains the concept, it's called "conceptual." It explains that we're measuring "spread about the mean" (Problem 1.2.10). The computing formula, by contrast, doesn't even mention the mean. You would never know from looking at that mysterious formula what it measures. It looks like some blind calculation. It does have its conceptual uses, however. You'll see an example of this in part (e) of this problem and other instances later and in more advanced courses.

(d)

$$\begin{aligned}\sigma^2 &= (1/N)\Sigma(X-\mu)^2 f\\ &= (1/N)\Sigma X^2 f - 2\mu\Sigma X f + \mu^2\Sigma f\\ &= (1/N)\Sigma X^2 f - 2N\mu^2 + N\mu^2, \qquad \text{since } \Sigma X f = N\mu, \Sigma f = N\\ &= (1/N)\Sigma X^2 f - N\mu^2.\end{aligned}$$

Now multiply the numerator and denominator by N and you get the required formula because $N^2\mu^2 = (N\mu)^2 = (\Sigma X f)^2$. For sample data, just change the notation.

You can gain practice with these computing formulas by redoing the problems of Chapters 1 and 2 where before you used the conceptual formula. Your answer will sometimes differ slightly due to round-off error and, if different, is more accurate.

(e) $E(X^2) = \mu^2 + \sigma^2$. This is just the computing formula all over again.

3.2.2

(a) Let X be the number of dots on the top face of a die. It will be uniformly distributed if the die is FAIR. Otherwise, of course, you'll not have all values equally likely. The parameter is $n = 6$.

(b) Toss a fair coin. Let X be the number of heads (zero or one). Because the coin is fair, the two values are equally likely. Parameter: $n = 2$.

(c) When you roll a pair of dice, the most obvious random variable is the total number of dots on the two top faces. However, even if the dice are both fair, the values are not equally likely. Suppose we have two fair dice, one red and one blue. Then for example, $P(X = 2) \neq P(X = 3)$. Here's how you show that:

$$\begin{aligned}1/36 = P(X=2) &= P(\text{one dot on each die})\\ &= P(\text{ one on red }) \times P(\text{ one on blue })\\ &= (1/6) \times (1/6)\end{aligned}$$

but,

$$
\begin{aligned}
1/18 = \mathrm{P}(X = 3) = \mathrm{P}(&\text{ one dot on the red die and two dots}\\
&\text{on the blue die } \textit{or} \text{ vice versa })\\
= \mathrm{P}(&\text{ one on red } \textit{and} \text{ two on blue })\\
&+ \mathrm{P}(\text{ two on red } \textit{and} \text{ one on blue })\\
= (1/6)& \times (1/6) + (1/6) \times (1/6)
\end{aligned}
$$

So the probabilities for $X = 2$ and $X = 3$ are NOT the same. Because the values are not all equally likely, X is not uniformly distributed.

(d) There can be from 2–12 dots on the top faces of two dice. Because the situation is completely symmetric, we can guess the mean to be seven. Before making up the probability distribution, recall the possible outcomes of one roll. If you follow the pattern of Chapter 1, you might describe the outcomes as "the pair of dice lying on the table top in some position." In the interest of simplicity, however, now that you understand the distinction between "outcomes" and "values," we'll take a simpler less physical view of the outcomes. In fact, the outcomes can be identified by a pair of integers from one to six:

$$
\begin{array}{cccccc}
(1,1) & (1,2) & (1,3) & (1,4) & (1,5) & (1,6), \\
(2,1) & (2,2) & (2,3) & (2,4) & (2,5) & (2,6), \\
(3,1) & (3,2) & (3,3) & (3,4) & (3,5) & (3,6), \\
(4,1) & (4,2) & (4,3) & (4,4) & (4,5) & (4,6), \\
(5,1) & (5,2) & (5,3) & (5,4) & (5,5) & (5,6), \\
(6,1) & (6,2) & (6,3) & (6,4) & (6,5) & (6,6).
\end{array}
$$

Because the dice are fair, we have 36 equally likely outcomes each with probability 1/36. It seems obvious on intuitive grounds that these outcomes are equally likely, but if you had to verify that by a computation, could you do it? Try. You'll find the answer in part (e).

Here's the probability distribution:

X	$P(X)$	$XP(X)$	$X^2P(X)$
2	1/36	2/36	4/36
3	2/36	6/36	18/36
4	3/36	12/36	48/36
5	4/36	20/36	100/36
6	5/36	30/36	180/36
7	6/36	42/36	294/36
8	5/36	40/36	320/36
9	4/36	36/36	324/36
10	3/36	30/36	300/36
11	2/36	22/36	242/36
12	1/36	12/36	144/36
	1	252/36	1974/36

$\mu = 7$
$\sigma = 2.4152$

(e) Take the outcome (3, 5). This is described verbally as "three on the red die *and* five on the blue." Because the rolls are independent and the dice are fair, the probability is

$$\begin{aligned} \text{P(outcome (3,5))} &= \text{P(three on red)} \times \text{P(five on blue)} \\ &= (1/6) \times (1/6). \end{aligned}$$

Because the dice are fair, the probability on each die separately is 1/6. We get to use the rule $P(A \text{ and } B) = P(A)P(B)$ because in our case A and B are independent.

(f) To be uniformly distributed, the *values* of the random variable must be equally likely. Here it's the outcomes that are equally likely, but some outcomes yield the same value for the random variable. Some values are more likely than others because those values arise from MORE different outcomes. The value two arises from only one outcome, but the value seven arises from six different outcomes. Because the *values* are not equally likely, this is not a uniformly distributed random variable.

3.2.3 You cannot guess the standard deviation without some calculations. After all, the standard deviation is a *comparative* measure of spread, but here there's nothing to compare with.

X	$\mathrm{P}(X)$	$X\mathrm{P}(X)$	$X^2\mathrm{P}(X)$
18.2	0.2	3.64	66.248
18.7	0.2	3.74	69.938
19.3	0.2	3.86	74.498
19.7	0.2	3.94	77.618
20.1	0.2	4.02	80.802
	1.0	19.20	369.104

$\mu = 19.2$,
$\sigma = 0.6818$.

3.2.4

(a) Take the example in Problem 3.2.2 [see level II part (c)]. When you roll a pair of fair dice, there are 36 equally likely outcomes (because the dice are fair), but the obvious random variable—the number of dots on the two uppermost faces—is NOT uniformly distributed.

(b) The key to such an example is that several different outcomes may correspond to the same value. Here, only one outcome gives the value two, but the value three corresponds to two different outcomes. The value four corresponds to three outcomes (what are they?). And so on.

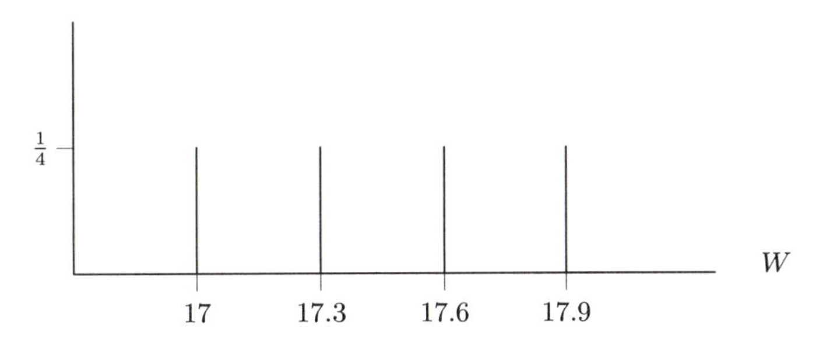

3.2.6

You require the values, nothing more. If there're 22 of them, the probabilities are all 1/22. If there are n values, then the probabilities are all $1/n$. So, if you know the values, you also know all the probabilities. That's the whole story! Everything else is calculated from this information.

3.2.7

$$\begin{aligned}\mu_X &= \Sigma X\mathrm{P}(X)\\ &= \Sigma(X \times (1/n))\\ &= (1/n)\Sigma X \qquad \text{since } n \text{ is constant}\\ &= \text{the average of the values of } X.\end{aligned}$$

3.3.1

(a) There are $n!$ ways to arrange n objects. Here's how to see that: There are n ways to choose an object for the first position. For the second position, there are $n-1$ objects left to choose from and so $n-1$ ways to fill that position. And so on. From the fundamental principle of counting, we get the number of ways to complete the arrangement:

$$n(n-1)(n-2)\cdots(3)(2)(1) = n!.$$

Note why it ends with 1. After filling the first $n-1$ positions, you have used up $n-1$ objects. So there's exactly one way to fill the last position.

(b) There are $n(n-1)(n-2)\cdots(n-x+1)$ ways to choose x objects and arrange them. Note that you stop after choosing the xth object. When you get ready to choose that object, there are $n-(x-1) = n-x+1$ objects left to choose from. Because we have two expressions for the number of ways to choose x objects from n objects and arrange them, those two expressions must be equal:

$$n(n-1)(n-2)\cdots(n-x+1) = x!C(n,x).$$

So

$$C(n,x) = \frac{n(n-1)(n-2)\cdots(n-x+1)}{x!}.$$

To obtain the required formula, use this more compact way to write the numerator:

$$n(n-1)(n-2)\cdots(n-x+1) = n!/(n-x)!.$$

(c) $C(7,1) = 7,\ \ C(7,7) = 1$ and

$$C(120,118) = \frac{120 \times 119 \times 118!}{118! \times 2!} = 60 \times 119 = 7140.$$

(d) Here are the answers. How do you get them?

$$C(n,0) = 1, \quad C(n,1) = n, \quad C(n,n) = 1, \quad C(n,n-1) = n.$$

(e) 7!

(f) $\frac{58!}{3!\ 55!} = \frac{58\times57\times56}{3\times2} = 29 \times 19 \times 56.$

(g) $6! = 720.$

(h) $11! = 39,916,800$. Factorials get large very fast. For most calculators the factorial key will give an error message at 70! and beyond.

(**i**) There are $C(14, 6)$ ways to pick 6 from 14 seats. Then there are 6! ways to seat 6 students in those 6 seats. The answer:

$$\begin{aligned}\frac{14!}{6!\,8!} \times 6! &= \frac{14 \times 13 \times 12 \times 11 \times 10 \times 9 \times 8!}{6!\,8!} \times 6! \quad \text{cancel 6! and 8!}\\ &= 14 \times 13 \times 12 \times 11 \times 10 \times 9\\ &= 2,162,160.\end{aligned}$$

Our two-step procedure is one way to think about this problem. Or you could just think in terms of the 6 seats with 14 ways to fill the first, 13 ways to fill the second, and so on. Both are correct and both are instructive. Make sure you understand BOTH analyses!

(**j**) $C(93, 3) = 32 \times 95 \times 47$ after canceling 6! and 93! giving 142,880 ways.

3.3.2

(**a**) The hypergeometric random variable models sampling without replacement from a *dichotomous* population. For the model to be valid, there must be some one characteristic of interest. In the example, "lifetime" is the issue but that doesn't divide the population of 50 components into two categories.

Suppose we're interested in a quality control criterion which specifies that each component should have a life of at least a 1000 hours. Then we're no longer interested in ALL the various possible lifetimes. We're only interested in "less than 1000 hours" versus "at least 1000 hours." Either of these might be considered the characteristic of interest, depending on the exact question. If we're trying to identify those components which do not meet the quality control criterion, "less than 1000 hours" becomes the characteristic of interest.

(**b**) OK, go ahead! Test the entire 50 components by burning them until they all burn out. Then report to the boss that none of the components are any good because they've ALL been destroyed!

This is an example of "destructive" testing. Any time you have a testing procedure which actually destroys the item being tested, obviously you don't want to test the entire population. We said sampling is used when the entire population is not accessible to you. Often that's because the population is too large. But there are other reasons why the entire population may not be accessible. Destructive testing illustrates this.

3.3.3

(**a**) For any sampling experiment, an outcome is a sample. Here it's a simple random sample chosen without replacement. The "doing" is the random sampling procedure. Such a procedure is designed to be repeatable and the random choice mechanism (like tossing fair coins or using a random number generator) guarantees that the outcomes cannot be predicted in advance. In other words, the procedure is a random experiment *by definition*.

(**b**) The smallest possible value is zero; maybe none of the elements in the sample have the characteristic of interest. The largest possible value is n (the sample size); maybe all the members of the sample have the characteristic of interest. Any integer in between is also possible. So the possible values are

$$0,\ 1,\ 2,\ 3,\ 4, \ldots, n-1,\ n.$$

3.3.4

(**a**) Yes, every candidate is either a woman (the characteristic of interest) or not.

(**b**) If the mayor is choosing the committee at random, she's choosing a random sample. So it's a sampling experiment. The sampling must, of course, be without replacement. Otherwise, if the same person can be chosen twice, after five selections we could end up with fewer than five members on the committee (unless we introduce clones!). So we're thinking about "simple random sampling WITHOUT replacement from a dichotomous population" as required by the hypergeometric model.

(**c**) Here, the committee, the outcome of the mayor's selection process, is to be regarded as a random sample. So a particular outcome in this instance is one particular choice of five persons from the 40 candidates. The committee actually chosen by the mayor is just one possible outcome. We're studying the question of prejudice by thinking (theory!) about ALL POSSIBLE outcomes, by thinking about all possible committees of five persons. This is necessarily theoretical—there are well over 600,000 such committees, it's impossible to deal with ALL committees, except in thought (in theory). The theoretical model is the hypergeometric random variable.

(**d**) This is the combinations of 40 things taken 5 at a time, $C(40, 5)$.

(**e**) To a particular choice of five candidates (an outcome), our random variable assigns the "number of women chosen." Calling it X,

$$X = \text{the number of women on the committee.}$$

(**f**) Yes. Because X arises from simple random sampling without replacement from a dichotomous population and because it *counts* the number in a sample having the characteristic of interest, it satisfies all the required conditions for the hypergeometric model.

3.3.5

(**a**) If you sample from the books themselves, obviously you'll never get a lost book! The point is to sample from the *catalog listing*. Thus, the population consists of catalog entries (cards or entries in a computer file, etc.). Furthermore, the population should be dichotomous; it should break down into exactly two categories, those having the characteristic of interest and those not. The characteristic of interest for a particular catalog listing is *lists a lost book*. Be careful how you express yourself. The characteristic is not "lost book." That's a characteristic of the books themselves, not of the catalog listings.

Note that multiple listings pose a problem. Any one book may be listed several times in a variety of ways: by author, title, subject, and so on. If there are several authors, there will be several author listings. A careful design of our sampling experiment will resolve this difficulty, but we're going to ignore it in our discussion. Classroom examples are never completely realistic; after all, the real world is much too complex!

(**b**) An outcome is a sample; it consists of 30 catalog listings, randomly chosen without replacement. Therefore, we're talking about a *sampling* experiment. By definition, it's a random experiment. The "doing" is simple random sampling without replacement from the catalog listing.

(**c**) The integers 0–30. The smallest number of lost books possible is zero. The largest is $n = 30$ (the sample size). Of course, it's ridiculous to imagine that ALL the books in the sample would be lost, but logically it's possible.

(**d**) $X = 2$.

(**e**) The population is not small. It's not wrong to use the hypergeometric model for large populations, but there's a BETTER way. See the next section.

3.3.6

It's a short list, there's only one characteristic: "all the values are *equally likely*."

3.3.7

(**a**) $C(50, 3)$, the "combinations of 50 things taken three at a time":

$$\frac{50!}{3!\,47!} = \frac{50 \times 49 \times 48}{6} = 50 \times 49 \times 8.$$

Don't multiply this number out. Leave it factored like this so you can see what will cancel when we put this denominator with the numerator.

(**b**) Step 1: There are $C(4, 2)$ ways to select 2 from the 4 defective components. Step 2: You've already selected $X = 2$ defective components. The rest of the sample (one more component) comes from the 46 components which are NOT defective. There are $C(46,1)$ ways to select them. So the numerator is

$$C(4,2) \times C(46,1) = \frac{4!}{2!\,2!} \times \frac{46!}{1!\,45!} = 6 \times 46.$$

(c)

$$\mathrm{P}(X=2) = \frac{6 \times 46}{50 \times 49 \times 8} = \frac{3 \times 23}{25 \times 49 \times 4} = 0.0141.$$

(d)

X	$\mathrm{P}(X)$	$X\mathrm{P}(X)$	$X^2\mathrm{P}(X)$
0	0.7745	0.0000	0.0000
1	0.2112	0.2112	0.2112
2	0.0141	0.0282	0.0564
3	0.0002	0.0006	0.0018
	1	0.2400	0.2694

$\mu = 0.2400$,
$\sigma^2 = 0.2118$.

3.3.8 (a)

$$\mathrm{P}(X=0) = \frac{C(15,0) \times C(25,5)}{C(40,5)}$$

and after some cancellation

$$= \frac{5 \times 23 \times 11 \times 7}{2 \times 39 \times 38 \times 37} = 0.0807.$$

(b) All you've shown is that the mayor's selection is unlikely to have been made randomly. But everybody knows she didn't choose randomly! Maybe she followed some perfectly legitimate nonrandom selection criterion that just happened by accident to produce an all-male selection. An argument against the mayor based on our analysis alone is incomplete. Our analysis is only the first step, it justifies further investigation into the accusation against the mayor.

(c)

$$\mathrm{P}(X=1) = \frac{5 \times 25 \times 23 \times 11}{39 \times 38 \times 37 \times 2} = 0.2884.$$

which together with our answer from part (a) gives

$$\mathrm{P}(X \leq 1) = 0.0807 + 0.2884 = 0.3691.$$

With a probability of more than 35% for such a result, one can hardly be suspicious of the mayor. Her choice looks reasonable.

This is much more conclusive than the argument in part (a). Now the accusation of bias against the mayor is demolished! She can unequivocally claim "no prejudice" because the number of women selected is consistent with a *random* selection. On the other hand, when the choice is NOT consistent with a random selection, as happened in part (a), the argument is not over. In this case, we've only established a reason to investigate further; we haven't proven bias. After all, we know the mayor doesn't select the committee randomly. She uses some systematic criteria, although GENDER shouldn't be among the criteria.

3.3.9

(a) The uppercase letter is the name of the random variable. It refers to the random variable as a whole. The lowercase letter stands for one of the particular values—an unknown or unspecified value.

(b) The denominator is the total number of ways to draw a sample of n from this population of N members. Note that this does not depend on any value of the random variable. So the formula does not contain the symbol x. The formula is $C(N, n)$.

(c) The numerator is the product of

> the number of ways to choose x members of the sample from the R members of the population which have the characteristic of interest: $C(R, x)$

and

> the number of ways to choose $n - x$ members of the sample (the rest of the sample) from the $N - R$ members of the population which do NOT have the characteristic: $C(N - R, n - x)$.

(d) Did you get this? Try again. We won't give the formula here. YOU try (don't worry, we'll give it to you later).

3.3.10

(a) We're randomly drawing four without replacement—simple random sampling without replacement—from a dichotomous population of ten objects, four of which have the characteristic of interest.

(b) Because the sample size is four, you should guess the mean to be $4 \times 0.4 = 1.6$. After all, with forty percent of the population having the characteristic, it would seem reasonable to guess that forty percent of the sample—*on average* anyway—also ought to have the characteristic. And that's true. The variance, on the other hand, is only a *comparative* measure of dispersion; its value does not have any intrinsic intuitive meaning. So, we do not attempt to guess it. There's nothing to compare it to.

(c) and (d) All the denominators are the same, the number of ways to choose four of ten objects: $C(10, 4)$, which is 210. Here's the probability distribution:

X	$P(X)$	$XP(X)$	$X^2P(X)$
0	0.0714	0.0000	0.0000
1	0.3810	0.3810	0.3810
2	0.4286	0.8572	1.7144
3	0.1143	0.3429	1.0287
4	0.0048	0.0192	0.0768
	1.0001	1.6003	3.2009

$\mu = 1.6003$,
$\sigma = 0.8000$.

3.3.11 (a)

$$P(X \geq 1) = 1 - P(X = 0) = 1 - (780/1218)$$
$$= 1 - 0.6404 = 0.3596.$$

(b) The mean, $\mu_H = 0.4$ and the variance, $\sigma^2 = 0.3227$. The finite population correction factor is 0.9310. That's relatively close to one, reflecting the fact that our sample is not particularly large compared with the population size. About 93% of the population is not in the sample.

(c) We'll leave $P(X = 1) = 10\%$ to you (sorry!). The mean of X is 0.1, the variance 0.09. Because, compared with parts (a) and (b), there are now fewer burnt out bulbs, we should not be surprised that the mean number of such bulbs in the sample is lower. Also, there should be less variability with only one bad bulb. This is reflected in the smaller variance. The finite population correction factor is the same as in part (a) because it's not affected by how many in the population have the characteristic of interest. The finite population correction factor is determined entirely by N and n with no reference to R.

(d) How could all three of the lightbulbs you selected be burnt out when there were only two burnt out lightbulbs in the first place?! Impossible! From the formulas, the numerator is the number $C(2, 3)$. By definition that's the number of ways to choose three from a collection of two. Clearly this number is zero.

(e) Solve $(30 - n)/(30 - 1) = 0.9$. So, $30 - n = 0.9 \times 29$. Now, what is it for 80%?

3.3.12 **(a)** X is the number of hearts in a six-card hand. The possible values of X are 0, 1, 2, 3, 4, 5, 6 . The underlying random experiment is "simple random sampling without replacement." In this case, "shuffle the deck well and take the top six cards." An outcome for any sampling experiment is "one sample"—in this case, a six-card hand. The "doing" is clearly repeatable and if we shuffle the deck thoroughly, the outcomes cannot be predicted in advance.

X is hypergeometric because it's a COUNT of how many in the sample have the characteristic. It counts how many cards in the hand are hearts. The expected value of X is $np = 6 \times 1/4 = 1.5$. One could guess this by thinking along these lines: The "typical" hand should look more or less like the entire deck, so about one-fourth of the hand should be hearts. Of course, this reasoning is only one step away from using the formula which says you should expect the same *proportion* of hearts in the sample as in the entire population. ON AVERAGE, that is.

(b) X is the number of black cards in a six-card hand. The possible values, the experiment, and the outcomes are all exactly the same as in part (a). And just as in part (a), X is hypergeometric because it's a *count* of how many in the sample have the characteristic; it counts how many cards in the hand are black. The expected value is $np = 6 \times 1/2 = 3$. You should expect the same proportion (one-half) of black cards in the "typical" hand as in the entire population. Here, "typical hand" means the "on average" hand.

(c) Because you use the mean for predicting (that's why it's also called the "expected value"), the "predictability" of the values of a random variable is measured by the variance. We can think of the two variances (for "color" and for "suit") without the finite population correction factor (46/51) because it's the same for both random variables and so will not affect the relative sizes of the two variances. So using npq as if it were the variance, we get $\sigma^2 = 9/8$ for the number of hearts. For the number of black cards we get $\sigma^2 = 12/8$. Of course, to get the true variance, you multiply by 46/51 each time. Thus "suit" is more predictable than "color." The finite population correction factor is 0.9020, so the actual variances are, respectively, 1.0147 and 1.3530.

(d) They should look something like this . . .

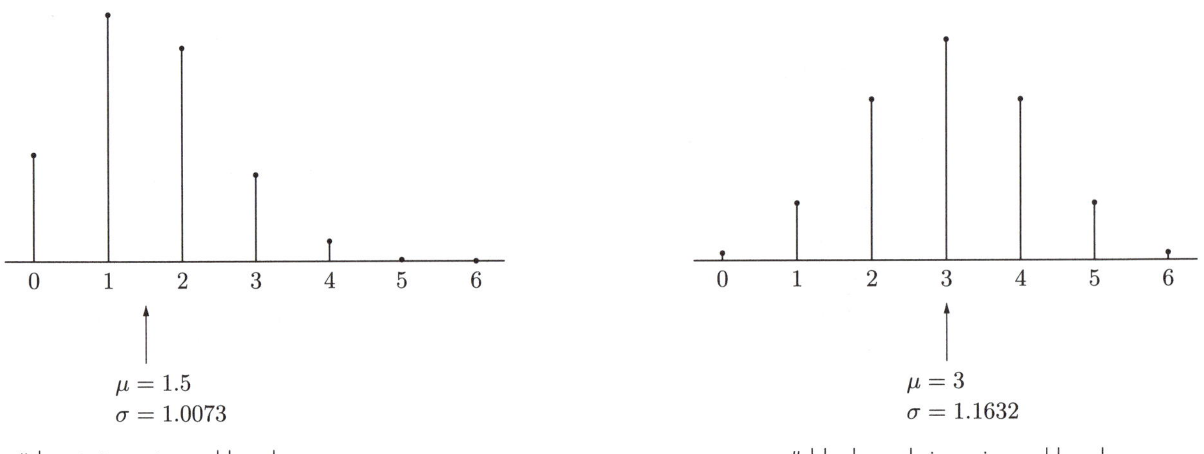

hearts in a six card hand

black cards in a six card hand

The number of black cards, note, is completely symmetric about its mean, reflecting the fact that there's a 50/50 chance of drawing such a card. Note also that five or six hearts is virtually impossible. Those values are more than three standard deviations away from the mean. By Chebyshev's Theorem, for any random variable there's less than one chance in nine ($1/3^2$) of values more than three standard deviations from the mean. Six is actually more than FOUR standard deviations from the mean. By Chebyshev's Theorem, that occurs with less than one chance in 16 ($1/4^2$). In fact, if you actually compute the probability of six hearts, it's less than one chance in 10,000. Chebyshev's Theorem is very conservative; it covers too much ground to yield very precise information (it's a theorem valid for *all* random variables).

(e) The finite population correction factor for both parts (a) and (b) is approximately the proportion of cards remaining in the deck after the six-card hand has been dealt, 46/52. The exact value is 46/51. We find 51 in the denominator rather than the intuitively expected 52 for technical reasons only. That should not affect your underctanding of what it means.

3.3.13 Up to the point where the sample becomes half the population, n, the sample size, predominates, making the variance larger as n gets larger. Once the sample is so large as to encompass MORE than half the population, the variance begins to get smaller. At that point the "space" in the sample (large n) which allows for variability is overwhelmed as a factor in the variance by the amount of information contained in the sample, the principle being: more information, more certainty; more certainty, LESS variability. At the extreme when $n = N$, there's no variability at all. After all, there's only ONE "sample of size N" (namely, the whole population). At the other extreme, when $n = 0$, you have no sample! So again there's no variability and the variance is zero.

3.4.1

(a) The "doing" for the random experiment is "simple random sampling with replacement" from a dichotomous population.

(b) We think of a sample as just a string of yes's and no's. Because the sample contains n members, we represent it as, for example,

yes, yes, no, yes, . . . , no, no   (n altogether).

(c) The smallest possible value is zero. What's the largest possible number in a sample which could have the characteristic? ALL!! Because the sample size is n, the possible values are

$$0,\ 1,\ 2,\ 3,\ \ldots, n-1,\ n.$$

3.4.2

(a) Because we select with replacement, individual selections are *independent*. What happens on the second draw will not be affected by the result of the first draw. Because our events are independent, we can use the simple product rule

$$\begin{aligned} \mathrm{P}(A \text{ and } B) &= \mathrm{P}(A) \times \mathrm{P}(B) \qquad A, B \text{ independent} \\ \mathrm{P}(\text{yes and yes}) &= \mathrm{P}(\text{yes}) \times \mathrm{P}(\text{yes}) \\ &= p \times p \\ &= p^2. \end{aligned}$$

The probability of a "yes" is p because the probability of getting the characteristic on any draw is just the proportion of the population having that characteristic. So,

$$\begin{aligned} \mathrm{P}(\text{yes, yes, yes, no, no, no, no, no, no, no}) &= pppqqqqqqq \\ &= p^3q^7. \end{aligned}$$

(b) For $X = 3$, the three yes's could appear anywhere; they need not have been the first three.

(c) $C(10, 3) = 120$.

(d) List systematically all the outcomes of the event $X = 3$. The probability of this event requires the "or" rule for mutually exclusive events.

	yes, yes, yes, no, no, no, no, no, no, no,	call it A_1
OR	yes, yes, no, yes, no, no, no, no, no, no,	call it A_2
OR	yes, yes, no, no, yes, no, . . .	call it A_3
	. . . . . .	
	and so on.	

There are 120 of these events: $A_1, A_2, \ldots, A_{120}$. And each of them has probability p^3q^7. Because the events A_k are all *mutually exclusive*, we get the probability by adding p^3q^7 to itself 120 times. Therefore, $P(X = 3) = 120p^3q^7$.

(e) $120 \times 0.74^3 \times 0.26^7 = 0.0039$.

(f) The probability of part (e) requires knowing that $n = 10$ and $p = 0.74$, nothing more.

3.4.3

(a) First we choose x of the n blanks to have yes's. There are $C(n, x)$ ways to do this. Then the probability for any one of these choices is just the product of x p's and $n - x$ q's because the selections are *independent*. So the probability for one such choice is $p^x q^{n-x}$. Now any one such arrangement of yes's and no's is *mutually exclusive* with any other. So we must add $p^x q^{n-x}$ to itself $C(n, x)$ times to get the probability $P(X = x) = C(n, x)p^x q^{n-x}$.

(b) The only information required is the values of n and p. That's one reason this model is more useful than the hypergeometric model. To work out probabilities for sampling without replacement (the hypergeometric model), we needed much more specific information. It was not enough to know p; we needed the two pieces of p, namely R and N ($p = R/N$, recall).

3.4.4

The condition means that at most 10% of the population is contained in the sample. The proportion of the population contained in the sample is given by n/N and the condition says it's at most $1/10$.

3.4.5

(a) Our sampling process will be much easier to implement if we don't bother trying to eliminate repetitions. This would seem reasonable given the large number of catalog listings because there's a very small chance the same listing will be chosen twice. If we proceed in that manner, the model is sampling WITH replacement.

Or, maybe we do, in fact, sample with*out* replacement. Still, the sampling WITH replacement model is a good approximation because it will be virtually identical to the WITHOUT replacement model (the hypergeometric model). For such a large population—all of several hundred thousand catalog listings—the probability of drawing the same listing twice in 200 draws is virtually zero.

(b) The finite population correction factor would be (approximately) the proportion of library listings not in our sample of 200. Because there are presumably thousands of books and we have only 200 listings in our sample, this proportion is essentially one. If the library has 200,000 volumes listed in the catalog, the finite population correction factor is

$$\frac{200,000 - 200}{200,000 - 1} = 0.999$$

which means 99.9% of the population is not in the sample.

(c) $p = 0.01$ and $q = 0.99$ so

$$\begin{aligned} P(X \geq 3) &= 1 - P(X \leq 2) \\ &= 1 - \left[q^{200} + 200pq^{199} + 100 \times 199p^2q^{198}\right] \\ &= 1 - q^{198}\left[q^2 + 200pq + 19,900p^2\right] \\ &= 1 - 0.6767 \\ &= 0.3233. \end{aligned}$$

(d) The expected value of this random variable is $np = 200 \times 0.01$. Thus we should expect about two of our 200 listed books to be lost.

(e) Here $n = 30$ and $p = 0.01$:

$$P(X = 2) = C(30, 2)p^2q^{28} = 15 \times 29p^2q^{28} = 0.0328$$

and

$$\begin{aligned} P(X > 2) &= 1 - P(X \leq 2) \\ &= 1 - q^{28}\left[q^2 + 30pq + (15 \times 29)p^2\right] \\ &= 1 - 0.9967 \\ &= 0.0033. \end{aligned}$$

(**f**) The expected value here is $np = 30 \times 0.01 = 0.3$. That is, we should expect about one percent of the sample (three-tenths of a book) to be lost because, in fact, one percent of all the listings are of lost books. The "three-tenths" answer means that theoretically (on average), out of ten samples of 30 catalog listings we should find three listings for books that are lost. Note the phrase "should find," not "will find." The word "should" refects the theoretical nature of the average.

(**g**) Here $n = 30$ and $p = 0.01$.

3.5.1

(**a**) Toss a coin, S = heads, F = tails, $p = q = 0.5$. Or another example: roll a die, S = six dots on top face, F = not six dots, $p = 1/6$, $q = 5/6$. Or draw one card from a well-shuffled deck of 52, S = spades, F = non-spade, $p = 0.25, q = 0.75$. You could give numerous other examples.

(**b**) You did this problem long ago! See Problem 1.2.7(d). You can also obtain the variance using the computing formula. Try it. You'll need to use: $p - p^2 = p(1 - p) = pq$.

3.5.2

Here $n = 1$, and there's no distinction between "with" and "without" replacement (there's no "second draw" to require replacement). So technically, we have THREE models. But all three models are the same. If you look carefully at the definitions, you'll see that for $n = 1$ you're saying the same thing in each of the three cases. So the formulas had better give the same result! For the sampling models, the mean is np and, with $n = 1$, this is just p as it should be. The variance for the hypergeometric random variable is $npq(N - n)/(N - 1) = 1pq1 = pq$. For sampling with replacement, the variance is $npq = 1pq = pq$.

3.6.1

(**a**) The "doing" is to test persons for blood type until you find one who has the type you seek.

(**b**) An outcome of this process would look like a string of no's followed by *one* yes:

no, no, no, no, no, no, no, no, no, no, no, no, yes.

(**c**) Here $X = 13$, the number of persons tested before finding one with the blood type you seek.

(**d**) Another string of no's followed by one yes:

no, no, no, no, no, no, no, no, no, no, no, no, no, no, no, yes.

Here you had to test 16 persons before finding the desired blood type. So, $X = 16$.

(e) If you could predict in advance whether a given person had the desired blood type, there would be no need for testing in the first place.

3.6.2

(a) The "doing" is to test ONE person for the desired blood type.

(b) The two possible outcomes can be recorded as either "yes" or "no," depending upon whether the person tested does or does not have the desired blood type.

(c) This is part (e) of the previous problem.

(d) The parameter p is the probability of success on the Bernoulli trial. Here it's the probability you find the desired blood type. So, p is the proportion of the population having that blood type. This assumes, of course, that you choose your test subjects randomly from the whole population; otherwise p may very well have some other value.

3.6.3

Blood type is partly a matter of heredity. But by assuming you choose your test subjects randomly from a large population, it's not likely you'll get two persons who are blood relatives. So it's not likely you'll have any significant dependency to worry about.

3.6.4

(a) The first success could occur on the first trial. It could hardly occur sooner. In that case, $X = 1$. So one is the smallest possible value. On the other hand, THERE IS NO MAXIMUM VALUE. If you can imagine that it might require seven million trials to obtain the first success, why not seven million and one? So the possible values are

$$1, 2, 3, 4, 5, 6, 7, \ldots \text{ ad infinitum.}$$

(b) There's only one parameter, p. It's the parameter for the associated Bernoulli trial: $p =$ the probability for success on one repetition of the trial. So, once you specify a value for p, you have picked out a particular geometric random variable from the class.

3.6.5

(a) It's just P(no, no, no, no, no, no, no, no, no, no, no, no, no, no, no, no, yes)

$$\begin{aligned} &= P(\text{no}) \times P(\text{no}) \times P(\text{no}) \times \cdots \times P(\text{no}) \times P(\text{yes}) \quad \text{(by independence)} \\ &= (1-p)^{16} \times p \\ &= q^{16} \times p \\ &= 0.92^{16} \times 0.08 = 0.0211. \end{aligned}$$

(b) It's just 0.08 because eight percent of the population have that blood type.

(c) Because eight percent of the population have the blood type you seek, eventually you will find one of them.

(d) $0.92^2 \times 0.08 = 0.0677$.

(e) $0.92^6 \times 0.08 = 0.0485$.

(f) This is the same as part (c).

3.6.6

The model assumes the repetitions of the Bernoulli trial are *independent*. By assuming there are no persons in the population who are blood relatives, you probably have assured that blood type is independent from one member of this population to another. Of course, it's a very restrictive assumption. In a realistic situation with a very large population from which you choose randomly, you may be able to assume the independence assumption to be approximately valid if blood relatives are fairly rare in the population. Or possibly you dismiss from your study anyone who turns out to be a blood relative of someone already tested.

3.6.7

There are an infinite number of possible values for X. I don't care how much time you have, you won't have enough time to list them all!

3.6.8

(a) You test people randomly, but the result is not a random sample as we have defined it. Our definition requires that you choose a FIXED number of population members for your sample, but here you do not test a fixed number of persons. This is a technicality, of course, but it's useful to think about if only to get clear what we're talking about.

(b) No.

3.6.9

(a) $\mu = 1/p = 1/0.08 = 12.5$.

(b) $P(X = 15) = q^{14}p = 0.92^{14} \times 0.08 = 0.0249$.

(c) $P(X \leq 15) = 1 - q^{15} = 0.7137$.

(d) $P(X \geq 16) = 1 - 0.7137 = 0.2863$.

(e)

$$\begin{aligned} P(X = 1) &= \text{probability that one person from this population has the desired blood type} \\ &= 8\% \qquad \text{(by the problem's assumption).} \end{aligned}$$

(f) Not possible; see Problem 3.6.5(c).

(g) $P(X \geq 11) = q^{10} = 0.4344$.

(h) $P(X \geq 3) = q^2 = 0.8464$.

(i) $P(X = 7) = q^6 p = 0.0485$.

(j) $P(X \geq 0) = P(X \geq 1) = 100\%$. Because eight percent of the population have the desired blood type, you're bound to find one eventually. Note that, in fact, X cannot be zero, so $P(X = 0)$ is zero. X doesn't take values less than one, so this is the certain event (probability 100%).

(k) $P(X \leq 22) = 1 - q^{22} = 0.8403$.

3.6.10 (a)

$$\sigma^2 = q/p^2 = 0.92/0.0064 = 143.75, \text{ so } 3\sigma = 35.9687.$$

The mean of X is $1/p = 12.5$ and so there are no values of X more than three standard deviations *below* the mean. Thus we need only compute the probability that X is more than three standard deviations *above* the mean:

$$P(X \geq \mu + 3\sigma) = P(X \geq 49) = q^{48} = 0.0183.$$

Note that we make use of the following relation:

$$P(X \geq x) = 1 - P(X \leq x - 1) = 1 - [1 - q^{x-1}] = q^{x-1}.$$

(b) We know the specific distribution of X. It's a geometrically distributed random variable with parameter $p = 0.08$. Chebyshev's Theorem makes no assumption whatsoever about the distribution of the random variable. Chebyshev's Theorem is very general, it's valid for any random variable whatsoever and so it cannot yield very precise information.

3.6.11

(a)

Class	$P(X\in$ class)
1–10	0.5656
11–20	0.2457
21–30	0.1067
31–40	0.0463
41–50	0.0201
	0.9844

0.9844 — not 1.0000 because of rounding *and* because $X \geq 51$ is omitted.

(b)

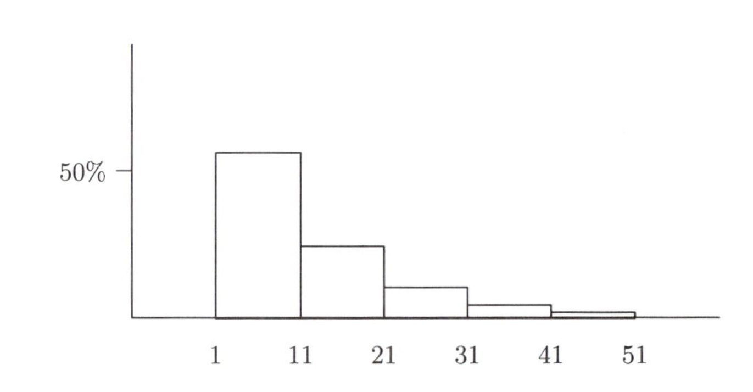

In fact, this graph is typical of what happens with the geometric distribution. It's easy to show that $X = 1$ is the most likely value, $X = 2$ is the next most likely value, and so on. In other words, the probabilities are strictly decreasing as X increases. Here's how you show that: Observe that if $x < y$, then $q^x > q^y$ because $q < 1$. Thus,

$$pq^x > pq^y \quad \text{from which we obtain } P(X = x) > P(X = y).$$

3.6.12

(a) For any geometric random variable, $X = 1$ is the most likely value, but the mean is $1/p$ which is always larger than one. In our blood type example, the mean is 12.5.

The mean is the "expected" value in the sense of "middle," not in the sense of "most likely." It's characterized by the fact that the average deviation from the mean is zero. That is, if N is a number such that $\Sigma(X - N)P(X) = 0$, then $N = \mu$. Would you like to prove that? It's not TOO hard. Watch:

$$0 = \Sigma(X - N)P(X) = \Sigma XP(X) - N\Sigma P(X) = \mu - N, \quad \text{so } N = \mu.$$

This algebraic characterization of the mean has a physical interpretation. If you think of the distinct values of X along a number line as if they were physical points of varying masses and if the probabilities are proportional to the masses, then the mean is the point on the line at which this system of masses would BALANCE. For engineers, this is the "center of mass" interpretation of the mean.

(b) $P(X < \mu) = P(X < 12.5) = P(X \leq 12) = 1 - q^{12} = 0.6323$.

(c) The mean is a "middle" value, but not in the sense of splitting the distribution 50/50; that's the median. The mean is the "middle" of the distribution in the sense that it's the one and only number for which the deviations all cancel out on average.

3.6.13

(a) You would predict the number of persons you must test by the *expected* number, the mean of the random variable. So the "predictability" is measured by how far away from that mean you are on average. We measure that with the variance.

(b) If p is 12%, then $\sigma^2 = 0.88/0.12^2 = 61$, compared with a variance of about 144 when p is eight percent. Because 61 is much smaller than 144, we conclude that the new situation is much more predictable than the old one.

3.6.14

It's just the geometric series

$$P(X \leq x) = \Sigma pq^{k-1}, \quad k = 1, 2, \ldots, x.$$

The "geometric series," you will recall, is the sum of powers of some number (here q) with some coefficient (here p). The cumulative probability formula, $P(X \leq x) = 1 - q^x$, is obtained from the formula for the sum of a geometric series. This analysis also explains the name of the model.

3.6.15

(a)

$$\begin{aligned} P(X = x + 1) &= pq^x = qpq^{x-1} \\ &= qP(X = x) \quad \text{since } P(X = x) = pq^{x-1} \end{aligned}$$

(b)

X	$P(X)$
1	0.0800
2	0.0736
3	0.0677
4	0.0623
5	0.0573
6	0.0527
7	0.0485
8	0.0446
9	0.0411
10	0.0378
	0.5656

(c) In Problem 3.6.11(a), you calculted $P(X \leq 10)$ using the cumulative probability distribution function. Note that you get the same answer, of course.

(d) If $x > 10$, $P(X = x) < P(X = 10) = 0.0378$, if $p = 0.08$.

3.6.16 Note that each entry of the table below is obtained from the preceding one if you just multiply by q. This follows easily from the recursion formula.

Frequency of capture	0	1	2	3	4	5	6	7
Expected # of rabbits	59.7	33.3	18.6	10.4	5.8	3.2	1.8	1

Most of these theoretically expected numbers are relatively close to the observed numbers. So the model might seem reasonable. Later (Problem 6.2.33), we'll have a more precise way to check the "goodness of fit" of this geometric model to Edwards and Eberhardt's data.

3.7.1 (a) . . . and then repeat it $n-1$ more times for a total of n repetitions. Those n repetitions constitute ONE execution of the binomial experiment. We can repeat this by doing n more repetitions of the Bernoulli trial. This constitutes ONE REPETITION of the binomial experiment. We have now repeated the Bernoulli trial $2n$ times, yielding two repetitions of the binomial experiment. This can go on indefinitely and so the binomial experiment is, indeed, repeatable.

(b) 90 repetitions. One execution of the binomial experiment consists of 15 repetitions of the Bernoulli trial, repeat that and you've done the Bernoulli trial 30 times. After six repetitions of the binomial experiment, you will have done the Bernoulli trial 90 times.

(c) A string of n successes and failures.

(d) Because the Bernoulli trial is a random experiment, its outcomes (the S's and F's) cannot be predicted in advance. So, of course, the outcome of the binomial experiment (a string of S's and F's) cannot be predicted in advance.

(e) For coin tosses, n is the number of tosses and p is the probability of heads on one toss (assuming "success" means "heads").

(f) From the n blanks, choose k to contain the S's. That can be done in $C(n, k)$ ways. With $n = 20$ and $k = 8$, this is 125,970.

(g)

$$FFFF$$

$$SFFF \quad FSFF \quad FFSF \quad FFFS$$

$$SSFF \quad SFSF \quad SFFS \quad FSSF \quad FSFS \quad FFSS$$

$$SSSF \quad SSFS \quad SFSS \quad FSSS$$

$$SSSS$$

3.7.2

(a) We may not have repetitions of the SAME Bernoulli trial—these are not the same coin, after all. If the probability for heads varies so that p is not constant, this is not a binomial experiment. On the other hand, if all the 20 coins have the SAME probability to come up heads—if the coins are "identical" in this sense—then the model is valid and we do, indeed, have 20 repetitions of the same Bernoulli trial. A Bernoulli trial is completely characterized by S and p .

(b) We require a fixed probability for one of these electronic components to be defective. Otherwise, we're not repeating the SAME Bernoulli trial each time. So if that probability increases as the day goes on, the model is not valid.

(c) We don't know. Ask an engineer who knows how defective components arise.

3.7.3

(a) The Bernoulli trial can be described as "observe one birth in this maternity ward during the week in question." To "repeat" the trial might mean "observe the next birth in this maternity ward during the week in question." Each birth is either a stillbirth or not, so we have exactly two possible outcomes.

Can stillbirths be predicted in advance? Yes, sometimes. But maybe no such information is available for your study. For example, you may be trying to model what will happen in this maternity ward for various weeks over the coming year.

Then there's no information about the mothers-to-be, let alone their babies. So yes, in appropriate circumstances "observe one birth in this maternity ward" can be taken as a Bernoulli trial.

(**b**) "Success" is "this birth was a stillbirth." Note that the word "success" is purely an abstract convention, it identifies the outcome of interest. It certainly does not mean you like the result!

(**c**) The Bernoulli trial is "observe one birth," you do it as many times as there are births over the one week in question. If there will be 34 births that week, then $n = 34$.

(**d**) There are two more considerations: Are you repeating the SAME Bernoulli trial each time (is p constant?) and are the repetitions independent?

The probability of a stillbirth from one birth to the next must be constant. Otherwise you're not repeating the same Bernoulli trial. If you consider these mothers as typical of their community, the probability of a stillbirth should be taken as the relative frequency of stillbirths in that community. Records may suggest that two percent of births in that community have in the past been stillbirths, so you may take that OBSERVED two percent figure as a THEORETICAL number, giving you the theoretical relative frequency of stillbirths for the community. A theoretical relative frequency is just a probability, so $p = 0.02$.

The independence assumption is satisfied if one mother's stillbirth makes it neither more nor less likely for another mother to have a stillbirth. That seems to be a reasonable assumption.

3.7.4

(**a**) The Bernoulli trial is: Make one drilling for oil. Success, S, means "we struck oil!" This clearly is a random experiment: It's something we do which is repeatable with clearly specified outcomes (two of them: strike oil or not) which cannot be predicted in advance. We repeat the trial $n = 112$ times. We must assume a fixed probability, p, for any one drilling to strike oil; the same p for all drillings so we have the same Bernoulli trial each time.

For the trials to be independent, some realistic physical assumptions would be required. Your local geologist will no doubt have something interesting to say about this (and you should consult her before accepting this model as valid), but at the very least the drillings would have to be spaced apart. Obviously, if we drill three feet away from a dry hole, we will just get another dry hole!

The parameters are $n = 112$ and $p =$ "the probability for one drilling on this tract of land to strike oil," a measure of how oil-rich the land is (note how we identify the real-world meaning of the parameter, "a measure of . . . ."). The value of p is not given here and might very well be the question we want to answer with our model. Our specific 112 drillings will yield a report (outcome) such as

dry, dry, dry, dry, oil, dry, oil, dry, dry, dry, dry, dry, . . . and so on.

(**b**) The Bernoulli trial: Throw the dart once. Clearly it's a random experiment (verify). Suppose we're interested in whether we hit a bull's eye or not. Then we have two possible outcomes: S = hit the bull's eye, F = miss the bull's eye. Clearly, these cannot be predicted in advance. We repeat this trial $n = 25$ times. To say we're repeating the same Bernoulli trial, we would have to assume the same player throws the dart each time, otherwise p would change depending on who throws the dart. The repetitions will be independent if we can assume that fatigue and psychological "holdover" (intimidation from missing the bull's eye or elation from hitting it) are negligible.

The parameters are $n = 25$ and p = "the probability of hitting the bull's eye on any one throw," a measure of the thrower's skill. Again, the value of p is not given and might actually be the question of interest. An outcome will be a record of hits and misses (25 of them):

hit, miss, hit, miss, miss, miss, ...

(**c**) The Bernoulli trial: one telephone contact by the saleperson. Clearly, it's a random experiment (verify) with two possible outcomes: S = make a sale, F = suffer a refusal (of course, there are other possiblilities for identifying S and F, depending on the marketing goals). We repeat this trial $n = 15$ times. If the probability of making a sale is constant from call to call (it may NOT be), then you're repeating the same Bernoulli trial. The repetitions will be independent if fatigue and "psychological holdover" for the salesperson are negligible and if the calls are placed to prospective customers who are unrelated.

The parameters are $n = 15$ and p = "the probability of making a sale on any one call," a measure of the salesperson's effectiveness. Once again, p is not given and might actually be the question of interest. An outcome will be a record of sales and refusals (15 of them):

sale, sale, sale, refusal, sale, sale, sale, refusal, ...

(**d**) The Bernoulli trial: Put the question to one interviewee. Clearly, that's a random experiment (verify) with two possible outcomes: S = get a truthful answer, F = get a deceptive answer. We repeat this trial $n = 1500$ times. The repetitions would seem to be independent if we can assume the persons being interviewed are unrelated in any way and are interviewed separately.

The parameters are $n = 15$ and p = "the probability of a truthful answer." However, the probability of a truthful answer would certainly change from person to person. What do we even mean by "the probability of a truthful answer" from one person? Such probabilities do not make sense from a relative frequency point of view. On the other hand, if you're studying a well- defined population, you might think in terms of the whole population and take p to be the proportion of the population which will answer the question truthfully. Then p becomes the "theoretical relative frequency" of a truthful answer for that population. With this interpretation, p remains the same from person to person and we can say we're repeating the same Bernoulli trial. Then the binomial model models "truth of reponse" for that population.

An outcome can be described as a record of yes's and no's.

3.7.5 $X = 0, 1, 2, 3, \ldots, n$. There could be no successes at all or every trial could result in a success. Or anything in between.

3.7.6 (a) By just multiplying: $pqpq^3p^4q^2pq^3pq^3 = p^8q^{12} = 9.0813 \times 10^{-7}$.

(b) $qp^2q^3p^4q^2pq^3pq^3 = p^8q^{12}$.

(c) Because the events are mutually exclusive, we can just add:

$$\begin{aligned} \mathrm{P}(A \text{ or } B) = \mathrm{P}(A) + \mathrm{P}(B) &= p^8q^{12} + p^8q^{12} \\ &= 2p^8q^{12}. \end{aligned}$$

(d) Exactly $C(20, 8) = 125,970$.

(e) $\mathrm{P}(X = 8) = 125,970p^8q^{12} = 0.1144$.

(f) If you round p^8q^{12}, you get ZERO for answer in part (e). That's a pretty serious error when the true value is over 11%!

Caveat Calculator:[5]

During the process of calculating, avoid intermediate rounding whenever possible.

3.7.7 About six, 30% of 20.

3.7.8 (a) $X = \Sigma X_k$. In the outcome shown below, there are eight successes and each one corresponds to a one. The failures contribute nothing to the sum because

[5] The latin word "calculator" means "one who calculates."

they correspond to zeros. So, $\Sigma X_k = 8$. But that's exactly the value of X (the number of successes):

$$\begin{array}{cccccccccccccccccccc} F, & S, & S, & F, & F, & F, & S, & S, & S, & S, & F, & F, & S, & F, & F, & F, & S, & F, & F, & F \\ 0 & 1 & 1 & 0 & 0 & 0 & 1 & 1 & 1 & 1 & 0 & 0 & 1 & 0 & 0 & 0 & 1 & 0 & 0 & 0 \end{array}$$

(b) The mean of each X_k is just p (Problem 3.5.1). So μ_X is the sum of p added to itself n times: $\mu_X = np$. In the example, $\mu_X = 6$.

The variance of each X_k is pq (Problem 3.5.1), so σ_X^2 is just pq added to itself n times: $\sigma_X^2 = npq$. The example: $\sigma_X^2 = 4.2$. This is valid because the X_k's are all independent, their random experiments are just the repetitions of the Bernoulli trial and those repetitions are required to be independent.

3.7.9

(a) $P(X = x) = C(n, x)p^x q^{n-x}$. Here's how:

(i) each outcome with x successes has probability $p^x q^{n-x}$ because the Bernoulli trials are independent,

(ii) there are $C(n, x)$ such outcomes,

(iii) the event $X = x$ is composed of the events in (i) connected by the word "or";

(iv) because the events of (i) are mutually exclusive, by (iii) we just add $p^x q^{n-x}$ to itself $C(n, x)$ times to obtain $P(X = x)$.

(b) We get the recursion formula if we multiply the following equation by $P(X = x)$:

$$\frac{P(X = x + 1)}{P(X = x)} = \frac{n - x}{x + 1}\frac{p}{q}.$$

Now, this equation is obtained by appropriate cancellation in

$$\frac{P(X = x + 1)}{P(X = x)} = \frac{x!(n - x)(n - x - 1)!}{n!}\,\frac{n!}{(x + 1)x!(n - x - 1)!}\,\frac{p^{x+1}q^{n-x-1}}{p^x q^{n-x}}.$$

3.7.10

(a) The X_k's are just an "ordered set of n values" of the Bernoulli random variable generated by "n independent repetitions of the random experiment" for that Bernoulli random variable. So, the n values constitute a simple random sample from that Bernoulli distribution (see Chapter 2 for "sampling from a probability distribution"). Because $X = \Sigma X_k$ is the sum of those values, X is the sum of the numbers in our sample as we were required to show.

(b) We're trying to model sampling with replacement from a dichotomous population. For any sampling experiment an outcome is "one sample." Here, a sample can be thought of as a string of yes's and no's. Obtaining ONE of these sample elements is the result of "drawing one at random from the population." That's the Bernoulli trial. Because the population is dichotomous, we have two possible outcomes: $S =$ "yes, has the characteristic of interest," $F =$ "no, does not." To get our sample of size n, we repeat that Bernoulli trial n times. The repetitions are independent because we're sampling WITH replacement. So we have "n independent repetitions of a Bernoulli trial." That's a binomial experiment.

Now for the random variable: The binomial random variable is the number of successes (here, it's the number of yes's) which is exactly our $X =$ "the number of observations in the sample having the characeristic of interest." So X is binomial as we were to show.

Finally the parameters n and p: The parameter n for the binomial model is "the number of repetitions." Here, that's exactly the sample size. The parameter p for the binomial model is "the probability of success." Here, it's the probability for one member of the population to have the characteristic of interest. Or you can interpret p as the proportion of the population having the characteristic of interest.

(c) Because it's a special case of the binomial model. That's why all the formulas for the binomial model are exactly the same as the formulas in Section 3.4.

(d) If we're sampling from a large population, the "with/without replacement" distinction doesn't make any significant difference in the models. If the population is large, the chances of drawing the same member twice is so negligible for sampling with replacement that it looks just like (approximately) sampling without replacement. Of course, for SMALL populations, sampling without replacement requires the hypergeometric model and in that case, the binomial model would give erroneous answers.

3.7.11

(a) The "predictability" of a player's game is measured by the variance for the Bernoulli random variable. For Shu Wen, it's 0.1411 and for Juan, it's 0.1056 [see Problem 3.5.1 (b)]. Because Juan has the smaller variance, his game is more predictable. Shu Wen is more likely to be successful on any one attempt, however, because her p is larger.

(b) She keeps repeating her Bernoulli trial until the first success. The number of attempts required to make the first basket is a geometric random variable with mean $1/p = 5.8824$. So BEFORE making a successful attempt, she should expect five unsuccessful attempts. Of course, you could give the theoretical figure by saying, "She should expect to require 5.9 attempts, on average, in order to sink a basket." That gives a more precise answer to the question and emphasizes the theoretical nature of the question because 5.9 is not actually possible.

Note that we're careful not to say 5.9 attempts *before* sinking a basket. Those words would refer to the number of UNsuccessful attempts which, on average, would be 4.9. If X is the geometric random variable, the number of failures before the first success is $Y = X - 1$. So the expected number of failures is $\mu_Y = 1/p - 1$.

(c) Here X is binomial with $p = 0.12$ and $n = 10$:

$$P(X \geq 3) = 1 - q^8 \left[q^2 + 10pq + 45p^2\right] = 1 - 0.8913 = 0.1087.$$

(d) Predictability is measured by the variance of the Bernoulli random variable, $\sigma^2 = pq$ (see Problem 3.5.1). This is maximum when $p = 0.5$. You can see this by looking at the graph of the quadratic polynomial $p - p^2$ [that's just $pq = p(1-p)$].

(e) If $p = 0$, the player never sinks a basket. That's very predictable, predictably BAD! If $p = 1$, the player sinks a basket on every attempt. Again that's very predictable. The most unpredictable case is when the player sinks a basket exactly half the time. Then you never know what to expect.

3.7.12 $P(X = 0) = q^{60} = 0.046$. Of course, the selection was withOUT replacement, but $N \geq 60$ and presumably $N \geq 10n$ so the sampling WITH replacement approximation is valid. That's a special case of the binomial. Because N is not known, the hypergeometric calculation is not possible.

3.7.13 (a)

X	P(X)	XP(X)	X^2P(X)
0	0.0313	0.0000	0.0000
1	0.1563	0.1563	0.1563
2	0.3125	0.6240	1.2480
3	0.3125	0.9375	2.8125
4	0.1563	0.6252	2.5008
5	0.0313	0.1565	0.7825
	1.0002	2.4995	7.4991

$\mu = 2.4995$,
$\sigma^2 = 1.2516$.

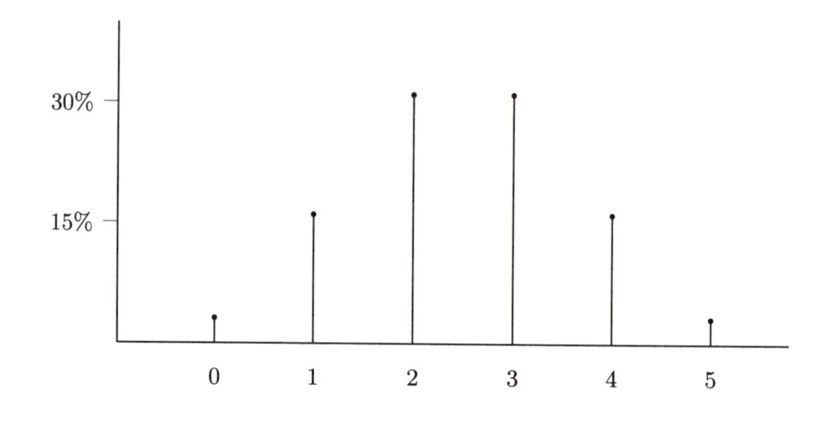

The formulas, of course, give $\mu = 2.5$ and $\sigma^2 = 1.25$. These values are exact, the values from the table suffer from rounding in the calculations.

(b)

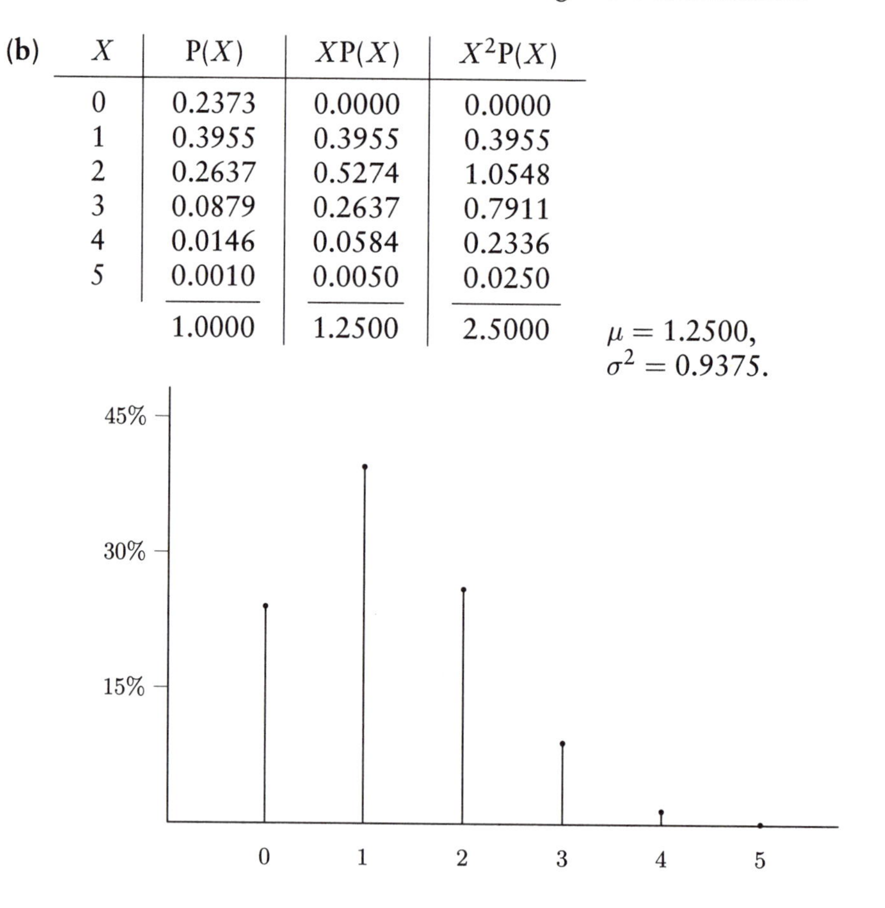

X	$\mathrm{P}(X)$	$X\mathrm{P}(X)$	$X^2\mathrm{P}(X)$
0	0.2373	0.0000	0.0000
1	0.3955	0.3955	0.3955
2	0.2637	0.5274	1.0548
3	0.0879	0.2637	0.7911
4	0.0146	0.0584	0.2336
5	0.0010	0.0050	0.0250
	1.0000	1.2500	2.5000

$\mu = 1.2500$,
$\sigma^2 = 0.9375$.

The formulas also give $\mu = 1.25$ and $\sigma^2 = 0.9375$.

(c)

X	$\mathrm{P}(X)$	$X\mathrm{P}(X)$	$X^2\mathrm{P}(X)$
0	0.5905	0.0000	0.0000
1	0.3281	0.3281	0.3281
2	0.0729	0.1458	0.2916
3	0.0081	0.0243	0.0729
4	0.0005	0.0020	0.0080
5	0.0000	0.0000	0.0000
	1.0001	0.5002	0.7006

$\mu = 0.5002$,
$\sigma^2 = 0.4504$.

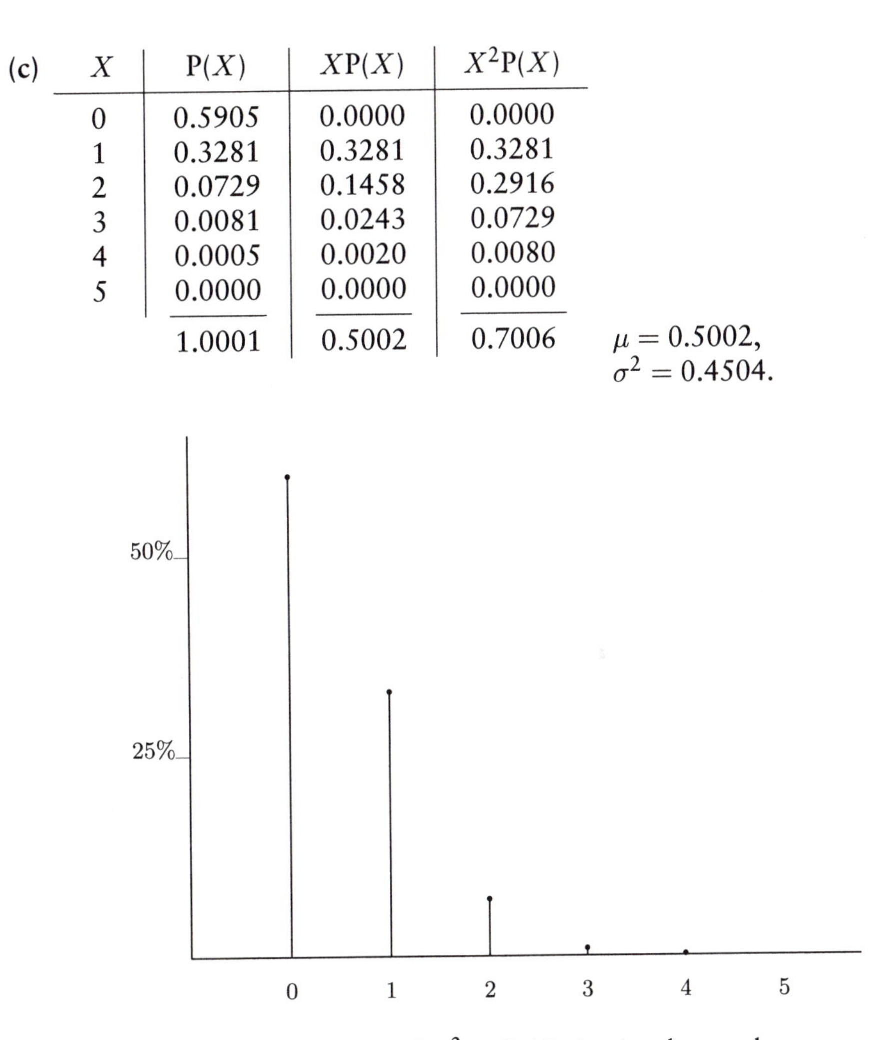

The formulas give $\mu = 0.5$ and $\sigma^2 = 0.45$. Again, these values are exact, the values from the table suffer from rounding.

3.8.1 (a)

$$\begin{aligned}\mathrm{P}(X < 2) = \mathrm{P}(X \leq 1) &= e^{-4.2}(4.2)^0/0! + e^{-4.2}(4.2)^1/1! \\ &= e^{-4.2}[1 + 4.2] \\ &= 0.0780.\end{aligned}$$

with the recursion formula

$$\begin{aligned} P(X \le 1) &= e^{-4.2} + [\lambda/(0+1)]e^{-4.2} \\ &= 0.014995577 + 0.062981423 \\ &= 0.0780. \end{aligned}$$

Note, you need not record the intermediate calculations. And certainly you should not round them and then calculate with the rounded values. We have recorded intermediate values here just so you can check your calculation. You should do calculations continuously within the calculator, storing intermediate values in memory if necessary. One of the principal advantages of the recursion formula is that it facilitates continuous computation without rounding.

(b) This is the same calculation, but now $\lambda = 2.1$:

$$\begin{aligned} P(X < 2) = P(X \le 1) &= e^{-2.1}[1 + 2.1] \\ &= 0.3796. \end{aligned}$$

Note that it's always true that $P(X \le 1) = e^{-\lambda}[1 + \lambda]$. With the recursion formula,

$$P(X \le 1) = 0.122456428 + 0.257158499 = 0.3796.$$

3.8.2 (a)

$$\begin{aligned} P(X > 2) &= 1 - P(X \le 2) \\ &= 1 - e^{-2.3}[1 + 2.3 + 2.3^2/2] \\ &= 1 - 0.5960 \\ &= 0.4040. \end{aligned}$$

(b) You could just imitate the calculation of part (a), but then you'll be repeating a lot of work. So we'll use the value from part (a). Here's how to do that:

$$\begin{aligned} P(X > 5) &= P(X > 2) - P(X = 3) - P(X = 4) - P(X = 5) \\ &= 0.4040 - e^{-2.3}[2.3^3/3! + 2.3^4/4! + 2.3^5/5!] \\ &= 0.4040 - 0.3740 \\ &= 0.0300. \end{aligned}$$

(c) Recursion requires a starting place. Start with

$$P(X = 0) = e^{-2.3} = 0.100258843.$$

Store this in memory and do NOT clear it from the display. This sets the situation up for recursion by calculating the initial probability which you feed into the recursion formula. Now, comes the recursion formula:[6]

- The current value of X is zero, so the next value is one. To get the probability of the next value, multiply the displayed number by 2.3 and divide by the next value:

  $$P(X = 1) = (2.3/1)P(X = 0) = 0.23059534.$$

  Add this to memory (there's a button on your calculator that adds the displayed value to memory). DO NOT CLEAR THE DISPLAY!
- Now, do it again. It's described just as before: To get the probability of the next value (the current value of X is one, so the next value is two), multiply the displayed number, $P(X = 1)$, by 2.3 and divide by the next value:

  $$P(X = 2) = 2.3/2P(X = 1) = 0.265184641.$$

  Add this to memory.
- Do it again: To get the probability of the next value, multiply the displayed number by 2.3 and divide by the next value:

  $$P(X = 3) = 2.3/3P(X = 2) = 0.203308225.$$

  Add this to memory.
- Continue this process two more rounds until you've finished with the "next value" being $X = 5$. Then add that last probability to memory. The memory now contains the sum of the probabilities from $X = 0$ through and including $X = 5$. That sum is $P(X \leq 5) = 0.970024306$.

That completes the calculation with recursion for $P(X \leq 5)$. We get the final answer as $1 - P(X \leq 5) = P(X > 5) = 0.029975693$. This rounds to the same answer we got above, 0.0300.

When we write all this out, it looks very formidable and certainly takes up a lot of space. But you aren't supposed to write ANY of this down. When you do the calculation continuously in the calculator, the whole calculation requires only 40 seconds! I just now did it and timed it with my favorite and only plastic Casio wristwatch which I carry in my pocket but is now sitting on this desk. No, THIS desk!

[6] This process will not work on some calculators.

(d) The Poisson recursion formula says: Multiply the probability of the current value by λ and divide by the next value. That gives the probability of the next value.

3.8.3

(a) Like the geometric distribution, the Poisson distribution takes on an infinite number of possible values. It's a count for which there is no upper limit—theoretically, that is.

(b) If you can imagine 15 accidents at an intersection in a given year, you can certainly imagine 16, and if you can imagine 16 you can imagine 17. So there's no limit, you can always imagine one more accident. Yet, it's hard to imagine 30,000 such accidents, say, or 30 million! There certainly cannot be more accidents then there are automobiles passing through that intersection. So there's a practical limit, but no theoretical limit. The theoretical model says that large values are possible, but highly unlikely.

3.8.4

You would have to look at the derivation of the formulas, but we don't study the very technical, mathematical derivation of the Poisson formulas. Furthermore, there's nothing in the three "rules of thumb" to help us. That the variance is the same as the mean should be considered a technical feature of the Poisson model. We won't attempt to understand these technicalities.

3.8.5

The second and third. Clearly, the second condition would not be satisfied. Furthermore, the incidence of contagious disease no longer looks random, so the third condition might well fail also.

3.8.6

No, not necessarily. There are other ways of deriving the Poisson model. The derivation from the binomial distribution—letting n go to infinity while p goes to zero—is only one possibility. If the rules of thumb hold, the model might work. But the model might still work from some very different justification.

3.8.7

(a) Over the 20 years from 1875 to 1895 for the ten Prussian army corps in question, there were 109 cases of no horsekick fatalities at all for some one of the corps over a one-year period.

(b)

B	CY	BCY
0	109	0
1	65	65
2	22	44
3	3	9
4	1	4
	200	122

(**c**) In 200 corps-years, 122 accidents were observed, that's a per-year average of $122/200 = 0.61$.

(**d**) If this data really "fits" the Poisson model, the mean and variance should be close to $\lambda = 0.61$, as indeed they are:

B	rf	B(rf)	B^2(rf)	
0	0.545	0.000	0.000	
1	0.325	0.325	0.325	
2	0.110	0.220	0.440	
3	0.015	0.045	0.135	
4	0.005	0.020	0.080	
	1.000	0.610	0.980	$\overline{B} = 0.6100$, $\sigma^2 = 0.6079$.

(**e**)

B	P(B)	BP(B)	B^2P(B)	
0	0.5434	0.0000	0.0000	
1	0.3314	0.3314	0.3314	
2	0.1011	0.2022	0.4044	
3	0.0206	0.0618	0.1854	
4	0.0031	0.0124	0.0496	
	0.9996	0.6078	0.9708	$\mu \approx 0.6078$, $\sigma^2 \approx 0.6014$.

Note that $\mathrm{P}(B > 4) = 0.0004$. So the missing values occur with very small probability. For example, show that

$$\mathrm{P}(B = 5) = 0.00038,$$

$$\mathrm{P}(B = 6) = 0.00004.$$

(**f**) Compare the distribution of observed data with the theoretical distribution. As you can see, the probabilities are very close. Such a comparison is not always so obvious. Later, we'll introduce a statistical technique to determine in a more precise way if observed data seems to fit a specific model. This is one of the major problems of statistics—to find "goodness-of-fit" tests, as they're called.

3.8.8

(**a**) $X \sim B(10, p)$. That is, X is binomial where $n = 10$ and p is the probability of YES. That means $p = \mathrm{P}(\text{YES}) = \mathrm{P}(Y > 1)$, where Y is "the number of horsekick fatalities in one corps-year." So Y is the Poisson random variable of the previous problem. From that Y we see that $p = 1 - 0.8748 = 0.1252$ and so $\mu_X = np = 1.2520$.

(b) YES is the same as in part (a). Here X is a geometric random variable with $p = 0.1252$. Again, just as in part (a), we have to reinterpret the question as asking for the mean of X. It's $1/p = 7.9872$.

In parts (a) and (b), note how we suppress irrelevant detail into the word "YES." By using just the word "YES" and ignoring the details of YES, we cast the question into a form where we can recognize the model. Once we see it's "repetitions of a Bernoulli trial" and once we've decided on the model (binomial or geometric), we turn to the probability question:

> What's the value of $p = \mathrm{P}(\text{YES})$? Now we focus on the meaning of YES and suppress all the other detail. What before was relevant (the other detail) is now suppressed; what before was suppressed (YES) is now relevant. The value of p is a separate question (Poisson) which must be answered before we can continue with the original question (binomial or geometric).

So we have a question within a question:

the original question, binomial

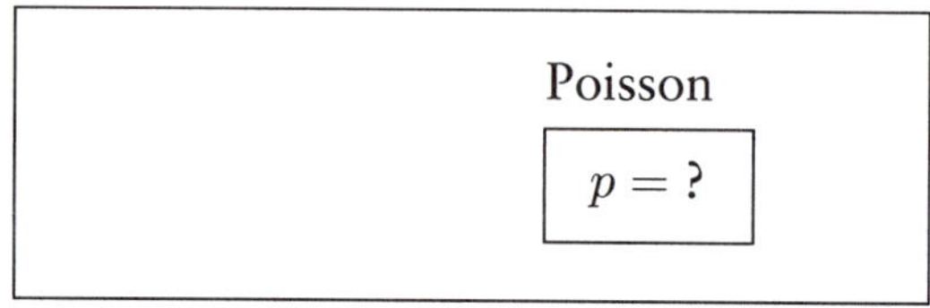

Don't get the two questions confused! Note that p is a probability, so there are numerous possibilites for answering the question "what is p?" Here p was calculated with the Poisson model but for some other situation p might be obtained in a very different way. The ability to sort a complex question into its component parts without confusing those parts is one of the important skills you'll develop as this course continues. It gets possible but it never gets easy! Sorry.

(c) X is hypergeometric, but that model is intractable here because it requires five very intricate calculations. Instead, because $N \geq 60$, in other words $N \geq 10n$, the sampling WITH replacement model is a good approximation. Of course, that's just a special case of the binomial. So we take X to be $B(15, p)$ where $p = 109/200$. Then using recursion, $\mathrm{P}(X > 4) = 1 - \mathrm{P}(X \leq 4) = 1 - 0.0279 = 0.9721$.

3.8.9 $\ln(14) = 2x\ln(e)$, so $x = 2.6391/2$.

3.8.10

(a) The Poisson distribution is determined by the value of its one and only parameter, λ. In general, you should use the Poisson distribution having the *same mean* as the binomial you are trying to approximate. That says you should choose the Poisson distribution which has $\lambda = np$. To approximate $B(300, 0.01)$, you should use the Poisson distribution with $\lambda = 3$.

(b) In each one-minute interval, you execute the Bernoulli trial only once. That says there can be at most one success observed in that interval. So no simultaneous occurrences.

The observance of a success in any one-minute interval will be independent of any other success because the Bernoulli trials are independent. So any two occurrences are independent.

Over the entire hour, you should expect np successes, the expected value of the binomial random variable. Now, in each subinterval, if you observe a success, it comes from one execution of the Bernoulli trial, with expected value p. So in any one interval, you expect p successes. This is an "expected" value don't forget, so it does not need to be an integer. In two intervals, you would be looking at two repetitions of the Bernoulli trial with expected value $2p$. The time is twice as much and the expected number of successes is twice as much. So you see the pattern: It says your expected number of occurrences in any interval is proportional to the length of the interval.

(c) $P(X = 2) = C(300, 2)p^2q^{298} = 0.2244$. If Y is the approximating Poisson with $\lambda = 3$, then $P(Y = 2) = 0.2240$.

(d)

	Binomial	Poisson
μ	np	λ
σ^2	npq	λ

For the approximating Poisson to have the same mean AND the same variance as the given binomial,

$$\lambda = np \quad \text{AND} \quad \lambda = npq$$

would both have to be equal to two *different* numbers. IMPOSSIBLE!! Unless $q = 1$, which we assume is not true. Why?

(e) $\lambda = np$ and $\lambda \approx npq$ together say $np \approx npq$ so that $q \approx 1$. This means $p \approx 0$. In other words, for the two variances to be approximately the same, *p must be small.* Now you see why we say p should be small. Of course, that corresponds to our idea that the Poisson distribution will be appropriate when we are looking at occurrences of some accidental (or rare) event.

3.9.1

(a) A specific outcome for this experiment will look like a string of S's and F's ending in an S, with exactly $k = 8$ S's .

(b)

$$F,\ S,\ F,\ S,\ S,\ S,\ S,\ S,\ F,\ F,\ F,\ S,\ S$$

or

$$F,\ F,\ F,\ S,\ S,\ S,\ F,\ F,\ F,\ S,\ F,\ F,\ F,\ S,\ F,\ S,\ S,\ S$$

or

$$S,\ S,\ S,\ S,\ S,\ S,\ S,\ S$$

(c) $X = 13,\ 18,\ 8.$

(d) The smallest possible value of X is eight (why?). There is no largest possible value. So, $X = 8, 9, 10, 11, 12, \ldots$ to infinity.

(e) The outcomes in part (d) have probabilities p^8q^5, p^8q^{10}, p^8, respectively. In other words, because the trials are independent you just multiply the probabilities.

3.9.2

(a) $X = 10$ means we observed our eighth success on the tenth repetiton of the Bernoulli trial. Because the tenth blank must contain an S, we need think only about the first nine blanks. How many ways can those nine blanks be filled with SEVEN S's, the other two to be filled with F's? That means we must choose seven of nine blanks for the S's. There are exactly $C(9, 7) = 36$ ways to do that.

(b) $C(11, 7) = 330$; (c) $C(8, 7) = 8$ (d) $C(7, 7) = 1$ (e) $C(x - 1, 7)$

3.9.3

Part (c) first:

(c) $X = 9$ means it takes us nine repetitions to obtain our eight successes. So we have exactly one failure. Here are the $C(8, 7) = 8$ ways we could have $X = 9$:

$$\underline{\text{F}},\ S,\ S,\ S,\ S,\ S,\ S,\ S,\ S$$
$$S,\ \underline{\text{F}},\ S,\ S,\ S,\ S,\ S,\ S,\ S$$
$$S,\ S,\ \underline{\text{F}},\ S,\ S,\ S,\ S,\ S,\ S$$
$$S,\ S,\ S,\ \underline{\text{F}},\ S,\ S,\ S,\ S,\ S$$
$$S,\ S,\ S,\ S,\ \underline{\text{F}},\ S,\ S,\ S,\ S$$
$$S,\ S,\ S,\ S,\ S,\ \underline{\text{F}},\ S,\ S,\ S$$
$$S,\ S,\ S,\ S,\ S,\ S,\ \underline{\text{F}},\ S,\ S$$
$$S,\ S,\ S,\ S,\ S,\ S,\ S,\ \underline{\text{F}},\ S$$

But, it's impossible for two of these to occur at the same time. In other words, THESE ARE MUTUALLY EXCLUSIVE EVENTS. So, we get the probability just by adding $p^8 q$ to itself $C(8,7)$ times:

$$C(8,7)p^8 q = 8 \times 0.42^8 \times 0.58 = 0.004492751 \text{ (round to 0.0045)}.$$

Do you understand the coefficient

$C(8,7)$

↑ This 8 is really $x - 1 = 9 - 1$. There are x blanks, but the last one MUST be filled with S. So, from the remaining $x - 1$ blanks, choose seven to put S's in. There are $C(x-1,7)$ ways to do this.

Now let's do the other parts of this problem.

(a) $C(9,7)p^8 q^2 = 36 \times 0.42^8 \times 0.58^2 = 0.0117.$

(b) $330p^8 q^4 = 0.0362$; **(d)** 0.0010; **(e)** $C(x-1,7)p^8 q^{x-8}$.

3.9.4 $C(x-1, k-1)p^k q^{x-k}$.

3.9.5 With $k = 1$ we get

$$\begin{aligned} P(X = x) &= C(x-1, k-1)p^k q^{x-k} \\ &= C(x-1, 0)pq^{x-1} \\ &= pq^{x-1} \quad \text{CORRECT!} \end{aligned}$$

And

$$\begin{aligned} P(X = x+1) &= \frac{xq}{x-0} P(X = x) \\ &= qP(X = x). \qquad \text{. . . CORRECT!} \end{aligned}$$

Show that the formulas for the mean and variance give the "correct" result.

3.9.6 **(a)** $\mu = k/p = 6/0.08 = 75$.

(b) $C(14,5)p^6 q^9 = 2002 \times 0.08^6 \times 0.92^9 = 0.0002$.

(c)

$$\begin{aligned} P(X \leq 15) &= P(X = 6) + P(X = 7) + \cdots + P(X = 15) \\ &= 0.0007 \quad [0.000695202]. \end{aligned}$$

The point here is to do a continuous calculation without stopping to write down intermediate results, losing accuracy by rounding (all the intermediate results would round to zero and you would get zero for an answer!). If you do a continuous calculation, the procedure is fast. Otherwise it's very slow, tedious, and inaccurate.

(d) $P(X \geq 16) = 1 - P(X \leq 15) = 0.9993$.

(e) Here, $k = 1$ and so you revert to the geometric random variable. You are asked for its expected value which is just $1/p = 12.5$.

(f) Here, $k = 2$ and the expected value is $2/p = 25$.

3.9.7 For us, $3\sigma = 88.1051$. So, there are no values more than three standard deviations below the mean. The values three standard deviations above the mean are all values bigger than $75 + 88 = 163$. Chebyshev says there's less than one chance in nine for a value larger than 163, but the true probability is virtually zero. After all, from part (d) of the previous problem, there's less than one chance in 1000 of a value larger than 16. Chebyshev's Theorem is not appropriate here because we have much more information than Chebyshev's Theorem assumes.

3.9.8 (a) The Bernoulli trial is "receive one bolt" for which "success" is "more than one defect." Thus, $p = P(\text{more than one defect in one bolt})$.

(b) Y is Poisson, a very typical case of the Poisson distribution. The parameter is $\lambda = 0.62$. Thus for the problem at hand, you find

$$p = P(Y > 1|\lambda = 0.62) = 0.1285.$$

(c) $P(X \geq 5|p = 0.1285) = 0.9923$, obtained as *one minus*

$$\begin{aligned} P(X < 5) &= P(X = 3) + P(X = 4) \quad [\text{3 is the smallest possible value}] \\ &= p^3 + 3p^3q \\ &= p^3(1 + 3q) \\ &= 0.0077. \end{aligned}$$

(**d**) You must assume that the bolts arrive "independently." That is, you must assume that whether one bolt has "more than one defect" is unrelated to whether another does, so that repetitions of the Bernoulli trial for X are independent. You must also assume that occurrences of defects satisfy the three rules for a Poisson distribution so that Y can be considered Poisson. But this is a very typical Poisson situation. Experience shows the Poisson model is usually valid. Both assumptions would seem reasonable under ordinary circumstances.

3.9.9

The number of rabbits caught in three traps, let's say, would be the sum of the numbers caught in each individual trap. But a sum of three independent geometric random variables with the same p is negative binomial. See the last comment in the text just before this set of "Try Your Hand" exercises.

3.10.1

(**a**) Geometric, $\mathrm{P}(X = 5|p = 0.1) = 0.0656$.

(**b**) Geometric, $\mathrm{P}(X \geq 5) = 1 - (1 - q^4)$; or binomial, $\mathrm{P}(X = 0|n = 4)$.

(**c**) Binomial, $\mathrm{P}(X \geq 4|p = 0.1, n = 10) = 1 - q^7[q^3 + 10pq^2 + 45p^2q + 120p^3]$.

(**d**) Geometric, $p = 0.7$, $\mu = ?$

(**e**) Both are binomial, compare variances: 0.9 with 0.475.

(**f**) Negative binomial, $p = 0.05$, $k = 4$, $\mu = ?$

(**g**) Uniform distribution of scores, $p = 0.25$, $\mu = ?$

(**h**) Compare variances of scores: 2.36, 8.1875. Either you set up probability distributions for your score and your opponent's score or you just enter the distributions into the statistical mode of your calculator.

(**i**) Negative binomial, $\mathrm{P}(X = 16|p = 0.25, k = 4) = 455p^4q^{12}$.

(**j**) $\mathrm{P}(A \text{ or } B) = \mathrm{P}(A) + \mathrm{P}(B) - \mathrm{P}(A \text{ and } B)$, A and B independent (reasonable?)

(**k**) Binomial, $\mathrm{P}(X = 4|p = 0.325, n = 10) = 210p^4q^6$.

(**l**) Binomial, $\mathrm{P}(X \geq 1|p = 0.1, n = 6) = 1 - \mathrm{P}(X = 0) = 1 - q^6$.

(**m**) This means your fourth miss occurs within the first five throws. So, negative binomial, $\mathrm{P}(X \leq 5|p = 0.05, k = 4) = \mathrm{P}(X = 4 \text{ or } X = 5) = p^4(1 + 4q)$.

(**n**) Let A = you don't get eliminated before that, B = at least one bull's eye in the first six throws. So: $\mathrm{P}(A \text{ and } B) = \mathrm{P}(B|A)\mathrm{P}(A) = 0.468559 \times 0.99997$.

Did you organize the given information for this problem appropriately? Otherwise, you waste a lot of time rereading the problem. Here's how:

Board	Points	P(your points)	P(opponent's points)
B	6	0.10	0.25
2nd	3	0.60	0.25
O	2	0.25	0.25
M	−2	0.05	0.25

Eliminated: more than three misses. One round: one throw for each player.

3.10.2

(a) Hypergeometric,

$$\begin{aligned}\mathrm{P}(X \le 1|N = 80, R = 30, n = 5) \\ &= \frac{50 \times 49 \times 48 \times 47 \times 46}{5!} \quad \frac{5!}{80 \times 79 \times 78 \times 77 \times 76} \\ &\quad + \frac{50 \times 49 \times 48 \times 47}{4!} \quad \frac{30 \times 5!}{80 \times 79 \times 78 \times 77 \times 76} \\ &= 0.088134717 + 0.287395815\end{aligned}$$

However, because $N \ge 60$ and $N \ge 10n$, the sampling WITH replacement model is a reasonable approximation (that's a special case of the binomial model, don't forget). So if you wanted to save some effort and were willing to accept an approximate answer,

$$\mathrm{P}(X \le 1|p = 0.375, n = 5) = q^5 + 5pq^4 = q^4(q + 5p) = 0.3815.$$

(**b**) $q^{47}\left[q^3+50pq^2+1225p^2q+19600p^3\right] = 0.0137$. What unstated assumption does this analysis require?

3.10.3

(**a**) If we can assume you won't find two potholes located on top of each other (no "simultaneous occurrences"), that they occur independently and that the number in any stretch of road is proportional to the length of the stretch, then the Poisson model is appropriate. $\mu = \lambda = 8 \times 0.2$ because two miles is eight "quarter-mile" stretches.

(**b**) Binomial, $\mathrm{P}(X < 3|n = 10, p =?)$, where $p = \mathrm{P}(Y > 2) = 0.2166$ for a Poisson Y. So (without rounding p),

$$\mathrm{P}(X < 3) = q^8[q^2 + 10pq + 45p^2] = 0.6272.$$

(c) Geometric, $P(X \geq 6|p = 0.7834) = q^5 = 0.0005$. Here X is the number of the first day when they find two or fewer potholes. So S is "two or fewer potholes" and $p = 1 - 0.2166$ [from part (b)].

(d) Binomial, let X = # NONbeach days:

$$P(X \geq 4|n = 20, p = 0.2166) = 1 - q^{17}[q^3 + 20pq^2 + 190p^2q + 1140p^3].$$

(e) Exactly: $\mu = 15.668$.

3.10.4

(a) $P(X = 5|\lambda = 3) = 3^5(e^{-3})/120 = 0.1008$.

(b) $X = 0, 1, 2, 3, 4, 5, 6$. The condition in the problem gives all integer values between -0.4641 and 6.4641. Note that $\sigma^2 = \lambda = 3$ and so $\mu \pm 2\sigma = 3 \pm 3.4641$.

3.10.5

(a) Expect a loss of about \$0.64 per roll. Set up a probability distribution table for X = # dots on the top face. You'll find that $\mu_X = 3.36$ and $\sigma_X^2 = 2.1504$. So if G is your net gain/loss random variable, $G = X - 4$ and $\mu_G = 3.36 - 4 = -0.64$. Or you can make a table for G directly.

(b) The game with the loaded die is more predictable because it has the smaller variance. Either by setting up a distribution table or by using the statistical mode of your calculator, you can find that with a fair die the variance of your gain/loss is 2.9167 which is larger than for the game with the loaded die. This means with the loaded die you're likely in the long run to come out closer to your expected loss of \$0.64.

(c)

G	$P(G)$	$GP(G)$	$G^2P(G)$
-3	0.12	-0.36	1.08
-2	0.12	-0.24	0.48
-1	0.40	-0.40	0.40
0	0.12	0	0
1	0.12	0.12	0.12
2	0.12	0.24	0.48
	1.00	-0.64	2.56

$\mu_G = -0.64$,
$\sigma_G^2 = 2.1504$.

(d) Done in part (a).

(e) $\mu_X - 4 = \mu_G$ and $\sigma_X^2 = \sigma_G^2$.

(f) $\mu = 1/p = 1.5625$, where $p = 0.64$.

(g) X is geometric, $P(X = 3) = q^2 p = 0.0829$.

(h) $P(X > 3) = q^3 = 4.67\%$.

(i) X negative binomial with $k = 5$. $P(X = 15) = C(14, 4)p^5 q^{10} = 0.0039$.

3.10.6 (a) $P(X \leq 1 | n = 12, p = 0.03) = q^{12} + 12pq^{11} = 0.9514$.

(b) $\mu = np = 0.36$.

(c) Before, the variance for "# defectives" was 0.3492, now it's 0.0952.

(d) Geometric, X = # boxes inspected to find one with a defective part. We need $\mu_X = 1/p$. Here $p = 0.3062$. How?

$$p = 1 - P(\text{no defectives in box}) = 1 - P(Y = 0) = 1 - q^{12} = 0.3062.$$

(e) Same X as in part (d), $P(X = 5) = q^4 p = 0.0710$.

3.10.7 (a) $\mu = np$.

(b) The variance is npq times the finite population correction factor. But that's the same for both situations, and so is $n = 5$. Thus, we need only compare the pq's. If you're looking for spades, $pq = 3/16$ which is smaller than $pq = 1/4$ for black versus red. So spades versus nonspades is more predictable. The exact variances are 0.8640 and 1.1520.

(c) $\mu = np = 2.5$.

(d) Hypergeometric, $P(X = 2 | N = 52, R = 13, n = 5) = 0.2743$.

(e) $P(X = 2 | N = 52, R = 26, n = 5) = 0.3251$.

(f) $1/p = 3.0760$.

(g) Geometric, with X as the number of deals required to get a YES hand. So $P(X \geq 4) = q^3 = 0.3074$. Or interpret it as binomial with Y as the number of YES hands in the first three deals. So you want $P(Y = 0) = q^3$, giving the same answer as with X. Of course, you have to get the same answer! With both interpretations, a YES hand is a hand with two red cards [so $p = 0.3251$ from part (e)].

3.10.8 (**a**) First, write out all the possible matches of balls with the court and players, where we number the balls as 1, your ball; 2, your friend's ball; 3, court's ball. Then record X = # players with their own ball:

You	Friend	Court	X
1	2	3	2
1	3	2	1
2	1	3	0
2	3	1	0
3	1	2	0
3	2	1	1

So

X	P(X)	XP(X)	X^2P(X)	
0	3/6	0	0	
1	2/6	2/6	2/6	
2	1/6	2/6	4/6	
	1	2/3	1	$\mu = 2/3$, $\sigma^2 = 5/9$.

Here σ^2 is $1 - (2/3)^2$.

(**b**) P(I take my ball)= (# balls that are mine)/(total # of balls) $= 1/3$. This is just the "theoretical relative frequency" definition of probability.

(**c**) $\text{P}(X \geq 1) = 2/6 + 1/6 = 3/6$.

(**d**) $\text{P}(A \text{ and } B) = \text{P}(A)\text{P}(B) = 1/9$, because evidently the events are independent. Otherwise you would need to know $\text{P}(A|B)$.

(**e**) Binomial, $\text{P}(X \geq 4|n = 7, p = 1/6) = 1 - \text{P}(X \leq 3) = 0.0176$.

(**f**) $\mu = np = 1.1667$.

(**g**) Geometric, $\mu = 1/p = 6$.

(**h**) Now,

X	$P(X)$	$XP(X)$	$X^2P(X)$
0	7/12	0	0
1	4/12	4/12	4/12
2	1/12	2/12	4/12
	1	1/2	2/3

$\sigma^2 = 2/3 - 1/4 = 5/12.$

Because $5/12 < 5/9$, this situation is MORE predictable. You might have guessed that by thinking of an extreme case. Suppose there were several million balls on the court. Then it's HIGHLY predictable that you and your friend both go home with a different ball than you brought.

(i) Yes?

3.10.9 (a) Binomial, $n = 17, p = 0.1$, $E(X) = 1.7$.

(b) $P(X > 2) = 1 - P(X \leq 2) = q^{15}[q^2 + 17pq + 136p^2] = 0.2382.$

(c) You should guess MORE unstable because there is more chance for a problem. The stability is measured by the variance for the number of breakdowns in the first month after the maintenance check (our X above). That variance is npq. In the second case, it's larger because n is larger.

(d) You're looking for the occurrence of a breakdown in the operation in the course of one week immediately following a maintenance check. It's Poisson, $P(X \geq 1|\lambda = 0.18) = 1 - 0.8353 = 0.1647.$

(e) $P(X > 1|n = 9, p = 0.1647) = 1 - q^8[q + 9p] = 0.4507.$

(f) There are at least three ways you could model this. Two are binomial, the third is geometric:

(i) $P(X = 0|n = 6, p = 0.1647) = q^6$, here X = # months with at least one breakdown.

(ii) $P(X = 6|n = 6, p = 0.8353) = p^6$, here X = # months with NO breakdown.

(iii) $P(X > 6|p = 0.1647) = q^6$, now X = # of the first month for which there's at least one breakdown.

In each case, you're computing $(0.8353)^6 = 0.3397$.

(g) $\mu = 1/p = 6.0716.$

3.10.10 (a) Poisson (lambda is 5.19 for one bolt), $E(X) = 5.19$.

(b) $1 - e^{-5.19}[1 + 5.19 + 13.4681 + 23.2997] = 0.7606$.

(c) $\mu = np = 12 \times 0.8227 = 9.8726$. Here $p = 0.8227$ is calculated as Poisson, $P(X = 0|\lambda = 1.73)$.

(d) $\mu = 1/p = 1.2155$.

3.10.11 It's possible for you to make a frequency distribution, but because relative frequencies are given, you should make up a *relative* frequency distribution. You'll find: $\Sigma X(\text{rf}) = 1.09$; $\Sigma X^2(\text{rf}) = 2.33$. From this distribution you get:

(a) median = 1; $\mu = 1.09$; $\sigma = 1.0686$; mode = 1; range = 4.

(b) the answer to this question is ΣXf, but from the relative frequency distribution you only get $\Sigma X(\text{rf})$. If you multiply by N you will get ΣXf. Answer: 120×1.09. Note that this question asks for the "total of all the data." That total divided by N is the mean, so "total of all the data" $= N \times \mu = \Sigma Xf$.

(c) Guess that the variance is lower because you're giving more weight to a value relatively close to the mean. Of course the mean also becomes smaller, so this is not obviously correct—still, the mean should not be lowered by much. Now take a specific case: Suppose 7% of employees take three days and none take four days: $\sigma = 0.8942$. Don't waste a lot of time with these calculations. They can be done quickly with the statistical mode on your calculator. And a more sensitive case: Suppose only one-tenth of one percent took three days. Now, $\sigma = 1.0663$ and still it is lower than when none took three days.

(d) $P(X \geq 2|n = 10, p = 0.07) = 1 - 0.8483 = 0.1517$.

(e) $\mu = 1/p = 1/0.07 = 14.2857$.

(f) Geometric, $P(X > 20) = (0.93)^{20} = 0.2342$.

3.10.12 (a) 5/12; (b) 4/3.

(c) The variance in each case is $2pq$ times the finite population correction factor, $7/8$. For the second situation, $pq = 20/81$ is larger. Hence, that's the less predictable situation.

3.10.13 Each of the three random variables counts the number of tails for its random experiment.

(a) $\mu_X, \mu_Y = 1/2$; $\mu_{X+Y} = 1$. It is not possible to guess the variance because it's only a comparative measure of spread about the mean and you have nothing to compare with.

(b) Z should be more variable just because it has a larger range of values, although that doesn't GUARANTEE more variability. You have to think about the concentration of probability. It's hard to make a meaningful comparison between two random variables which do not take the same values.

(c) The variances of X and Y are each $1/4$; the variance of Z is $1/2$. You can do this directly from the probabiliity distribution of Z or from the formulas relating Z to X and Y. Here are the general formulas: If $Z = aX + bY$ then $\mu_Z = a\mu_X + b\mu_Y$; and if X and Y are independent, $\sigma_Z^2 = a^2\sigma_X^2 + b^2\sigma_Y^2$.

(d) Set up a probability distribution and compute.

3.10.14 (a) min $= 1.2$, max $= 1.5$. $M \neq 1.4$, that's the SAMPLE median. M is the population median. Or, if you're sampling from the probability distribution of a random variable X, M is the value of X such that $P(X < M) = 0.5$. The value of M is not known; we're talking about estimating that unknown number from the sample.

So M is the QUESTION. As a first guess, we're going to estimate M to be somewhere between min and max, somewhere in the interval $(1.2, 1.5)$. That interval is our "confidence interval estimate" for M. The "confidence coefficient" is the probability that M is actually in this interval. In part (f), we'll show that the confidence coefficient is essentially 100%.

(b) Y is $B(n, 0.5)$.

(c) If $M <$ max, there must be at least one observation above the median. So,

$$P(M < \text{max}) = P(Y \leq n - 1) = 1 - P(Y = n) = 1 - 0.5^n.$$

(d) $P(M > \text{min}) = P(Y \geq 1) = 1 - P(Y = 0) = 1 - 0.5^n$.

(e) Let A be the event that min $< M$ and let B be the event that $M <$ max.

Then the confidence coefficient is

$$\begin{aligned} P(A \text{ and } B) &= 1 - P(A^c \text{ or } B^c) && (A \text{ and } B)^c = A^c \text{ or } B^c \\ &= 1 - [P(A^c) + P(B^c)] && A^c, B^c \text{are mutually exclusive} \\ &= 1 - [0.5^n + 0.5^n] && \\ &= 1 - 0.5^{n-1} && 1 - (2 \times 1/2^n) \end{aligned}$$

Note how the confidence coefficient depends only on the sample size. For a sample of size three, we would be only 75% sure. It increases from there.

(f) $1 - 0.000000238 = 0.999999761$. We've used the formula $1 - 0.5^{n-1}$ from part (e) with $n = 23$. So we can be essentially 100% sure the true median is somewhere between 1.2 and 1.5.

(g) It's virtually certain that the median weight of U.S. pennies is somewhere between 2.99 and 3.21 grams. Note that $0.5^{99} = 1.5772 \times 10^{-30}$, a number with 29 zeros before the first significant digit after the decimal. This confidence coefficient is very wasteful. A confidence coefficient of 90% or 95% would usually be considered adequate and will yield a much more useful estimate for M. In Chapter 4, we'll show you how to get that better estimate [see part (h)].

(h) We can be 99% sure that the median weight of U.S. pennies is somewhere between 3.11 and 3.13 grams. However, this statement is subject to misunderstanding. The median weight of U.S. pennies is a fixed number (a parameter) and is NOT subject to probability statements. It's the INTERVAL which varies here—it varies from sample to sample. A more careful interpretation of our confidence interval would say, "We are confident that the median weight of U.S. pennies is somewhere between 3.11 and 3.13 grams because, using our technique, 99% of all samples of 100 pennies would yield an interval containing the median weight." Of course, one percent of such samples would give an erroneous interval. We cannot know if our interval is one of the erroneous ones or not. That's why we say "confident" and not "certain." That's also why these are called "confidence intervals"—we're "confident" the interval contains the parameter, but not certain.

(i) M is a fixed number. It can't be sometimes one place and sometimes another! The 99% confidence coefficient refers to the INTERVAL, not to the parameter. It's the INTERVAL that's sometimes here and sometimes there—sometimes it contains M and sometimes not. You see how this is true in the calculations of parts (b) and (c) where the probabilities were calculated from a binomial random variable associated with the sampling experiment, counting how many observations in the sample are greater than the median. So the probability—the confidence coefficient—goes with the SAMPLE. It's the sample that varies, and because the interval is calculated from the sample, the interval varies. See part (h).

(j) It's virtually certain the sample median is less than the true median, M. After all, the sample median is 3.11 and there's a 99% chance M is between 3.11 and 3.21 [see part (h)].

Chapter **4**

4.1.1 There are infinitely many values. If there were n equally likely values, each would have to have probability $1/n$. As n gets larger and larger, this probability gets smaller and smaller. So it seems reasonable to think that when n becomes "infinite," the probability would have to be zero. For a continuous distribution, there are so many possibilities that any one value has probability zero. This intuitive argument is helpful, but in fact, mathematically the situation is more sophisticated than that. After all, you already know examples of random variables with an infinite number of values where the probabilities of individual values are NOT zero. When the values are discrete, this is possible. The geometric, Poisson, and negative binomial distributions are examples.

But when the possible values of the random variable constitute a continuous interval, it's not possible for specific individual values to have nonzero probability. There are just too many possible values. We're asking you to believe this, it's not obvious and it's not easy to prove.

You'll get some idea of what's going on here if you recall that our discrete models which are infinite—the geometric, Poisson and negative binomial distributions—are all counts. So, although all counts are theoretically possible, after some point, all the probabilities are so small they become zero from a practical point of view. That leaves only a finite number of values with significant probabilities. This is what does not happen for a continuously distributed random variable.

So any specific value of a continuously distributed random variable has probability zero. It's a little like throwing a dart at a dart board. There are infinitely many points on the board where the dart might lodge; the board is a continuous two-dimensional "interval." Imagine marking one tiny point with a very sharp pencil. What is the likelihood you will be able to hit that exact point with the dart? ZERO!

4.1.2 . . . some SUBINTERVAL. Although any specific value of a continuously distributed random variable is virtually impossible, it IS possible to fall within some specified range of values, a subinterval. It is virtually impossible to have the dart lodge in an exact tiny point. But if we encircle a small—or not so small—region of the dart board, then there is a genuine possibility (nonzero probability) of throwing the dart into that region. That's why the bull's eye on a dart board is always a small region. It's never just a point.

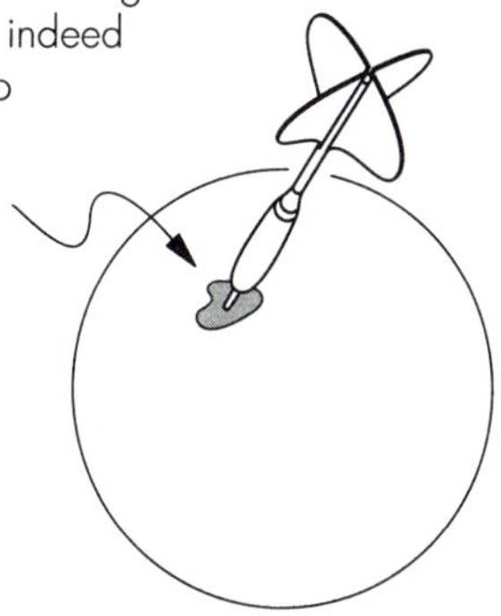

If you encircle a small region of a dart board, it's indeed possible (non-zero probability) to throw the dart into that region

Now, if I'm throwing darts at a one-dimensional interval of real numbers, any given point would be virtually impossible to hit, but a small subinterval would be possible. In other words, there's a nonzero probability to succeed in throwing the dart within some specified range of values. So within a specified range of values, a continuously distributed random variable can be thought of as a mechanism for "throwing a dart at an interval of real numbers" with a significant possibility to fall within any given subinterval. Thus, the nonzero probabilities for a continuous distribution are associated with subintervals—with ranges of values—not with specific values.

4.1.3

(**a**) The shaded area represents $P(0 < X < 100)$. That probability is necessarily ONE because X only takes on values in that interval. In other words, it is certain that X will fall within the interval $(0, 100)$.

(**b**) Obviously, the shaded area is $1/2$ because it is $P(X < \text{median})$.

(**c**) By Chebyshev's Theorem, the probability that X is within three standard deviations of its mean is at least $1 - 1/3^2 = 8/9$. Because that's the indicated range of values, the shaded area must be $8/9$ or more.

4.1.4

The total probability is 1 and there are $n + 1$ segments. So each segment of the distribution represents a probability of $1/(n + 1)$. Now, between $x_{(h)}$ and $x_{(k)}$ there are $k - h$ segments. So the probability of being between these two values is $1/(n + 1)$ added to itself $k - h$ times.

By direct counting, there are $n + 1$ ways for a future observation to distribute itself into a current sample of size n. Because all the observations are independent—that is, the n observations in the sample and the one future observation—these $n + 1$ ways are equally likely. So the denominator of your probability is $n + 1$. Now, among those $n + 1$ ways, how many correspond to the future observation being smaller than the the ith member of the sample? Answer: i. So, how ways can one future observation distribute itself into the sample so that it's larger than the hth and smaller than the kth observation in the sample. Answer: $k - h$. That's the numerator of your probability. In both arguments, the continuity of the distribution from which you're sampling is not essential; it just allows us to avoid dealing with the possibility that two observations might be equal.

4.1.5

(a) Each of these intervals has length one-fourth. Because they all have the same length, they must all have the same probability. That's the definition of the uniform distribution. If that probability is denoted by p, then

$$p + p + p + p = 1, \qquad \text{in other words, } 4p = 1, \text{ so } p = 1/4.$$

From this you can see that $\mathrm{P}(2.25 \leq X \leq 2.5) = 1/4$.

(b)

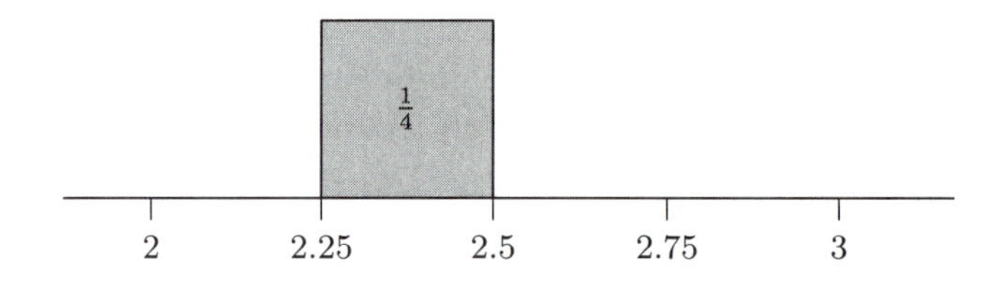

(c) Because X is bound to take a value somewhere between two and three,

$$\mathrm{P}(2 \leq X \leq 3) = 1.$$

But the "event" that $2 \leq X \leq 3$ can be split into the four MUTUALLY EXCLUSIVE events:

$$(2 \leq X \leq 2.25) \text{ OR } (2.25 \leq X \leq 2.5) \text{ OR } (2.5 \leq X \leq 2.75) \text{ OR } (2.75 \leq X \leq 3),$$

and so the simple addition rule for mutually exclusive events holds. Note that the events actually have endpoints in common, so technically they are not mutually exclusive. But because the probability of a specific value—an endpoint—is zero, we can ignore that. With a continuous distribution, you can always ignore an equal sign!

4.1.6 The curve will have to be a HORIZONTAL LINE!! Nothing else will work. In terms of $f(x)$, this means the density function must be a constant function.

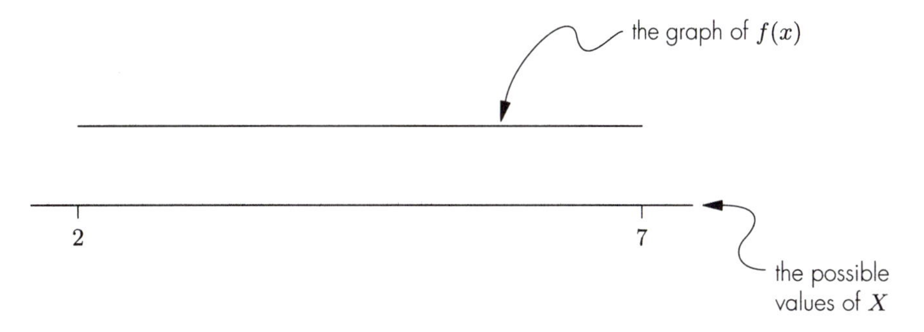

Note how any two subintervals of the same length will have the same probability:

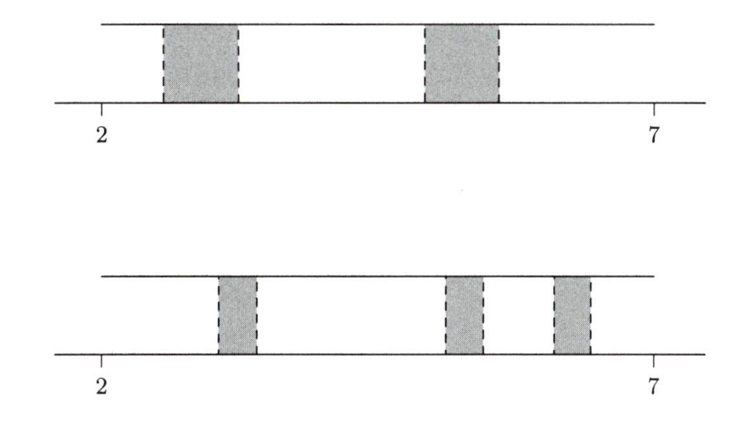

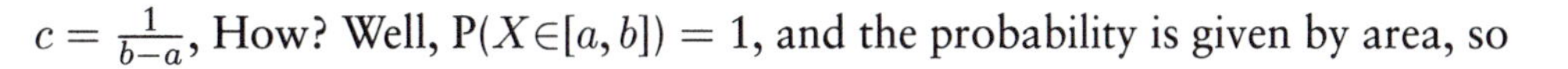

4.1.7 $c = \frac{1}{b-a}$, How? Well, $\mathrm{P}(X \in [a, b]) = 1$, and the probability is given by area, so

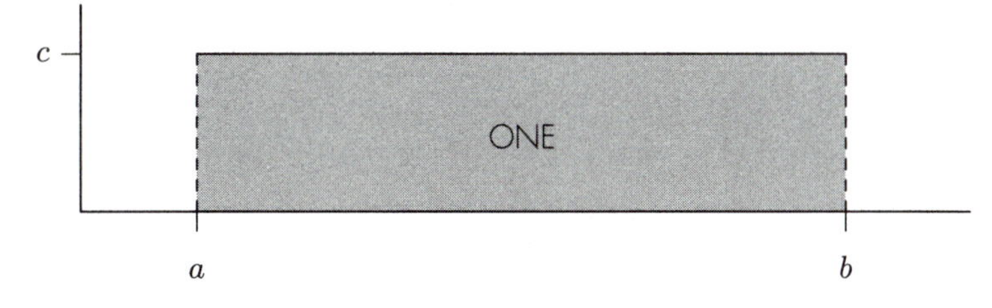

Thus the area of this rectangle is ONE, but the area of a rectangle is "length times height." The length here is $b - a$ and the height is c. So

$$(b - a)c = 1$$

which just says that c is $1/(b - a)$.

4.1.8 (a) From the graph you can see that the median, the point on the x-axis that divides the distribution 50/50, is the half-way point between a and b. Thus $\mu_X = (a+b)/2$.

Some people forget why the average of two numbers—here $(a+b)/2$, the average of a and b—is exactly the half-way point between them. Well, half the distance between a and b is just $(b-a)/2$, so the half-way point is just a plus that: $a + (b-a)/2$. Here's the picture:

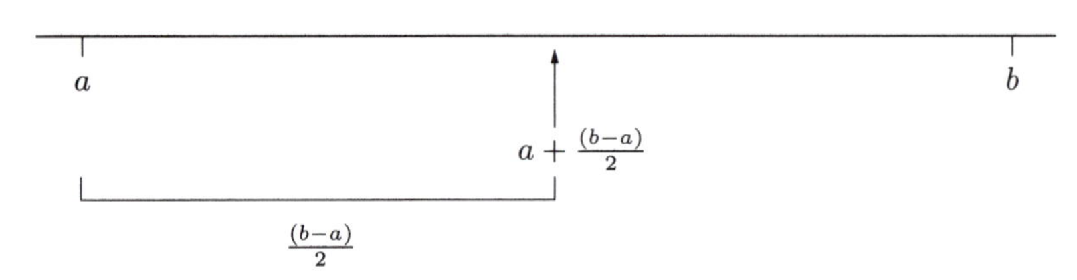

Now just simplify:

$$a + \frac{b-a}{2} = \frac{2a+b-a}{2} = \frac{a+b}{2}.$$

(b)

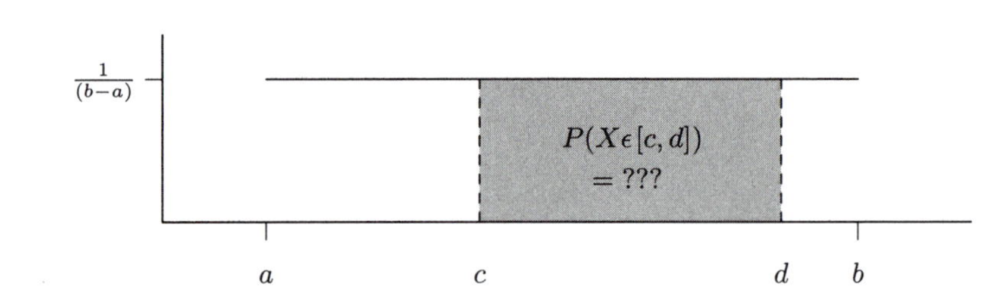

$P(X\in[c,d])$ is just the area of the shaded rectangle. Its height is $1/(b-a)$, as you saw in Problem 4.1.7 (just because the total area is ONE). The length of the base is $d-c$. So the area is the product of those two numbers:

$$P(X\in[c,d]) = (d-c)/(b-a).$$

(c) The probability that X takes on a value between c and d is proportional to $d-c$. In other words, the probability that a uniformly distributed random variable falls within a certain interval is proportional to the length of the interval.

4.1.9 (a) $X \leq x$ means X takes a value somewhere less than or equal to x. But $X \geq a$ because all the values fall between a and b. So $X \leq x$ means $X\in[a,x]$. Now, applying the probability formula with $x_1 = a$ and $x_2 = x$ gives $P(X \leq x) = P(X\in[a,x]) = (x-a)/(b-a)$.

(b)

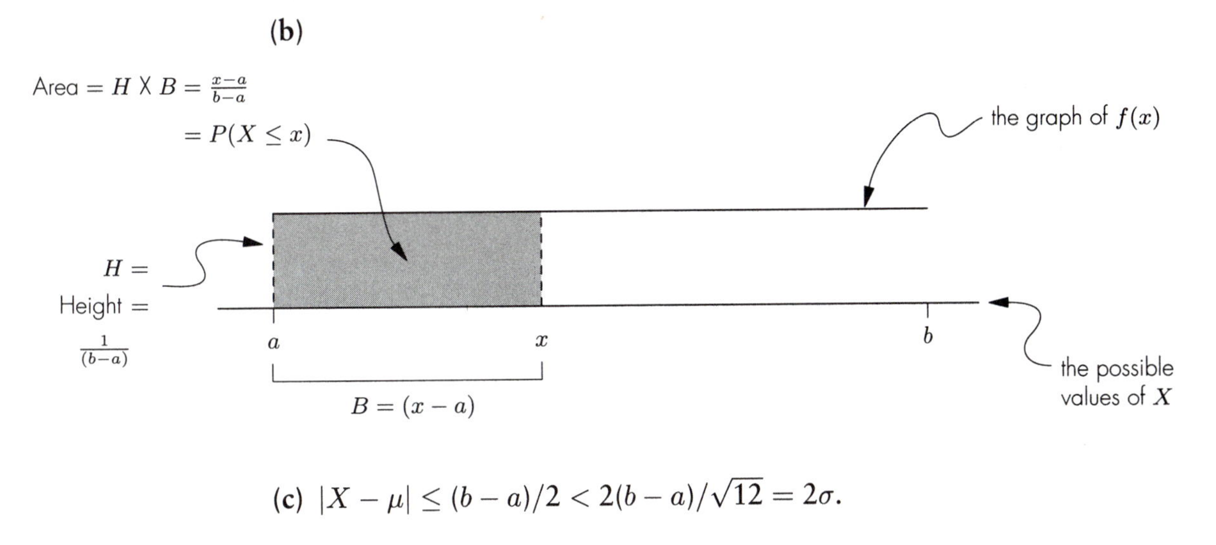

(c) $|X - \mu| \le (b - a)/2 < 2(b - a)/\sqrt{12} = 2\sigma$.

4.1.10

(a) Here's the picture:

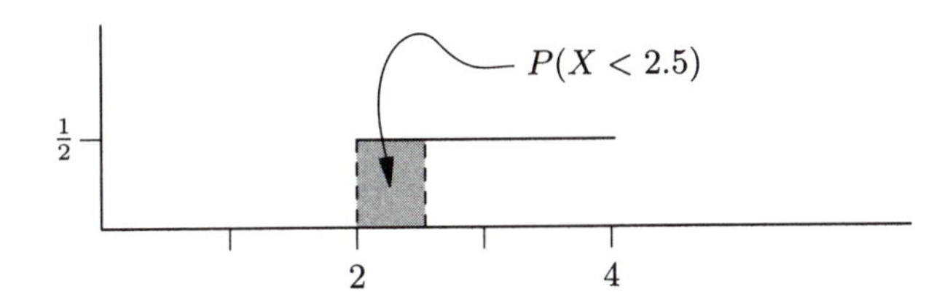

Here $f(x) = 1/(b - a) = 1/(4 - 2) = 1/2$, and so its graph is just the horizontal line $y = 1/2$ over the interval $[2, 4]$. Because the shaded area is a rectangle with sides of length one-half, clearly we get

$$P(X < 2.5) = \frac{1}{2} \times \frac{1}{2} = \frac{1}{4}.$$

This answer also comes from the formula of the previous problem:

$$P(X < 2.5) = (2.5 - 2)/(4 - 2) = 0.5/2 = 0.25.$$

Note that we need not be concerned that the formula in the previous problem is for $X \le x$ instead of $X < x$ because the probability that $X = x$ is zero.

(b)

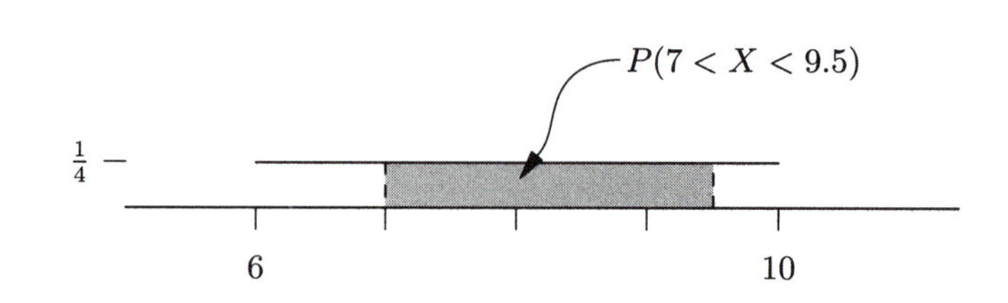

Here it's obvious from the picture that the shaded area is $2\frac{1}{2} \times \frac{1}{4} = 5/8$. Or instead of the picture, look at the formula

$$P(X \in [7, 9.5]) = (9.5 - 7)/(10 - 6) = 2.5/4 = 0.625.$$

(c)

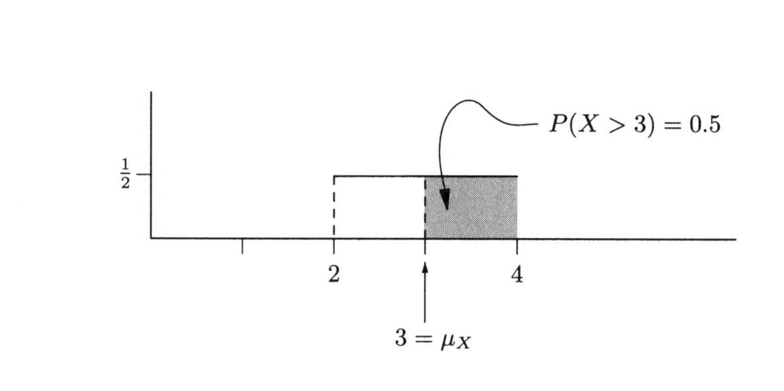

(d)

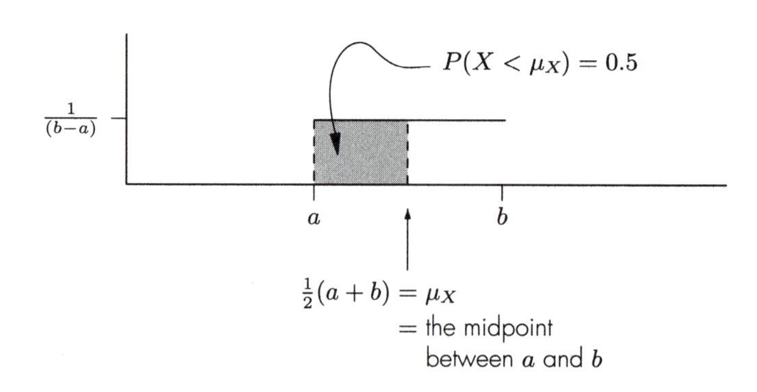

From the formula, $\frac{b-\frac{1}{2}(a+b)}{b-a} = 1/2$.

(e) Here $\mu = 8$ and $\sigma^2 = (6 - 10)^2/12 = 16/12 = 4/3$ and so $\sigma = 1.1547$. Thus we need

$$P(6.2679 < X < 9.7321) = 0.8661.$$

This is a probability of the form $P(X\in[c, d])$ where the endpoints are just

$$8 \pm 1.5\sigma = 8 \pm 1.7321.$$

4.2.1

(a) Here are the three rules of thumb:

(i) Two *simultaneous failures* are not possible because together they would simply constitute one failure—the system fails.

(ii) Again, when we say that failures are due to external causes such as random overloads, and so on, we are implicitly assuming the *independence* of two failures. This independence assumption would not necessarily be true, but often it is. If it's true, our Poisson model will be appropriate. Think about a household fuse blowing as a result of a current overload. Typically, two such events would be independent of each other. Similarly, if you think about other examples we have mentioned (or will mention), you will see that this assumption of independence during the period of useful life would often be satisfied.

(iii) The third rule of thumb, *the proportionality rule*, simply says that during half of the period of useful life, we should expect half as many failures, during one-third of the period, one-third as many failures, and so on. That kind of porportionality assumption would be very reasonable in many situations. When it's reasonable, our Poisson model will be valid.

(b) $P(X = 0) = \frac{e^{-t\lambda}(t\lambda)^0}{0!} = e^{-t\lambda}$.

4.2.2

(a) The real-world "doing" for the underlying experiment is the same as X in Problem 4.2.1, "operate the system during the period of useful life." However, the outcomes are different. Here, an outcome would be a time period, the period of time before the first failure. T assigns a number to that period of time, namely, its length. Again, as above, the occurrence of a failure cannot be predicted in advance. Thus, T is indeed a random variable.

(b) A continuous random variable is a random variable which takes on any value in some interval of real numbers. Once our system enters its period of useful life, any length of time whatsoever is theoretically possible as a value of T. Because any value whatsoever in an interval of time is possible as a value of T, it should be considered to be continuously distributed as a random variable.

(c)

$$\begin{aligned} P(T \le t) &= 1 - P(T \ge t) && \text{the equals sign is irrelevant, why?} \\ &= 1 - P(X = 0) && \text{with } X \text{ as the number of failures on the interval 0 to } t \\ &= 1 - e^{-t\lambda}. \end{aligned}$$

4.2.3

(a) The number of calls received in a period of time is reasonably modeled by the Poisson distribution with λ as the expected number of calls per minute. So, $\lambda = 1/5 \times 2.3 = 0.46$ because for the Poisson distribution we assume the expected number of occurrences in any interval to be proportional to the length of the interval. Now for the exponential distribution, the expected value is $1/\lambda = 1/0.46 = 2.1739$.

(b) Poisson with $\lambda = 3 \times 2.3 = 6.9$. So using recursion,

$$P(X > 7) = 1 - P(X \le 7) = 1 - 0.6136 = 0.3864.$$

(c) $P(T > 10) = e^{-0.46 \times 10} = 0.0101$. Note that we're using the fact that

$$P(T > t) = e^{-t\lambda}$$

because

$$P(T \le t) = 1 - e^{-t\lambda}.$$

4.2.4

(a) $\mu_T = 2.5$ (in thousands of hours). Here $\lambda = 0.4$ (notice that $\lambda = 1/\mu_T$) and we're asked for $P(T \le 1) = 1 - e^{-0.4} = 1 - 0.6703 = 0.3297$.

(b) The number of burn outs has a Poisson distribution. The expected number of burn outs per 1000 hours is $\lambda = 1/2.5 = 0.4$. So in 10,000 hours the expected number should be $10 \times 0.4 = 4$ because the expected number should be proportional to the length of time (assuming the Poisson model valid). Because we expect four components to burn out, we will need five components to keep our system running for 10,000 hours.

Or think in terms of the expected life of the system when you have M components available. When $M = 1$, the expected life is given to be 2500 hours. You require a value of M for which the expected life is greater than 10,000 hours: $\mu_\lambda = 2.5M > 10$. This implies $M > 4$, so buy $M = 5$ components.

(c) Let T_1 be the time to failure for the first component and T_2 the time to failure for the second component. So, we have

$$\begin{aligned}
\text{P(system runs for more than 1000 h)} &= \mathrm{P}(T_1 > 1 \text{ AND } T_2 > 1) \\
&= \mathrm{P}(T_1 > 1) \times \mathrm{P}(T_2 > 1) \\
&= e^{-0.4} \times e^{-0.4} \\
&= 0.4493.
\end{aligned}$$

But, to make use of the multiplication rule, we must assume the two components of the pair operate INDEPENDENTLY of each other. That might or might not be reasonable; you'd better ask the engineers. Without independence, you would need the conditional probability of a burn out for one component, given that the other one burns out. But we don't know that.

4.2.5

(a)

$$\begin{aligned}
\mathrm{P}(T > t + s | T > t) &= \frac{\mathrm{P}(T > t + s \text{ AND } T > t)}{\mathrm{P}(T > t)} \\
&= \frac{\mathrm{P}(T > t + s)}{\mathrm{P}(T > t)}.
\end{aligned}$$

The numerator simplifies as we've shown here because $T > t + s$ AND $T > t$ represents redundant information. If the first condition holds, then of course the second condition holds. You need not assert it separately.

But you have formulas for the numerator and denominator here:

$$\begin{aligned}
\mathrm{P}(T > t + s | T > t) &= \frac{P(T > t + s)}{P(T > t)} \\
&= \frac{e^{-\lambda(t+s)}}{e^{-\lambda t}} \\
&= \frac{e^{-\lambda t} e^{-\lambda s}}{e^{-\lambda t}} \\
&= e^{-\lambda s} \\
&= P(T > s).
\end{aligned}$$

(b) The probability of no failures before 2 P.M. given no failures from noon until 1 P.M. is the same as the probability of no failures from 1 P.M. to 2 P.M. given no other information whatever. In other words, if you know there are no failures in the first hour, you might just as well start the model over again beginning at 1 P.M., with no memory of what happened from noon to 1 P.M.

4.3.1

(a) According to the "fundamental fact" of the normal distribution, it's virtually certain for a normally distributed X to be between $\mu - 3\sigma$ and $\mu + 3\sigma$. In other words, it's virtually certain X will fall within three standard deviations of its mean. Because for Z the mean is zero and the standard deviation is one, it's virtually certain for Z to fall between -3 and 3, as asserted.

(b) Z is the symbol for the STANDARD normal random variable. By definition, that's the normally distributed random variable with mean zero and standard deviation one. Because the standard deviation is one, so is the variance. Thus, $Z \sim N(0, 1)$.

(c) $P(Z < -2) = 0.025$. Here's why: According to the "fundamental fact," there's about a 95% chance for Z to fall within two standard deviations of the mean (to fall between $\mu \pm 2\sigma = \pm 2$). So there's about a five percent chance of falling outside this interval. Because the picture is symmetric, this five percent divides equally into the two tails:

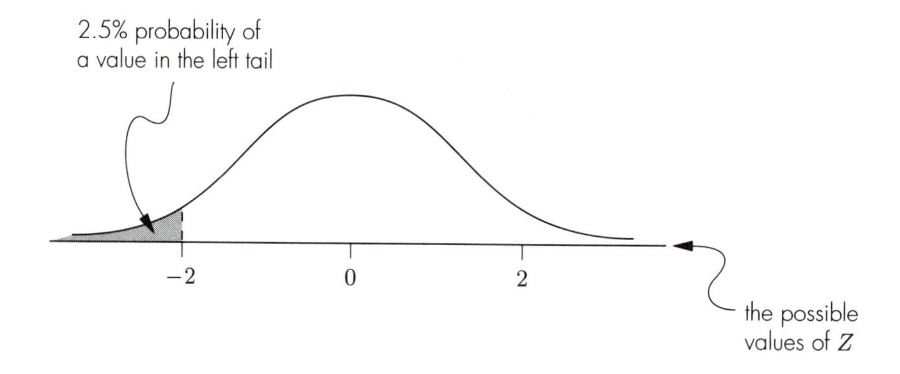

Here's the analysis: $P(-2 < Z < 2) = 95\%$ and so $P(Z < -2 \text{ OR } Z > 2) = 5\%$. But this probability involves two MUTUALLY EXCLUSIVE events: Z cannot be both smaller than -2 and bigger than 2 at the same time. So,

$$0.05 = P(Z < -2 \text{ OR } Z > 2) = P(Z < -2) + P(Z > 2).$$

Now, by symmetry these two probabilities are equal, which just says

$$P(Z < -2) = 0.025.$$

(d) The pictures tell it all:

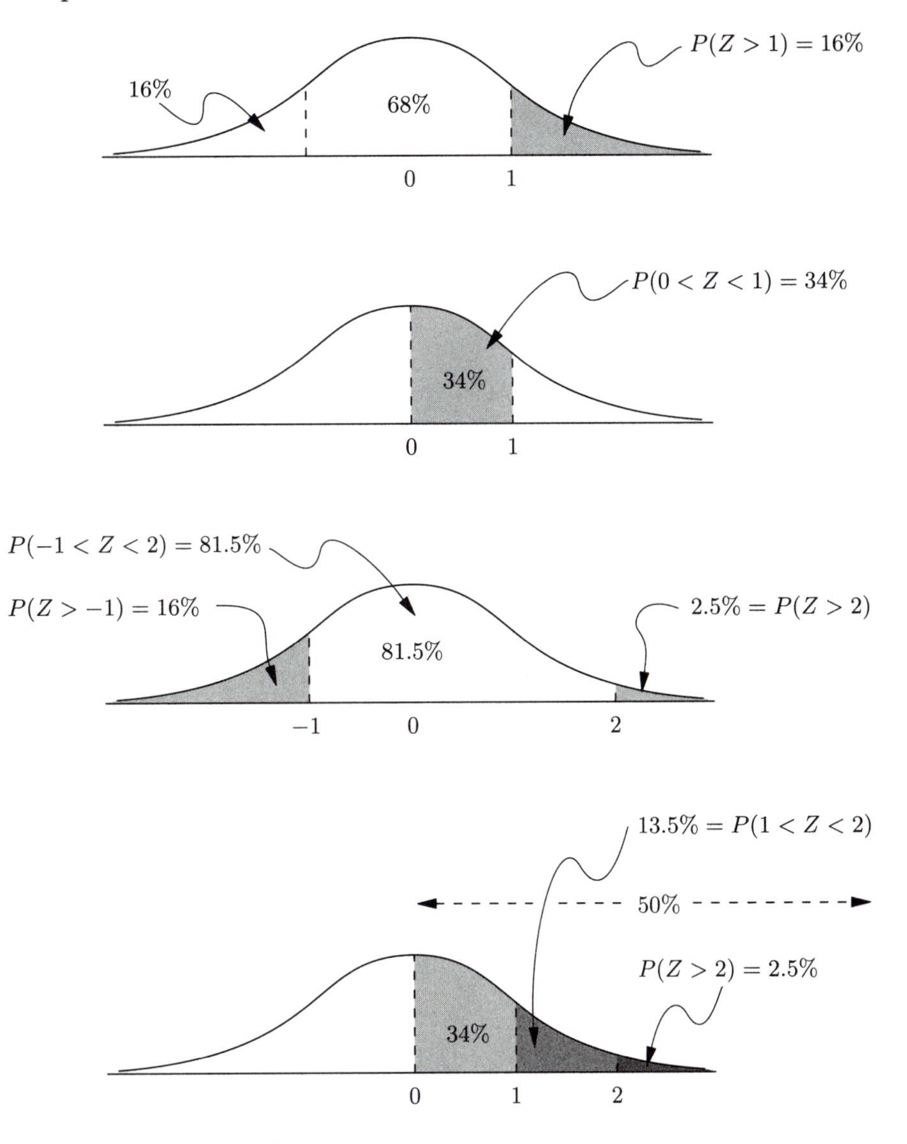

For this last picture, subtract out the relevant areas:

$$50\% - 34\% - 2.5\% = 50\% - 36.5\% = 13.5\%.$$

4.3.2

(a) $Z = 1$; (b) $Z = 1$; (c) $Z = -1$; (d) $Z = 1$; (e) $Z = 2$; (f) $Z = -3$;

(g) $Z = -3$ and 3 are possible answers, but many other pairs are also. For example -3.84 and 4.96. Or -3 and 22.

(h) $Z = -1$; (i) $Z = -1$.

4.3.3

Assume there's no systematic source of error. Then you'd expect the errors to cancel out on average. For the numbers suggested in Level I, the errors are

$$0.18, \ -0.91, \ -0.63, \ 0.26, \ 0.08;$$

some are positive and some are negative. Now, they do NOT average to zero, but, after all, these are five specific observations. You were asked to explain why ALL POSSIBLE errors should (theoretically!) be zero, on average. This is like the difference between saying a coin is fair (theory: half the time you should get heads) and noting that on a specific sequence of ten tosses (real-world observation), you don't get five heads.

So if there's no systematic source of error, the errors "in excess" and the errors "in defect" (the positive and negative errors) ought, theoretically and on average, to cancel out. Here's what Simpson said:

> That there is nothing in the construction, or position of the instrument whereby the errors are constantly made to tend the same way, but that the respective chances for their happening in excess, and in defect, are either accurately, or nearly, the same.

4.3.4

(a)

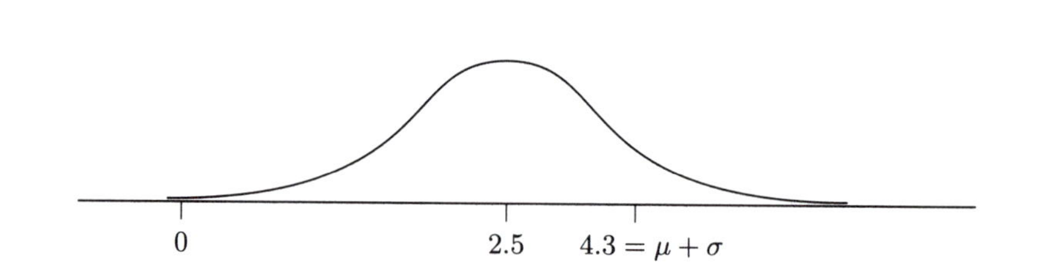

(b) Because zero is roughly 1.5 standard deviations below the mean, zero should be about as shown in part (a).

4.3.5 (a)

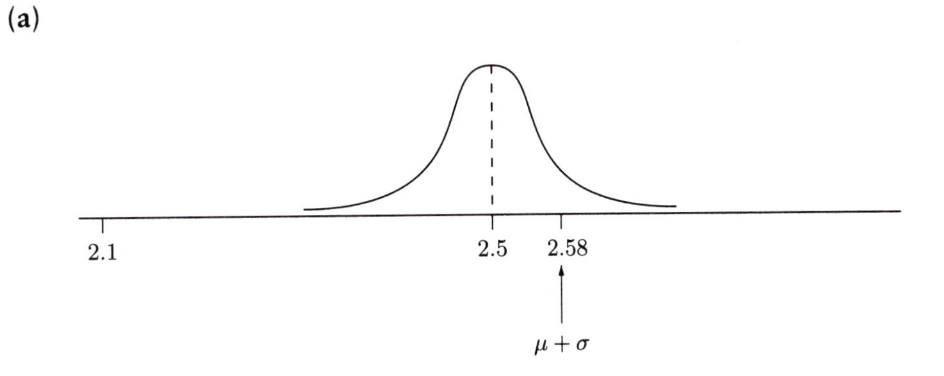

(b) Zero is how many standard deviations below μ?

$$\underbrace{0 - \mu}_{\text{the distance of zero below } \mu_x} = 0 - 2.5 = k\sigma, \quad \text{so } k = -31.25.$$

Zero is more than 30 standard deviations below the mean. The leftmost point on the graph looks like it might be four or five standard deviations below the mean. In other words, zero is way out of the picture to the left. It's so far from the mean that it's not even in the picture.

4.3.6

a) Clearly *less* than 50% of the area is between -1 and 1. It should be *more* than 50%—about 68% as in the picture...

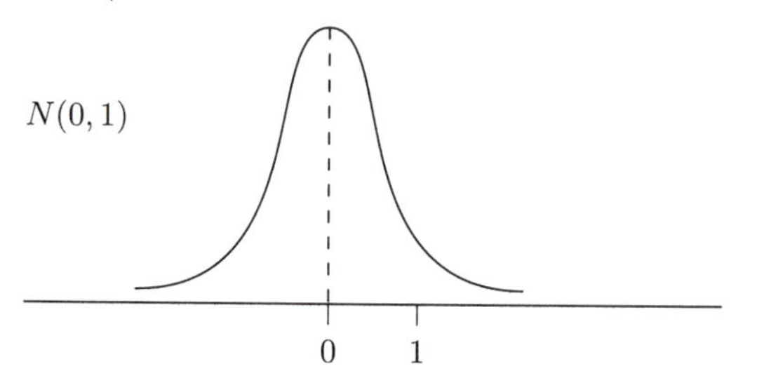

b) The picture was centered at zero but in fact the mean is one. Here is a correct picture...

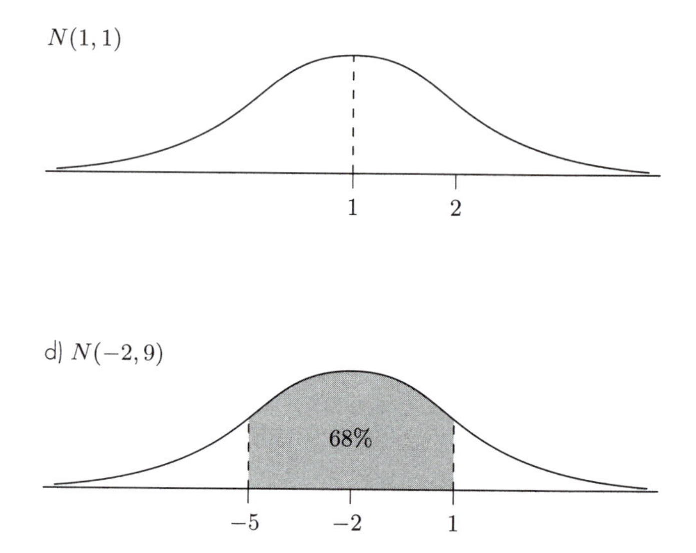

c) $N(1, 1)$

68%

0 1 2 3

No change—this looks OK. Notice that a little more than two-thirds the area is centered between 1 and 2 as required.

d) $N(-2, 9)$

Again, no change—this looks OK also. Here $\mu + \sigma = 1$ and '1' is located to cut off a bit more than two-thirds (about 68%) of the area between it and $\mu - \sigma = -5$.

(e) This picture was OK.

(f) This picture was NOT OK. The point $\mu + \sigma = 4$ was located way off in the right tail with clearly less than the required area to the right of it—less than 16%. There should be about 32% in the two tails because the area between zero and four (within one standard deviation of $\mu = 2$) should be about 68% of the total area. Here's a correct picture:

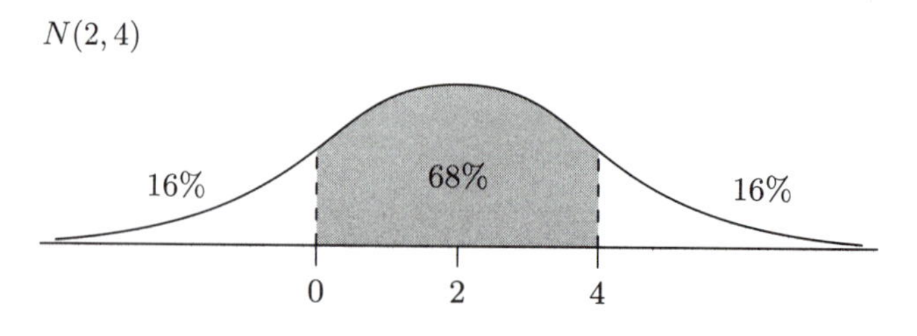

4.3.7

For Z the parameters are $\mu = 0$ and $\sigma = 1$. With these values, the formula for the density function becomes

$$\begin{aligned} f(x) &= \frac{1}{\sqrt{\sigma^2 2\pi}} \exp\left(-\frac{(x-\mu)^2}{2\sigma^2}\right) \\ &= \frac{1}{\sqrt{1^2 2\pi}} \exp\left(-\frac{(x-0)^2}{2}\right) \\ &= \frac{1}{\sqrt{2\pi}} \exp\left(-\frac{x^2}{2}\right), \quad \text{as required.} \end{aligned}$$

4.3.8

(a) The DOING is to "make a measurement." Clearly, we can REPEAT that process. An OUTCOME is the resulting physical situation when you do the experiment once (make one measurement). It can be described as a "state of the measuring device." For example, you observe a measuring rod placed up against the edge of a table, or you observe the dial in a certain position on a scale.

Are the outcomes PREDICTABLE? No, measurement always involves some error. There's no such thing as a 100% exact measuring process. One time when you place the measuring rod it's 1 mm to the left of where you placed it the last time. Or at a very slight angle. When you weigh produce at the grocery store, watch the dial on the scale—it wavers ever so slightly. And when you read that dial, you get a slightly different reading depending on the position of your head. In fact, the phenomenon of measurement error is common to all attempts at measuring, that's why it's such an important random variable for us to understand. And that's also one reason the normal distribution is so important as a model. If this discussion sounds vaguely familiar, take another look at Problem 2.1.8.

So far we've verified that the measuring process is a random experiment. Now, what's the random variable? What's the rule which assigns numbers to the outcomes? As usual, if the outcomes are clearly specified, the rule will be clear. Here, when you look at the "state of the measuring device" (an outcome), you assign the observed measurement to that outcome. Here's the RULE which assigns numbers to outcomes:

> Read the measuring device and report your observed measurement (a number).

(b) Intuitively, if there's nothing systematically wrong in the measuring process, repeated measurements ought to give, on average, the true measurement of the object being measured. Analytically, measurement error is the difference between the observed measurement and the true value: $E = M -$ true value. So it's a linear function of the observed measurement:

$$E = a + bM \qquad a = -(\text{true value}),\ b = 1.$$

Now unless there's something systematic wrong in the measuring process, the "mean error" should be zero: $\mu_E = 0$. This means $\mu_M =$ true value.

(c) If $Y = a + bX$, then $\sigma_Y^2 = b^2\sigma_X^2$. For us, $b = 1$ as we saw in the previous part of this problem. That means the measurements and the errors have the same variance. This conclusion makes perfect sense: The only reason the measurements vary at all is because of the error!

(d) There are two sources of variability: the accuracy of the measuring device and the skill of the person doing the measurement. Here's the quote from Simpson:

> . . . there are certain assignable limits between which all these errors may be supposed to fall; which limits depend on the goodness of the instrument and the skill of the observer.

4.3.9

(a) The "numbers" in this case are the various possible values the measurements could take on. The "situation" giving rise to these numbers is a random experiment, the measuring process (see Problem 4.3.8).

What determines the difference between any two measurements of the same object? Well, because it's the SAME object each time, the only reason you don't get the same value, the "true value" being measured, is because of measurement error. There are two categories of cause for error in a measuring process. Shades of Simpson (see Problem 4.3.8):

(i) the accuracy of the measuring device is not absolute,
(ii) the skill of the observer is not absolute.

But why is the measuring device not absolutely accurate? If you think this through, you'll realize there are going to be "many independent random factors" accounting for this: environmental factors such as temperature, humidity, and so on. Factors inherent in the measuring device are stress in some small component, how well-lubricated the moving parts are, and so on. Of course, if the measuring device is not operating properly, there will also be systematic error. But then you'll no longer expect to see a normal distribution because now some of the differences in measurements are due to something systematic—not due solely to "many independent RANDOM factors."

And why is the observer's skill not absolutely accurate? We'll leave this to you. In brief: "many independent random factors." Spell out a few of the specific possibilities—they're legion.

Thus, if the measuring device is operating properly, the difference between any two measurements seems to be accounted for by "many independent random factors." That means the measuring process is a "situation" like what's required by our rule. Therefore, the measurements should be approximately normally distributed. The word "approximately" is required because no real-world situation ever fits an abstract model exactly.

(**b**) Choose one number from the population at random. This is something you do which is repeatable. Outcomes: Usually we try to have our outcomes be real-world objects, but, here, it's more natural to describe an outcome as simply the "chosen number." You cannot predict in advance which number you'll get (why?), so we've verified that this is a random experiment. Now, to an outcome (the number chosen) the random variable assigns the number itself. Thus, we have a random variable whose values are just the numbers of our numeric population. To see if a normal distribution seems appropriate, you would apply our difference-between-any-two-values criterion to these numbers.

4.3.10

(**a**) Specification error is just a form of random error. In other words, the error will be due to many independent random factors.

(**b**) If there's something systematically wrong in the manufacturing process, the difference between the actual dimension of the object and the specified dimension would be determined in part by that systematic, nonrandom factor.

4.3.11

(**a**) There are categories of such factors. For example, there are environmental factors: one student took the SAT test in San Diego with beautiful sunny weather, the other in New York City where she had to fight a snow storm! One student was stuck in a testing room which was too warm. You name some others.

Then there are factors connected to background of the student: One student comes from a supportive family of highly educated intellectuals, the other from a dysfunctional, economically disadvantaged family where she had to struggle with part-time jobs and strife at home. Another student comes from a supportive family, but of a minority culture and the test is subtly biased in favor of the majority culture. You name other such "background" factors which would account in part for some of the differences in scores.

There are personal factors: One student is smart, the other is not so smart. One student had a good night's rest and a good breakfast before coming to the test. Another student was up all night at a terrific party, woke up late, barely got to the testing center on time, and took the test on an empty stomach. Name some other such "personal" factors.

Because the difference in two test scores will be due to "many independent random factors"—note the independence in all our examples—you would expect to see a distribution of scores that is approximately normal. Note that the scores are, in fact, discrete (they're all integers) and there're finitely many of them, so they cannot fit the normal model exactly—it's a continuous model after all.

(**b**) Maybe someone got the test in advance and shared it with a few of her friends. Of course, that's always unpleasant to think about.

Maybe a small group of students have been meeting regularly and studying together, encouraging each other and putting in a lot of time. As a result, they might very well be significantly better as a group than the rest of the class. Note that for this to work, you must be describing some systematic factor which accounts for some of the differences in the scores. That's what makes our criterion fail. It's no longer true that ALL differences are due to independent random factors because there is this one systematic factor—the "study group"—which accounts for SOME of the differences.

If only one student saw the test in advance, that would just become one of the many random factors determining differences. It would not affect the approximate normality of the distribution. Similarly, if you say, "Some students studied harder than others," that would not disturb the normality; that seems simply to say that "effort" is just one of the many random factors determining differences in score. To make the distribution bimodal (non- normal), you must identify something SYSTEMATIC which identifies a clear-cut group of students who outperform the rest of the class.

Note that the bimodal distribution is really a combination of two normal distributions: the scores for the study group and the scores for the rest of the class. The difference in scores for two students in the study group is not affected by "study group" because both students are in that group. Similarly, "study group" is irrelevant for the scores of the rest of the class—none of them were in the study group. So, for each group separately, the difference in two scores is accounted for by "many independent random factors" with nothing systematic, the systematic effect of "study group" being irrelevant. Here's the picture:

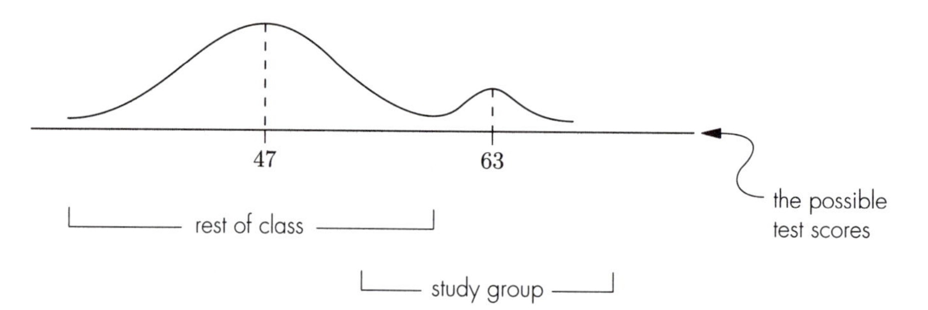

4.3.12 (a) $S = T + E$.

(b) Think of the population of students as a numeric population—each student is associated with a certain value of T. Then the various values of T are values of a random variable as in Problem 4.3.9(b). So, T should be approximately normally distributed because the difference in two values will be due to many independent random factors [analogous to Problem 4.3.11(a)].

(c) E is a form of measurement error. So it's normally distributed.

(d) Look at the difference in two values of $X + Y$:

$$(X_1 + Y_1) - (X_2 + Y_2) = X_1 - X_2 + Y_1 - Y_2.$$

Because each of X and Y is normally distributed, each of $X_1 - X_2$ and $Y_1 - Y_2$ is due to many independent random factors. Therefore, the criterion holds for the difference in two values of $X + Y$.

4.3.13 (a) What accounts for the value of $Y_1 - Y_2$? Well,

$$\begin{aligned} Y_1 - Y_2 &= (a + bX_1) - (a + bX_2) \\ &= b(X_1 - X_2). \end{aligned}$$

Because b is just a constant, the difference $Y_1 - Y_2$ will be accounted for by exactly those factors which account for the difference in the X values. But X is normally distributed, so the difference in two X values is determined by "many independent random factors."

Because the difference between any two values of Y is determined by "many independent random factors," the conditions of our rule of thumb are satisfied and we should expect Y to be normally distributed.

(b) A binomial random variable takes on all the integer values between zero and n . But $Y = 2X$ will take on all the even values between zero and $2n$. It doesn't take on odd values as it must if it were binomial.

One example is enough to show that no general rule holds, but two examples may make the point more clear: If $Y = X + 1$, Y never takes the value zero; therefore, Y is not binomial.

(c)

Y	$P(Y)$	$YP(Y)$	$Y^2P(Y)$
1	q	q	q
3	p	$3p$	$9p$
	1	$2p+1$	$8p+1$

$\mu = 2p + 1$,
$\sigma^2 = 4pq$, how?

$$\text{Or}\quad \mu_Y = 2\mu_X + 1 = 2p + 1.$$
$$\sigma_Y^2 = 2^2\sigma_X^2 = 4pq.$$

Note that, of course, you get the same answer both ways.

(d) The point of this entire problem is to see that a linear function of a normally distributed random variable is again a normally distributed random variable. Part (b) shows there's no general rule like that: The rule fails for the binomial model. Part (c) just invites you to recall in detail facts about binomial random variables and linear functions thereof.

(e) The difference in two values will just reduce to the difference in two values of X and then of Y. Because each of X and Y are normally distributed, each of their differences are "due to many independent random factors." So the difference in two values of aX+bY is also. See the solution to Problem 4.3.12(d) where you applied the criterion to $S = T + E$.

4.3.14

(a) Remember, in this problem μ and σ refer specifically to the random variable X. Now if $X = a + bZ$, then

$$\mu = \mu_X = a + b\mu_Z = a \quad \text{since } \mu_Z = 0,$$
$$\sigma^2 = \sigma_X^2 = b^2\sigma_Z^2 = b^2 \quad \text{since } \sigma_Z^2 = 1.$$

Thus, if X is a linear function of Z, it must be that $a = \mu$ and $b = \sigma$.

Now prove that $\mu + \sigma Z$ is indeed X. In other words, show that it's $N(\mu, \sigma^2)$. That it's normally distributed was shown in Problem 4.3.13(a). And we already know it has the right mean and variance; that's how we found a and b. Once again: the mean of $\mu + \sigma Z$ is $\mu + \sigma\mu_Z = \mu + 0 = \mu$; the variance is $\sigma^2\sigma_Z^2 = \sigma^2 1 = \sigma^2$.

Because there's only one normally distributed random variable with mean μ and variance σ^2 and because we have two candidates (X and $\mu + \sigma Z$), these two must, in fact, be equal.

(**b**) Look at the equation in part (a): $X = \mu + \sigma Z$. Solve for Z:

$$Z = \frac{X - \mu}{\sigma} = \frac{1}{\sigma}X + \left(-\frac{\mu}{\sigma}\right)$$
$$= dX + c.$$

So, $d = 1/\sigma$ and $c = -\mu/\sigma$. And d is positive because σ is.

4.3.15 (**a**) An outcome here is a "state of the measuring device." To that outcome, M assigns the observed measurement, whereas E assigns the ERROR in the observed measurement. That's a different number, of course. That's exactly the difference between M and E. The experiment is the same and so the outcomes are the same, but the values assigned by each of these two random variables differ. Therefore, the random variables are different.

(**b**) The error is the difference between the measurement and the true value: $E = M-$ (true value). Thus, $\mu_E = \mu_M-$ (true value). But $\mu_E = 0$, so $\mu_M =$ (true value). What about the variances? Because the only variable part of M is the error, M should be exactly as variable as the error. That's true: $E = 1 \times M-$ (true value) and so $\sigma_E^2 = 1^2 \times \sigma_M^2 = \sigma_M^2$.

(**c**) M is normally distributed. This is the content of Problem 4.3.9. The normal curve for M looks exactly like the normal curve for the errors, E, except that the curve for M is centered at the true value being measured instead of being centered at zero like E. Suppose, for example, the true value of the object being measured is 1.830 cm and the standard deviation of the measurement error is 0.025 cm. Then,

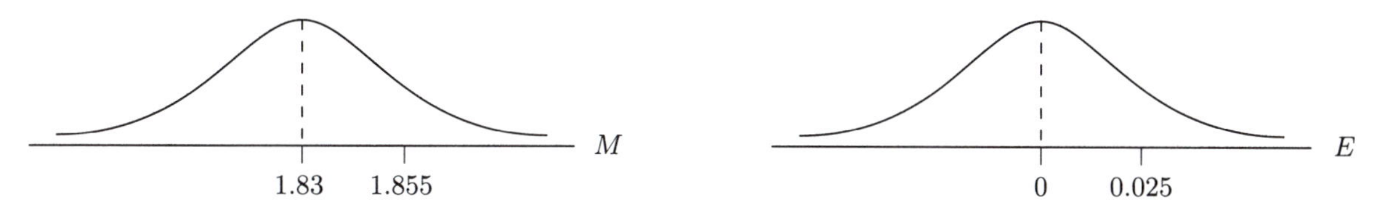

These pictures have exactly the same shape. Only the center is different.

(**d**) For E and M, an outcome is a "state of the measuring device." For X and Y, an outcome is the "die sitting on the table top." Further, E and M assign different values to their outcome. E assigns the error in the measurement, M assigns the measurement itself. Similarly, X and Y assign different values to their common outcome, the "sitting die."

Each case provides an example of a pair of random variables defined for the same random experiment where the only difference is the rule which assigns numbers to the outcomes. Further, in each case, the relationship is linear: $E = M-(\text{true value})$ and $Y = 7 - X$. There are differences between the two cases: X and Y are discrete and take on the same possible values, the integers from one through six. E and M are continuous and take on values from different ranges. E takes values centered on zero; M takes on values clustered near the true value of the object being measured.

4.3.16 (a) Done.

(b) $P(X < 2) = 0.5$. Because X is symmetric about its mean of two, there's a 50/50 chance of a value less than two. Any normally distributed random variable is symmetric about its mean. Draw the picture!

(c) Okay, once more we'll give you the picture:

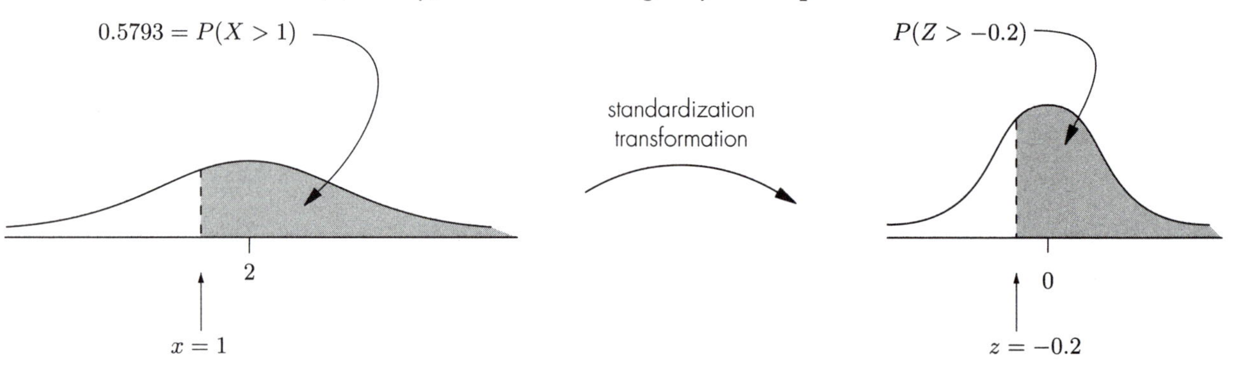

The standardizing transformation here is $\frac{(1-2)}{5} = -1/5 = -0.2$. But, by symmetry, the shaded right-tail area in the last picture is the same as the shaded left-tail area in

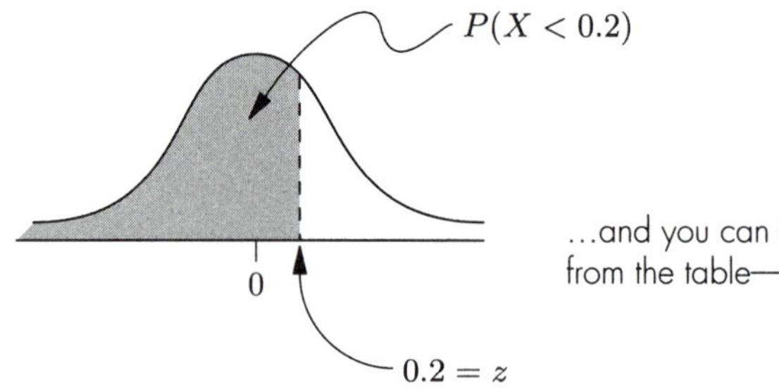

...and you can find this probability from the table—it's 0.5793

(d) The standardizing transformation is $[(-1) - 2]/5 = -3/5 = -0.6$. And by symmetry $P(Z > -0.6) = P(Z < 0.6)$.

(e) This comes from $\mathrm{P}(X > -1)$ which you just computed in the previous part:

$$\mathrm{P}(X < -1) = 1 - \mathrm{P}(X > -1) = 1 - 0.7257 = 0.2743.$$

Remember, we can ignore any equal signs because X is continuously distributed.

(f) The answer is 0.3446. Here's how:

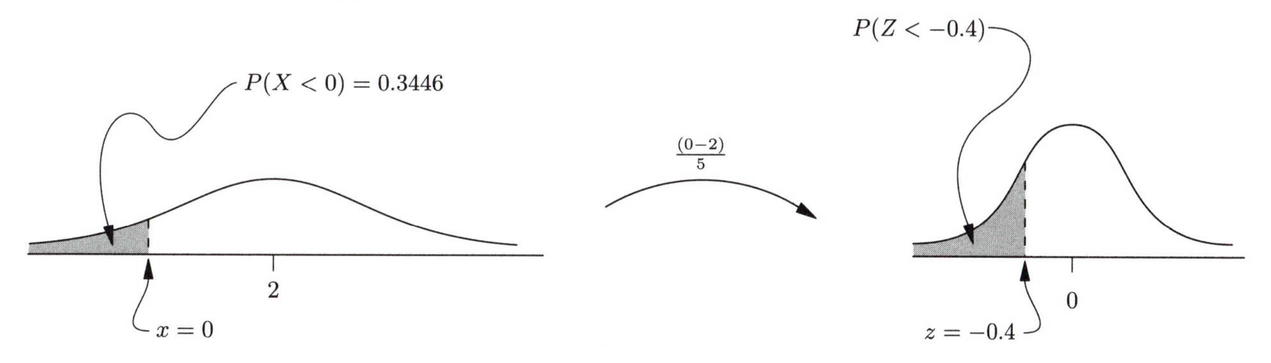

This left-tail area is the same as the symmetric right-tail area:

$$\mathrm{P}(Z > 0.4) = 1 - \mathrm{P}(Z < 0.4) = 1 - 0.6554 = 0.3446.$$

In a calculation of this kind, you should remember what sort of answer you're looking for, in this case LESS THAN 50%. It's easy in your thinking to flip-flop back and forth between "right-tail area" and "left-tail area" too many or too few times. If you remember you're looking for a probability of less than 50% you won't make this mistake! Of course, you can't "remember" unless you keep the PICTURE in mind.

(g) From the picture, clearly you've got to take the area in the right tail above $X = 1$ and subtract off the area above $X = 7.4$.

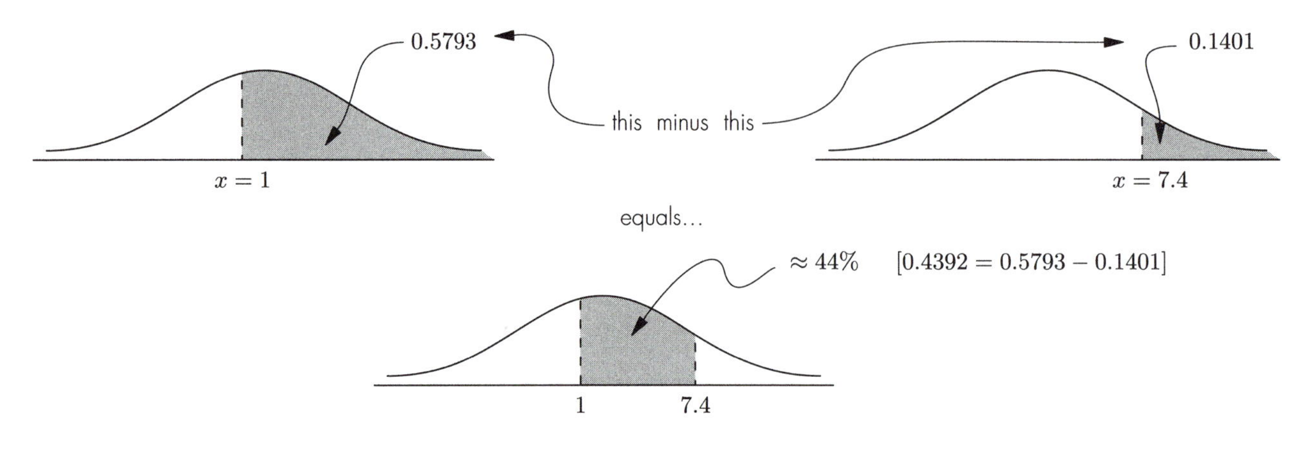

The area above $X = 1$ was computed in part (c) and the area above $X = 7.4$ is $1 - \mathrm{P}(X < 7.4)$ from part (a). So,

$$\mathrm{P}(1 < X < 7.4) = 0.5793 - 0.1401 = 0.4392.$$

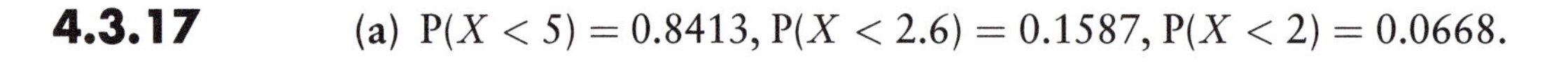

4.3.17 **(a)** $\mathrm{P}(X < 5) = 0.8413$, $\mathrm{P}(X < 2.6) = 0.1587$, $\mathrm{P}(X < 2) = 0.0668$.

(b) $\mathrm{P}(X < 14) = 0.8413$, $\mathrm{P}(X < 10) = 0.1587$, $\mathrm{P}(X > 9) = 0.9332$.

(c) $\mathrm{P}(X < 0) =$ ZERO, $\mathrm{P}(X < 0.17) = 0.1587$, $\mathrm{P}(X > 0.16) = 0.9772$.

(d) This X is just Z, of course. Here are the answers—you've computed all of them in the previous parts:

$$\mathrm{P}(Z < 1) = 0.8413, \quad \mathrm{P}(Z < 2) = 0.9772, \quad \mathrm{P}(Z > 1.5) = 0.0668.$$

(e) $\mathrm{P}(X > 1) = 0.0032$, $\mathrm{P}(X > 0) = 0.0344$, $\mathrm{P}(X > -3) = 0.8186$.

(f) $\mathrm{P}(4 < X < 9) = 0.4996$, $\mathrm{P}(3 < X < 5) = 0.4972$, $\mathrm{P}(X > 3) = 0.7486$.

(g) Take $4 < X < 9$ for instance. The trick here is to observe that X less than nine can be split into the two events A, B defined as

$X < 9$ if and only if $A = X$ is between nine and four
OR $B = X$ is less than four
(these are mutually exclusive)

so

$$P(X < 9) = P(A \text{ or } B) = P(A) + P(B)$$
$$= P(4 < X < 9) + P(X < 4)$$

since A and B are mutually exclusive. Now subtract $P(X < 4)$ from both sides to get the required equation.

4.3.18

(a) Solve the following equation to get $X = 0.6091$:

$$(X - 1.3)/0.42 = -1.645.$$

(b) Find the value of Z from the table which puts 0.18 in the right tail (0.92 is as close as you can get). Then solve for X in $(X - \mu)/\sigma = 0.92$.

(c) From the table, $P(Z > 0.77) = 0.2206$.

(d) $(X - \mu)/\sigma = (X - 5)/1 = -0.77$.

(e) $[X - (-2)]/2.2 = 0.61$ from which $X = 1.342 - 2 = -0.6580$.

(f) Here $Z = 0.025$ because the value you're looking up falls exactly halfway between the two values in the table.

(g) First, you need the area in the left tail below ten: $0.0409 = P(X < 10)$ found from $Z = -1.74$. Now find X so that $P(X <?) = 0.38 + 0.0409 = 0.4209$. DRAW THE PICTURES! The closest you can get is $Z = -0.20$ giving $X = 13.54$.

(h) Because 12 is roughly one standard deviation below the mean ($\sigma = 2.3$), you would have the following impossible picture:

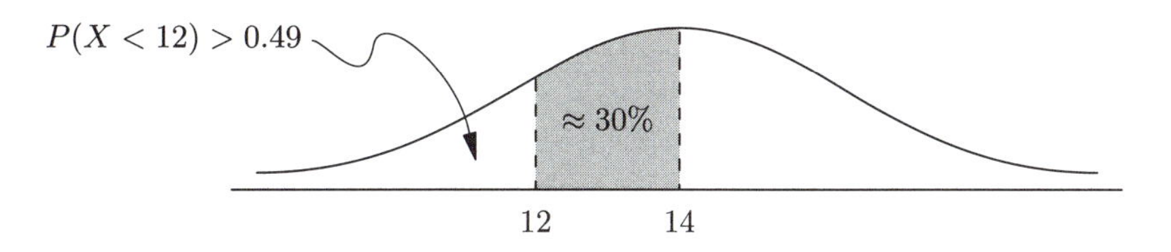

Or just calculate $P(X < 12) = 0.1922$. It's NOT greater than 0.49.

(i) $P(X > -30) = P(Z > 1.2) = 0.1151$ and so

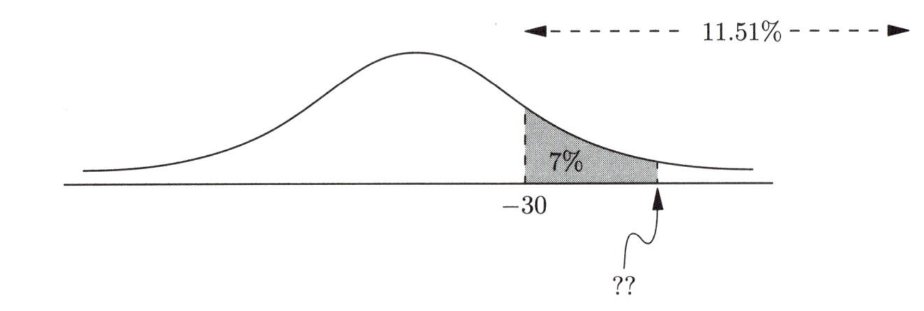

So $P(X > x) = 0.1151 - 0.07 = 0.0451$ and $Z = 1.695$ giving $X = -27.525$.

(j) This is the same as $P(X <?) = 84\%$:

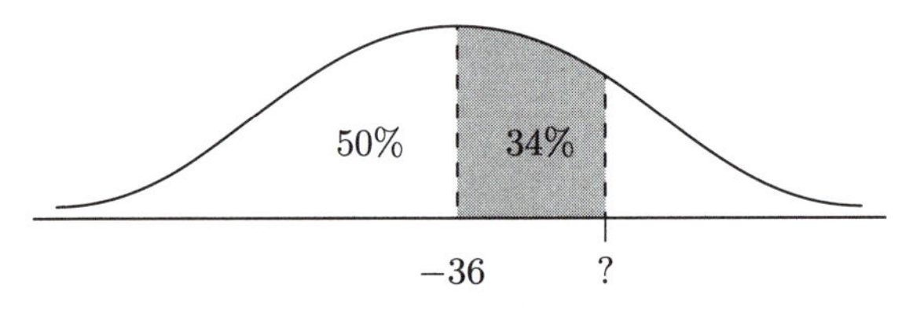

4.3.19

(a) Nine values of Z are required to cut the distribution into ten regions. These nine values of Z are $0, \pm 1.28, \pm 0.84, \pm 0.52, \pm 0.25$. For example, 40% of the Z distribution is below $Z = -0.25$ and 70% is below $Z = 0.52$. If this is not clear, you didn't draw the pictures. (Don't blame me! I told you long ago: DRAW THE PICTURES!)

(b) The "ideal" sample is evenly spread in its distribution. However, the phrase "evenly spread" does not refer to the observations in the sample, but rather to the probabilities. Now if you had a uniform distribution, "evenly spread" probabilities would mean evenly spread observations along the axis. But we're talking about a NORMAL distribution where the probability is concentrated near the center. So an area (probability) near the center representing ten percent will lie over a shorter interval than an area of ten percent out in the tails:

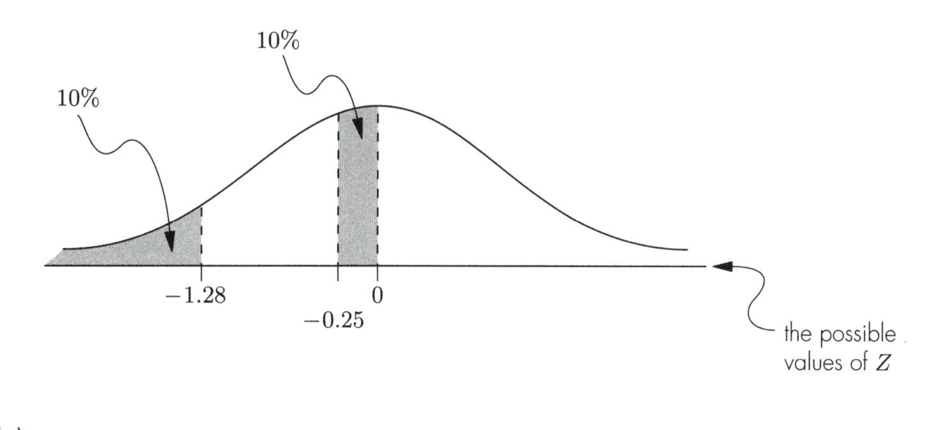

(c)

$$(-1.28, 0.2),\ (-0.84, 0.4),\ (-0.52, 1.2),\ (-0.25, 1.8),\ (0, 2.5),$$
$$(0.25, 2.7),\ (0.52, 3.5),\ (0.84, 4.8),\ (1.28, 5.2).$$

(d) For the "ideal" sample—a sample that's distributed exactly like $N(\mu, \sigma^2)$—the normal probability plot would be a line. If the sample is distributed more or less like $N(\mu, \sigma^2)$ (but not exactly), the normal probability plot will lie more or less along a line.

Of course, a random sample might not look anything at all like the distribution it's drawn from. Atypical samples are possible. In that case, we'll be led astray by the normal probability plot—it doesn't take the possibility of atypical samples into account. That's why earlier we referred to the normal probability plot as a "relatively weak" check on normality. There are more powerful tests for normality which do take into account the possibility of an atypical sample, but those tests are too sophisticated for a first course. You might take some comfort in remembering that although atypical samples are possible, they're not probable. We'll justify that statement in the next chapter.

(e)

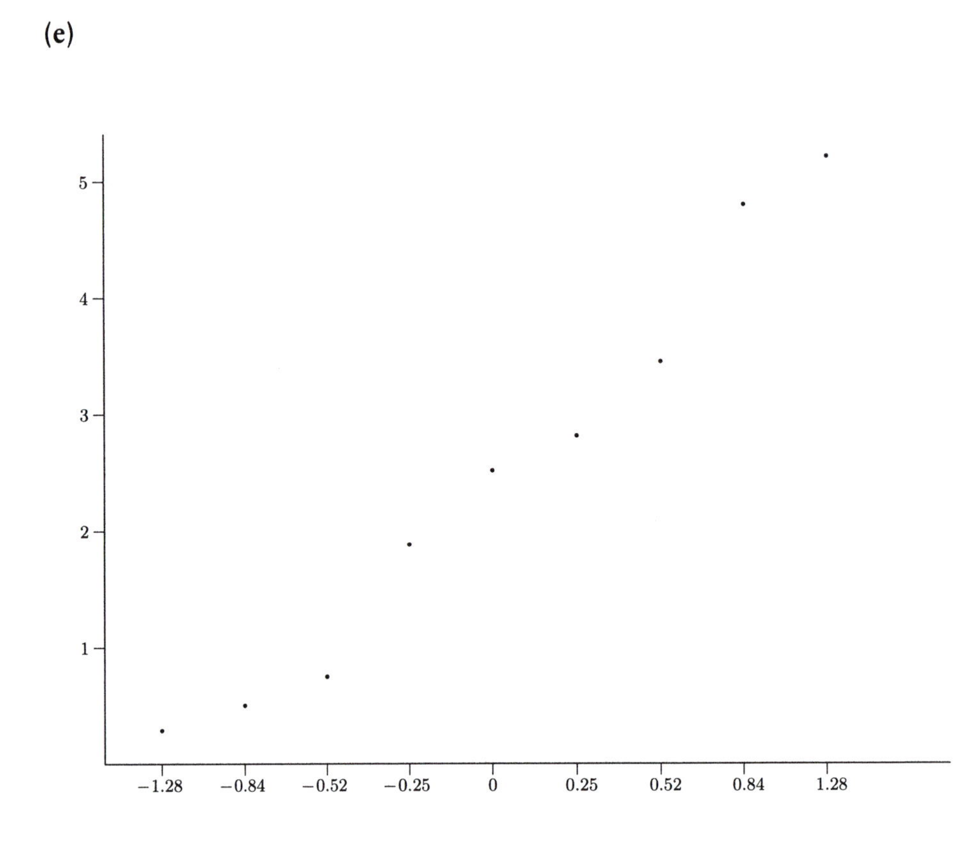

(f) It would seem our sample is not from a normal distribution because the normal probability plot strays away from a line at the extreme ends. This kind of plot—one with a vague S shape—suggests a symmetric distribution that's thicker in the tails than a normal distribution. A certain amount of experience is required to become good at interpreting probability plots. Any book on Exploratory Data Analysis will give you further insight into this and many other graphical techniques for exploring your data.

You can gain some experience on your own by playing with normal probability plots using any computer statistical package. Use the package to generate samples from various distributions and then have the package graph a normal probability plot. You'll learn a lot from this exercise.

(g) The second value of Z was the 20th percentile, determined by $2/(n+1) = 0.20$. The third was the 30th percentile given by $3/(n+1) = 0.30$. In other words, the formula $k/(n+1)$ tells what percentage of the distribution should be "cut off" by the percentile in question. Other formulas that are sometimes used are $(k-0.5)/n$ or $(k-0.3)/(n+0.4)$. There are others! Minitab uses the function: $(k-3/8)/(n+1/4)$.

4.3.20

When you draw your normal probability plot, you'll see that the plot veers upward away from a line. That suggests that "time between earthquakes" is not normally distributed. In fact, a better fit can be found in the family of Weibull distributions (the exponential distribution is one member of this family) which is often used to model "time until the next occurrence" in a series. A Weibull distribution is skewed to the right. A Weibull probability plot for Sieh's data—it plots the data against percentiles of the Weibull distribution—lies very tightly along a line, suggesting a reasonable fit to that distribution. So it would seem that "time between earthquakes" might better be modeled by a Weibull rather than a normal distribution. The Weibull model can be used to estimate the probability of an earthquake at Pallett Creek in, for example, the coming year. The Weibull plot for Sieh's data along with further discussion is given in the interesting article "Estimating the Chances of Large Earthquakes by Radiocarbon Dating and Statistical Modeling" by David R Brillinger of the University of California at Berkeley (Tanur et al.). Of course, ordinarily you would use a computer package such as Minitab to generate a normal probability plot.

4.3.21

(**a**) The normal distribution is a CONTINUOUS probability distribution. So our discrete random variable must look approximately continuous. For that, it must have many possible values.

The example in the text was defined on the interval from 0 to 10,000. A continuous random variable on this interval would take on any value from 0 to 10,000. Our discrete random variable takes on only 10,001 values in that range. It doesn't take the value 1.2 for example, or 1.8, or any other fractional value. Thus, it's not really continuous, but with so very many possible values (10,001 of them) it's close!

(**b**) The normal distribution is symmetric about its mean. So anything that's approximately normal would have to be approximately symmetric about its mean.

(**c**) The normal distribution is symmetric about its mean, so the mean and median must be the same—50% of the area to the left and 50% to the right. And there's only one "peak" point (it's unimodal), occurring at the mean. So to be approximately normal, a discrete random variable must be at least approximately like that.

4.3.22 Choose the normal distribution which has the same μ and σ^2 as our X:

$$X \approx \widetilde{X} = N(32.8, 106.09).$$

Now calculate the required probability:

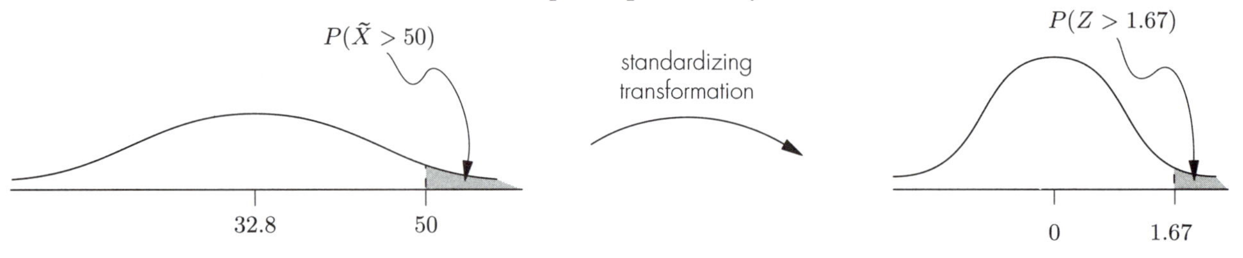

The standardizing transformation is $Z = (50 - 32.8)/10.3 = 1.6700$. From this, you find

$$\mathrm{P}(Z > 1.67) = 0.0475 \quad \text{and so} \quad \mathrm{P}(X > 50) \approx 0.0475.$$

4.3.23 (a) Because 7 is not included in $X > 7$, the rectangle over 7 must be excluded:

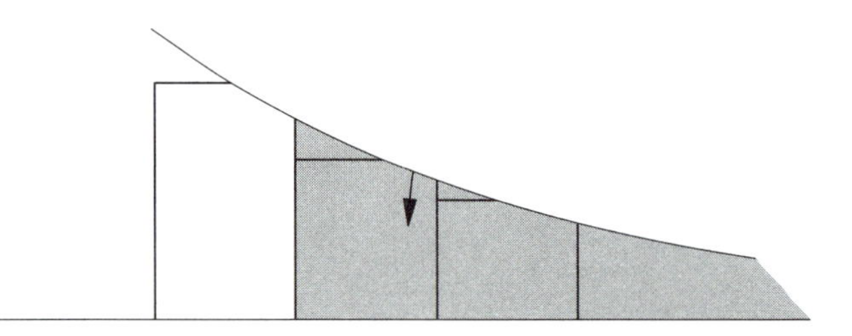

(b) With a continuous random variable an equals sign will make no difference because the probability of any one particular value is zero. Thus, $\mathrm{P}(6.5 < \widetilde{X}) = \mathrm{P}(6.5 \leq \widetilde{X})$. For $X \geq 7$, the value 7 is included and 6 excluded, so the rectangle over 7 must be included:

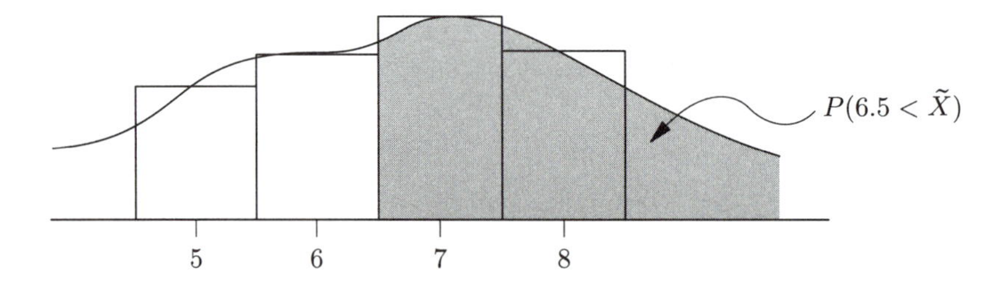

You can ignore the equals sign for $\widetilde{X}$ because it's continuous but you CANNOT ignore the equals sign for X. It's discrete.

(c) The value 2 is to be excluded, 8 included:

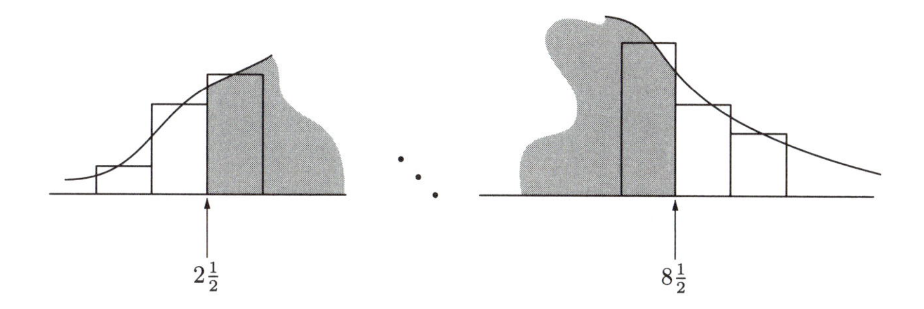

4.3.24

(a) $n \geq 10$; **(b)** $n/4 \geq 5$; **(c)** $np = 0.05n \geq 5$, so $n \geq 5/0.05$.

(d) $np = 0.002n \geq 5$.

(e) nq is smaller than np. Because BOTH must be greater than or equal to five, you check the smaller one. If it's big enough, of course the larger one is also. So, $nq = n/4 \geq 5$ says $n \geq 20$.

(f) $np = 12p \geq 5$ says $p \geq 5/12$ which is just 0.4167.

(g) $np = 100p \geq 5$ says $p \geq 0.05$.

(h) $150p \geq 5$ says p is at least 0.0333; **(i)** 0.005; **(j)** 0.0005.

4.3.25

(a) Because $np = 2.8$ which is LESS THAN FIVE, the normal approximation is not valid. Besides, you shouldn't use an approximation when the true value is easily available.

(b) $0.0262 = \mathrm{P}(\widetilde{X} > 136.5 | \mu = 120, \sigma = 8.4853)$.

(c) Because $np = 6$, the criterion for the normal approximation holds and the approximation should be valid (just barely!). But it's not appropriate to use it. Why should you settle for an approximate value when the exact value is easily available?! So your answer should be 0.8062.

(d) You want to approximate the area of the rectangle centered at $X = 24$. Here's the picture:

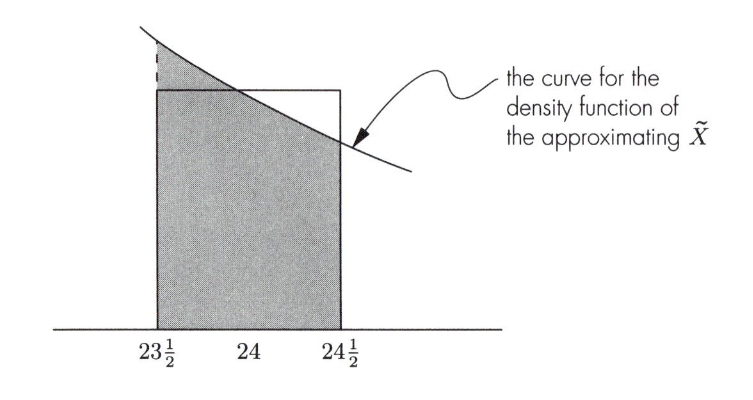

and so

$$P(23.5 < \widetilde{X} < 24.5|\mu = 24, \sigma = 4.5957) = P(-0.11 < Z < 0.11) = 0.0876.$$

4.3.26

(a) The characteristic is $D > 2.35$, where D is the diameter of a randomly selected machine part. It's reasonable to assume D normally distributed. You should be able to justify that assumption, of course. Remember, if an assumption is required to work a problem, make that assumption and then go ahead and solve the problem. You should be prepared to discuss the appropriateness of the assumption if asked. Assuming D is normally distributed, we find that

$$\begin{aligned} p &= \text{P[one machine part has the characteristic]} \\ &= P(D > 2.35|\mu = 2.3, \sigma = 0.1) \\ &= 0.3085. \end{aligned}$$

With this value for p, we can answer the original question:

$$P(X \geq 5|n = 12, p = 0.3085) = 1 - 0.7009 = 0.2991.$$

You get the value 0.7009 using either the binomial recursion formula, or calculating directly

$$P(X \leq 4) = q^8[q^4 + 12pq^3 + 66p^2q^2 + 220p^3q + 495p^4] = 0.7009.$$

The recursion formula is much faster, of course! I do it both ways just as a double check. Here's the real-world answer:

> There's about a 30.32% chance that when twelve of our machine parts are chosen at random at least five will be more than 2.35 cm in diameter.

(**b**) Note that in six months there are 4380 hours (365/2 times 24). Now, assuming the lot of machine parts is chosen randomly, X is binomial, just as in part (a), but with a different characteristic of interest. The basic question is

$$\mathrm{P}[X \geq 250 | n = 500, p = ??] = ???$$

The characteristic of interest is that one part "lasts six months," meaning that it operates at least six months. Let T be the time to failure for one of these parts. Then "having the characteristic" means $T \geq 4.38$ (in thousands of hours). It's reasonable to assume an exponential distribution for T, a typical application of that distribution. So,

$$\begin{aligned} p = \mathrm{P}(T > 4.38 | \lambda = ?) &= e^{-t\lambda} \\ &= \exp[-4.38/6] \qquad \text{in fact } \lambda = \tfrac{1}{6} \\ &= 0.4819. \end{aligned}$$

We found the parameter λ for the exponential distribution from the formula $\mu_T = 1/\lambda$. Because $\mu_T = 6$ (thousand hours), $\lambda = 1/6$. So now the original question becomes

$$\mathrm{P}[X \geq 250 | n = 500,\ p = 0.4819] = ???$$

Use the normal approximation. It's valid here because np and nq are at least five. In fact, $np = 240.9545$ and nq is even larger. So we must evaluate

$$\mathrm{P}[\widetilde{X} > 249.5 | \mu = 240.9545, \sigma = 11.1730] = \mathrm{P}(Z > 0.76) = 0.2236.$$

For the binomial $\mu = np$ and $\sigma^2 = npq$, so we took these values as the mean and variance of $\widetilde{X}$. The real-world answer is

> There's about a 22.36% chance that at least half of our customer's lot of five hundred machine parts will last six months or more.

(c) The pattern is the same as for part (a) because the characteristic of interest concerns the diameters, but now the value of p is

$$\begin{aligned} p &= \mathrm{P}[\text{one machine part has the characteristic}] \\ &= \mathrm{P}(D > 2.55 | \mu = 2.3, \sigma = 0.1) \\ &= 0.0062. \end{aligned}$$

We need to find $P(X > 5|n = 500, p = 0.0062)$. A direct computation would be tedious (six calculations with some large numbers), but unfortunately the normal approximation is not valid because $np = 3.1 < 5$. However, we CAN use the Poisson approximation. Because $p \leq 5\%$ and $n \geq 20$, that approximation is valid (see Problem 3.8.10).

Let $\widetilde{X}$ be the approximating Poisson random variable. Using the Poisson recursion formula, we find that $P(\widetilde{X} \leq 5) = 0.9057$ which gives our answer: $P(\widetilde{X} > 5|\lambda = 3.1) = 0.0943$. So, the real-world answer is

There's about a 9.43% chance that more than five of a lot of five hundred machine parts will be unusable.

Of course, the binomial recursion formula itself is not much harder. And it gives a somewhat more precise answer: 0.0937. Try if you like.

You might have saved yourself some work with the Poisson approximation if there had been a Poisson table handy. Of course, a handbook with such a table would also have a binomial table, but usually binomial tables don't go as high as $n = 500$. They don't need to. When n is large, either p is small, so the Poisson approximation is valid, or p is not so small, in which case the normal approximation is valid. Note how the normal and Poisson approximations are complementary to each other.

4.4.1

(a) The answer is 2.5%. There's a 97.5% chance for χ^2 to take a value *less than* 23.336 and you need the probability to be *greater than* that.

(b) Here's the picture of χ^2 with 19 degrees of freedom:

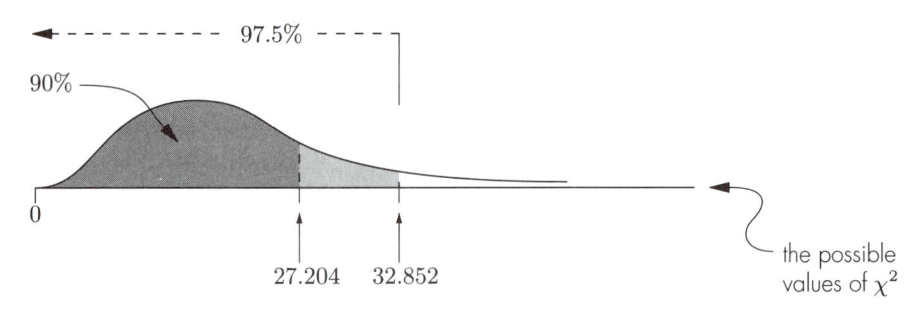

To the left of 32.852 is 97.5% of the area and to the left of 27.204, 90%. So *between those two* is 97.5% − 90% = 7.5%.

(c) You want the probability that $\chi^2 > 40$, with $d = 20$. Answer: 0.005.

(d) Look at the definition of sampling from a probability distribution. You have a simple random sample of size 20 from the distribution of Z. That means the values were generated by *independent* repetitions of the underlying random experiment.

If the values of Z are from a numeric population and you're not thinking in terms of a random experiment, then the sample should have been chosen with replacement, making the Z's independent. If the sample was chosen withOUT replacement, but the population was quite large compared with the sample size, the distinction between with and without replacement doesn't much matter, so still the Z's would be at least approximately independent. Unless otherwise stated, we assume our populations large.

(e) While the scores would be normally distributed, they do not have mean zero and standard deviation one. In other words, you do not have a sum of squared Z's.

4.4.2

(a) $\chi^2 = 36.191$.

(b) 97.5% of the area under the curve is to the left of $\chi^2 = 30.191$. You have to find the point to the left of 30.191 which removes the first 95% of the 97.5%. It's 27.587.

(c) Assuming these Z's are independent, we want the value of χ^2 which cuts off 90% in the left tail (to put 10% to the right). Here $d = 6$ because there are six Z's . The answer is $\chi^2 = 10.645$.

4.4.3

(a) χ^2 is a sum of squared quantities. So it can't be negative.

(b) With more squared Z's in the sum, the expected value should be *larger*. This also follows from the formula for the mean of χ^2—its just d. So of course, if d (the degrees of freedom) increases, then the mean (it's just d) will increase.

(c) With $d = 4$, $\chi^2 = Z^2 + Z^2 + Z^2 + Z^2$. Now for this to take a large value (so it falls in the "far right tail") at least some of these Z's must be quite extreme. That's possible but unlikely. That it's possible says all values in the far right tail are possible, so the curve must cover the entire positive axis all the way to infinity. That it's unlikely says the area in the far right tail is small.

(d) Here $\mu = 14$ and $\sigma = 5.2915$ ($\sigma^2 = \sqrt{28}$)

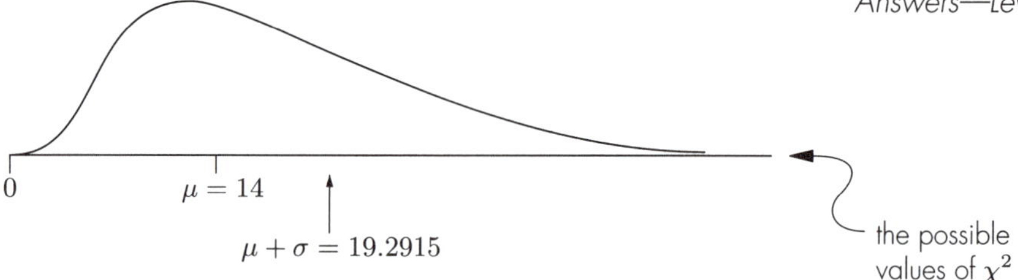

(e) $E(Z^2) = \sigma^2 + \mu^2 = 1 + 0 = 1$. Then since the mean of a sum is the sum of the means, $E(\chi^2) = d$.

4.5.1 (a) Draw the picture for the normal distribution of diameters and you'll see that 1.3 must standardize to $Z = -0.52$ and 1.71 to $Z = 0.25$. That will give you two equations (the standardizing transformation) in the unknowns μ and σ. Solve for μ and σ.

(b) The assumption should be valid unless there is something systematically wrong with the machine. See Problem 4.3.10.

(c) Draw the normal distribution for diameters. You want to put an area of 0.0001 in the right tail. Then diameters falling in the right tail will occur with a probability of one in 10,000. The cutoff for that area is the desired maximum. It must standardize to $Z = 3.86$. Solve the standardizing transformation to get the maximum.

4.5.2 So, $0.045 = \text{P}(T > 10) = e^{-10\lambda}$. The trick for getting variables out of the exponent is to take logarithms

$$\ln(0.045) = -10\lambda$$

because for any a, $\ln(e^a) = a$ [recall: $\ln(e^a) = a\ln(e)$, but $\ln(e) = 1$]. Thus, the mean of T, the average waiting time, is $1/\lambda = 3.2247$ minutes. Now 0.2247 minutes is about 13.48 seconds.

4.5.3 (a) $Z = 1.2$ so 88.49% of the students pass. That's 28,983 students.

(b) $p = 0.1151$ from part (a). The exact probability is $C(100, 10)p^{10}q^{90}$, but that coefficient will be difficult to calculate. Use the normal approximation; it's valid here because $np > 11 \geq 5$ and nq is even bigger. So, with $\mu = 11.51$, $\sigma = 3.1914$,

$$\text{P}(9.5 < \widetilde{X} < 10.5) = \text{P}(-0.63 < Z < -0.32) = 0.1102.$$

4.5.4 $p = 1/35$, so $\text{P}(X > 0) = 1 - \text{P}(X = 0) = 0.5155$.

4.5.5 (a)

Size	P (size)
I	0.0228
II	0.1359
III	0.6826
IV	0.1359
V	0.0228
	1

(b) The mean of the hypergeometric is np, so

Size	I	II	III	IV	V
Expected # in a carton	0.2736	1.6308	8.1912	1.6308	0.2736

(c) The simple addition rule holds because the sizes are mutually exclusive. Thus, 15.87% of all stones are of size IV or V. You will have a sample of $n = 12 \times 110$ stones. The expected number in a sample to have the characteristic of interest is $np = 209.484$.

(d)

$$\begin{aligned} P(X > 120 \mid n = 1320,\ p = 0.1587) &\approx P(\widetilde{X} > 120.5) \\ &= P(Z > -6.70) = 1. \end{aligned}$$

Here $\mu = 209.484$, $\sigma = 13.2755$.

(e) Draw the picture! $P(Z > 1.28) = 0.9$, so $(24 - \mu)/\sigma = 1.28$.

$$\begin{aligned} 24 &= 1.28\sigma + \mu \\ &= 1.28\sqrt{npq} + np \end{aligned}$$

You know p here, so you have an equation to be solved for the one unknown, n. However, the equation involves the square root of n. Introduce the substitution $u = \sqrt{n}$. This gives the following quadratic equation in u which you solve using the quadratic formula

$$24 = 0.1911u + 0.0228u^2$$

$$\text{or} \quad 0 = 0.0228u^2 + 0.1911u - 24.$$

You'll find $u = 28.5230$ (ignore the negative root because u cannot be negative). And so $n = u^2 = 814$, rounding UP to be conservative. Now if you need 814 stones, you'll need 68 cartons (67 cartons would give you only 804 stones which is not enough).

4.5.6

(a) $\mu_T = 2.5$, so you would expect $\lambda = 1/2.5$ malfunctions in one hour. Thus, 3.2 in eight hours (if the model is valid).

(b) $\mathrm{P}(X = 0) = e^{-3.2} = 0.0408$.

4.5.7

If $p = 0.154$, $\mathrm{P}(X \leq 15) \approx \mathrm{P}(\widetilde{X} < 15.5) = \mathrm{P}(Z < -6.45) = 0$. If $p = 0.057$, $\mathrm{P}(X \leq 15) \approx \mathrm{P}(\widetilde{X} < 15.5) = \mathrm{P}(Z < -1.63) = 0.052$. The case was remanded to a lower court to determine what the pool of candidates should be.

4.5.8

(a) So it's the "twelfth order-statistic," denoted by $x_{(12)}$. And $x_{(6)} = 1.3$.

(b) $M < x_{(h)}$ means $Y < h$ because $h - 1$ or fewer observations in the sample are BELOW M. And $M > x_{(k)}$ means $Y \geq k$ (Y is discrete, we cannot ignore equals signs!), so

$$\mathrm{P}(M < x_{(h)}) + \mathrm{P}(M > x_{(k)}) = \mathrm{P}(Y < h) + \mathrm{P}(Y \geq k).$$

(c) Taking the continuity correction into consideration and taking $\widetilde{Y}$ as the approximating normal, you want

$$\mathrm{P}\left(\widetilde{Y} > h - \frac{1}{2}\right) = 5\% \quad \text{and} \quad \mathrm{P}\left(\widetilde{Y} > k - \frac{1}{2}\right) = 5\%.$$

From the pictures of $\widetilde{Y}$ and Z, you'll see that $h - \frac{1}{2}$ must standardize to -1.645 and $k - \frac{1}{2}$ must standardize to 1.645. Write down those two standardizing transformations and solve the first for h and the second for k. Remember that the mean is $n/2$ and the variance $n/4$. Letting $z_o = 1.645$, you'll get the formulas given in the problem statement.

(d) Because $n = 10$, h and k are given by $\left(11 \pm z_o\sqrt{10}\right)/2$. Thus, $h = 2.9$, $k = 8.1$ and the endpoints are $x_{(3)}$ and $x_{(8)}$. The order statistics, $x_{(3)}$ and $x_{(8)}$, are determined from the ordered sample

$$6.2,\ 6.3,\ 6.3,\ 6.4,\ 6.4,\ 6.4,\ 6.5,\ 6.5,\ 6.7,\ 6.8.$$

(e) You must assume your observations really are a simple random sample from the distribution of "fill." They will be if the "fill" is independent from one cup to the next. Ask an engineer who knows the fill mechanism!

(f) The median is a fixed number (a parameter), it can't be sometimes here and sometimes there! The 90% probability refers to the INTERVAL. A different sample would yield a different interval. So it's the interval that's sometimes here and sometimes there—sometimes it contains M, sometimes not. The 90% confidence coefficient means that 90% of all intervals obtained in this way would contain M. Of course, we'll never know whether our particular interval is one of the 90% good intervals or not. So there's a ten percent risk of error in our procedure.

(g) In the problem statement just before part (a), the confidence coefficient was analyzed schematically this way: $0.9 = 1-0.1 = 1-(0.5+0.5)$. In part (b), this analysis led you to choose a value for z_o to locate the 90% confidence coefficient in the middle of the Z-distribution. Now we have a 95% confidence coefficient. We should locate this 95% in the middle of the Z-distribution. That requires $z_o = 1.96$. So h and k are still given by $(11 \pm z_o\sqrt{10})/2$, but with the new z_o. Thus, $h = 2.4$, $k = 8.6$, and the endpoints are $x_{(2)} = 6.3$ and $x_{(9)} = 6.7$.

(h) Now h and k are given by $(101 \pm z_o\sqrt{100})/2$. So $h = 37.6$ and $k = 63.4$. Thus, $x_{(38)} = 3.11$ and $x_{(63)} = 3.13$ are the endpoints.

(i) "With a one percent risk of error, we can be sure that U.S. pennies have a median weight of between 3.11 and 3.13 grams." This is the formal conclusion. However, it's subject to misinterpretation. See Problem 3.10.14(h), (i) and see also part (f) of this problem.

Chapter **5**

5.1.1

(a) Populations are either dichotomous or numeric. Because you're asked about a MEAN, clearly the population must be numeric —there's no way to speak of a mean unless you have a set of numbers which could be averaged together.

(b) If you're going to compute the mean score of a test, you need all the test scores. In general, to compute a population mean requires that you have the entire population available to you.

To compute the mean of a random variable, you need some mathematical description of the entire distribution of that random variable. For example, to compute the mean length of all rods from a manufacturing process, you require a description of the lengths and the corresponding probabilities. That's much harder to know than just to find the mean itself. Note that there's no question of computing the mean length from "all the rods that will every be manufactured." The whole point is to know the mean length BEFORE we manufacture them!

(c) You're assuming the random sample is exactly like the original distribution and so gives the same mean.

(d) You are assuming that the random sample is *more or less* like the original distribution and so gives approximately the same mean.

(e) A random sample is NOT necessarily typical of the distribution it's drawn from—even approximately! In fact, it could be quite atypical, and so NEITHER of the assumptions in (c) or (d) is acceptable. Do you remember this discussion concerning populations from Chapter 2? It's in the subsection of Chapter 2 entitled "The Statistical Questions."

(f) You don't. For example, suppose you're sampling from a population. The only way you would know if a sample is typical of the population it's drawn from would be to actually compare it with the entire population. But if you knew the population completely enough to make such a comparison you'd have enough information to just compute the value of μ—that's exactly what we said was not possible. In this chapter, we're asking what you can do if it's not possible to actually compute the value of the parameter.

Similarly, if you're sampling from the distribution of a random variable, to know if a sample is typical would require some complete description of the distribution so you could compare the sample with the distribution as a whole. But as a practical matter, no such complete description will exist.

(g) One particular value of $\overline{X}$ by itself tells you NOTHING. At this point the situation looks hopeless. But it's not—READ ON!

5.1.2

(a) This is a *dichotomous* population. It's true that the population consists of numbers, but we are not interested in the actual values of these numbers, we're only interested in the fact that some have a certain characteristic of interest and that others do not. It's a dichotomous population which happens to consist of numbers.

(b) If you wanted to know the proportion of students on a test who had a passing score, you would need two pieces of information—how many students took the test and how many passed. The ratio of these two numbers is your answer. If 30 took the test and 22 passed, then $p = 22/30$.

In our problem, you would need to know (1) how many persons reside in the geographic area in question and (2) how many of them are over 50 years of age. In other words, you would need to have some very specific information about the total population. If the population as a whole is not readily accessible to you, this information would not be available.

(c) Wrong!! You are not justified in saying that the true proportion is even approximately the same as your sample proportion. Why not?

(d) Your sample might be very unlike the population as a whole. Furthermore, you'll never know whether it's a typical sample or not. That would require comparing the sample with the whole population. If you could do that—if the whole population were that accessible to you—there would be no need for sampling or for estimating the value of p, you could just *calculate* the true value of p from the whole population.

So what can you say about the population based on one simple random sample alone? NOTHING! If that makes the situation sound hopeless—KEEP READING!

5.1.3

(a) There are 21 of these samples. They are:

$$
\begin{array}{cccccc}
\{0,1\} & \{0,2\} & \{0,3\} & \{0,\overline{3}\} & \{0,4\} & \{0,5\} \\
 & \{1,2\} & \{1,3\} & \{1,\overline{3}\} & \{1,4\} & \{1,5\} \\
 & & \{2,3\} & \{2,\overline{3}\} & \{2,4\} & \{2,5\} \\
 & & & \{3,\overline{3}\} & \{3,4\} & \{3,5\} \\
 & & & & \{\overline{3},4\} & \{\overline{3},5\} \\
 & & & & & \{4,5\}
\end{array}
$$

(b) Here are the corresponding values of $\overline{X}$:

$$
\begin{array}{cccccc}
0.5 & 1.0 & 1.5 & 1.5 & 2.0 & 2.5 \\
 & 1.5 & 2.0 & 2.0 & 2.5 & 3.0 \\
 & & 2.5 & 2.5 & 3.0 & 3.5 \\
 & & & 3.0 & 3.5 & 4.0 \\
 & & & & 3.5 & 4.0 \\
 & & & & & 4.5
\end{array}
$$

(c) $\overline{X}$ computes a number for each sample—its mean. So, $\overline{X}$ is just a random variable. The random sampling process is indeed a random experiment, and so the samples are *outcomes of a random experiment*. The sample mean is an assignment of a number to each of these outcomes. Don't forget, by the way, for us the sample size in random sampling must be fixed; in this case, it's fixed at $n = 2$.

(d) There are 21 samples, each occurring with probability 1/21. So if a value of $\overline{X}$ occurs three times—if it's the mean for three different samples—it has probability 3/21. Here's the probability distribution for $\overline{X}$:

$\overline{X}$	$P(X)$
0.5	1/21
1.0	1/21
1.5	3/21
2.0	3/21
2.5	4/21
3.0	3/21
3.5	3/21
4.0	2/21
4.5	1/21
	1

(e)

X	$P(X)$	$XP(X)$	$(X-\mu)^2P(X)$
0.5	0.0476	0.0238	0.2043
1.0	0.0476	0.0476	0.1176
1.5	0.1429	0.2143	0.1641
2.0	0.1429	0.2857	0.0467
2.5	0.1905	0.4762	0.0010
3.0	0.1429	0.4286	0.0262
3.5	0.1429	0.5000	0.1232
4.0	0.0952	0.3810	0.1943
4.5	0.0476	0.2143	0.1770
	1.0001	2.5715	1.0544

so, $\mu_{\overline{X}} = 2.5715$, $\sigma^2_{\overline{X}} = 1.0544$.

(f) It's the answer to the original question! It's the value of the unknown μ. Except for the last digit, which is off due to rounding, the mean of the sample means is the same as the mean of the original population . . .

$$(0+1+2+3+3+4+5)/7 = 2.5714.$$

General Fact: The "mean of the sample means" is just the mean of the original population—not just for this one example; it's a general fact. This makes sense too if you think about it —the average value of the numbers in a sample is *on average* the same as the average value of the population of numbers it's drawn from.

(g) The sample means are less variable than the numbers making up the population. Here are three ways of thinking about this:

> (i) *The sample means compared with the population*: The sample means are concentrated near their center—four of them take on the value 2.5. SIXTEEN of the 21 samples give means within a range of only two points, 1.5 to 3.5. By contrast, the population is spread fairly evenly over its entire range, with only one value duplicated. (Of course it's true the *range* is smaller—the sample means have a range of only four, whereas the population has a range of five. But this is just a fluke of our unrealistically small population. For a large population, this difference in the range would be negligible.)

or

> (ii) *a general principle*: You could observe that in looking at the means of samples you are "averaging out" some of the variability by looking at several population members at once.

or

> (iii) *variances compared*: You can just compare the population variance with the variance of the sample means: $\sigma^2 = 2.5306$, whereas $\sigma^2_{\overline{X}} = 1.0544$, a smaller number.

(h) The true population mean is $\mu = 2.5714$. The typical samples are the ones in the box . . .

	$\{0,1\}$	$\{0,2\}$	$\{0,3\}$	$\{0,\overline{3}\}$	$\{0,4\}$	$\{0,5\}$
		$\{1,2\}$	$\{1,3\}$	$\{1,\overline{3}\}$	$\{1,4\}$	$\{1,5\}$
If "typical" means $\overline{X}$ is			$\{2,3\}$	$\{2,\overline{3}\}$	$\{2,4\}$	$\{2,5\}$
within one point of μ,				$\{3,\overline{3}\}$	$\{3,4\}$	$\{3,5\}$
the "typical" samples are $\longrightarrow$					$\{\overline{3},4\}$	$\{\overline{3},5\}$
						$\{4,5\}$

Let's say, in trying to estimate this unknown μ, you had drawn the sample $\{3,5\}$, with a mean of four. Based on this sample alone, you'd get a very wrong idea of the true population mean. You can see that there are eight such atypical samples and they're *entirely possible* as outcomes of your sampling experiment.

So remember, with simple random sampling from a population, each sample is equally likely to be drawn. If you draw one sample and have nothing else to go on, you could be very misled about the true nature of the population!

(i) By itself, absolutely NOTHING!

5.1.4

(a) That random sample by itself would tell you NOTHING about p.

(b) NUMERIC population?! Ouch! That's not correct—it's a *dichotomous* population which just happens to consist of numbers. We're not interested in the actual values of those numbers, rather we're interested in a certain characteristic the numbers may or may not have. Some of the numbers are "positive even numbers," some are not.

(c) $\hat{p}$ looks at a random sample and assigns a number to it. In other words, $\hat{p}$ is a random variable—it's an assignment of a number to each of the possible outcomes (the samples) of a random experiment. What's the random experiment?

(d) $\hat{p} = 0, \frac{1}{2}$ or 1.

(e) The proportion of a sample having the characteristic of interest should *on average* be the same as the proportion of the population having that characteristic. Watch:

$\hat{p}$	$P(\hat{p})$	$\hat{p}\,P(\hat{p})$	$\hat{p}^2\,P(\hat{p})$
0	10/21	0	0
1/2	10/21	5/21	5/42
1	1/21	1/21	2/42
	1	6/21	7/42

so $\mu_{\hat{p}} = 2/7$,
$\sigma^2_{\hat{p}} = 0.0850.$

So the mean of $\hat{p}$ is $2/7$ which equals the proportion of the population having the characteristic of interest. In other words, $\mu_{\hat{p}} = p$. The variance of $\hat{p}$ is

$$7/42 - (6/21)^2 \quad \text{which is just } 0.0850.$$

(f) The word "typical" now refers to a different aspect of the population. Before, we were thinking of it as a numeric population and asking about the mean. Now it's a dichotomous population and we're asking about the *proportion* which have a certain characteristic of interest—being positive and even. So different samples look typical because we've changed the meaning of "typical."

The word "typical" now means the sample proportion is close to the population proportion of $2/7$. Because $2/7 \approx 30\%$ and the closest sample proportion is 50%, you might want to say that none of the samples is typical! But this happens only because we've taken such a small sample. Let's not be too restrictive; let's agree that a sample is typical if $\hat{p}$ is within 25 percentage points of the true p. So a sample is typical if $\hat{p}$ falls between $2/7 - 1/4$ and $2/7 + 1/4$. In that case, those samples for which $\hat{p} = 1/2$ are the "typical" samples—there are ten of them:

$\{0,1\}$	$\{0,2\}$	$\{0,3\}$	$\{0,\bar{3}\}$	$\{0,4\}$	$\{0,5\}$
	$\{1,2\}$	$\{1,3\}$	$\{1,\bar{3}\}$	$\{1,4\}$	$\{1,5\}$
		$\{2,3\}$	$\{2,\bar{3}\}$	$\{2,4\}$	$\{2,5\}$
			$\{3,\bar{3}\}$	$\{3,4\}$	$\{3,5\}$
				$\{\bar{3},4\}$	$\{\bar{3},5\}$
					$\{4,5\}$

If "typical" means $\hat{p}$ is within $1/4$ of p, the "typical" samples are the ten in the boxes.

(g) On the basis of one random sample by itself with nothing else to go on, you can say NOTHING AT ALL about the population.

(h) In fact, one random sample *by itself, with no other information to go on*, tells you NOTHING about the population as a whole. So what's the use of sampling experiments? Isn't this whole discussion just a waste of time? NOT AT ALL—READ ON!

5.1.5

Statistics can involve a "three valued logic." Statements taken abstractly are either true or false, yes. But taken in the light of *given information* there are not just two possibilities, there are three: justified, inconclusive, or unjustified. In light of the information $\overline{X} = 2.37$, with nothing else to go on, the statement $\mu \approx 2.37$ is inconclusive.

So the defect we're discussing is to think some conclusion must be possible based just on our one sample. You're drawing a conclusion where no conclusion is possible. It's true that μ is either close to 2.37 or it's not. But the argument goes further—it tries to decide between these two alternatives. That's NOT possible; no such conclusion can be justified based on one sample alone.

5.1.6

(a) An estimator is a random variable. The underlying random experiment is "random sampling." An outcome is one random sample. To that outcome (the sample), the estimator assigns a number—the value of the estimator computed from that sample. So the estimator is "an assignment of a number to each of the possible outcomes of some random experiment."

(b) . . . its probability distribution. For any random variable, we would want, at the very least, the mean and variance. Beyond that, we would want some way (formulas, tables, pictures, etc.) to get hold of the probabilities. In other words, we would want the probability distribution for that random variable.

5.1.7

(**a**) The variable, X, is either hypergeometric or binomial depending upon whether you sample with or without replacement. But even if the sampling is without replacement, the binomial model is still a reasonable approximation if the population is large. Let's assume the binomial model for X. Here's how you obtain $\hat{p}$ from X . . .

$$\hat{p} = 1/nX.$$

↑ count how many have the characterisitic

↑ divide by the sample size

Now use your familiar formulas

If	$Y = a + bX,$
then	$\mu_Y = a + b\mu_X$
	$\sigma_Y^2 = b^2\sigma_X^2$

Here $a = 0$ and $b = 1/n$. So the formulas for the mean and variance of $\hat{p}$ become

$$\mu_{\hat{p}} = (1/n)\mu_X = (1/n)np = p,$$

$$\sigma_{\hat{p}}^2 = (1/n^2)\sigma_X^2 = (1/n^2)npq = pq/n.$$

This result is consistent with the answer to Problem 5.1.4(e) where, for that specific example, you calculated that $\mu_{\hat{p}} = p$. Now you see there was nothing special about that example, it's a general fact:

> When sampling from a dichotomous population, the expected value of $\hat{p}$, $\mu_{\hat{p}}$, is just p, the proportion of the population having the characteristic of interest.

(**b**) As above, $\hat{p} = (1/n)X$ for a binomial X. X counts how many in the sample have the characteristic of interest. Its possible values are

$$0,\ 1,\ 2,\ 3,\ \ldots,\ n-1,\ n.$$

So the possible values of $\hat{p}$ are

$$0,\ 1/n,\ 2/n,\ 3/n,\ \ldots,\ (n-1)/n,\ 1.$$

Here, because $n = 100$, $\hat{p}$ takes the values

$$0,\ 0.01,\ 0.02,\ 0.03,\ \ldots,\ 0.99,\ 1.00$$

There are 101 values, too many to attempt drawing a line graph.

But the normal approximation is appropriate for X because np, $nq \geq 5$. Because $\hat{p}$ is a linear function of X, $\hat{p}$ is itself approximately normally distributed (using Problem 4.3.13). Here's the picture:

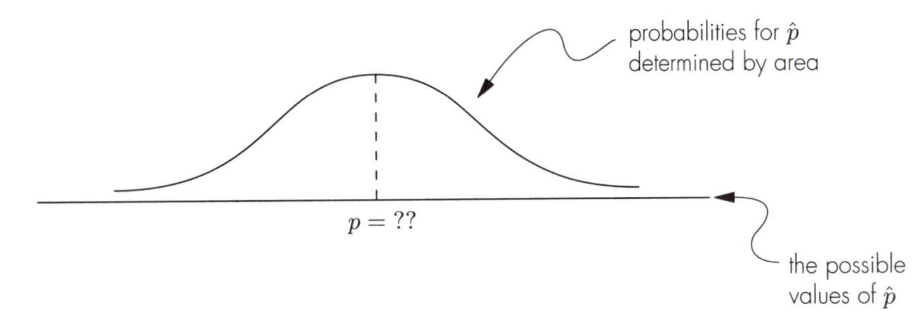

In the picture, you see p as the mean and you see that its exact value is unknown. We didn't label $\mu + \sigma$ because it's algebraically too complicated to carry much intuitive meaning $(p + \sqrt{pq/n})$.

(c) X is the binomial random variable! Everytime you observe the characteristic of interest ("success"), you record a one, otherwise a zero. So X, "the number of ones," is the number of successes. That's the binomial X. Thus, once again, $\hat{p} = (1/n)X$, with X binomial. Now the argument is exactly the same as before.

5.1.8

(a) The "total context" of the estimator is its probability distribution as a random variable. You can summarize that distribution in a picture such as the one for $\hat{p}$ given just above. So the "whole picture" becomes a picture in the literal sense!

The important, subtle point here is that a random variable—in this case, an estimator—is much more than just a number, or even a set of numbers. Random variables are very sophisticated theoretical models. You've spent a lot of time learning all about them! The "total picture" or "entire context" of a random variable is summarized in its probability distribution which displays all its possible values together with the associated probabilities and other relevant information (the mean, etc.).

(b) Look at the total context of the estimator $\hat{p}$. In other words, look at the normal distribution of $\hat{p}$ given in the previous problem. You see the unknown parameter p—it's the center (the mean) of that probability distribution.

The relationship of $\hat{p}$ to p is the sampling distribution pictured in the previous problem together with the pair of equations

$$\mu_{\hat{p}} = p,$$
$$\sigma^2_{\hat{p}} = pq/n.$$

5.1.9

(a) You don't know and never will know where to locate $\hat{p} = 0.18$ in the picture because the picture is centered on p and you don't know its value. Furthermore, in any practical situation, the true value of a parameter like p is *not knowable*! But if you don't know the center of the distribution, you can't locate a specific value.

Think about this: The population is much too large to actually compute the true proportion of persons over 50 years of age. Even if you had enough money to survey the entire population, such a large survey would contain unknown and uncontrollable errors of various kinds. Furthermore, the survey requires time to complete, so the proportion of the population over 50 would certainly change before you were finished. People are born and die and people move in and out. Many practical problems will prevent your ever knowing the center for the probability distribution of an estimator.

(b) (i) Done.

(ii) This picture is entirely possible. The only difference between it and the first is the spread; the second is a little less spread about its center. But that's possible because we don't know the standard deviation of $\hat{p}$ (it requires knowing p).

(iii) Because $\mu_{\hat{p}} = p$ and this picture is centered on the observed $\hat{p}$, we would have to accept that $p = 0.18$. Exactly 18% of the entire population is over 50 and we observed a random sample also having exactly 18% over 50! This is *possible* alright, but we'd be very surprised if it did, in fact, happen. It's like tossing a coin 1002 times and getting exactly 501 heads!

(iv) This picture is silly—it implies that 0.18 is a negative number!

(v) This picture is not possible either, for essentially two reasons. First of all, it implies that $p = 0$, which is not likely. If none of the population have the characteristic of interest, you wouldn't have been asking about it in the first place. In this problem, however, we KNOW that p is not zero! How?

Second, even if p could be zero, this picture is not possible because it's symmetric about its mean. That implies there's a 50/50 chance for a sample proportion which is negative, but proportions are necessarily non-negative. Could you have MINUS three percent over 50 years of age?

(vi) This picture is not possible because it's not centered on p, contradicting the fact that $\mu_{\hat{p}} = p$.

(vii) This picture is entirely possible.

(viii) This picture is impossible. It implies $\hat{p}$ could be larger than one. Could more than 100% of your sample be over 50 years of age? Of course not, proportions can't be larger than one!

5.1.10

(a) If $p = 0$, then none of the population are over 50 years of age. But you obtained a sample with 18% having that characteristic, so even if nobody else is over 50, those people are! For example, if $n = 200$, then at least 36 people in that geographical area are over 50—you've seen them and interviewed them for your sample!

(b) The estimator $\hat{p}$ is "essentially" a binomial random variable. It's $\hat{p} = (1/n)X$, a linear function of a binomial X. If you're sampling without replacement, X is really hypergeometric. But even so, it's approximately binomial *if the population is quite large* so that the "with–without replacement" distinction doesn't much matter.

Now if $np, nq \geq 5$, the binomial random variable is approximately normally distributed. But this is tricky because you don't know the value of p and the condition $np, nq \geq 5$ involves p. To guarantee validity for the normal approximation, you would explore "worst-case" scenarios: If you're sure p is at least 80%, for example, then $q < 0.2$, so you would have to choose $n = 25$. In general, the further p is from a half, the larger n would have to be to guarantee the condition $np, nq \geq 5$. This means *the sample would have to be large*.

Here's how you analyze this: $5 \leq nq$ and $q < 0.2$ together give

$$5 \leq nq < 0.2n,$$

$$\text{and so} \quad 5/0.2 < n, \quad \text{where} \quad 5/0.2 = 25.$$

5.2.1

(a) The question we're trying to study in this chapter is of the form "What's the value of this unknown parameter?"—"What is p" for example, or "What is μ?" When the estimator is unbiased, its "on average" value (its mean) is the parameter in question. That's the connection between the question being asked and the estimator being used to answer the question. For $\hat{p}$ and p, the connection is the equation $\mu_{\hat{p}} = p$; for $\overline{X}$ and μ, it's the equation $\mu_{\overline{X}} = \mu$. See Problem 5.1.8.

Of course, there are other important connections between the estimator and the unknown parameter—through the equations for the variance, for example. These various "connections" of p to $\hat{p}$ are aspects of the "whole context" of the estimator which we talked about in Problem 5.1.8. That "context" is the sampling distribution taken as a whole. By the way, note that we've not yet seen how the sampling distribution actually allows us to say something meaningful about the parameter—in the very next section, we'll do that!

(b) The standard error is $\sqrt{pq/n}$; it measures the *accuracy* of $\hat{p}$ as an estimator. Because n is in the denominator, as n increases, the standard error gets smaller. This fact is what controls variability in the model from sample to sample: When n is large, there's less variability from sample to sample; when n is small, there's relatively more variability from sample to sample.

To say "most samples are typical" is just to say that most $\hat{p}$ values are concentrated near p . Because we have a normal distribution centered at p, that's true. Most of the probability is concentrated near the center, but that center: $\mu_{\hat{p}} = p$. Suppose, for example, a "typical" sample is a sample which gives a value of $\hat{p}$ that's within, say, ten percentage points of the true p. The picture will be:

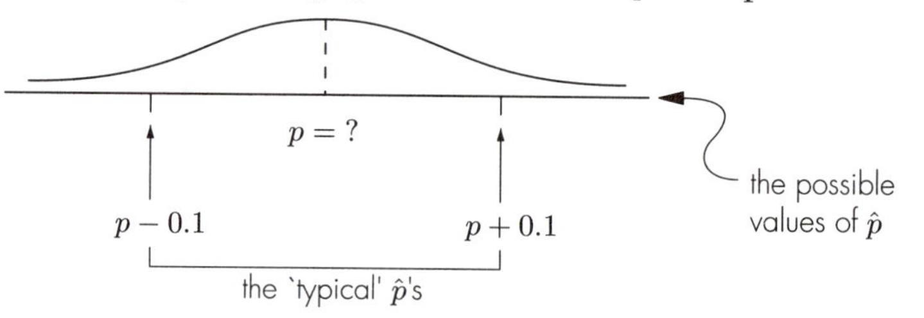

Suppose the standard error is smaller:

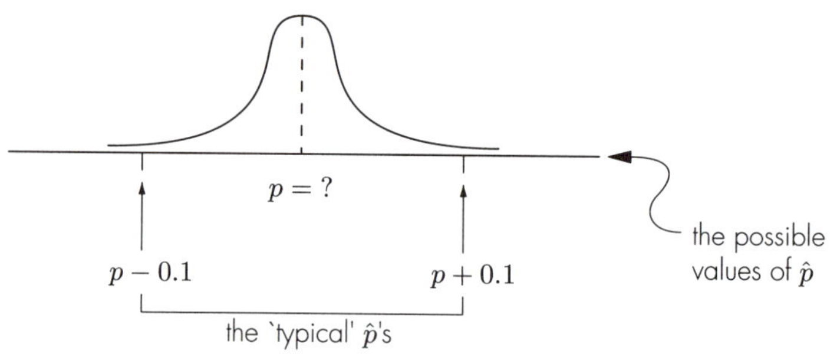

The second picture shows that a much larger percentage of the $\hat{p}$ values fall within the range of "typical" $\hat{p}$'s. So we see that

SMALLER STANDARD ERROR . . .

IMPLIES . . . more values of $\hat{p}$ close to p,

IMPLIES . . . more samples are typical,

IMPLIES . . . $\hat{p}$ is a more accurate estimator of the unknown p.

This gives the conclusion we stated above—the standard error measures the accuracy of the estimator. Here's another way to think of all this: The estimator will become more accurate if we take larger samples because

- "larger sample" means "more information,"
- "more information" means "more accurate estimate."

(c) The smallest value, zero, is too far below the true value of p (one percent). And ten percent, the *very next value of* $\hat{p}$, is already too far ABOVE the true value! Even when $n = 100$, the situation is not much improved.

(d) Choose n to make equality hold: $pq/n = (0.02)^2$. With $p = 0.5$, we get $0.25/0.0004 = n$ so let $n = 625$. A larger n would make the standard error even smaller, because n is in the denominator. So we've made the standard error be "no more than two percentage points," as required.

(e) Here's the picture:

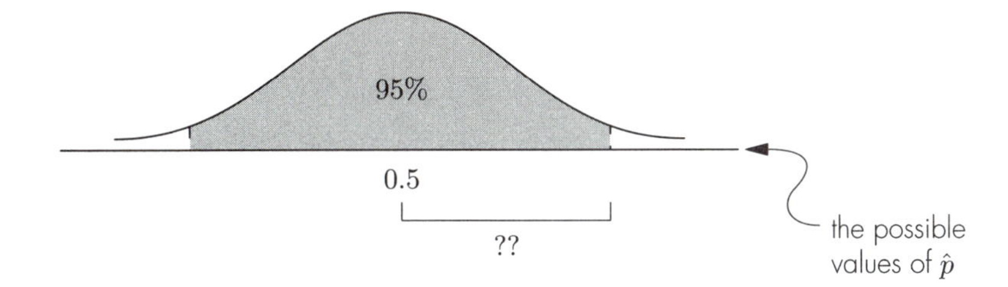

For this picture to hold, you must be looking at all values of $\hat{p}$ within about two standard deviations of the center (two "standard errors'). That means 2×0.02 (four percentage points).

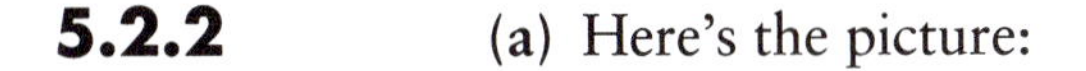

5.2.2 **(a)** Here's the picture:

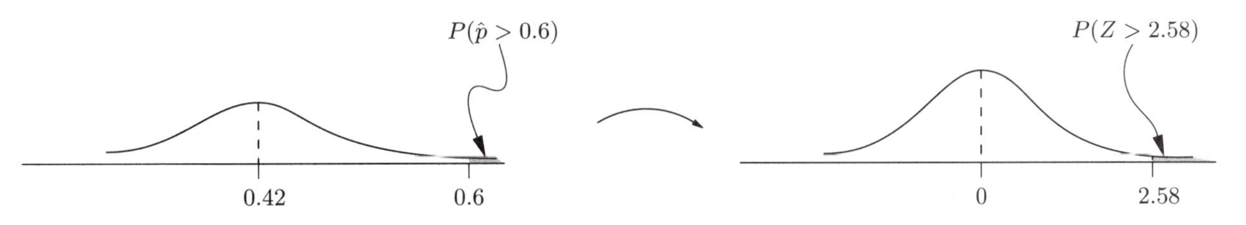

The square of the standard error (the variance of the sampling distribution) is $pq/n = 0.42 \times 0.58/50 = 0.004872$ (you should not round until you have completed your calculations). So $\hat{p} = 0.6$ standardizes to

$$\frac{0.6 - 0.42}{\text{s.e.}}$$

which is just $z = 2.58$. Then $P(Z > 2.58) = 0.0049$.

Real-world conclusion:

> There's a 0.5% chance of a sample of 50 morning commuters where more than 60% say they use public transportation.

(b) The *number* of possible "yes" responses increases with the sample size. In a sample of 50, you could have 50 yes's, in a sample of 100, you could have 100. On the other hand, the *proportion* of yes's in a sample is relative to the sample size. No matter what the sample size, the proportion can be anywhere between zero and one (or, as a percentage, between 0% and 100%).

In each case, the appropriate measure of variability is the variance. For the number of "yes" responses, the variance is npq (binomial model). It *increases* with the sample size, so the number of yes's is more variable for a larger sample. For the proportion of yes's, the variance is pq/n (the variance of $\hat{p}$). That *decreases* with the sample size.

(c) Real-world problems in this chapter ask for the value of an unknown parameter such as p or μ. But in this problem, we KNOW p—it's 0.42. We're exploring what could be said about the estimator $\hat{p}$ based on a random sample. So the population seems to be known quite well (we know what proportion use public transportation) and we ask a question about what *ought, in theory*, to be true of a random sample. That's theoretical and "unrealistic," but it's very useful for understanding how the theory works.

(d)

$$P(15 < X < 20) = P(0.3 < \hat{p} < 0.4) = P(-1.72 < Z < -0.29)$$
$$= 0.3859 - 0.0427 = 0.3432.$$

(e) You know exactly what proportion of morning commuters, if asked, would say they don't use public transportation. You were given that, it's 58%!

(f) If $\hat{p}$ continues to be the proportion of the sample who say "yes" as we've been doing it, then $1 - \hat{p}$ is the proportion who say "no." So

$$1 - \hat{p} > 2/3 \quad \text{if and only if } 1/3 > \hat{p}.$$

The original question is $P(\hat{p} < 1/3) =??$, and the answer is

$$P(\hat{p} < 1/3) = P(Z < -1.24) = 0.1075.$$

Now change the problem around. Do it in terms of the parameter q, the proportion of all morning commuters (58%) who would say "no." With this new parameter, $\hat{p}$ is the proportion of your sample who say "no." So now, $P(\hat{p} > 2/3) = P(Z > 1.24) = 0.1075$.

(g) The question here asks for $P(0.4 < \hat{q} < 0.46)$. Let's stick with $p = 0.42$ and convert the question to a question about $\hat{p}$. First, multiply by -1 (inverting the inequality): $-0.4 > -1 + \hat{p} > -0.46$, then add one: $0.6 > \hat{p} > 0.54$. This converts the inequality in $\hat{q}$ to one in $\hat{p}$. So you need

$$\begin{aligned} P(0.54 < \hat{p} < 0.6) &= P(1.72 < Z < 2.58) \\ &= 0.9951 - 0.9573 = 0.0378. \end{aligned}$$

Or you can do this problem with $q = 0.58$. You have to find the probability: $P(-2.58 < Z < -1.72)$. It's exactly the same probability, of course: 0.0378. It must be the same because it answers the same question, just taking another point of view.

(h) Here's the picture of $\hat{p}$:

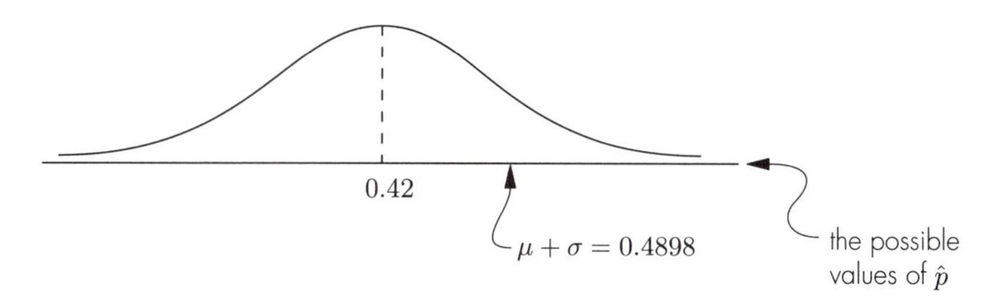

Knowing there's about a 68% chance that $\hat{p}$ falls within one standard error (one standard deviation) of the mean, we locate the point $\mu + \sigma$. Between it and the symmetrically placed point, $\mu - \sigma$, we should find a bit more than two- thirds the area.

5.2.3

(a) If all values (call them X) are the same, the mean takes that same value also. So $X = \mu$ for all X. But then the deviations from the mean, $X - \mu$, are all zero. So $\sigma^2 = 0$, because it's the average of the squared deviations from the mean.

(b) The normal approximation is justified because $np, nq \geq 5$. The total context of $\hat{p} = 15/24$ is the probability distribution of $\hat{p}$:

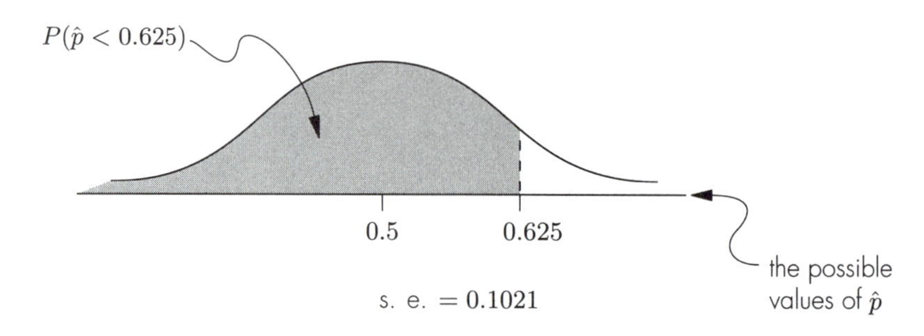

The standard error is the square root of $pq/n = 0.5 \times 0.5/24$. Applying the standardizing transformation to 0.625, we obtain our answer from the Z table: $P(Z < 1.22) = 0.8888$.

Why is $p = 0.5$? The parameter p is the probability of having the characteristic of interest. It's the probability that any ten-year-old is less than 129 cm tall, shorter than average: $P(H < \mu_H)$. It's should be about one-half. You can obtain this from the probability distribution of H (height).

(c) $p = 0.1736$ is obtained as follows:

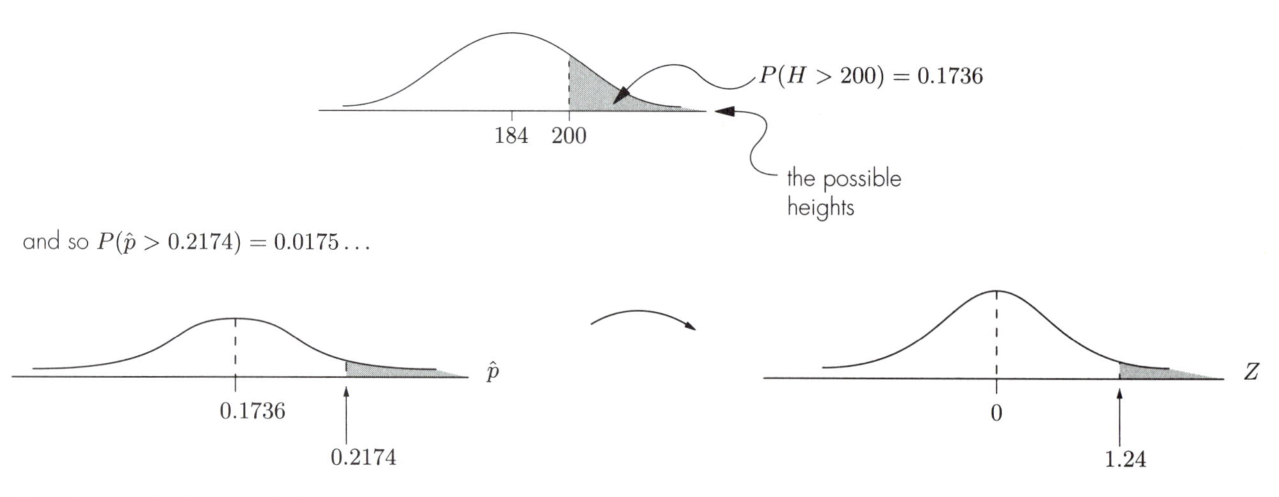

here the standard error is 0.0353

(d) In part (b) the normal distribution for height is *symmetric about its mean* of 129. That means there's a 50–50 chance ($p = 0.5$) for a ten-year-old to be shorter than 129 cm. In part (c) you use the normal distribution to calculate $p = 0.1736$.

A variable like height will often be normally distributed—the difference between any two heights would be determined by many independent random factors with nothing systematic going on. But it would not be true, for example, if you were speaking of the ten-year-olds *and their parents*. In such a case, there's a relevant systematic consideration—adult versus child—which accounts for some of the differences in height. The population of children and parents should have a bimodal distribution.

In addition to the normality assumption, in parts (b) and (c) you had to assume the students were "like" a random sample. Otherwise, the theory you used is not valid. This assumption might fail in many ways. Name some.

5.2.4

We are given: (1) a *range of possible values* for p:

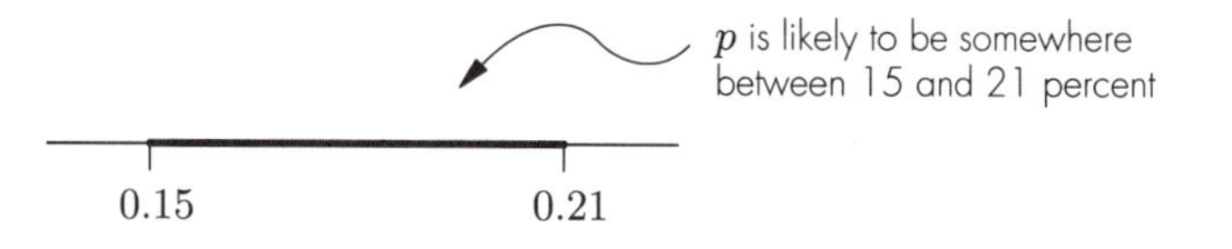

and (2) a 95% *probability* that the range of values actually does include the unknown p. These are the two "parts" of a confidence interval—the range of possible values and the probability for the parameter to fall within that range.

5.2.5

(**a**) The "total situation" consists of a question about an unknown parameter p together with the tool for answering that question—the probability distribution for $\hat{p}$. This shows what's variable: the possible values of $\hat{p}$ (a random *variable*) as determined by the various samples. The center of the interval and its width are determined by the observed value of $\hat{p}$ (in our example, 18%). With a different sample you'll have a different value of $\hat{p}$ and, so a different interval (different center and different endpoints).

Therefore, for a confidence interval, IT'S THE RANGE OF VALUES THAT VARIES—the range of values is determined by the sample and varies from sample to sample. Exactly how all this happens can't be clear to you yet. We need to see how the range of values for a confidence interval is actually determined by the sample data.

(**b**) It's impossible for p to be sometimes here and sometimes there! After all, p is not variable. It's some fixed quantity which, in the case we're looking at, happens to be unknown.

(**c**) It's the interval itself which varies [see part (a)]. So the probability of 95% means

> Of 100 samples, about 95 on average will yield a confidence interval containing the unknown p and about five will generate an interval that fails to contain p.

Of course, in any real-world situation, you get one sample only, so this probability remains *theoretical*. It's a way of measuring your "risk of error" in using this estimation technique.

So of 100 intervals, 95 should be "good" and five "bad." You probably could guess that the five bad intervals are the ones generated by the atypical samples. This should sound familiar—you saw in Problem 5.2.1(b) that most samples are typical. In other words, most $\hat{p}$'s are close to p. This is because $\hat{p}$ is a normally distributed random variable centered on p: The probability is concentrated near p, the mean. Here's the picture:

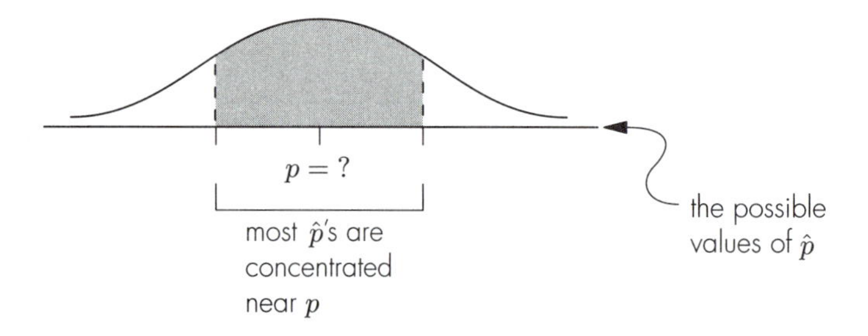

Now you see the tremendous significance of having a sampling distribution which is symmetric and "mound-shaped" and which is centered on p: Very atypical samples giving $\hat{p}$ values far from the true p are *possible* but not likely.

5.2.6

(a) Different samples would give different intervals for p. As a practical matter, you obtain only one such interval. There's a 95% chance that your particular interval, in fact, does contain p. So, you can be *confident* that p is in the interval—it may not be, but it probably is. The usual terminology is to say, "We are 95% confident that p is in the interval (0.15, 0.21)."

(b) More certainty in your conclusion would require more *information* on which to base the conclusion. Information is never free. In the present context, "more information" means you would have to take a LARGER SAMPLE. And that costs—sampling is an expensive process!

5.2.7

(a) Instead of an exact value for p, you are given a range of possible values. You never know which of these possible values is the true value of p. So there is inexactness in the *value* of p .

But not only that, there's a small probability (5% in this case) that the true value of p isn't within that range at all. That's true if you were unlucky enough to obtain an atypical sample. So another source of inexactness in the answer is the possibility of highly atypical samples which generate "bad" confidence intervals. Here, we use the word "bad" as we did above in the answer to Problem 5.2.5(c)—a confidence interval is "bad" if the true value of the parameter is not in the interval at all. This is not standard terminology by the way; we use it here to help focus on the effect of very atypical samples.

Of course, in any realistic situation there are many sources of error, not just two! Principal among them are human errors of various forms—improperly collected or recorded data, computational errors in summarizing the data, and so on! All we've done in this problem is to identify the two sources of error that are *formally* accounted for and controlled in our procedure for constructing a confidence interval.

(b) Use the midpoint of the interval; that's the observed value of $\hat{p}$ which you obtained from your sample. In the example, it's 18% because that's halfway between 15% and 21%.

So, in fact, you'll be acting as if p were approximately 18%—as if $p \approx \hat{p}$. But you know exactly how it could be wrong. There are two sources of error and you exercise control over the those two possible errors.

(c) First of all, there is the range of possible values, 0.18 ± 0.03. The observed 18% which you will use for p could be off by three percentage points either way. Second, there's a chance (in this example, a 5% chance) you could have been seriously misled by an atypical sample so that the true value of p is nowhere near 18%! The true value of p may not be in the "range of possible values" at all.

You control both of these possible errors. The first you control by knowing the upper and lower limits of the range of possible values for p (21% and 15%, respectively). You control the second by choosing an acceptable probability (5% in our example) that your range of values doesn't contain the true value of p at all.

5.2.8

(a) The confidence coefficient is a *probability*. A probability assertion about $\hat{p}$ requires its probability distribution! You will need the following picture:

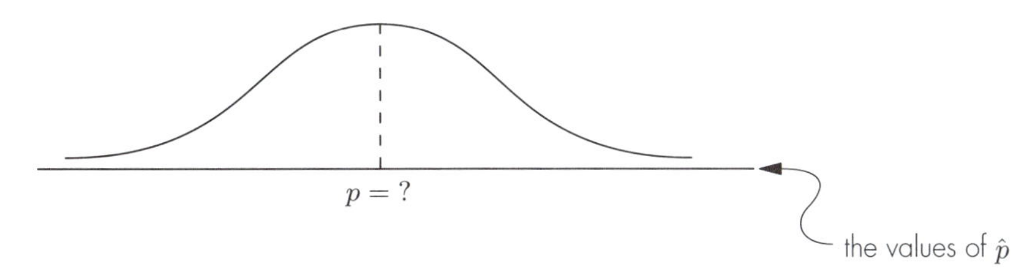

In other words, to construct a confidence interval for an unknown p, you require more than just one sample and one value of $\hat{p}$. You require the "entire context" of $\hat{p}$—its sampling distribution.

(b) When you say, "Based on this sample we believe, $p \approx 18\%$," you're trying to draw a conclusion from one sample alone. That's not possible. Complete the conclusion by giving (1) the *probabilty* that the estimate is valid (in our example, 95%) and (2) the *maximum error of the estimate* (3% in our example). Here's a corrected version of the conclusion:

> Based on this sample, with a five percent risk of being wrong, we believe p to be approximately 18% with a margin of error of three percentage points.

5.2.9 **(a)** Here's the picture:

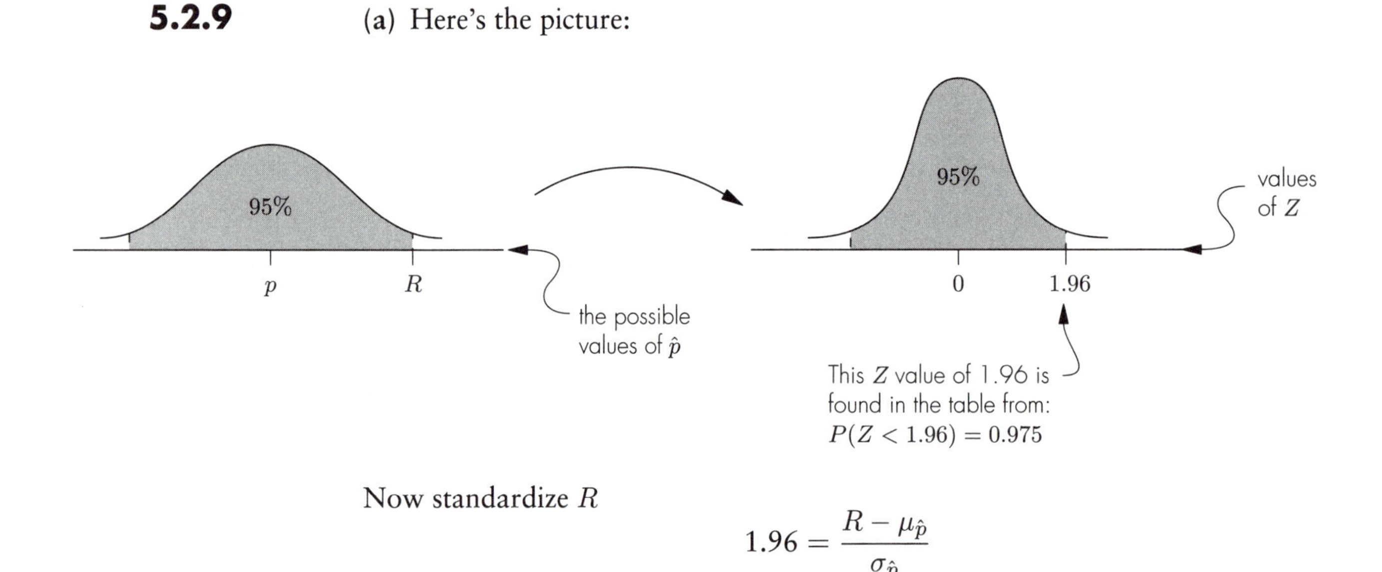

Now standardize R

$$1.96 = \frac{R - \mu_{\hat{p}}}{\sigma_{\hat{p}}}$$

and solve to find

$$R = \mu_{\hat{p}} + 1.96\sigma_{\hat{p}}$$
$$= p + 1.96\sigma_{\hat{p}} \quad \text{because } \mu_{\hat{p}} = p.$$

Similarly find

$$L = p - 1.96\sigma_{\hat{p}}.$$

So, $\delta = 1.96\sigma_{\hat{p}}$.

(b) The standard error of $\hat{p}$ is $\sqrt{pq/1000}$. If you can evaluate the standard error, you must already know the value of p. But if you already know p, why would you want to estimate it?

(c) Here's the parabola, $p - p^2$:

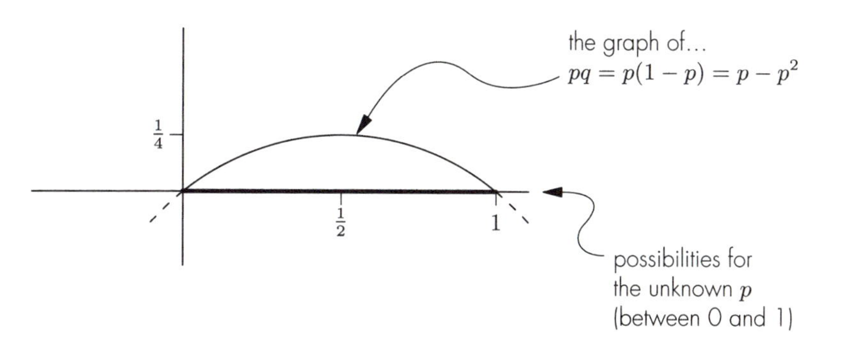

When $p = 0.5$, pq is as large as possible. Then, of course, $q = 0.5$ also. If we assume this "worst-case" scenario, the standard error for $\hat{p}$ will be approximated by 0.0158. This is the square root of

$$pq/n = 0.5 \times 0.5/1000.$$

5.2.10 A confidence interval answers a question of the form: "What is the value of this unknown parameter?" Because a parameter is just a number, you would expect a very simple answer: a number! But a confidence interval is much more complex than that. It's based on partial information in the form of ONE random sample, information that could be very misleading. To account for the possibility of a misleading sample, a *theory* is required—the theory of ALL random samples. That theory provides the sampling distribution of the estimator. Using the sampling distribution, we can obtain (1) a range of possible values for the parameter and (2) the probability that that range of values actually does contain the unknown parameter value. These two items constitute the confidence interval. It's not just a number! And it's derived from some very sophisticated considerations.

5.2.11 (a) If the confidence coefficient is 95%, then you should expect 9.5 out of ten intervals *on average* to contain the unknown p. If the confidence coefficient is 80%, then eight out of ten intervals should contain the unknown p, on average. Note that you cannot make any assertion about a SPECIFIC ten intervals—you can say what *ought* to happen on average. That's what the confidence coefficient is all about. It's a probability, a theoretical relative frequency.

(b) In the real world, ONE SAMPLE ALONE is all that's practical.

5.2.12 (a) $\sigma_{\hat{p}}^2 = pq/n \leq 0.25/600 = 0.000416667$ (giving all the accuracy of the calculator), so that the standard error is 0.0204. From the Z table we find $Z = 1.28$ will give the required $1 - \alpha$:

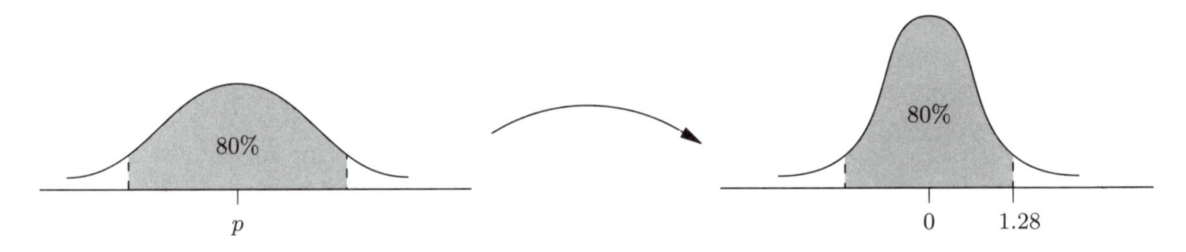

and so the endpoints of our interval are

$$\hat{p} \pm 1.28\sigma_{\hat{p}} = 0.43 \pm 1.28 \times 0.0204,$$

where we are *approximating* the standard error by assuming the worst case. The worst case occurs when $p = 0.5$—that's when the standard error will assumes its largest possible value.

(b) Here the picture to determine Z is

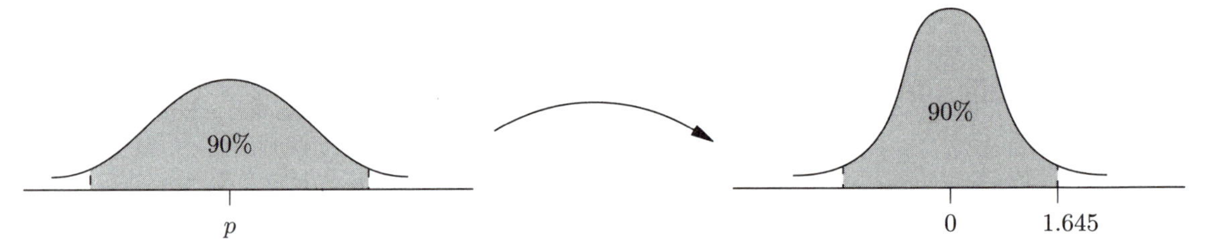

The probability we must find in the body of the table is exactly halfway between the probabilities for $Z = 1.64$ and $Z = 1.65$. Thus, we interpolate in the table by assuming the required value of Z to be exactly halfway between 1.64 and 1.65.

Now the standard error is obtained from $\sigma_{\hat{p}}^2 \leq 0.25/450$. So the endpoints of our interval are

$$\hat{p} \pm 1.645\sigma_{\hat{p}} = 0.71 \pm 1.645 \times 0.0236.$$

(c) $Z = 1.08$ and $\sigma_{\hat{p}}^2 = 0.25/280$ giving endpoints

$$\hat{p} \pm 1.08\sigma_{P} = 0.45 \pm 1.08 \times 0.0299.$$

(d) $Z = 1.75$ and $\sigma_{\hat{p}}^2 = 0.25/40$.

5.2.13 This is like Problems 5.2.1(b) and 5.2.5(c). The normal distribution is "mound shaped" with most of the probability concentrated at the center. Because $\hat{p}$ is unbiased, that center is the unknown p. So most of the $\hat{p}$'s are close to p, meaning most samples are more or less typical.

Atypical samples ($\hat{p}$ far from p) are possible, but their $\hat{p}$'s are in the tails of the distribution and only a small part of the probability is in the tails—again the "mound" shape of the normal distribution. Atypical samples are possible, alright, but they're not probable! Here's the picture:

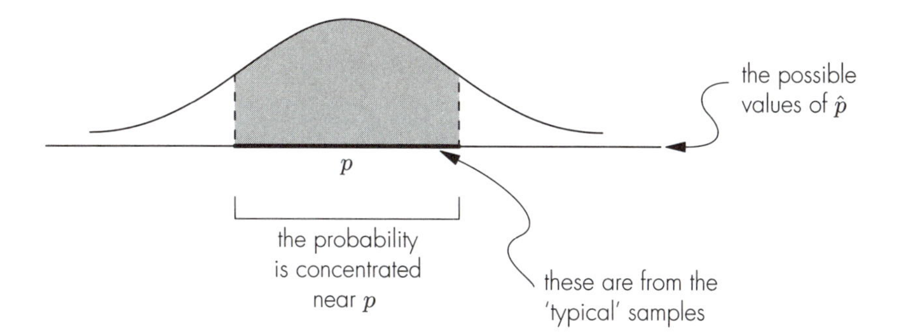

5.2.14 (a)

$$\frac{0.18 + 1.645^2/480 \pm (1.645/\sqrt{240})\sqrt{0.18 \times 0.82 + (1.645)^2/960}}{1 + (1.645)^2/240}$$

$$= \frac{0.185637552 \pm 0.041182367}{1.011275104}$$

(b) $0.18 \pm 1.645 \times 0.0323$, approximating $\sigma^2_{\hat{p}} = pq/n$ by $0.5 \times 0.5/240$.

(c) $0.18 \pm 1.645 \times 0.0248$, approximating $\sigma^2_{\hat{p}}$ by $0.18 \times 0.82/240$.

(d) The conservative and the less conservative approaches are always centered on the observed value of $\hat{p}$; here 18%. The exact endpoint approach will have a center slightly off from that, the larger n is the more slight is that difference. Here, the "exact" interval is centered at 18.355%.

The center of an interval (a, b), by the way, is just the average of its endpoints. Draw a number-line picture: The center is $a+$ "half the width of the interval," $a + (b - a)/2$, which is easily seen to be $(a + b)/2$.

The error tolerance of the first interval is 0.0408, of the second, 0.0526, and of the third, 0.0408. The error tolerances for the first and third intervals look the same, but they're not really the same. The difference is in the negligible term

$z^2/4n = 0.0028$—this term is so negligible that in the first four decimal places of the standard error we don't even see the difference!

The second interval has a larger error tolerance, reflecting the conservative approach of approximating the standard error by its largest possible value. It's good for the error tolerance to be *as small as possible*—that means we locate the unknown value of p somewhere within a very narrow range of values. So, of course, a conservative estimate (a "worst-case" estimate) will allow a relatively large error tolerance.

(e) The "exact" formula is derived using as model the normal *approximation* for the distribution of $\hat{p}$.

5.2.15

(a) (0.1914, 0.2515); (b) (0.1837, 0.2563); (c) (0.1899, 0.2501).

(d) The error tolerances: 0.0301, 0.0363, and 0.0301. Again, the second interval is the widest (it has the largest error tolerance), reflecting a conservative estimation procedure—it has the greatest uncertainty as to the actual value of p. The other two are centered at 22%. Of course, all three intervals have a 92% probability of containing p.

5.2.16

The conservative approach gives an interval which is very crude. You observed 8% having the characteristic, but all you're able to say is that p is less than about 18%. And you get no information at all about the smallest value because the calculated left endpoint is a negative number, but p cannot be negative. That's why we give zero as the left endpoint.

The less conservative approach gives an interval of more or less the same length as the exact interval, but it's centered 1.55 percentage points too low—it's centered at 8%, but the exact interval is centered at 9.55%. That could be a significant distortion.

This sample of n=100 is really too small because $z^2/n = 0.0384$, that's almost four percentage points and it's not negligible when the sample proportion is 8%!

5.2.17

(a) You might assume these 318 printheads form a random sample of all printheads from your company—a simple random sample chosen WITHOUT replacement from a dichotomous population. Because, presumably, you produce thousands of printheads, the population is very large compared with the sample size of 318 and so our normal model of $\hat{p}$ should be approximately valid. But you would have to look carefully at the process by which these 318 printheads were chosen to see if "random choice" seems at all reasonable. If not, the real-world conclusion given in part (b) could be very wrong.

A second way to model this problem is to think of the production process as a Bernoulli trial with "success" being "produce a printhead which will fail within the warranty period." If "fails within warranty period" seems to be independent from one printhead off the production line to the next, you have a simple random sample of 318 from the distribution of that Bernoulli random variable, the variable being $X = 1$ if "failure within warranty," $X = 0$ otherwise.

This second way of modeling the problem focuses the unstated assumption more clearly. You should consult with one of your engineers to see if "fails within warranty" could be reasonably considered independent from one printhead to the next as they are produced.

On the other hand, if the engineer says, "No, that's not reasonable. That kind of problem tends to come in runs," then the second way of modeling the problem is not valid. But note that these 318 printheads almost certainly were not taken from the production line successively one after the other. If you look carefully at exactly how they were selected, you may feel satisfied that "random choice" seems a reasonable assumption. Then you are back to the first model.

Note how these two ways of modeling the sampling process are not equivalent—they are in a sense complementary. If one is not reasonable, the other may be!

(b) For $1 - \alpha = 0.99$, we look up 0.995 in the body of the Z table. It's exactly halfway between two values in the table: 0.9949 and 0.9951. So select the midpoint between the two corresponding Z's, select $Z = 2.575$. Because you were not told otherwise, use the less conservative approach. Then, the *real-world* answer is

> We believe there is somewhere between a 1.44% and a 7.36% chance for one of our printheads to fail during the warranty period. There is a one percent risk of error in this conclusion.

Or you could say, "We are 99% sure that . . . ." Any formulation of the real-world answer is fine as long as it accounts for both the range of possible values of p and the probability that that range of values actually does contain p.

5.2.18

(a) Ham interviewed 563 people. The only meaningful answer would be a confidence interval: "We can be about 95% sure that between 18% and 25% of residents of Tallahassee consider themselves to have stuttered at some point." We do not claim they are "stutterers." We only say that they considered themselves to have stuttered at sometime in the past (maybe only once!). The standard error is 0.0176.

(b) "With a 5% chance of error, somewhere between about nine and fifteen percent of Tallahassee residents considered themselves to be stutterers at the time of the interview." The interval: (0.0939, 0.1477).

(c) "Accepting the judgments of the interviewers, we can be about 90% sure that between two and five percent of Tallahassee residents were true stutterers at the time of the interview." Note that 48 respondents were judged to have normal speech with only "normal dysfluencies," so only 20 were judged to be true stutterers. The standard error: 0.0078.

5.2.19 (a) If $\hat{p} = 0.18$ and $1 - \alpha = 0.95$, there is a 95% chance the confidence interval actually does contain p, so the confidence coefficient measures our certainty that the interval estimate is really valid. Here's how it works: If we need a value of p for some computation, we'll use 18%, the observed $\hat{p}$—the center of the interval. The maximum error of the estimate is the maximum distance between our estimate and any point in the interval—between $\hat{p} = 0.18$ and any point which we think might be the true value of p:

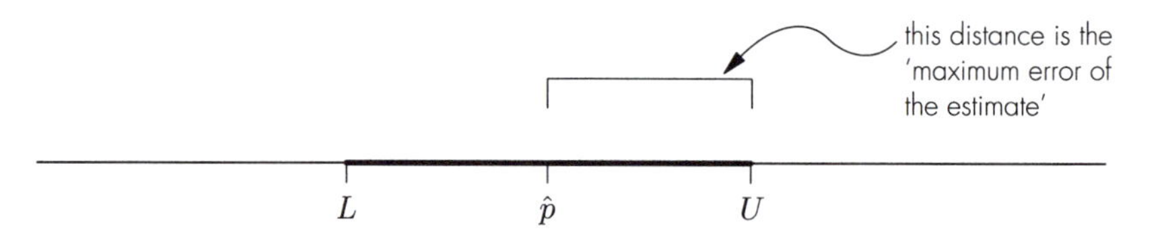

(b) (i) The "worst case" interval was (14.9%, 21.1%), so the maximum error of the estimate was 3.1 percentage points:

$$\frac{0.211 - 0.149}{2} = 0.031.$$

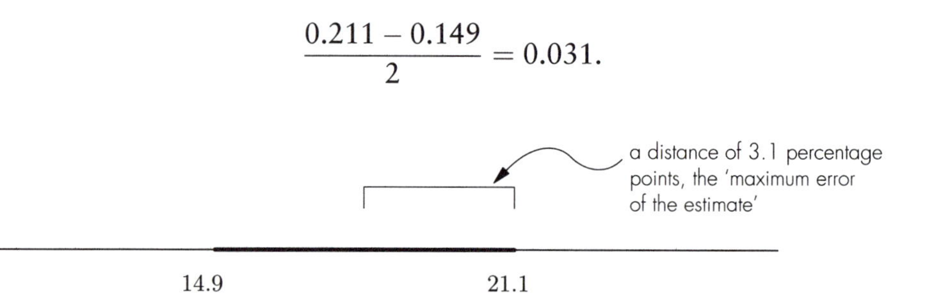

Because the endpoints of the interval are $\hat{p} \pm z\sigma_{\hat{p}}$, it's clear that the general formula for the maximum error of the estimate is just $z\sigma_{\hat{p}}$. Here Z is 1.96 and the standard error is estimated by the square root of 0.25/1000, so we get $1.96 \times 0.0158 = 0.0310$.

(ii) The less conservative estimate of the standard error is obtained by using $\hat{p} = 0.18$ in place of p in the standard error formula:

$$\sqrt{0.18 \times 0.82/1000}.$$

With this, the maximum error of the estimate is $1.96 \times 0.0121 = 0.0238$, about 2.4 percentage points (compared with 3.1). This seems to be a better error tolerance (maximum error of the estimate) but it comes at the cost of less certainty—we know the true standard error is no larger than our estimate in the conservative case. In the less conservative case, we can't be sure.

(c) The conservative estimate for the standard error takes the largest possible value it could ever have. That occurs when $p = 0.5$. So the true standard error is smaller than our estimate of it. But the standard error should be small; so if our estimate is always bigger than the true value, the estimate is always valid.

5.2.20 A 95% confidence interval for the specificity: (0.9985, 0.9995). The standard error here is estimated to be 0.000241661, calculated using $\hat{p} = 0.999003108$. Don't calculate with a rounded value, these numbers are all too sensitive to loose accuracy like that! Note that $p = 17/17053$.

5.2.21 (a) We've shown that pq is maximum when $p = 0.5$. So, we know for sure that

$$2401pq \leq 2401 \times 0.25 = 600.25.$$

But n must be an integer. Here you should round UP to 601 just to make absolutely sure your sample is large enough.

(b) You know that $pq \leq 0.21$. Here's how: We know that $pq = p - p^2$ is a parabola where the maximum value occurs at $p = 0.5$:

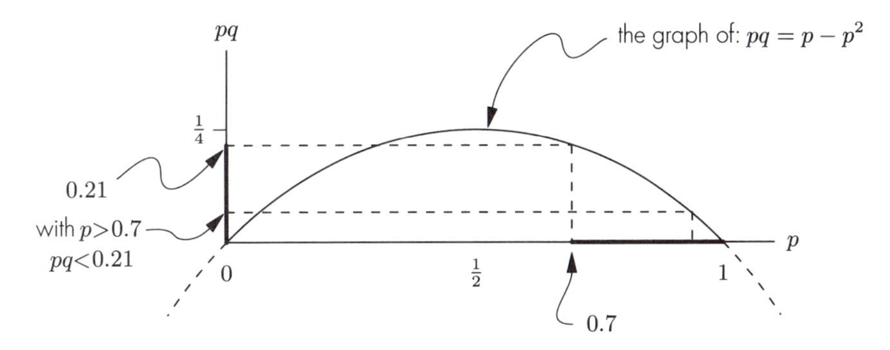

You were given that $p > 0.7$. Now from the picture, it's clear that pq (on the vertical axis) gets smaller and smaller as p moves above 0.7. So $pq \leq 0.21$ and

$$2401pq \leq 2401 \times 0.21 = 504.21.$$

With $n = 505$, you can be sure n is larger than $2401pq$, as required. Now q is the smaller of p and q and you know it's AT MOST 30%, but you have no limit on how small it could be (the larger p, the smaller q). So to use the normal approximation, you would require q to be at least about 1% ($505/5 = 0.0099$).

5.2.22

(a) You require that $z^2pq/n \leq (0.005)^2$ and so n must be at least $(2.575)^2 \times 0.25/(0.005)^2 = 66,306.25$.

(b) $n \geq (1.75)^2 \times 0.25/(0.005)^2 = 30,625$.

(c) $n \geq (2.575)^2 \times 0.0099/(0.005)^2 = 2625.7275$.

(d) $n \geq (2.575)^2 \times 0.000999/(0.005)^2 = 264.9598$. If you do this with your calculator as a continuous calculation, it's not necessary to round. That's why we show more than four decimals in the intermediate steps.

But there are two conditions to be met here. One is the specified error tolerance—you've just taken care of that. The other is the condition for validity of the normal approximation which you are using, that's where the 2.575 came from! For it to be valid, you must also have $np \geq 5$. With $n = 265$ and with the worst case of p ($p = 0.001$), np would be less than one. Taking the worst case of p, you find that n must also be at least 5000. To satisfy both conditions, you'll have to choose $n = 5000$. In part (c), this problem does not arise because there, $2626 \times 0.01 \geq 5$.

5.3.1

(a) First, sampling from the distribution of the random variable X: Because X_k is a value of X, the "on average" value is μ_X, which we're writing more simply as μ. Similarly, its variance is $\sigma_X^2 = \sigma^2$.

Suppose X_k is the number obtained on the kth draw from a numeric population. Then its "on average" value is just the mean of the population it's drawn from, μ, and its variance would be σ^2, the variance of that population.

(b) You must show that $\mu_{\overline{X}}$ is just μ, the parameter to be estimated. But because

$$\overline{X} = (1/n)(X_1 + X_2 + X_3 + \cdots + X_n),$$

you get

$$\mu_{\overline{X}} = (1/n)(\mu + \mu + \mu + \cdots + \mu).$$

This is true because the mean of each X_k is just μ by part (a) and because "the mean of a sum is the sum of the means." Now, μ added to itself n times is just $n\mu$ and so

$$\mu_{\overline{X}} = (1/n)(n\mu) = 1\mu = \mu.$$

(c) Now, the *variance* of a sum is the sum of variances only if the random variables are independent, but our X_k's are all independent. That's guaranteed by the definition of a "simple random sample." So,

$$\begin{aligned}\sigma^2_{\overline{X}} &= (1/n)^2(\sigma^2 + \sigma^2 + \sigma^2 + \cdots + \sigma^2)\\ &= (1/n)^2(n\sigma^2)\\ &= \sigma^2/n.\end{aligned}$$

Thus, the standard error is $\sigma/\sqrt{n}$.

5.3.2

(**a**) $\overline{X} = (1/n)\Sigma X_k$ is a linear function of the X_k's which are normally distributed random variables (they're "versions" of X). But linear functions preserve normality [see Problem 4.3.13(e)] and so we get our required result, namely: $\overline{X}$ is normally distributed because it's a linear function of normally distributed random variables. A linear function, recall, is one which involves only sums and multiplication by constants—that's true for $\overline{X}$.

(**b**) Use this notation: $\overline{X}_1 = \frac{1}{2}(X_1 + X_2)$, $\overline{X}_2 = \frac{1}{2}(X_3 + X_4)$, so

$$\overline{X}_1 - \overline{X}_2 = \frac{X_1 + X_2 - X_3 - X_4}{2}.$$

Clearly, the difference between these two $\overline{X}$'s is determined by the differences among the four numbers in the numerator. But those four numbers come from the underlying population. Because that population is normally distributed, the differences among those four numbers are determined by "many independent random factors." So, $\overline{X}_1 - \overline{X}_2$ is determined by the same "many independent random factors." Note how the criterion transfers from the population to the sample means.

Now think again about part (a). If those X_k's are values of a normally distributed random variable X, still the differences among the numbers in the sum are accounted for by many independent random factors because X is normally distributed—the same argument. So, part (a) can be done two ways.

But sampling from a population is a special case of sampling from the distribution of a normally distributed random variable; so the argument in part (a) can be made to work in part (b). In short, there are two arguments for each part of this problem: the linear function argument and the "rule of thumb" criterion. Both are instructive and helpful in understanding the sample mean as a normally distributed and unbiased estimator for an unknown mean.

5.3.3 Here are the two sample means:

$$\overline{X}_1 = 1/n(X_1 + X_2 + X_3 + \cdots + X_n),$$
$$\overline{X}_2 = 1/n(X_1 + X_2 + X_3 + \cdots + X_n).$$

For simplicity, we write them with the same notation, but don't forget that they come from different samples and so the X_k's are different.

We have MANY RANDOM FACTORS:

> The difference between these two sample means is determined by the different X_k's from which they're computed. There are $2n$ of these X_k's. Now, we're assuming n large; suppose, for example, $n = 100$. Then there are 100 X_k's in the first sample and 100 in the second; 200 altogether. That means there are MANY X_k's. And they're randomly generated because these are *random* samples.

which are INDEPENDENT:

> The values of any two of the X_k's are independent. That's how the sample is chosen: either through (1) *independent* repetitions of the experiment for the random variable X or through (2) sampling from a numeric population, *with replacement* (i.e., the choices are *independent*).

So if n is large, the sample mean should be approximately normally distributed (our criterion holds): The difference between any two sample means is "determined by many independent random factors." Note the critical use of the condition that n be large; it gives us the required *many* of "many independent random factors." In Problem 5.3.2, that "many" came from the fact that the underlying situation was itself normally distributed.

5.3.4 (a) There's a simple and very basic relationship between the total of a sample and its mean: $\Sigma X_k = n\overline{X}$. It's a *linear* relationship. So ΣX_k is just $a + b\overline{X}$, where $a = 0$ and $b = n$.

Now, a linear function of a normally distributed random variable is itself normally distributed (Problem 4.3.13). Here, $\overline{X}$ is approximately normally distributed because we have a large sample (the Central Limit Theorem). And ΣX_k is a linear function of $\overline{X}$. So ΣX_k must also be approximately normally distributed.

By the way, you could also obtain this result using our heuristic criterion for normality, but you were asked to use the Central Limit Theorem instead.

(b) Here's the matching

RANDOM ERROR IS	AND CAN BE THOUGHT OF AS
• the result • of many small influences • from many different sources beyond our control	• the sum • of a large sample of numbers • independently and randomly generated

But the right-hand column is just what we need: Random error can be thought of as the sum or total of many independently and randomly chosen numbers—the sum or total of a large simple random sample.

(c) Random error is "like" the sum of a large sample. By part (a) such a sum is approximately normally distributed.

5.3.5 The "true" measurement is a constant (although unknown). It's only the errors that vary. So M is a linear function of random error:

$$\begin{array}{ccccc} \text{one measurement} & = & \text{true measurement} & + & \text{random error} \\ M & = & a & + & \in \end{array}$$

By the previous problem, random error should be approximately normally distributed. Because M is a linear function of random error, M also should be approximately normally distributed (Problem 4.3.13).

5.3.6 We estimate $\sigma_{\overline{X}}$, the standard error, by $s/\sqrt{n} = 0.0618$. Then we compute the endpoints as $2.37 \pm 1.645 \times 0.0618 = 2.37 \pm 0.1017$.

5.3.7 **(a)** Cancel the n in the denominator:

$$s^2 = \frac{n}{n-1}\hat{\sigma}^2 = \frac{n}{n-1}\left(\frac{1}{n}\Sigma(X-\overline{X})^2 f\right) = \frac{1}{n-1}\Sigma(X-\overline{X})^2 f.$$

(b) $\Sigma(X-\overline{X})^2 f = \Sigma\left[X^2 f - 2X\overline{X}f + \overline{X}^2 f\right] = \Sigma X^2 f - 2\overline{X}\Sigma X f + \overline{X}^2 \Sigma f.$
Now use $\Sigma f = n$ and $\overline{X} = (1/n)\Sigma X f$:

$$\Sigma\left(X-\overline{X}\right)^2 f = \Sigma X^2 f - \frac{2}{n}\left(\Sigma X f\right)^2 + n\left(\frac{1}{n}\Sigma X f\right)^2 = \Sigma X^2 f - \frac{1}{n}\left(\Sigma X f\right)^2.$$

Now when you multiply by $1/(n-1)$ you get the required formula for s^2.

(c) First set up a frequency distribution for the data:

X	f	Xf	X^2f
4	2	8	32
5	2	10	50
7	3	21	147
8	1	8	64
	8	47	293

Now, $s = 1.5526$ is the square root of $s^2 = 2.4107 = (8 \times 293 - 47^2)/(8 \times 7)$.

(d) A value of s is calculated from the numbers in a random sample, so the random experiment, the "doing," is just "random sampling." There's no need to verify that it's a random experiment—that's true by definition: "random sampling" is any random experiment which produces a sample as outcome. This also identifies the outcomes: An outcome is one sample. Now, s assigns a number to an outcome according to the rule given by the formula in part (a). So s satisfies the definition of "random variable"—it's an assignment of numbers to the outcomes of some random experiment. Clearly, from that formula, s is a measure of spread about the sample mean exactly in the same sense that σ is a measure of spread about the population mean, so s is an "estimator" for the parameter σ.

(e) Note that s^2 is a linear function of $\hat{\sigma}^2 : s^2 = [n/(n-1)]\hat{\sigma}^2$: You just multiply $\hat{\sigma}^2$ by the constant $n/(n-1)$. So the *expected value* of s^2 is that same function applied to the expected value of $\hat{\sigma}^2$:

$$\mu_{s^2} = \frac{n}{n-1}\mu_{\hat{\sigma}^2} = \frac{n}{n-1}\left[\left(\frac{n-1}{n}\right)\sigma^2\right] = \sigma^2.$$

5.3.8

(a) For proportion problems, your original question is, "What is the value of this UNKNOWN proportion, p?" But that same unknown p appears in the standard error formula (the squared standard error is $\sigma_{\hat{p}}^2 = pq/n$).

(b) You can get around the unknown p entirely by using the "exact" formula for the endpoints of the confidence interval. That formula does not involve p at all. Then there's the conservative estimate that replaces p by 0.5. This is conservative in the sense that the standard error is larger for that value of p than for any other possible value. Finally, there's the less conservative approach based on the "exact" formula that replaces the unknown p by the observed $\hat{p}$. This depends on the fact that for large samples, the term z^2/n in the "exact" formula is negligible.

(c) In each case, you replace the unknown parameter by a value calculated from the sample itself. This is not obviously valid—in each case it requires careful justification.

For proportions, the justification comes from looking at the "exact" formula and observing that all the z^2/n terms are negligible if n is large. When you omit these negligible terms, you get a formula for the endpoints which just replaces the unknown p in the standard error formula by $\hat{p}$. In the case of means, we use s in place of σ when n is large. So far we have not justified using the sample in this way—we will see that later when we study the small sample case.

5.3.9

(a) $z = 2.33$, $\sigma_{\overline{X}} = 1.21/\sqrt{250} = 0.0765$ and so the endpoints are determined by $6.4 \pm 2.33 \times 0.0765$.

(b) Now $\sigma_{\overline{X}} = 0.0071$ and the endpoints are: $0.32 \pm 2.33 \times 0.0071$. Note that you're given the population variance and so there is no need to estimate the standard error (The squared standard error: $\sigma^2/n = 0.02/400$).

(c) $z = 1.75$, $\sigma_{\overline{X}} = 0.6$ giving endpoints of $122.51 \pm 1.75 \times 0.6$.

(d) $z = 1.60$, $\sigma_{\overline{X}} = 0.0115$.

(e) Here $\sigma_{\hat{p}}$ is to be estimated by $\sqrt{\hat{p}\hat{q}/n} = 0.0144$ with endpoints determined by $\hat{p} \pm 2.33\sigma_{\hat{p}} = 0.23 \pm 2.33 \times 0.0144$.

5.3.10

(a) If $1 - \alpha = 0.85$, then $z = 1.44$. The standard error here is approximated by 1.1314, calculated from $s/\sqrt{n} = 8/\sqrt{50}$. So the endpoints of the 85% confidence interval expressed in months are

$$32.3708 \quad \text{and} \quad 35.6292.$$

Now, 32.3708 months is 2.6976 years—two years plus 69.76% of a year. Well, 69.76% of a year is $0.6976 \times 12 = 8.3708$ months—but you were to round to the nearest month. So you get two years and eight months. Similarly, 35.6292 rounds to two years and 11.6292 months which rounds to two years and 12 months, in other words, three years.

(b) "We can be about 85% sure that between 14% and 30% of the children coming to the daycare center are under one year in age." The endpoints to four decimal places are 0.1356 and 0.3044, and the standard error is 0.0586.

(c) Your sample came from the population of children brought to the ski resort by their parents even though no daycare facility was available. Now with the daycare center set up, it's entirely possible, even likely, that more small children come to the resort. In that case, your question is about a different population than the population from which you drew your sample. Of course, the manager may feel her question is not important enough to warrant the time and expense of obtaining a new sample. That decision is her responsibilty.

5.3.11 There is no need to estimate the standard error because you know σ to be 0.0341. The sample standard deviation ($s = 0.0090$) is irrelevant. So your standard error for this data is $\sigma/\sqrt{n} = 0.0341/\sqrt{76} = 0.0039$.

(a) The endpoints of the interval are $7.2449 \pm 1.645 \times 0.0039$. The final conclusion of the problem is

> We can be about 90% sure that the average fill per cup after resetting the fill mechanism is at least 7.2384 ounces and not more than 7.2513 ounces.

(b) This is exactly like part (a), but with $z = 1.96$. You were asked about the "fill" not the "average fill," but evidently the person asking the question means fill for the "typical" cup. This situation is very common in real-world questions where the *actual* question is not exactly the *intended* question. So the question must be interpereted. So far, the only interpretation possible for us is "typical fill" in the sense of average. So we give a confidence interval for the average fill and state the answer accordingly

> We can be about 95% sure that the AVERAGE fill per cup after resetting the fill mechanism is at least 7.2428 ounces and not more than 7.2491 ounces.

Later, in section 5.5 we'll see how to answer the question where it's interpreted as asking for the "predicted" fill of one particular cup. For that we give a so-called "prediction interval" as answer.

(c) Again, this is exactly like part (a), but now $z = 2.33$. A question of, "How much drink is dispensed per cup?" must be taken to mean "What's the average fill per cup?"

> We can be about 98% sure that the average fill per cup after resetting the fill mechanism is at least 7.2358 ounces and not more than 7.2540 ounces.

(d) You cannot provide the required answer! The only information available is the observation of 76 random cups of drink dispensed after resetting the fill mechanism. Because this is only a random selection of all possibilities and could be misleading, any conclusion you draw is open to error. The most you can do is control the error by specifying in advance an acceptable NON-ZERO probability of that error.

(e) "We can be about 90% sure that somewhere between six and eighteen percent of all cups will overflow after resetting the fill mechanism." The endpoints to four decimal places are 0.0575 and 0.1794. The standard error is 0.0371.

(f) You have no information about fill before resetting the fill mechanism.

(g) Assume the fill mechanism of the dispensing machine to be a *random* mechanism and assume repetitions to be independent. The data is a random sample from the probability distribution of the random variable "fill': We have an ordered set of 76 values of "fill" obtained from 76 independent repetitions of the random experiment.

(h) The "doing" of the underlying experiment is "operate the fill mechanism," clearly this is repeatable. An outcome is "one filled cup." Evidently one cup will have a bit more or less than another cup—we cannot predict the outcome in advance. This identifies the random experiment. Then "fill" assigns a number to such an outcome by measuring the amount of fill. That number, the amount of fill, is assigned to the outcome "filled cup." So "fill" is an assignment of a number to each outcome of a random experiment and is, therefore, a random variable.

(i) You don't know the value of p, but $np = 76p \geq 5$ implies $p \geq 6.58\%$. Because your observed proportion is close to 12%, it might seem you have nothing to worry about. But in fact, the lower endpoint of your confidence interval allows for p to be less than 6.58%. So, the normal approximation may well be invalid. To avoid this uncertainty, measure the fill of a few more cups.

Or, if you're willing to accept a bit more uncertainty, scale down your original requirements: Do an 80% confidence interval. Then the lower endpoint is above 7%, and there's no suggestion of a problem with the normal approximation. Now you report that with a 20% risk of error between seven and seventeen percent of cups will overflow.

Ignoring the problem and staying with the original 90% interval introduces an unmentioned and uncontrolled uncertainty. By reducing the confidence coefficient and recalculating the interval, you accommodate that uncertainty into your answer in an explicit, *controlled* way.

5.3.12

(a) The picture determines which samples are going to be considered "typical," but the range of "typical" $\overline{X}$'s is determined by $\mu \pm 1.34\sigma_{\overline{X}}$ and you don't know the value of μ.

(b) Still, you know that a particular value of $\overline{X}$ has an 82% chance to be within 1.34 standard errors of μ. By using this *theoretical information* about the sampling distribution of $\overline{X}$, you can construct an interval centered on your observed sample mean. That particular interval may or may not contain μ. However, you have controlled for this uncertainty—you know that 82% of all such intervals will contain the unknown value of μ.

5.3.13

(**a**) The "total context" is the sampling distribution of the estimator. If the estimator is unbiased, the sampling distribution is *centered on the unknown parameter*. If the sampling distribution of the estimator is normally distributed (at least approximately), then the value of the estimator computed from your one sample is found somewhere in a picture such as

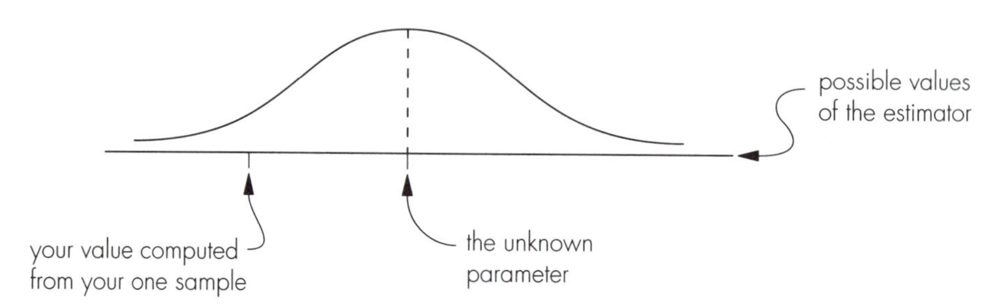

Finally, there will be a standard error formula which will measure the accuracy of our estimate. It measures how spread this picture is about the unknown parameter. But the more spread the picture, the less accurate the estimate—the more spread the picture, the greater the probability of observing values far from the unknown parameter.

(**b**) The endpoints will have the form 3.12 ± 1.645 s.e. To give actual values to these endpoints, you'll need the formula for the standard error of that particular estimator. The standard error formula will depend on which parameter and which estimator you're working with.

The standard error may itself involve some unknown parameter. If so, you have a problem. For proportions, the standard error involves the proportion itself—we've studied that case in detail. The standard error for means involves the population standard deviation, σ, which may or may not be known.

5.3.14

(**a**) The endpoints are $\overline{X} \pm 1.645\sigma_{\overline{X}}, \sigma_{\overline{X}} = 1.01/\sqrt{8} = 0.3571$.

(**b**) Here the endpoints are $\overline{X} \pm 1.96\sigma_{\overline{X}}, \sigma^2_{\overline{X}} = 3.6/50 = 0.0720$.

5.3.15

(**a**) The standard error is $1/\sqrt{12} = 0.2887$. You should not use $s = 2$ which was computed from the sample because σ is known. Why introduce an estimate for the standard error when you have enough information to compute it exactly?! Even if you didn't know σ you couldn't use s as an estimate here because the sample size is not large. The endpoints of the interval are $17 \pm 1.96 \times 0.2886$.

(b) "We can be 95% sure that the new mean of this population is something greater than 11.8741 but less than 12.9259." This is just part (b) of the previous problem. Don't use s here!

(c) We require a "maximum error of the estimate" of 0.4. That means $1.96^2\sigma^2/n \leq 0.4^2$. With $\sigma = 1$, $n \geq (1.96)^2/(0.4)^2 = 24.01$, so let $n = 25$. We need to select 13 more from the population to complete our sample. Note this will give a new, more accurate sample mean.

(d) 16.7833 ± 0.4. Note that you don't need any calculations—your sample size was chosen to force the 0.4 error tolerance.

(e) Note that for the interval to be half a point wide, the error tolerance (maximum error of the estimate) should be half that: it should be 0.25. So now we have $1.645^2\sigma^2/n \leq 0.25^2$, with $\sigma^2 = 3.6$.

5.3.16 It's a *probability distribution*—the distribution of the random variable t.

5.3.17 (a) The underlying random experiment is the random sampling experiment which produced our sample. An outcome of that experiment is just one particular sample. Now s assigns a number to that outcome. That number is calculated from the sample data according to either the conceptual or the computing formula as given in Problem 5.3.7.

(b) If X comes from a normal distribution (and only then), $(X - \mu)/\sigma = Z$. So if the sample comes from a normal distribution,

$$\Sigma\left[(X-\mu)/\sigma\right]^2 = \Sigma Z^2$$

is just a sum of squared Z's. That's chi squared if all the Z's are independent just from the definition of chi square. Because our sampling procedure obtains the X's through *independent* repetitions of the random selection process, these Z's are independent. Well s^2 is not equal to that sum of squared Z's, but it's closely related to it. So, as we claimed, s^2 is intimately related to the chi-squared distribution. Gosset made ingenious use of this relationship to derive the t-distribution and the table of values which we use.

5.3.18 (a) The standardizing transformation takes $\overline{X}$ to Z. Student's t-distribution arises when—not knowing the true population standard deviation, σ—we *estimate* the standard error. So the variance of Z accounts for uncertainty among values of $\overline{X}$. The variance of Student's t-distribution must account for still *more* uncertainty—in addition to the uncertainty about the value of μ, there is uncertainty about the value of σ which is estimated from the sample by s. So Student's

t-distribution is a model for a less precise situation than Z and, hence, has a larger variance.

(b) The true standardizing transformation should take $\overline{X}$ to Z. That transformation requires the true standard error in the denominator—here we have modified that. We have *estimated* the true standard error:

$$\sigma/\sqrt{n} \approx s/\sqrt{n}.$$

(c) With a larger sample, we have more information. With more information, the estimate s for σ should be more accurate. Thus, if we use Student's t we should get closer to Z (the "true" model) as the sample gets larger and larger.

(d) *Situation 1:* The sample size is large. The many independently and randomly chosen numbers in the sample provide the "many independent random factors" of our criterion.

Situation 2: The population is normally distributed. Between any two numbers from the population, there are "many independent random factors" accounting for the difference. These same factors will account for the difference between two values of $\overline{X}$.

See the solutions to Problems 5.3.2(b) and 5.3.3 for complete details.

5.3.19

(a) $1.2 \pm 2.1448 \times 0.17/\sqrt{15}$. Assume the population normally distributed in order to use t.

(b) A smaller sample yields less information and so there should be less certainty in the estimate—that means a wider interval. This arises from a *larger* standard error. The interval is (1.0693, 1.3307)—here you have $t = 2.3060$ and standard error 0.0567. Assume the population normally distributed in order to use t.

(c) $1.2 \pm 1.96 \times 0.17/\sqrt{45}$. The population need not be normally distributed! With a large sample, even though you have to estimate the standard error, the standardizing transformation takes you to Z, *approximately*.

In this situation, you might have tried to use the t-distribution. That's not wrong, but, for large samples, t is approximately Z. That's why you didn't find your required 44 degrees of freedom in the t table—the usual procedure is to approximate t by Z if $n \geq 30$. Still, if you use 45 degrees of freedom (for which $t = 2.0141$), you will get a slightly more accurate interval—that t is closer to the true t than our approximating $Z = 1.96$.

(d) $1.2 \pm 1.96 \times 0.17/\sqrt{9}$. With a small sample size, the population must be normally distributed so that $\overline{X}$ will be normally distributed—but then the standardizing transformation takes $\overline{X}$ to Z because the standard error is known exactly.

(e) $1.2 \pm 1.96 \times 0.17/\sqrt{150}$. No special assumptions are required—here you invoke the Central Limit Theorem. You are not estimating the standard error because σ is known.

(f) $3.4 \pm 1.7959 \times 1.21/\sqrt{12}$. The population must be normally distributed in order to use the t-distribution.

(g) $0.21 \pm 2.575 \times 0.03/\sqrt{8}$. Assume the population to be normally distributed, otherwise $\overline{X}$ is not normally distributed. Because you are not estimating the standard error here, the standardizing transformation takes $\overline{X}$ to Z.

(h) $0.21 \pm 3.4995 \times 0.03/\sqrt{8}$. Assume the population to be normally distributed in order to use Student's t-distribution. You require Student's t because you are estimating the standard error.

(i) The difference is that we don't know σ here, whereas in part (e) we did. But the sample size is large, so estimating the standard error using s as calculated from the sample in place of σ does not have a significant effect on the standardizing transformation—it still takes $\overline{X}$ to a random variable that is approximately Z. This is the "difficult theorem of mathematical statistics" we have referred to several times. You can think in these terms: The great amount of information contained in the large sample "swamps" the uncertainty introduced by estimating σ with s.

5.3.20

(a) When the population is normally distributed (Problem 5.3.2(b)), $\overline{X}$ is too. But when $\overline{X}$ is normally distributed, it standardizes to Z. However, if σ is not known, the standardizing transformation must be "modified" by estimating σ using s as calculated from the sample. This modification takes us to Student's t-distribution instead of to Z.

(b) *σ known:* If the population is normally distributed, $\overline{X}$ is normally distributed (Problem 5.3.2) and so standardizes to Z. So use Z.

If the population is not known to be normally distributed, the Central Limit Theorem tells us that $\overline{X}$ is approximately normally distributed and the standardizing transformation takes us to what is *approximately* Z. So use Z.

σ not known: If the population is normally distributed, use s instead of σ. This modification of the standardizing transformation then takes us to t, but because the sample size is large, t is approximately Z. So use Z.

If the population is not known to be normally distributed, because the sample size is large, the modified standardizing transfromation using s instead of σ still takes us to a model that is approximately Z (a "difficult theorem of mathematical statistics"). So use Z.

5.3.21

(a) If the population is not normally distributed, you will have to rely on the Central Limit Theorem which guarantees that $\overline{X}$ is approximately normally distributed for large samples. But then, the standardizing transformation transforms $\overline{X}$ into a random variable which is only *approximately* Z. If the population is exactly normally distributed, then $\overline{X}$ is exactly normally distributed and we don't have this uncontrolled approximation. In other words, our calculations with the Z-distribution are now exact and not approximate.

(b) If σ is known, then the standardizing transformation takes us to Z *exactly*. This is true because $\overline{X}$ is normally distributed (and that's true because the population is normally distributed).

If σ is NOT known, we use Student's t which requires the population to be normally distributed—again, this model is exact. Student's t-distribution was derived to account *exactly* for the uncertainty introduced by estimating the standard error formula using s instead of σ.

5.3.22

(a) $0.42 \pm 2.1318 \times 0.04/\sqrt{5}$. The population must be normally distributed in order to use t.

With a ten percent risk of being wrong, the true mean of this population should be at least 0.3819 but not more than 0.4581, provided we can assume the population normally distributed.

(b) $0.2 \pm 1.96 \times 0.0381$, the square of the standard error is $0.2 \times 0.8/110$.

There's a 95% chance that the true proportion of this population which have the characteristic of interest is at least 12.52% and not more than 27.48%.

(c) $23 \pm 1.645 \times 2.1/\sqrt{35}$. No special assumptions are required because the sample size is large. Furthermore, we have the exact standard error.

We can be 90% sure that the mean of this population falls somewhere between 22.4161 and 23.5839.

(d) $1.27 \pm 2.7969 \times 0.064$. The population must be normally distributed in order to use t.

With a one percent risk of error, we can assert that the mean of this population falls between 1.0910 and 1.4490.

(e) $87 \pm 2.575 \times \sqrt{16/14}$. The underlying population must be normally distributed, otherwise we have no technique for this situation. If the population is normally distributed, then $\overline{X}$ is normally distributed and the standardizing transformation takes us to Z.

We can be 99% sure that the mean of this population is at least 84.2472 but not more than 89.7528.

(f) The estimator $\hat{p}$ is approximately normally distributed *if the sample size is large*! When the sample size is small, you have to revert to interpreting $\hat{p}$ as X/n for a binomial X. But this makes the confidence interval problem difficult and we don't do this case.

(g) 0.49 ± 1.645 s.e. where the s.e. is estimated to be 0.0884.

We believe the unknown proportion of this population which have the characteristic of interest is somewhere between 34.46% and 63.54%. There's about a ten percent chance of error in this conclusion.

(h) $7.2 \pm 1.7341 \times \sqrt{2.73/19}$. The population must be normally distributed in order to use Student's t-distribution.

There's a 90% chance that the typical value to be expected from this population is at least 6.5427 and not more than 7.8573.

(i) 44 ± 2.2622 s.e., with the standard error estimated at 0.0095. To use t, the population must be normally distributed.

The mean of this population should be between 43.9785 and 44.0215 where we run a five percent risk of being wrong in this assertion.

5.4.1

(a) The lower endpoint for σ^2 is $7s^2/U = 7.5796/16.013 = 0.4733$. The upper endpoint is 4.4850. Now, there's a 95% chance that σ^2 falls between these two points. But that happens if and only if σ falls between the square roots of these points. So there's a 95% chance that σ falls between 0.6880 and 2.1178, as required.

(b) Because $(n-1)s^2/\sigma^2 = \chi^2_{n-1}$, solving, we find that s^2 is just chi-squared multiplied by $\sigma^2/(n-1)$.

(c) You control the contaminant best when the amount is predictable. It's twice as predictable for the first supplier as for the second. You wouldn't mind having *more* contaminant provided it's controllable. It's certainly not desirable to have less contaminant and be unable to control it!

(d) $a = 0$ and $b = (n - 1)/\sigma^2$.

(e)

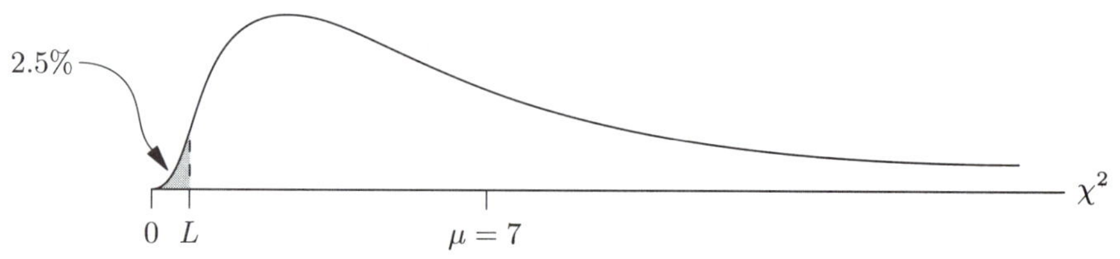

2.5%

0 L $\mu = 7$ χ^2

5.4.2

(a) $U = 23.685$, $L = 6.571$.

(b) Because not told otherwise, you make your own choice of confidence coefficient. We'll choose $1 - \alpha = 0.90$, giving $U = 19.675$ and $L = 4.575$. Here $(n - 1)s^2 = 11.59298316$ and so the endpoints for σ^2 are (0.5892, 2.5340).

(c) Using the statistical mode of your calculator, you find s^2 for your 76 cups to be 0.000071494. Don't round this number or you'll get zero, trivializing your answer. The endpoints for the *variance* of fill are $75s^2$ divided by $U = 100.826$ and $L = 52.9555$, respectively (with 75 degrees of freedom, we're exactly halfway between two values in the table: Take the average). For σ^2, this gives the interval (0.000053181, 0.000101256). Of course, for your solution you need not write down these large numbers; you obtain them in your calculator and then take the square roots. Only when you complete the calculation should you round the answer.

5.5.1

(a) Because $MD = 0.1X + 3$, $\mu_{MD} = 0.1\mu_X + 3$. But linear functions with positive slope preserve inequalties. So $a < \mu_X < b$ is valid if and only if $0.1a + 3 < \mu_{MD} < 0.1b + 3$ is valid. Thus, any probability statement involving such a sequence of inequalities must assign the same probability to each sequence. Of course, this is also true if it's a one-sided inequality.

(b) Let $Z = 1.645$. Note that $n = 64$, $\overline{X} = 8.3906$, $s^2 = 3.1307$, and so a 95% upper confidence interval for μ_X has the endpoint $8.3906 + 1.645s/8$, giving endpoint 3.8754 for μ_{MD}.

(c) Let M be the *population* median for all values of X. Let $Y = \#$ of observations in the sample above M. We must find $x_{(h)}$ to satisfy $95\% = \mathrm{P}(M < x_{(k)}) = \mathrm{P}(Y \geq k)$. Using the normal approximation for Y, we see that $k - \frac{1}{2}$ must standardize to 1.645. Thus, recalling the mean and variance of Y (respectively, $n/2$, $n/4$),

$$k = \frac{n + 1 + z_O\sqrt{n}}{2}.$$

Because $k = 39.08$, the endpoint of our confidence interval is $x_{(39)} = 9$. This is transformed from X to MD [justified as in part (a)] to give 3.9 as the endpoint of our interval for μ_{MD}.

(d) It's useless to have the box designed for either the average or "median" apple! The box must accommodate an INDIVIDUAL apple. We'll see how to deal with such a problem in the next part of this section—we need a "prediction interval," not a confidence interval.

(e) Now the endpoint is $\hat{p} - 1.28$s.e $= 0.0982$, with $\hat{p} = 0.1563$ and s.e. $= 0.0454$.

5.5.2

(a) $Z = 2.33$, $\overline{X} = 3.1078$, and $s = 0.0431$. The upper confidence limit for the mean weight of ONE penny is 3.117842644 g, so for 100,000 pennies, it will be 311,784.2644 g, which is 682.0281 pounds. Now add the weight of the cart.

(b) Here a two-sided confidence interval is required because if there's a miscount, it could mean either too many or too few pennies. So $Z = 2.575$ and the mean weight of one penny should be between 3.0967 and 3.1189 grams. Now, each bag has 1000 pennies. Note, however, that knowing the range for a "typical" bag is not relevant to determining whether a particular bag is miscounted. Again, we need a "prediction interval." See the next part of this section. Note that a confidence interval estimate for the average WAS appropriate for part (a) of this problem. The question was about "such a cart."

5.5.3

In the picture below, there's a 95% chance to obtain an $\overline{X}$ in the indicated range. Note that for the $\overline{X}$'s in that range, the confidence interval will contain μ, otherwise not:

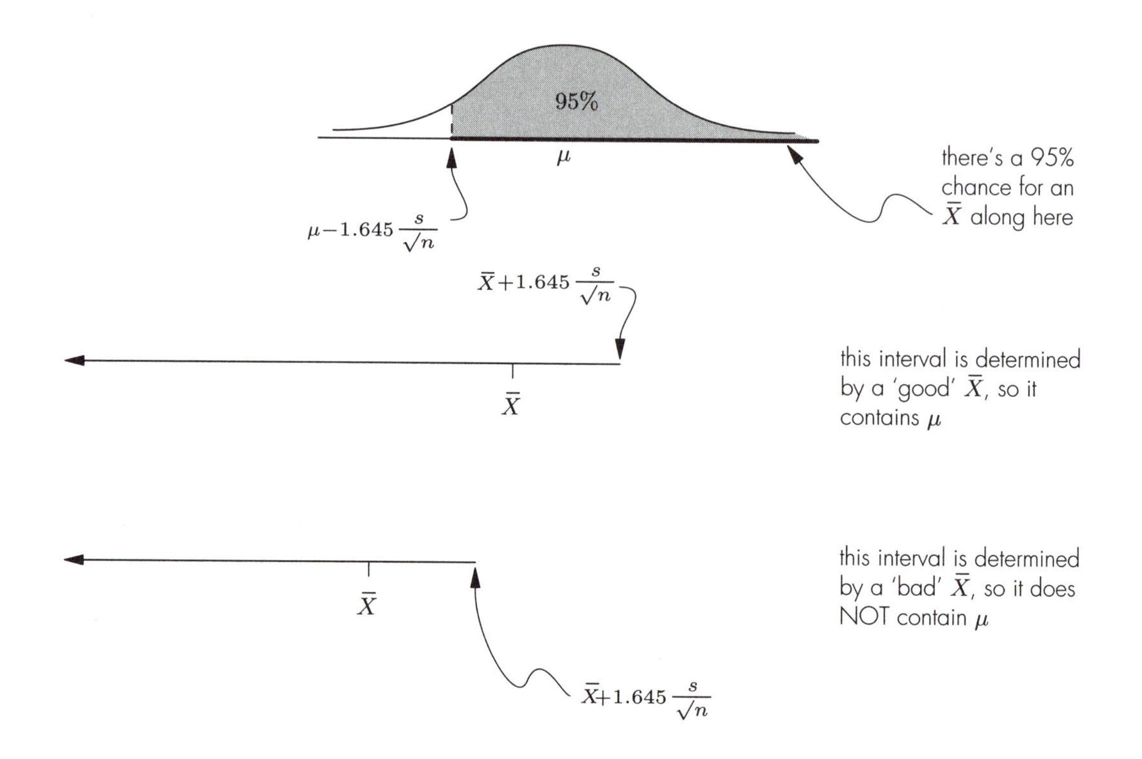

5.5.4 (**a**) Unless there's something wrong with the machine, it's reasonable to assume a normal distribution for "fill." Draw a picture of the distribution of "fill" and you'll see that the interval $\mu \pm 1.96\sigma$ encompasses 95% of all possible values of "fill." So there's a 95% chance to get a "fill" in the interval (5.7904, 6.8096).

(**b**) We've given a range of possible values for a numeric observation—my "amount of drink"—together with the probability that that range of values actually does contain the observation.

(**c**) Ordinarily, you would not know the mean and standard deviation of "fill." When we speak of an "interval estimate," we ordinarily are thinking of a situation where the context of the numeric quantity you're estimating is largely unknown. In other words, an interval estimate is ordinarily based on a sample in lieu of complete information about the process (or population) you're studying.

5.5.5 The model is $\overline{X}$ for samples of size $n = 3$ with s.e. $= 0.2942$. The interval is (6.0058, 6.5942). Of course, just as in the previous problem, we would ordinarily not know μ and σ for the drink machine. In that case, an answer to our question would require the results of a prior sample taken from the machine.

5.5.6

(a) Make a "random observation." For example, get one cup of drink from a drink machine (observe a process) or measure the diameter of a randomly chosen machine part (observe a population).

(b) The model was $\overline{X}$.

(c) The model is $X + (-\overline{X})$. In Problem 4.3.13(e), you saw that the sum of two normally distributed random variables is normally distributed. So our model is normally distributed. Now the mean of X is μ and the mean of $-\overline{X}$ is $-\mu$. So the mean of our model is $\mu - \mu = 0$. Finally, the variance of X is σ^2 and for $-\overline{X}$ is $(-1)^2\sigma^2/n$. Because we assume observations of X to be independent, we obtain the variance for the model. It's just $\sigma^2 + \sigma^2/n = \sigma^2(1 + 1/n)$.

(d) There's a 95% chance for $X - \overline{X}$ to take a value between L and U. So, standardizing, there's a 95% chance for

$$-1.96 < \frac{(X - \overline{X}) - 0}{\text{s.e.}} < 1.96$$

which just says there's a 95% chance for -1.96 s.e. $< X - \overline{X} < 1.96$ s.e. Or equivalently, there's a 95% chance for

$$\overline{X} - 1.96\text{s.e.} \; < X < \overline{X} + 1.96\text{s.e.}$$

(e) With $z = 1.96$, part (d) shows that the range of values from

$$\overline{X} - 1.96\text{s.e.} \qquad \text{up to} \qquad \overline{X} - 1.96s.e.$$

has a 95% chance to contain X. If we choose an appropriate value of Z, the range with endpoints $\overline{X} \pm z$s.e. will have a $1 - \alpha$ probability to contain X. This is exactly what's required for a prediction interval.

If σ is unknown, we should use s instead. If the sample size is small, that requires using the t-distribution with degrees of freedom $n - 1$, following the principle that each parameter in the standard error which must be estimated from the data decreases the degrees of freedom by one. Here, σ is estimated from the data by s.

(f) Because σ is unknown, we will have to use s instead. With nine degrees of freedom, $t = 2.2622$. Using the formula from part (c), the standard error is the square root of $0.27^2 \times 1.1$. The endpoints are 6.6 ± 0.6406. Of course, all this assumes (1) that the previous ten cups provide a simple random sample from the machine. They do if, when we operate this machine repeatedly, the amount of drink dispensed is independent from one cup to the next. This analysis also assumes (2) that "fill" from this drink machine is normally distributed, a reasonable assumption if the machine is not malfunctioning. For both assumptions, consult an engineer who knows the machine!

(g) Because $\overline{X}$ and A are independent, the variance of the sum is the sum of the variances. So the model $A-\overline{X}$ has variance $\sigma^2/m+\sigma^2/n$, estimated if necessary by $s^2(1/m+1/n)$. Note that if $m=1$, you get the "right answer." That is, you get the formula in part (c).

(h) We can be about 95% sure that the three of us will get an average of between 6.1979 and 7.0021 ounces each.

(i) It's not enough for $\overline{X}$ to be normally distributed. The model for one future observation is $X-\overline{X}$. This model is normally distributed if each term in the sum is normally distributed [see Problem 4.3.13(e)]. So if X is not normally distributed, the model won't be either and our analysis in parts (d) and (e) fails.

(j) The model is $X-\overline{X}$, where X models the future observation we want to predict and $\overline{X}$ models our data. And so a 95% confidence level involves both of these variables—the sample mean AND our future observation. For a confidence interval you are estimating a parameter, NOT a variable.

So for a 95% prediction interval five out of 100 times, on average, either our data or our future observation—one or the other (or both)—will be sufficiently atypical that our interval will not contain that future observation. Here, "atypical" means "far from the true value of μ". Note that our sample might be quite typical, with a sample mean quite close to μ, and still give an interval that misses our future observation. That would happen if our future observation is quite atypical (draw a picture of the sampling distribution of $\overline{X}$ to see this).

But "typicality" is not really the issue here because the true value of μ is irrelevant to our question. We're asking about a future observation not about μ. It's just a question of how far that future observation is from our sample mean. So here's a more direct interpretation of the 95%: There's a five percent chance that the sample mean and our future observation will be so far apart that our interval does not contain that observation.

5.5.7 Here the squared standard error is 3.1307×1.0156 and the right endpoint of an upper 95% prediction interval for X is 11.3239. So the maximum diameter will be estimated at 4.1324 cm. Add 0.5 cm for paper and straw. We must assume the maximum diameters of all the candy apples is at least approximately normally distributed, probably a reasonable assumption. Of course, this solution ignores the possibility that a particular apple might fit better by putting the maximum diameter at an angle in the box, thus requiring less than the estimated dimension. Note the interpretation given to the 95% confidence coefficient in level I: if the maximum diameter of an apple is far from the average of our sample, the box will be too tight for that apple.

5.5.8

(a) Here the standard error to predict the average weight of 100,000 pennies is 0.0043, the square root of $0.0431^2 \times 0.01001$. The right endpoint of an upper 99% prediction interval for the weight of the pennies in a cart will be 682.0292 pounds. Recall: $Z = 2.33$, $\overline{X} = 3.1078$, $s = 0.0431$. Now add 43 pounds, the weight of the cart.

(b) Here, a two-sided prediction interval is required because we would suspect a miscount if the bag is either too heavy or too light. So $Z = 2.575$ for a 99% prediction interval. But now we start with a prediction interval for the average weight of 1000 pennies ($10) and then convert that to an interval for the weight of a bag. The standard error is the square root of $(0.0431)^2 \times 0.011$. Do you see why the answer in level I is given in kilograms and not grams?

The "one percent risk of error" does NOT refer to how atypical this bag is, yet a recount is required when the bag itself is atypical, when it's heavy or light enough to suggest it contains too many or too few pennies. Note that if Youden's data really is atypical, you would begin to note after awhile that you would be constantly recounting bags which had the right number of pennies.

(c) We have to assume the weight of U.S. pennies is normally distributed, an entirely reasonable assumption. That assumption was not required in Problem 5.5.2 because there the model is $\overline{X}$ and the sample is large ($n = 100$). That model is normally distributed even if X is not. But here the model is $X - \overline{X}$ and this model is normally distributed only if X is.

5.5.9

(a) This is the result of Problem 4.1.4. Note that it makes no assumption at all about the shape of the distribution from which you are sampling. That's why it's a "non-parametric" technique.

(b) There's one segment above $x_{(n)}$, two above $x_{(n-1)}$, and so on. So there would be h above $x_{(n-h+1)}$.

(c) Solve the equation for h and you'll get

$$h = \frac{1}{2}\left[n + 1 - (1 - \alpha)(n + 1)\right] = \frac{1}{2}(n + 1)\alpha.$$

(d) $h = 5.05$ and so $k = 95.95$.

5.5.10

(a) Saying there's a 90% chance the interval $(6.4, +\infty)$ encompasses 93% of the values of X, as in the picture, is the same as saying there's a 90% chance

$(6.4, +\infty)$ contains the seventh percentile of the distribution of X. In other words, our 90% tolerance interval for 93% of the values of X is a 90% CONFIDENCE interval for the seventh percentile of X.

Of course, if we had constructed an UPPER tolerance limit for 93% of the values of X, say $(-\infty, 7.1)$, that would be a confidence interval for the 93rd percentile of the distribution of X. Draw the picture to see this.

(b) Let's do the upper endpoint first. Let P be the 100pth percentile of X. We want $x_{(k)}$ so that a proportion p of values of X is below $x_{(k)}$. Let Y = # of observations in the sample which are less than P. Then Y is binomial, $Y \sim B(n,p)$. We want to find k so that

$$1 - \alpha = \mathrm{P}(P < x_{(k)}).$$

For example, if $1 - \alpha = 0.9$ and $P = 93\%$, this means we'll be 90% sure that at least 93% of the values of X are below $x_{(k)}$. This is exactly what's required if $x_{(k)}$ is to be our upper endpoint. But the condition $P < x_{(k)}$ simply says "$k - 1$ or fewer observations in the sample are below p," so

$$\mathrm{P}(P < x_{(k)}) = \mathrm{P}(Y < k).$$

Using the normal approximation to Y (with continuity correction), we see that $k - \frac{1}{2}$ must standardize to Z,

$$\frac{k - \frac{1}{2} - np}{\sqrt{npq}} = z.$$

Solve this equation for k and you obtain the formula in the problem statement.

A similar argument will derive the lower endpoint. However, now you should let Y = # of observations in the sample which are less than Q, where Q is the 100qth percentile. To see why we switch from P to Q, look at the picture of X below. This Y has a $B(n,q)$ distribution. And the condition that a proportion p of all values of X are greater than $x_{(k)}$ translates into $Y \geq k$. Here's the picture of X with $100p = 93$ and $100q = 7$, so that Q is the seventh percentile:

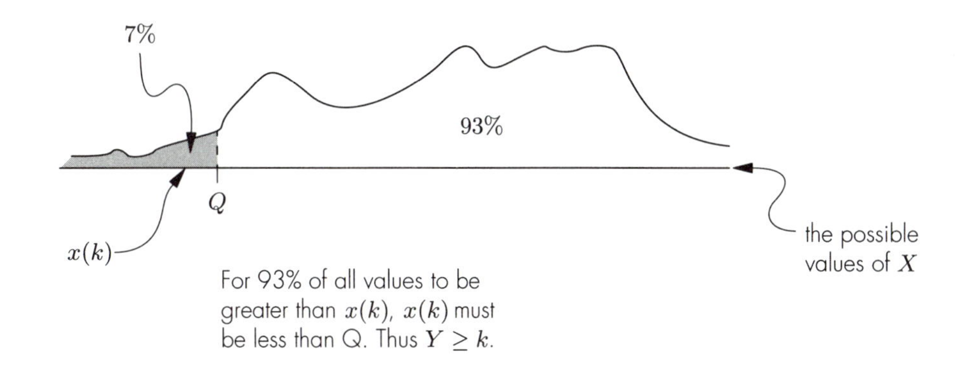

Because the mean of Y is now nq (NOT np), we see that $k - \frac{1}{2}$ (continuity correction!) must standardize to a *negative* value of Z:

$$\frac{k - \frac{1}{2} - nq}{\sqrt{npq}} = -z.$$

Now solve to obtain the required equation for the lower endpoint.

(c) Our analysis is based on the binomial random variable Y. We've used a normal approximation to Y to get our formula for k.

(d) The corresponding formula for the median is given in Problem 4.5.8(c). You get that formula if you set $p = 0.5$ in the formula from part (b) of this problem.

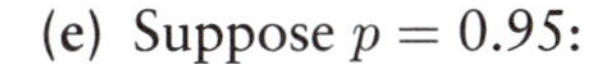

(e) Suppose $p = 0.95$:

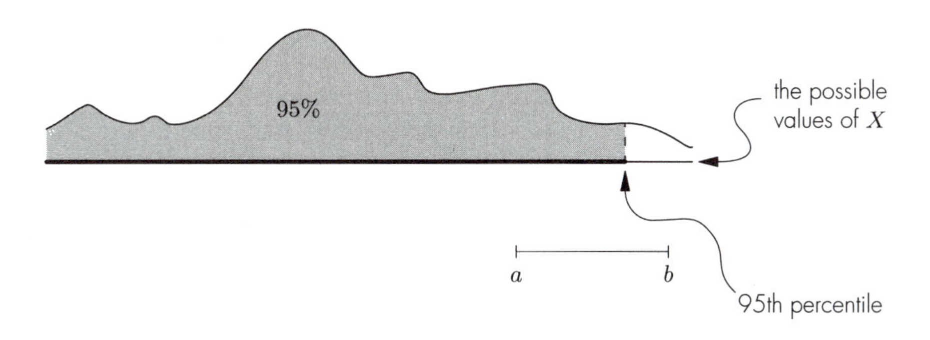

Here, (a, b) is a confidence interval for the 95th percentile, but it's NOT true that 95% of all values of X are encompassed within that interval.

5.5.11 Put $n = 100$, $p = 0.93$, and $Z = 1.28$ into the formula from Problem 5.5.10(b). Then, $k = 4.2$ and the lower limit is $x_{(4)} = 6.4$.

5.5.12 (**a**) This problem clearly is asking about the "typical" U.S. penny, so we've given a confidence interval for μ, the mean weight of all U.S. pennies.

(**b**) This requires a prediction interval for one observation of the variable $X =$ weight of a U.S. penny. The standard error is 0.0433, the square root of $1.01s^2$. The interval is (3.0229, 3.1927).

(**c**) Here, we require a tolerance interval. For an upper tolerance limit with a confidence coefficient of 95%, $Z = 1.645$. Thus, $\sqrt{npq} = 3$, $np = 90$, and, using the formula from Problem 5.5.10(b), $k = 95.4$. So the upper limit is $x_{(95)} = 3.17$.

(**d**) $h = 2.525$ and $k = 98.475$.

(**e**) Part (b) uses the normality of the weight of U.S. pennies (a reasonable assumption, but we have not actually tried to verify it). In other words, that technique uses information extraneous to the actual observed data. Part (d) makes no assumptions beyond the data but rather utilizes the information contained in the order statistics of the sample. This information is ignored in part (b).

That's the idea of a nonparametric technique: "no assumptions beyond the data." In fact, in our case we did make an the assumption that no two observations could be exactly equal. In Youden's data, we see many observations which are equal, but that's only because we rounded to two places. Youden, in fact, made measurements accurate to four places.

Which technique is better? Because they make no assumptions beyond the data—or in any case, very weak assumptions—nonparametric techniques might be preferred if they give reasonably strong conclusions. That's the case here because our nonparametric interval is virtually the same as the parametric one.

(**f**) $1-\alpha = (k-h)/(n+1) \leq (n-1)/(n+1)$; solve for n. Or, if the interval is to be centered in the sample, $h = (n+1)\alpha/2$ and so $n = (2h-\alpha)/\alpha \leq (2-\alpha)/\alpha$. This second analysis shows that when n equals $(2-\alpha)/\alpha$, your endpoints will be just the largest and smallest observations in the sample, thus making little use of the information contained in the order statistics. To have the endpoints be into the sample by, let's say, three observations would require an even larger sample: $n = (6-\alpha)/\alpha$.

(**g**) For example, a 95% nonparametric prediction interval requires a sample of at least 39 observations. And, with $n = 39$, the endpoints of your interval will be the smallest and largest observations, not making much use of information contained in the order statistics of the sample. So the nonparametric interval is not useful for small samples. The parametric interval is valid for any sample size, BUT you must be sampling from a normal distribution.

Chapter 6

6.1.1

(a) The hypothesis about the fairness of the die certainly is a simple test of significance. You're making a one-time inference: You roll the die 100 times, say, and try to determine if the observed result seems to suggest the die is unfair. You ask: Does the data seem to challenge the hypothesis (the fairness of the die)?

(b) Are you going to monitor your parts supplier's claim periodically as you receive shipments? That should be a hypothesis test. Or are you simply trying to find out if the claim is true (should you sign a contract with that supplier or not?)? In that case, you're making a one-time inference and so you should just do a test of significance.

(c) Almost certainly this is not a monitoring process. A test of significance is probably called for.

(d) This may or may not be a monitoring situation.

(e) This also may or may not be a monitoring situation.

6.1.2

When the data seems consistent with the hypothesis, you learn nothing because it will be consistent with many other hypotheses as well. We'll see this with some specific examples later.

6.1.3

(a) This is a point frequently misunderstood, so let's get it clear now. It's five percent of GOOD shipments that will be rejected. That's what "erroneously rejected" means. Thus, you need to know how many of the 500 shipments were good. Five percent of that number is the answer. But the answer can only be hypothetical because we don't know how many shipments were good. Suppose (hypothetically!) 418 of the shipments actually meet specifications. Then you would expect to reject about 20 or 21 of them. Note that the answer to the question is necessarily no more than than 25 (it's 5% of a number that's at most 500).

(b) For a confidence interval problem, making the procedure more exact requires increasing the sample size. And that costs MONEY! Sampling is an expensive process. Furthermore, a zero probability would require an infinite sample size! Similarly, while you can make the probability of erroneously rejecting a good shipment as small as you like (but is it worth the cost?), you cannot make it zero. Just as with a confidence interval, you balance the precision of your technique against the cost.

6.1.4

(a)

Test of Significance	Hypothesis Test
1. provides a numeric measure of consistency between . . . ?	—is a decision procedure which seeks to control . . . ?
2. hypothesis and data on the same footing (in what sense?)	—the data is secondary (in what sense?)
3. can test for randomness	—cannot test for randomness
4. hypothesis either true or false	—hypothesis sometimes true, sometimes false (usually . . . ?)
5. a case of inductive inference	—a monitoring procedure
6. replication not intrinsic	—replication intrinsic
7. meaning of probability rather tenuous	—meaning of probability very concrete
8. implies exact replicability	—replication not exact
9. often preferred for testing a scientific hypothesis	—often preferred for matters of public policy, business, and so on

This last is the weakest contrast by far. Both tests of significance and hypothesis tests are used in scientific investigations and in more practical areas as well. In analyzing complex random experiments, a scientist will, indeed, use the hypothesis test as a tool. However, if it's just a question of evaluating the consistency of data with a hypothesis, the test of significance will be more suitable.

(b) **1.** A test of significance computes a NUMBER from the data, the p-value. The p-value measures the CONSISTENCY of the data with the hypothesis. If the p-value is interpreted as "small," we conclude that the data does not seem to be consistent with the hypothesis (it seems to challenge the hypothesis). In our example, if the p-value is small, it appears the population is not normally distributed. If the p-value is not small, we get no conclusion (we'll see later why this is true).

The hypothesis test is more complicated. It monitors incoming shipments of chain links on a month by month basis, sometimes rejecting a shipment as not meeting specifications. It bases its DECISION each month on a sample of that month's shipments. The procedure seeks to CONTROL THE ERROR which we will make if the sample from that month's shipment happens to be very atypical of the shipment as a whole (sampling error!).

2. The question addressed by our test of significance asks if "this data" seems to challenge the hypothesis that "the population is normally distributed." So the data and the hypothesis stand in opposition to each other in the original question.

For the hypothesis test, there is originally no data. The original problem just presents a situation which requires monitoring on a monthly basis. The data is secondary: Each month we decide if that month's shipment of chain links should be rejected. The data comes in month by month—then only—as a basis for the decision of that month. The data is part of the solution; it's not part of the original problem.

3. Omit.

4. Either the population is normally distributed or it isn't! Sometimes a shipment of chain links will meet specifications, sometimes it won't (usually it will, otherwise we'd better change suppliers!).

5. In trying to challenge our hypothesis, we're trying to deduce a general fact ('non-normal population') from "specific cases" (the observed data). That's exactly inductive inference in the classic sense. And it's very different from the monitoring process of our hypothesis test with its monthly decision of whether or not to reject that month's shipment of chain links ('repeating decision').

6. If our study gets repeated by other investigators, fine. But that repetition is not part of OUR study! On the other hand, the month-by-month repetition of the hypothesis test (examining a sample from that month's shipment to see if the shipment seems to meet specifications) is an integral part of the test procedure.

7. Our hypothesis test will be run every month. So, for example, we can ask how many times in two years will we not have to reject a shipment? If the probability of that event is 2/3, the answer would be 16, on average (2/3 of 24). For a test of significance, a probability will have to take a more theoretical meaning because it is a one-time inference.

8. Other investigators at other times and places may obtain other data on this SAME population by way of repeating our study. So if they obtain their data through the same procedure we used, they're repeating our study exactly.

By contrast, with the month by month repetition of our hypothesis test, when we take a sample of next month's shipment of chain links, we're sampling a DIFFERENT shipment than we did this month. Here, the circumstances change on each repetition. Some months the shipment meets specifications, some months not. It's not the same shipment from month to month, after all. So next month, we are NOT repeating this month's process exactly.

9. Often a scientific hypothesis takes the form of a simple statement just like ours ("this population is normally distributed"). By contrast, a monitoring procedure, as opposed to a detached scientific question, often entails practical interests just like the concern our engineers have about short chain links. Let's remember, however, that this contrast is not a clear-cut distinction. It's at best a very rough "rule of thumb."

6.2.1

(**a**) 0.0548.

(**b**) The test of significance does not tell you the probability of heads! But it's obvious that the coin seems biased in favor of heads. How biased? To answer that question you should compute a confidence interval, of course.

(**c**) You can't conclude the coin is fair! With a p-value of zero, all the test shows is that it seems wrong to think heads comes up 40% of the time. The correct conclusion is: "Based on our observations of this coin, one would think heads should come up on average more than four times in ten tosses."

(**d**) The p-value is $P(\hat{p} \leq 0.44 | p = 0.5, n = 100)$. Thus $z = -1.2$ for a p-value of 0.1151. By no criterion is this a small p-value (it's bigger than 10%), so the test is inconclusive. If the kids had good reason to believe the coin fair, that reason stands unchallenged. In that case they should accept the coin as fair. But NOT because of the data, rather because their "good reason" stands unchallenged by the data. If they thought the coin unfair, or just didn't know, all they can say is, "Well, we found no evidence to suggest the coin is biased. Maybe it is. We don't know. Maybe it isn't."

6.2.2

The hypothesis was "40% chance of heads" in Problem 6.2.1(c). But we don't just say the data supports "not 40% chance of heads." In fact, the data suggests heads MORE than 40% of the time because 58% was observed. If the kids had observed only 10 heads in 100 tosses, they should say the data supports LESS than 40% chance of heads.

6.2.3

(**a**) Here's one possible explanation: It can be the data that's discrepant. Randomly generated data need not be typical of the situation from which it is drawn. As we said in the text: "The hypothesis might be true and the data extreme just because of sampling error. All heads on a hundred tosses of a fair coin is *possible* however unlikely it may be!"

Or: Maybe the hypothesis is false. That could certainly explain why the data looks extreme—you're comparing it with the wrong hypothesis! From the text: "The hypothesis might be wrong and the data not really extreme at all. Suppose the coin really is biased with a probability of heads of 60%. Then, 58 heads out of 100 tosses is certainly not extreme."

(**b**) Randomly generated data can be discrepant (atypical), but it's not likely to be! The smaller the p-value, the less likely that seems. On the other hand, from the very beginning, we thought the hypothesis might be false. So, having to choose between these two, it's reasonable to believe the hypothesis false.

But don't forget: As always in a statistical analysis, THERE'S A POSSIBILITY OF ERROR. If the first explanation is the valid one—if the data really is atypical (even though that's not likely)—our conclusion will be erroroneous.

(c) The data is summarized into an unbiased estimator which is normally distributed. "Unbiased" says the mound in the normal curve is centered on the parameter being estimated. So although atypical values of the estimator (values in the tail of the distribution) are possible, they are not likely. Most of the probability is concentrated near the true value of the parameter, meaning most samples are more or less typical:

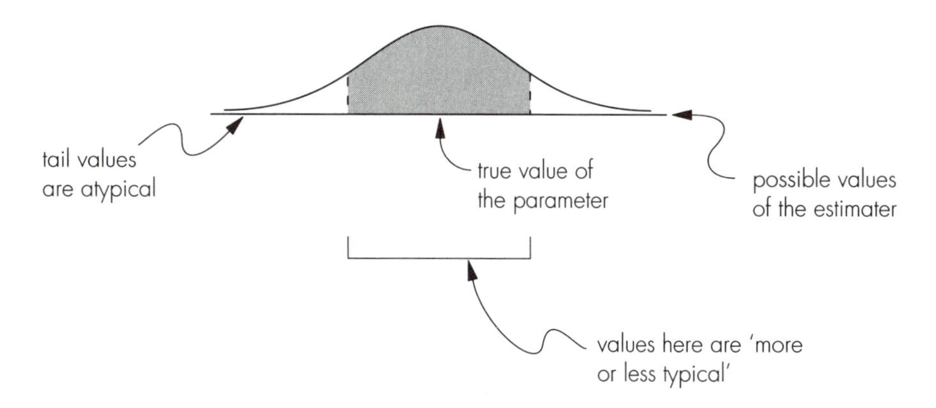

Now you see what the mound of the normal distribution means: It assures that most values of the estimator are concentrated near the center. Because that center is the true value of the paramter, those values (and so "most data") are more or less typical. Even when the estimator is only approximately normal, this qualitative understanding is still valid.

(d) There's a seven percent chance of obtaining 58 or more heads on 100 tosses of a fair coin.

(e) The p-value is NOT the probability the hypothesis is true, nor that it's false. The p-value as a probability concerns the *data*—it's the probability of data like ours or worse if the hypothesis is true. See part (d).

6.2.4 "Consistent with the hypothesis" does not mean "proves the hypothesis"; it means "the test is inconclusive." In this problem, you see that data consistent with one hypothesis will always be consistent with many other hypotheses as well. On the other hand, when data is INconsistent with an hypothesis (small p-value), that hypothesis is ruled out. So we do get a conclusion.

6.2.5 **(a)** It seems the mean is smaller than 70.18, but how much smaller? The test of significance does not answer that. That question asks for the unknown value of your parameter and the answer should be a confidence interval.

(b) For a mean of 68.42, $t = 1.12$ and the p-value is the *same*—greater than ten percent. For a mean of 68.7, $t = 0.762$, so you get a still larger p-value. For a mean of 69.3, $t = 0$ and the p-value is as large as a p-value can get: 50%. After all, a sample mean of 69.3 is more consistent with a population mean of 69.3 than it is with any other possible value!

(c) The sample in part (a) is five times larger than the sample in part (b). Furthermore, in part (b), you're estimating the variance from the sample. So there are two ways in which part (b) has lost information. Not surprisingly, in part (b), with little information, your analysis is inconclusive. In part (a), with much more information, you get a conclusion—that your mean is less than 70.18.

6.2.6 If they're playing with the same coin every day, the coin is either fair or not fair—the hypothesis "fair coin" is either true or false. If it's a different coin each day, the hypothesis is sometimes true and sometimes false because the coin is sometimes fair and sometimes not. When they play with the same coin every day, data from one day can be meaningfully compared with any such data collected on other days because they're always talking about the same coin. Otherwise, data from one day has no relevance whatsoever for other days because they have different coins. If the question is, "Is our *coin-tossing game* fair?," the answer in the first case is either "yes" or "no"; in the second case, the answer is sometimes "yes" and sometimes "no."

Suppose every day before they play, the kids do a series of ten tosses, recording the results. If they always play with the same coin, at the end of the summer they'll have accumulated the results of over 1000 tosses. If it's a different coin each time, at the end all they have is ten tosses on each coin—there's no accumulation of evidence.

The role of repetition is entirely different for the two procedures. An hypothesis test is a monitoring procedure and so the repetition is part of the procedure. You're monitoring an hypothesis that's sometimes true and sometimes false. The point of the procedure is to "flag" the occasions when the hypothesis is false. There's no accumulation of evidence from one repetition to the next just as there would not be for the kids if they play with a different coin each day.

6.2.7 If you're calculating a p-value for a single proportion p, that calculation is carried out assuming the hypothesis. But the hypothesis gives you a value of p, and that's the value you use. So you're not estimating the standard error from the sample. This will not work for a difference of two proportions. Yes the hypothesis gives a value of $p_1 - p_2$, but that doesn't provide a value for the individual proportions. So you have to estimate them from the sample.

6.2.8

(**a**) We estimate the standard error as 1.0510, the square root of $(4.8)^2/50 + (5.2)^2/42$. So a 90% confidence interval is (-6.5289, -3.0711), where the endpoints are obtained from -4.8 ± 1.645 s.e.. The parameter is $\mu_1 - \mu_2$ where μ_1 is the average height of second graders from the inner city and μ_2 from the suburbs. You could just as well have taken this in the reverse order obtaining the interval (3.0711, 6.5289). Of course the interpretation is the same.

(**b**) The assumption of "no difference" implies that the mean of the estimator is zero ($\mu_1 - \mu_2 = 0$), so our observed difference in means of -4.8 standardizes to $(-4.8 - 0)/1.051 = -4.56$:

$$p\text{-value} = \mathrm{P}(\overline{X}_1 - \overline{X}_2 < -4.8 | \mu_1 = \mu_2) = \mathrm{P}(Z < -4.56) = 0.$$

This zero p-value is not evident from the confidence interval of part (a).

6.2.9

In the standard error formula put $53/200 = 0.265$ in place of each of p_1 and p_2. You get an estimated standard error of 0.0624 obtained from the square root of (0.265)(0.735)(1/50). So,

$$p\text{-value} = \mathrm{P}(\hat{p}_1 - \hat{p}_2 < 0.22 - 0.31 | p_1 = p_2) = \mathrm{P}(Z < -1.44) = 0.0749.$$

Or if you took the difference in the other order,

$$p\text{-value} = \mathrm{P}(\hat{p}_1 - \hat{p}_2 > 0.31 - 0.22 | p_1 = p_2) = \mathrm{P}(Z > 1.44) = 0.0749.$$

As you can see, it doesn't matter which way you do this, you'll get the same answer. Note how the observed values standardize: Because the mean of the estimator is zero ($p_1 - p_2 = 0$), the standardizing transformation is just $[(0.22 - 0.31) - 0]/$ s.e. .

So the p-value is about seven percent. If you think that's a small p-value, then based on our observations it seems the two proportions are indeed different. If you think the p-value is NOT small, you have to say the data is inconclusive.

6.2.10

First, because $Y = 7 - X$, $\mu_W = \mu_{X+Y} = \mu_X + \mu_Y = \mu_X + (7 - \mu_X) = 7$. The variance formula is not applicable here because it requires that X and Y be INDEPENDENT. Clearly, they are not.

6.2.11

(**a**) Each $\hat{p}_i$ is unbiased, so the mean in each case is p_i. Thus, the mean of $-\hat{p}_2$ is $-p_2$. Adding, we find the mean of $\hat{p}_1 - \hat{p}_2$ is $p_1 - p_2$. Thus, the estimator $\hat{p}_1 - \hat{p}_2$ is unbiased. Next, $\hat{p}_1$ is approximately normally distributed and so is $-\hat{p}_2$; thus, their sum $\hat{p}_1 - \hat{p}_2$ is also approximately normally distributed. Finally, each $\hat{p}_i$ has variance $p_i q_i / n_i$, so we get the variance formula for $\hat{p}_1 - \hat{p}_2 :\ p_1 q_1/n_1 + p_2 q_2/n_2$.

For this, we use the fact that the samples are independent and, therefore, the two random variables $\hat{p}_i$ are independent.

(b) This argument is identical to part (a); just change the symbols!

6.2.12

(a) Here we observed $\hat{p}_1 - \hat{p}_2 = (0.5417 - 0.7954) = -0.2537$, with a standard error of 0.0675 (we pooled the samples, see Problem 6.2.9). Now if the test is not biased in some way, there should be no difference in pass rates. So we calculate

$$p\text{-value} = \mathrm{P}(\hat{p}_1 - \hat{p}_2 < -0.2537 | p_1 = p_2) = \mathrm{P}(Z < -3.76) = 0.$$

Note how we stated the conclusion of this analysis in level I. Any interpretation going beyond the conclusion not "due just to chance" is exactly that, interpretation. For example, without further information, you cannot conclude that the employer is biased. It's the TEST that seems to be biased. Even in the absence of biased intention, subtle bias can creep into a test. Our point: The real-world interpretation of a p-value is not a STATISTICAL question, it's a judgment call.

(b) Here, we cannot estimate the standard error by pooling the samples because we have no reason to think the two proportions—the two pass rates—are the same. We estimate the standard error to be 0.0659, from the square root of $\hat{p}_1\hat{q}_1/n_1 + \hat{p}_2\hat{q}_2/n_2$. The endpoints of a 95% confidence interval are given by -0.2537 ± 1.96 s.e.

Note that the left endpoint gives the condition $p_1 - p_2 > -0.3830$ which is the same as $p_1 > p_2 - 0.3830$. This gives our interpretation of the left endpoint: The pass rate for blacks is more than 38.3 percentage points below the pass rate for whites. We interpret the right endpoint similarly.

6.2.13

But this is not a case of statistical significance! The figures given are the total number of jobs lost, not a sample. Statistical significance compares the difference between what's observed in a sample and what is hypothesized to be true. If that difference is large enough (statistically significant), you will abandon the hypothesis.

This question can only refer to PRACTICAL significance. Is a loss of 300,000 jobs of any practical significance? Most people would say YES.

6.2.14

(a) The mistake is to confuse "practical significance" and "statistical significance." The p-value of seven percent is a measure of statistical significance. It says nothing about practical significance. The p-value says nothing about how serious the bias might be from a practical point of view if there really is a bias. If the little kid thinks there's no practical significance to some small bias, then she should reason as in part (b).

(b) For the question, "How biased?," they should construct a confidence interval for the probability of heads. It should be centered at $\hat{p} = 0.58$. For the second situation, they should test the hypothesis $p = 0.53$.

(c) It's not clear what the little kid would do! But at least it's clear what range to think about when she thinks about the bias. At worst, there's about a 68% chance for heads. Assuming, of course, the interval contains the parameter (there's a 5% chance it doesn't).

A p-value of 16% is not small by any standards. So the little kid should be willing to play! Note that, here, an *acceptable level of bias* has been specified very clearly in advance.

(d) In the first situation, we simply ask, "What's the bias?" In the second situation, we set an acceptable level of bias in advance. Then we look at the data to see if the bias seems greater than that. A very different question.

Is it, "What's the parameter?" or is it, "Here's what's acceptable, does our data suggest 'unacceptable'?" The first is a confidence interval, the second a test of significance ("does our data challenge the hypothesis?").

6.2.15

(a) The parameter is μ, the mean number of ounces of soft-drink dispensed by our machine into the cups.

(b) The hypothesis is $\mu = 7.4$.

(c) The p-value is $P(\overline{X} > 7.53|\mu = 7.4) = P(z > 4.27) = 0$.

(d) It would be virtually impossible to get 35 cups with an average of 7.53 ounces per cup if the resetting device is working properly. Note, once again, that the p-value is NOT the probability the hypothesis is true (it's not the probability the resetting device is working properly).

(e) Because the p-value is *very* small (as far as we can see from our table, it's zero) we should conclude that the hypothesis is false.

(f) Real-world conclusion: "Based on this data, it seems the resetting device is not operating properly. It seems to cause overfilling."

Note that the question concerns the *resetting device*, so your report must answer that question. If you said the *machine* is not working properly, then you didn't answer the question asked. Such confusion could have important consequences in the real world.

6.2.16

It's not necessary to think of the residents of the southern part of the city as a random sample. We think of an incidence of cancer as the outcome of a random process. That is, the randomness here does not come from sampling; it's inherent in the situation itself.

(**a**) The parameter is the mean (λ) of the Poisson random variable.

(**b**) The hypothesis is that $\lambda = 8.4$—that the rate of cancer in the southern part of our city is the same as what would be expected in any population of that size.

(**c**) The p-value is $P(X \geq 12|\lambda = 8.4)$, which with the recursion formula is $1 - 0.8571 = 0.1429$.

(**d**) There's about a 15% chance to observe 12 or more such cancers in a population like ours if there is nothing exceptional happening. It's NOT correct to say, "There's a 15% chance nothing exceptional's happening."

(**e**) A p-value of 15% is not small by any criterion. So the test is inconclusive.

(**f**) Real-world conclusion: "The incidence of 12 such cancers does not by itself suggest any extraordinary environmental cause of cancer in the southern part of this city. Twelve such cancers in a year appears to be consistent with what would be expected in any population of this size."

6.2.17

(**a**) The parameter is $p_1 - p_2$; the hypothesis, $p_1 - p_2 = 0.02$. Then, $\hat{p}_1 - \hat{p}_2 = 0.0575$ with an estimated standard error of 0.0371 (using the observed $\hat{p}$'s). This gives a p-value of

$$P(\hat{p}_1 - \hat{p}_2 > 0.0575) = P(z > 1.01) = 1 - 0.8438 = 0.1562.$$

By any standard, this is NOT a small p-value.

Real-world conclusion: "This data provides no evidence of a significant difference between the proportion of passengers on our New York to San Francisco flight requesting a vegetarian meal and the proportion on the New York to Chicago flight making such a request."

(**b**) You know estimates: $\hat{p}_1$ and $\hat{p}_2$. Multiplying each by their respective n's gives, respectively, four and one. Neither is at least five. And the true value could well be even smaller! Because the model you've used is not valid, your conclusion is meaningless. An exact test might give a very different result, but such a test is very complex.

(**c**) Now s.e $= 0.01059$ giving $z = (0.0761 - 0.02)/\text{s.e.} = 5.30$. So the p-value is zero. Real-world conclusion: "Based on this more complete data, it appears there is, indeed, a significant difference in the number of such requests on the two flights."

6.2.18 The p-value is: P(data like ours or worse | hypothesis);

it's NOT: P(hypothesis | the data).

As strict frequentists—that's our point of view: probability means "theoretical relative frequency"—we must say the second probability is either zero or one. In other words, either the hypothesis is true or it's false; there's nothing in between.

But for a Bayesian statistician, the second probability is, indeed, meaningful. It can be interpreted, for example, as a measure of our "degree of belief" in the hypothesis. And there are other possible interpretations. The Bayesians think p-values are a poor substitute for what's really required and have developed their own Bayesian test specifically to evaluate P(hypothesis | the data). This is better in two ways. First, because it's the *hypothesis* we're interested in, the probability should refer to that hypothesis. Second, what's the point of talking about data "worse than ours"? We didn't observe any such data! So (say the Bayesians), you've got the wrong probability and you're talking about totally irrelevant data which you never observe!

Ah! . . . the probability the hypothesis is true given the data! So much more satisfying than a p-value! BUT (guess what) . . . this Bayesian procedure is not without its own pitfalls. Unfortunately, to take a look at those interesting pitfalls and the wonderful flora and fauna which surround them would require more than we can do in this course.

Yes, you're right: None of the statistical testing procedures is entirely satisfactory. If that bothers you a lot, GOOD! It means you'll become a statistician yourself, or maybe a philosopher, and spend your career helping to find improved methods. Maybe you'll even discover a completely satisfactory way of testing statistical hypotheses! If this situation does NOT bother you much, that's also good. You'll find other interesting things to do. Thank goddess for diversity. We're not all alike! Anyway—for now—why don't you just go on to the next problem.

6.2.19 It could be that the hypothesis is really true even though the p-value is small. This could happen if our data is very atypical. One should never forget that statistical techniques based on randomly generated data can be misleading because the data can be very atypical of the distribution from which it was obtained. See Problem 6.2.3(b).

6.2.20 (a) However unlikely they may be, atypical samples are possible. Even if the mean life of these tires is around 28,000 miles, keep taking samples and eventually you'll get a sample with a much higher mean. Such repeated sampling is very deceptive and, unfortunately, has been known to be used by unscrupulous advertisers.

(b) It would seem the question is: What's the mean life of the tires? So you wanted a confidence interval estimate for that parameter, not a test of significance. But we're trying to illustrate the fallacy involved in looking for a specific predetermined result through repeated sampling. No matter what the technique—confidence interval, test of significance or any other statistical technique based on random data—that's a fallacy.

6.2.21 (a) Two contradictory conclusions! With the *negative binomial model* you conclude the coin is biased in favor of tails (p-value less than 5%). But with the *binomial model* you conclude the coin is fair—not based on the data of course, but rather on your prior belief which was not challenged by the data.

(b) This is very disturbing because the fairness of the coin has nothing to do with what was in the mind of the experimenter when the coin was being tossed. It has nothing to do with whether she was thinking of a binomial or negative binomial model. In other words, there seems to be a *subjective* element in the method of tests of significance. That's uncomfortable because in any branch of science we think conclusions drawn from data should be entirely *objective*. Bayesians love this example because it shows, so they claim, that Bayesian tests of significance are superior. But their tests have other weaknesses. We're all hoping your generation of statisticians will come along and help resolve some of these problems!

This example reveals a definite weakness for tests of significance. Still, the example seems to depend on having a p-value that was small but not unequivocally small. We've already pointed out that when the hypothesis is false, we ordinarily expect (or at least hope for) a very extreme p-value, one representing a probability of one chance in one billion, say. In such a case it seems unlikely some other model could yield a "not small p-value." This statement would be difficult to prove because it requires having a model for "all possible models."

The problem of this example will not arise for hypothesis tests because they involve data gathered repeatedly. Note that if the experimenter were to repeat the test many times, we would soon know *objectively from the data* which model was being used. You'd probably know by the first repetition! If it's 12 tosses with seven heads, you know she's using a binomial model. Why?

6.2.22 (a) The Court calculated $P(X = 7|n = 90, p = 0.27)$. Now it's alright to calculate the p-value from the binomial distribution (in fact, it's more accurate; our calculation with $\hat{p}$ uses the normal approximation); however, it's not appropriate to calculate the probability that X is EXACTLY seven. When n is large, the probability of any exact value will always be small—there're too many values.

That's why a p-value is the probability of data like ours OR WORSE. That's a more meaningful probability.

(b)

$$\begin{aligned} P(X = 45) &\approx P(44.5 < \widetilde{X} < 45.5) \quad \mu = 45, \sigma = 4.7434 \\ &= P(-0.11 < Z < 0.11) \\ &= 0.0876. \end{aligned}$$

6.2.23

(a) Let p be the probability a person dies in the three months prior to their birthday. In the absence of some special power to postpone death, $p = 0.25$. But $\hat{p} = 0.08$, giving a p-value of zero ($z = -10.7302$). That's small! Here's the real-world meaning of the p-value: It's virtually impossible as a purely random phenomenon to observe obituaries of 747 randomly chosen persons and find that only eight percent or fewer died in the three months prior to their birthdays.

So it does, indeed, appear from this data that "people" are less likely to die in the three months prior to their birthday than in the other nine months. There is a problem, however: Exactly to what "people" is this conclusion applicable? From exactly what population was the sample drawn? The *Newsweek* article does not tell us that. Still, without that information, the data does suggest that, at least for some people, there exists a power to postpone one's death until after a birthday.

If this effect is real and not just due to chance error, how is it brought about? That's NOT indicated by the test. That question—what *causes* the observed effect?—requires an interpretation having nothing to do with statistics. It's natural to speculate, of course. The most obvious speculation is that attitudes affect our physical state. Is it mind over matter? Does attitude have a significant effect on health in general? Hmmm! We'll look at this question again in later problems.

(b)

1. We don't require a decision procedure. We want a measure of consistency between our observed 8% and the hypothesized 25%.

2. The hypothesis and the data are on the same footing, our 8% is very much part of the question.

3. Omit.

4. Either there is or there is not a potential to hold onto life until you pass an upcoming birthday. Our hypothesis is either true or false. We're not monitoring a hypothesis which is sometimes true and sometimes false.

5. This is a classic case of inductive inference. We want to establish a general fact (people have a certain power to postpone death) on the basis of particular instances (our observed 747 persons).

6. We don't have any plans to repeat this test. You may see some such study later and so may I, but we won't be doing it together! Those "repetitions" are not part of our study. An hypothesis test, by contrast, is a monitoring procedure collecting data *periodically*. So, repetition is part of the procedure. In fact, we will see another study of the same question at the end of this chapter.

7. The p-value is the probability of data like ours, or worse if people have no special powers to postpone death. But our specific data (our 747 persons) is all we have and it's all we're ever going to have. We don't see any "or worse" data. So our probability is somewhat tenuous. It refers to something we've never observed and never will observe. For an hypothesis test by contrast, you gather data repeatedly on a periodic basis for purposes of monitoring. So probability has a very concrete meaning as "expected relative frequency."

8. If someone wants to repeat our "study," they'll be testing the same population (human beings in general) with the same hypothesis ($p = 25\%$ in the absence of special powers to postpone death). So if they obtain their data through simple random sampling (as we're *assuming* for our data) they'll be repeating our study exactly. For an hypothesis test, on some repetitions the hypothesis is true, on others false, and so the repetitions are not exact.

9. We are testing a scientific hypothesis. Only a few years ago our hypothesis would have been labeled "pseudoscientific"; however, there's an increasing awareness among scientists of the important connection between mental and physical states.

6.2.24 This is NOT a small p-value and so there's no reason to accuse the accounting firm of collusion. It's entirely believable their sample was chosen randomly. However, our analysis assumes simple random sampling. In fact, accountants ought to control for "size;" they should stratify their samples according to the invoiced amount. We don't know if they did that or not.

6.2.25

Face	O_i	E_i	$(O_i - E_i)^2/E_i$
1	6	16.6667	6.8267
2	61	16.6667	117.9267
3	9	16.6667	3.5267
4	6	16.6667	6.8267
5	10	16.6667	2.6667
6	8	16.6667	4.5067
	100	≈ 100	142.2802

The largest value in the χ^2 table at five degrees of freedom is 16.75, so our observed $\chi^2 \approx 142$ is far off the table. That means the p-value is essentially zero. The test does suggest bias in this die, very clearly and unambiguously.

6.2.26

(a) Accident free means $X = 0$. Theoretically, that should occur, according to the Poisson model, for 54.34% of the corps-years: $200 \times 0.5434 = 108.68$. Let's call the cell for $X = 0$ the 0th cell. So $E_0 = 108.68$. Similarly, $E_1 = 66.28$, $E_2 = 20.22$, $E_3 = 4.12$, $E_4 = 0.62$.

(b) $\chi^2 = 0.0009 + 0.0247 + 0.1567 + 0.3045 + 0.2329 = 0.7197$.

(c) Bortkiewicz' observed value of χ^2 is certainly not large because it's *below* the mean (with $df = 3$, $\mu_{\chi^2} = 3$). So there is no evidence for "lack of fit." The p-value for "lack of fit" is close to 95%, as you see from the χ^2 table because there's a 5% chance for χ^2 to fall below 0.352, and 0.7197 is only slightly above that. Note that unlike the tests we've done up to now based on the normal distribution, for χ^2 tests a p-value can be greater than 50%.

The fit of Bortkiewicz' data to the Poisson model seems to be exceptionally good. Of course, that should be true; the data had to call loud and clear to Bortkiewicz to make him think of a Poisson model. The Poisson distribution had never been used like that before, after all.

(d) Only about 5% of all simple random samples of size $n = 200$ from a Poisson distribution with $\lambda = 0.61$ would fit that distribution as well as or better than Bortkiewicz' data. In other words, of all simple random samples from the Poisson distribution with $\lambda = 0.61$, 95% would be less typical of the parent distribution than Bortkiewicz' data.

6.2.27

With $\chi^2 = 0.5$, three degrees of freedom, we see the fit appears to be very good. In fact, really TOO good. Thinking again of the meaning of the χ^2 probability, we realize there's only (roughly) a five percent chance of data so close to what Mendel's theory predicts [a more complete table would give $P(\chi^2 < 0.5)$ as approximately 0.08]. This is too good to be true.

Mendel's brilliant new theory which postulated the existence of genes and explained the laws of genetic transmission did not receive the recognition it deserved and lay dormant until 1900 when it was "rediscovered." Evidently, Mendel was aware that the radically new understanding embodied in his theory might not be appreciated, for, in fact, *clearly he altered his data*. This fact was discovered by Ronald Fisher (see [Mendel]) from the χ^2 test as we've used it here. Fisher showed that the fit of Mendel's various sets of data is literally too good to be true. Much too good to be true! In one case, his computation of the theoretically expected values is wrong; yet STILL his "observed" values are very close!

But don't be too hard on Mendel! He had a highly significant and important scientific discovery to give to the world. And the historical record shows that he was, indeed, right if he thought the general ignorance of the day would make it difficult for his theory to be recognized. Even with his "improved" data, it wasn't recognized. It took 35 years to be appreciated! In no way does this make Mendel comparable to those unscrupulous persons who've begun to show up in the newpapers these days; people who "cook" their data merely to advance their careers and to hide their own incompetence!

6.2.28

(a) $E_2 = 138.8564, E_3 = 281.6304$.

(b) $\chi^2 = 0.0247 + 1.0124 + 0.7332 = 1.7703$.

(c) The p-value is greater than ten percent. This data is consistent with our contention that the three precincts are pretty much alike as far as support for our candidate is concerned; the data provides no reason to think otherwise.

6.2.29

(a)

	Recovered	Died within five years
Male	81.48	15.52
female	81.48	15.52

(b) $\chi^2 = 25.4304 + 0.1411 + 19.1295 + 3.6437 = 48.3447$.

(c) The p-value is essentially zero and so our hypothesis looks very doubtful in the light of our data. We conclude that recovery for this type of cancer is dependent on gender.

(d) Because about 88% of the men in the sample versus 84% of the women recovered, clearly IF THE DATA SUGGESTS ANYTHING AT ALL, it suggests women are less likely to recover than men. We justifiably draw this conclusion because our zero p-value says the data does, indeed, seem to suggest something, namely, a dependence of recovery on gender.

If this analysis looks suspicious to you, you're right. After all, there are almost three times as many men in the sample as women, but our analysis was based on the rather dubious assumption that this disease affects men and women equally. See parts (e) and (f).

(e) Each parameter estimated from the data results in a loss of one degree of freedom. Thus, if the 84% recovery rate is unknown and estimated by the observed $169/194 \approx 87.11\%$, the degrees of freedom would be two. If you also have to estimate the proportion of females by the observed $50/194 \approx 25.77\%$, the degrees of freedom would be reduced to one.

(f) $\chi^2 = 0.0193 + 0.1306 + 0.0556 + 0.3761 = 0.5816$. With one degree of freedom, this gives a p-value above ten percent. So this data seems to provide no evidence for dependence of recovery on gender (if the recovery rate and the effect on gender are not known).

(g) The p-value is $P(\hat{p} > 0.7423|p = 0.5) = P(Z > 6.75) = 0$. Here, p is the proportion of men in the population at large who fall victim to this type of cancer. With such a small p-value, it would seem on the basis of this data that men fall victim to this type of cancer more often than women. This test for independence does not use χ^2 because we have an either/or situation which reduces to a simple proportion. As a technical consideration however, you really should not use the same data set to test two related hypothesis as we're suggesting here. Get some new data!

6.2.30 $E_1 = (307)(48/307)(232/307) = 36.2736$, and so on. So,

	Pass	Fail	Totals
Black	26 (36.2736)	22 (11.7264)	48
White	206 (195.7264)	53 (63.2736)	259
Totals	232	75	307

Note how there's a shortcut for the expected values. The 307 figure cancels so you get E_1 as just $48 \times 232/307$. The rule would be "multiply corresponding totals and divide by the grand total." Thus,

$$\chi^2 = 2.9098 + 0.5393 + 9.0008 + 1.6681 = 14.1179.$$

With one degree of freedom, we're off the table, giving a p-value of zero.

6.2.31 For a test of significance, you get a conclusion only when the p-value is small. Otherwise, the test is inconclusive—the data is consistent with the hypothesis, but it's always consistent with other hypotheses as well. So, for a goodness of fit test, if the p-value is small, we conclude we have the wrong distribution. That means the data does NOT fit the distribution. This is what we can conclude—we cannot show "fit," only "nonfit."

In the nineteenth century, the normal distribution was used for everything in sight. Gradually people began to understand that this crude approach was not working. In the 1890s, Karl Pearson developed many new distributions in the course of his work on the book *Mathematical Contributions to the Theory of Evolution*. In 1900, he derived the χ^2 test to see which of those distributions fit his data. But it was a process of elimination because the test only shows which distributions do NOT fit.

6.2.32 The expected number for each quarter year is 186.75. This gives $\chi^2 = 258.84$ which, with three degrees of freedom, is far off the table, giving a zero p-value. This analysis confirms our analysis in Problem 6.2.23. There appear to be too few deaths in the three months prior to the birthday and too many in the three months afterward.

6.2.33 (a) There are eight cells and one parameter which was estimated. In fact, there can be technical difficulties depending on how the parameter was estimated. Let's ignore that technicality. So we have six degrees of freedom for χ^2 and obtain a p-value greater than ten percent. Based on that analysis, there is no reason to doubt the model.

(b) MLE $= 1/1.8684$.

Frequency of capture	0	1	2	3	4	5	6	7
Expected # of rabbits	72.3	33.6	15.6	7.3	3.4	1.6	0.7	0.3

Here $\chi^2 = 12.8$ with six degrees of freedom, giving a p-value a bit less than 5%. This does not seem to challenge the model.

Still the model from part (a) with a much smaller p-value is not necessarily better: In a χ^2 test, the parameters should be estimated by the so-called "minimum chi-squared estimator" which is often well approximated by the MLE. But the estimator we used in part (a) is neither of these!

6.3.1 (a) The decision is based on a *random sample* of the day's output. That sample could be quite atypical of the entire day's output. If so, you will be misled into the wrong decision. Of course, although this is possible, it's not likely.

(b) You would hope the error could be controlled by specifying IN ADVANCE an acceptable probability for such error. That's how you control error for confidence intervals. In that case, even though the interval may not contain the parameter, you believe and act as if it does. That's an error! You "control" that error by choosing a confidence coefficient of, say, 95%. The error is still possible, but you avoid it 95% of the time. As you'll see, one of the two errors of an hypothesis test can be controlled this way. The other cannot.

(c) The normal distribution for $\hat{p}$ is centered on p and so the "mound" of the normal curve says the probability is concentrated near p. That says exactly that typical values of $\hat{p}$ are more likely than atypical ones. Look at the picture for $\hat{p}$ [see the level II solution to Problems 5.2.5(c) and 6.2.3(c)].

6.3.2

(a) Wrong! Even if $\mu = 3.15$, you don't expect $\overline{X}$ to be exactly 3.15. If it's just slightly below 3.15, that says nothing about μ. Only if $\overline{X}$ is SIGNIFICANTLY below 3.15 do you have "evidence" that $\mu < 3.15$. This, of course, is a case of *statistical* significance.

(b) You can't expect the mean diameter to attain an exact value! Evidently, the engineers thought a mean as low as 3.15 would not be a problem, that's why they said "too small" means less than 3.15 mm. In other words, they're saying a difference of less than 0.05 mm is of no *practical* significance.

(c) If the spread of diameters about the mean is too great, many diameters will be far from the mean even though the mean itself is "right on target." You would want to monitor the spread of diameters also.

(d) Monitor the proportion of parts with a diameter less than 3.15 mm.

6.3.3

(a) The evidence is the sample which is drawn periodically. In this problem, it's a sample drawn from an incoming shipment of machine parts.

(b) The diameters are "uncertain" if the spread of diameters about the mean is too great. That's measured by the variance of the diameters.

(c) $P(D < 3.15|\mu = 3.2, \sigma = 0.025) = P(Z < -2) = 0.0228$. That assumes a normal distribution for diameters. Under ordinary circumstances, that should be reasonable. Check with the engineers!

(d) With the same μ, less than 2% of the parts would be useless if $\sigma = 0.025$ as compared with nearly 16% if $\sigma = 0.05$. The point is: If there's too much *variability* in the diameters, a large proportion of the parts will be useless, even when the mean diameter is "in control" (even when μ meets the specification of 3.2 mm). See Problem 6.3.2(c).

(e) The estimator is s^2 and we use the chi-squared distribution for this test. Recall that $\chi^2 = cs^2$, where $c = (n-1)/\sigma^2$ (a constant). The hypotheses will be

$$H_o: \quad \sigma^2 = 0.000625 \quad (\sigma^2 < 0.000625 \text{ irrelevant}),$$

$$H_A: \quad \sigma^2 > 0.000625.$$

(f) You're not going to reject a shipment because the diameters are too close to the specified value!

(g) To provide evidence that $\sigma^2 > 0.000625$, s^2 must be *significantly* larger than 0.000625. Is this practical or statistical significance?

6.3.4

We quote from the text: " . . . (the null hypothesis) plays a purely logical role in the hypothesis test by giving us a value of the parameter to work with." On the other hand, the alternative hypothesis plays the practical role of specifying an "exceptional" situation calling for some alternative course of action. Action, after all, is a practical matter. The alternative hypothesis " . . . is "alternative" only because it flags an exceptional action. From the point of view of the test, the alternative hypothesis is, in fact, the main hypothesis. After all, it's precisely the exceptional action that the test wants to flag."

6.3.5

(a) Your concern is whether a particular part is functional or not. The mean says nothing about a particular parts. So you monitor the proportion of functional parts:

$$\begin{aligned} H_o: &\quad p = 0.02 \quad (p < 0.02 \text{ irrelevant}), \\ H_A: &\quad p > 0.02. \end{aligned}$$

Note that $p < 0.02$ is irrelevant because you're willing to discard as many as one in 50 parts. Only when more than 2% of the parts would have to be scrapped is the expense of "corrective action" justified. This kind of trade off is typical in any analysis seeking to minimize costs. Evidently, the cost of "corrective action" is greater than the cost of scrapping one in 50 parts. Abraham Wald's Decision Theory provides a more detailed and rigorous way to incorporate costs of possible decisions into the analysis of this kind of situation.

(b)

$$\begin{aligned} H_o: &\quad p = 0.81 \quad (p > 0.81 \text{ irrelevant}), \\ H_A: &\quad p < 0.81. \end{aligned}$$

Can you give a real-world description of p?

(c)

$$\begin{aligned} H_o: &\quad \mu = 1.2 \quad (\mu > 1.2 \text{ irrelevant}), \\ H_A: &\quad \mu < 1.2. \end{aligned}$$

The test attempts to flag an unacceptable shipment. That means you're "trying" to flag a shipment for which the mean length is too small. This is an example of "acceptance sampling." Give a real-world description of μ.

(d)

(i) It's reasonable to assume the lengths of chains approximately normally distributed. So $P(L > 1) = P(L > 3\sigma_L) \approx 99\%$, as required.

(ii) $L = \Sigma X$, the sum over the 92 links of a chain. Because $\sigma_L^2 = 92\sigma_X^2$ and because we want $\sigma_L^2 < 1/9$, $\sigma_X^2 < (1/9)(1/92) = 0.0012$.

(iii)

$$\text{H}_\text{o}: \quad \sigma_X^2 = 0.0012 \quad (\sigma_X^2 < 0.0012 \text{ irrelevant}),$$

$$\text{H}_\text{A}: \quad \sigma_X^2 > 0.0012.$$

(iv) We're assuming the links making up a chain form a simple random sample. The "monitoring procedure" of the hypothesis test uses the theory of simple random sampling. We'll see that later. And in (iii), we used the equation $\sigma_L^2 = n\sigma_X^2$ which is true only if the X's for the 92 links making up a chain are independent. That's true if the links making up a chain form a simple random sample from the distribution of X.

(e) The standard deviation for the length of links should be specified in the contract to be "no more than 0.0348 cm" (the square root of 0.0012).

6.3.6

(a) We control type I error by specifying the significance level in advance. That's the error which could be made when we act on H_A. The error for H_o is controlled only in a rather unsatisfactory sense.

Our comment in the text was, "the testing procedure is more conclusive when it decides in favor of the alternative hypothesis. In that case the probability of error is completely under our control. We cannot eliminate that error, but we can specify its probability in advance. This is why the alternative hypothesis is the principal hypothesis from a practical point of view."

(b) The significance level is the probability of type I error. You determine it in advance before you ever look at any data.

(c) The power of the test is $1 - \beta = \text{P}(\text{act on H}_\text{A}|\text{H}_\text{A} \text{ is true})$. But H_A does not provide a particular value of the parameter. So the power of the test depends on the various parameter values allowed in the alternative hypothesis. The power is a function of the parameter.

(d) A hypothesis test is trying to spot the times when H_A is true.

(e) A test of significance is trying to spot data which challenges the hypothesis. Note that a hypothesis test can be thought of as trying to challenge H_o.

(f) The table helps us understand the logical structure of the test. It sorts out the four situations which arise. Two involve correct decisions, two involve errors. Because the "state of the world" is unknown, we'll never know whether we're making a correct decision or an error. That means any discussion of correct decisions and errors is theoretical. This is always true in making a statistical inference based on sample data—you'll be misled into error if the data is atypical of the general situation, but YOU DON'T KNOW THE GENERAL SITUATION! So you don't know whether your data is typical or not. All you know is the data is *probably* typical. Thus, you cannot *avoid* error, but you can hope to control it. You control error by specifying the probability of the error in advance. In an hypothesis test, only one of the two errors can be controlled in a completely satisfactory way. See Problem 6.3.1.

(g) As a practical matter, no physically determined number ever takes *exactly* a predetermined value.

(h) The test tries to flag "too many defectives." "Too few defectives" doesn't make sense. So small values of p are irrelevant. Beyond the test itself is the entire production process. Certainly, from that perspective "few defectives" is of critical importance! Don't tell your stockholders that "few defectives" is irrelevant. You'll cause a Wall Street crash!

(i) From the text: " . . . the testing procedure is more conclusive when it decides in favor of the alternative hypothesis. In that case, the probability of error is completely under our control. We cannot eliminate that error, but we can specify its probability in advance."

(j) H_A gives a condition on the parameter, NOT on the estimator.

6.3.7

(a) Either by identifying the "alternative" action or by forcing the more serious of the two possible errors be type I error.

(b) The first error is inconsequential and the second quite serious! The first error is to use the new procedure even though it's ineffective. So no harm is done; after all, there's no risk. The second error is to do nothing, even though the new procedure is effective. In light of the seriousness of the disease, this is a crucial error. Of course, you could take an extreme case in the "other direction" and you would reverse the seriousness of the two errors.

(c)

$$\begin{aligned} H_o: &\quad R = 0.001 \quad (R > 0.001 \text{ irrelevant}), \\ H_A: &\quad R < 0.001. \end{aligned}$$

Here, R is the recovery rate for the new treatment. Note that believing and acting on H_A means believing the new treatment is not as good as the old, so you'll use the old procedure. Thus, type I error is "using the old procedure when, in fact, the new procedure is more effective." By part (b), that's the more serious error.

6.3.8 False. The five percent "error rate" refers to five percent *of those weeks in which quality was really in control* (it's conditional on that "state of the world'). Suppose quality was really in control for only 70 of the 100 weeks. Then we would expect to have unnecessarily halted production for corrective action about three or four times (5% of 70). We are not talking about five percent of all runs of the hypothesis test; we're talking about five percent of those runs for which, in fact, although we didn't know it, there was no need for corrective action. The five percent "error rate" refers only to those cases for which the null hypothesis was in fact (unbeknownst to us!), true. It's "conditional" on the null hypothesis. We'll see why this is true later when we discuss how the data is analyzed.

6.3.9 (a) **Reject H_o:** On the basis of our evidence, it appears we are producing too many useless parts. Take corrective action.
Fail to reject H_o: The test is inconclusive. There's no evidence we are producing useless parts at a rate of more than one in 50; continue in production.
Type I error: There's about a five percent risk that we interrupt production to take corrective action when, in fact, that's not necessary. Although the evidence suggests otherwise, we are not producing more than one in fifty useless parts.
Type II error: There's an unknown risk that we are continuing in production even though more than one in 50 parts is entirely useless.

(b) **Reject H_o:** It appears we have lost more than three points for the month in question. We must step up the campaign effort in that district.
Fail to Reject H_o: The test is inconclusive. There's no evidence our support in that district for the month in question has weakened. We will continue at the present level of campaign activity.
Type I error: We increase the campaign effort in that district when in fact, although we didn't know it, our level of support for the month in question has not fallen more than three percentage points. There's a five percent chance of this error.

Type II error: With an unknown risk of error, we continue at the present level of campaign activity when, in fact, our support in that district for the month in question has slipped more than three points.

(c) **Reject H_o:** Our evidence seems to indicate this month's shipment of chain links does not meet specifications. Following the terms of our contract, we return this shipment to the supplier.
Fail to Reject H_o: The test is inconclusive. Because there's no evidence, it does not meet specifications; accept this month's shipment of chain links.
Type I error: There's a five percent risk that we return the shipment of chain links this month when, in fact, it does meet the specified mean length per link of at least 1.2 cm.
Type II error: There's an unknown risk that we accept, for the month in question, a shipment of chain links which does not meet specifications.

(d) You'll have to change the reference to the specification in the statement of type I error. Here you not talking about "mean length per link."

6.3.10

(a) Formally, either you "reject H_o," which in real-world terms means "act on H_A," or you "fail to reject H_o" meaning you "act on H_o."

(b) Formally, you have either "type I error," which in the real world means you "act on H_A when, in fact, H_o is true." Or you have "type II error" which means you "act on H_o when, in fact, H_A is true."

(c) Logically, "H_o is true" means $p = 0.01$, practically (in real-world terms) it means $p < 0.01$. These are the "irrelevant" values, but they are NOT irrelevant from a practical point of view! See Problem 6.3.6(h).

(d) The alternative hypothesis plays a practical role in the test. So its "logical meaning" is exactly the same as its practical, real-world meaning. In the example, it says you're producing too many defectives: $p > 0.01$.

(e) Logically, to say "the test is inconclusive" means your data provides no evidence that H_A is true. In other words, "the data is consistent with H_o." But it will be consistent with many other hypotheses as well. So the data proves nothing. By contrast, in real-world terms some course of action is required. Because the test is inconclusive, that action cannot be based on the test and must, instead, be based on some prior information. In the quality control example, if there's no evidence of a problem, you "stay in production." That action, as we said in the text

... is based on the fact ("prior information") that you have a production process which is well designed, free of problems, and run by well trained workers. With no evidence to the contrary, it makes sense to let the process go forward.

6.3.11 It does unless you're also checking the VARIANCE of the diameters! See Problem 6.3.2(d).

6.3.12 The parameter is μ, the mean life of ALL the new supplier's part. The estimator is $\overline{X}$, the mean life of a SAMPLE of the new supplier's part.

Reject H_0: We should write a contract with the new supplier. It appears the average life of their part is significantly higher than for the present supplier.
Fail to reject H_0: The test is inconclusive; the sample provides no new information. In particular, the evidence obtained from our sample does not support switching to the new supplier. We should remain with our present supplier.
Type I error: There's a 1% risk that we switch to the new supplier when, in fact, the mean life of their part is not significantly greater than that of the present supplier.
Type II error: There's an unknown risk that we remain with the old supplier even though the new supplier's part has a significantly longer life.

Note the use of the word "significant" in stating the error. That's *practical* significance. It has nothing to do with the significance level of the test which is a form of *statistical* significance. For example, the present supplier's part has a mean life of 285 hours. Suppose the new supplier's part has a mean life of 286 hours. Then it has a longer life—by one hour! Evidently, in the judgment of the purchasing department, such a small difference would have no practical significance. That's why they gave you a criterion of "greater than 310 hours."

6.3.13 The parameter is p, the proportion of ALL registered voters who will see the spots. The estimator is $\hat{p}$, the proportion of a SAMPLE of registered voters who will see the spots.

Reject H_0: Your study suggests that fewer than 25% of registered voters will see the spots. So, with a 5% risk of error, you do not launch the series of television spots.
Fail to reject H_0: Your study is inconclusive. Therefore, you go ahead with the series of television spots as previously planned (NOT on the basis of the hypothesis test, but on the basis that "you have decided to go ahead with the series of spots, but . . . .").
Type I error: There is a 5% chance that you will fail to launch the series of television spots when, in fact, it would be effective (i.e., it would reach at least 25% of the registered voters).
Type II error: There is an unknown risk of launching the series of television spots even though they will reach fewer than 25% of the registered voters.

6.3.14

Reject H_o: Your study suggests that more than 25% of the registered voters will see the series of television spots. So with a 5% risk of being in error, you go forward with the series.
Fail to reject H_o: Your study is inconclusive. You tell your candidate's supporters that the preliminary study provides no evidence to support launching such an expensive television series. You do not launch the series.
Type I error: There is at most a 5% risk that you go forward with the series of television spots when, in fact, it would not be effective (i.e., when fewer than 25% of registered voters would see it).
Type II error: There is an unknown risk that you will fail to launch the series of television spots when, in fact, it would be effective.

6.3.15

It's a question of practical significance. The word "significant" refers to the difference between two population means. If the average SAT score at your school is one point higher than at Bad U, you hardly have anything to brag about. A difference of one point makes no *practical* difference. Evidently, only a difference of 15 or more points will be considered of practical significance.

In this situation, data is statistically significant only if it leads us to reject H_o. Statistical significance compares the data, as summarized in a value of $\overline{X}_1 - \overline{X}_2$, with the hypothesis, $\mu_1 - \mu_2 = 15$. Here, this means $\overline{X}_1 - \overline{X}_2$ is significantly far above 15, too far to explain the difference as due just to sampling error.

Reject H_o: Your data suggests that there is at least a 15-point superiority of your school over Bad U in SAT scores. So with a 10% chance of being wrong, you go ahead with publication of the "good news."
Fail to reject H_o: Your data is inconclusive. Your study revealed no evidence to support the contention that there is a significant superiority of SAT scores at your school compared to those of Bad U. Do not publish—there is no "good news." What does the word "significant" mean here?
Type I error: There is a 10% chance that you will go ahead with publishing what you take to be the "good news" about SAT scores at your school being significantly superior to those at Bad U when, in fact, that is not the case. What does the word 'significantly" mean here?
Type II error: There is an unknown risk of missing out on a chance to publish that there is a significant superiority of SAT scores at your school over those at Bad U when, in fact, such an article would be consonant with the facts.

Did you identify in real-world terms the parameter $(\mu_1 - \mu_2)$ and its estimator?

6.3.16

This is a left-tailed test because "missing a chance to boast" will have to be type I error and that's the error we control. Be sure to identify the parameter and its estimator! We leave it to you to give the real-world interpretation of the conclusions and errors.

6.3.17–6.3.20

For these problems, we leave it to you to give the real-world interpretation of the conclusions and errors.

6.3.21

(a) If in any week your sample gives more than nine useless parts, assume the process is producing too many useless parts. Take corrective action. For a 10% significance level, $Z = 1.28$ and $\hat{p}_c = 0.0303$ and the decision rule is the same.

(b) Step up the campaign effort in that voting district during any month in which you found fewer than 36 voters among the 50 interviewed who support your candidate. If $\alpha = 10\%$, $\hat{p}_c = 0.7390$. Now the decision rule says " . . . fewer than 37 voters."

(c) The decision rule is: "Each month, compute t, the test statistic. It's $(\overline{X} - \mu_o)$ divided by $s/\sqrt{10}$. Compare it with -1.8331. Reject that month's shipment if the number you compute is less than -1.8331." Of course, you'll have to explain to the person carrying out this rule how to calculate $\overline{X}$ and s by entering the data into a calculator in statistical mode. For $\alpha = 10\%$, reject the shipment if the computed value of t is less than -1.383.

When you know that $\sigma^2 = 0.1$, the decision rule can be more simply stated: "Reject the shipment in any month when the average length of the ten links you've measured is less than 1.0355 cm." For $\alpha = 10\%$, say, " . . . less than 1.072 cm."

(d) Reject the shipment in any month when s is greater than 0.0475 cm. Here's how you get this number: First, DON'T USE $\sigma = 1/3$. That was for the length of *chains* not links [see Problem 6.3.5(e)]. For the links, you test $\sigma^2 = 0.0012$ [see Problem 6.3.5(d)]. That means H_o specifies σ^2 to be 0.0012. Because you assume H_o true, from the chi-squared table with nine degrees of freedom you get

$$16.919 = (n - 1)s^2/\sigma^2 = 9s^2/0.0012.$$

Solve to find the value we gave above: $s = 0.0475$. For $\alpha = 10\%$, your decision rule should say, " . . . greater than 0.0442 cm."

6.3.22

(a) In the second and sixth weeks, you had to take corrective action because ten or more useless parts were observed. In the other weeks no action was required.

(b) In months two, six, and seven you found fewer than 36 voters among the 50 interviewed who supported your candidate. In those months you "stepped up the campaign effort in that district." In all other months, support for your candidate seemed adequate and no special campaign effort was required.

(**c**) Here are the observed values for t, your test statistic:

$$0.8433,\ 0.3814,\ -0.3,\ 0.6325,\ 0.2236,\ -1.8974,\ -0.9487,\ 0.1.$$

Only the sixth is below -1.8331, so only in that month did you return the shipment as unacceptable. Note that the five positive values of t were unnecessary because in those months the sample mean was ABOVE 1.2 cm which could hardly be evidence that the true mean was BELOW 1.2 cm..

(**d**) The largest observed s^2 was 0.11 ($s = 0.3317$), but that's not large enough to suggest a problem (s is NOT greater than 0.4570).

6.3.23

(**a**) It's not just the *mean* length of these chain links you want to monitor. The variability in length is also important. Because your observed $s = \sqrt{0.22}$ is too large—it's larger than 0.0475—you should reject this month's shipment.

(**b**) Even if the *average* length for the chain links is acceptable (1.2 cm or more), when the standard deviation is too big, the distribution for length (approximately normal) will be too spread, giving a large number of links that are too long. Draw the picture!

c6.3.12

The rejection region is $\{t | t > 2.624\}$. It's given in terms of t because it would change with each sample if given in terms of the sample mean. After all, the standard deviation for the life of these parts is not known. Be sure to recall the *real-world* meaning of your conclusions in each case.

(**a**) Fail to reject H_o, the observed mean is in the wrong tail of the distritution. There's no way an observed mean of 287 could be evidence that the true mean is MORE than 310!

(**b**) Fail to reject H_o. The observed value of the test statistic is less than 2.624; it's 2.2951.

(**c**) Reject H_o. The observed value of the test statistic is $t = 2.9508$.

(**d**) Reject H_o. The observed value of the test statistic is $t = 7.3939$.

(**e**) FAIL to reject H_o. Now $t = 2.5505$. Although the observed mean for this sample is greater than for all the others, the observed standard deviation is also much larger, degrading the value of the data. The standard deviation is so large we have to say "no evidence" for the true mean being above 310. Such a large standard deviation might suggest that, instead of worrying about the average life, you should look more carefully at the *unpredictability* of life for the new supplier's parts.

(f) "Lifetime" should be exponentially distributed. To have $\overline{X}$ be normally distributed, you would have to take a large sample.

c6.3.13 You need not look at the decision rule in part (a) because 43/130 is greater than 0.25. Fail to reject H_0. A value of $\hat{p}$ larger than 0.25 could hardly be evidence that p is *smaller* than 0.25!

The decision rule is in terms of "25 of the 130." That 25 is NOT the 0.25 of the hypothesis!! The 25 of the decision rule is 18.75% of 130. The decision rule: "Do not launch the series of television spots if fewer than 25 of the 130 registered voters interviewed say they will see the series of spots." For parts (a)–(c), fail to reject H_0. For parts (d)–(g), reject H_0. What do these conclusions mean in real-world terms?

c6.3.14 The rejection region is $\{\hat{p}|\hat{p} > 0.3125\}$. You state the decision rule in real-world terms.

(a) Reject H_0; **(b)–(g)** fail to reject H_0. Real-world meaning?

c6.3.15 The t-distribution is irrelevant here because you know the standard deviations of the two populations. The decision rule at the 10% significance level says to reject H_0 if the difference in sample means is greater than 55.3053. What's the real-world decision rule?

(a) A sample from your school which is worse than the Bad U sample could not possibly be evidence that your school as a whole is better!

(b) and (c) Fail to reject H_0; **(d)** Reject H_0. Real-world meaning?

c6.3.16 **(a)** Reject H_0; **(b)–(d)** fail to reject H_0. Real-world meaning? Parts (b)–(d) are not even candidates to reject H_0. Part (a) is the only case where your school's score was less than Bad U's.

c6.3.17 This, note, is the same test as in Problem 6.3.15.

c6.3.18 The decision rule is: "If more than 145 of the 1215 tax returns examined show evidence of attempts at tax evasion, then the IRS should audit all income tax returns which show these characteristics of evasion."

(a)–(d) Reject H_0; **(e) and (f)** fail to reject H_0. Real-world meaning?

c6.3.19 The decision rule is: "If more than 260 of the 1215 tax returns examined show evidence of attempts at tax evasion, then the IRS should audit all income tax returns which show these characteristics of evasion."

(a) reject H_o; (b)–(f) fail to reject H_o. Real- world meaning?

c6.3.20 We will reject H_o if $\chi^2 = 6s^2/0.3 > 12.592$. That says that s^2 must be greater than 0.6296 mm^2, so s must be greater than 0.7935 mm. So the decision rule is: "Any week in which the standard deviation of thickness for the seven observed pieces of paper from our production process exceeds 0.7935 mm, you should assume the process is no longer in control and take appropriate corrective action."

(a) **and** (b) Fail to reject H_o; (c) **and** (d) reject H_o. Real-world meaning?

6.3.24 Treating one-time decisions as hypothesis tests instead of tests of significance provides a clearer analysis of the possible errors. In fact, Problems 6.3.14, 6.3.16, 6.3.17, and 6.3.18 were all formulated in terms of "risk of error." So treatment as tests of significance would have entailed a translation from the hypothesis testing concept of error.

6.3.25 (a) Done!

(b) For a one-time decision, all you need is a conclusion. One conclusion! Calculate the p-value and interpret it to get the conclusion. Finished!

For a true hypothesis testing situation—a *monitoring* situation—the decision rule determined by the rejection region can be given in real-world terms to someone like a production line supervisor. It's a rule to decide between two possible courses of action depending on the data for the particular sampling period. Because the data changes from one period to the next, such a decision rule is NOT redundant.

6.3.26 (a) $P(Z > 0.26) = 0.3974$. You knew this would be smaller than 50% because you're calculating the power at a value of p which is between 0.01 and the crictical $\hat{p}$.

(b) $P(Z > -0.72) = 0.7642$. How could you see in advance that the power would be greater than 50%?

(c) $P(Z > -2.71) = 0.9966$.

(d) Necessarily the power is $\alpha = 0.01$. Draw the picture and you'll see why.

(e) The power is one. If $p = 1$, every item is defective, so you know for certain that any sample gives $\hat{p} = 1$, which is in the rejection region. Because you always reject H_o, the power is one:

$$\text{power} = P(\text{act on } H_A | H_A \text{ is true}) = 1.$$

The normal approximation for $\hat{p}$ is not valid because $nq \geq 5$ doesn't hold (it's zero!).

6.3.27

(a) The power is the probability of acting on H_A when you should have. That means H_A is the condition that's true. The power is a conditional probability, conditional on H_A being true

$$\text{power} = P(\text{act on } H_A | H_A \text{ is true}).$$

So to calculate power, we must choose a parameter value where H_A is true.

(b) In real-world terms, the test has decent power to detect too many defectives only if 2% or more are defective. In other words, even when we have an unacceptable defect rate above 1%, there's less than a 75% chance for the test to actually detect that fact when less than 2% are defective.

6.3.28

(a) $P(Z > 0.92) = 17.88\%$, $P(Z > -0.16) = 56.36\%$, and $P(Z > -1.33) = 90.82\%$.

(b) 30.5%, 84.61%.

(c) 20.33%, 56.75%.

(d) Because $9s^2/\sigma^2 = \chi^2$, the power is

$$\begin{aligned} P(s^2 > 0.0475^2 | \sigma^2 = 0.0065) &= P(\chi^2 > 9 \times 0.0475^2/0.0065) \\ &= P(\chi^2 > 3.124) \\ &\approx P(\chi^2 > 3.325) \\ &= 95\%. \end{aligned}$$

6.3.29

Sorry.

6.3.30

(a) The picture for $\hat{p}$ centered at $p = 0.015$ with 40% of the area in the left tail ($Z \approx -0.25$) gives $\hat{p}_c = 0.015 - 0.25$ s.e. With the picture centered at $p = 0.01$, we get $\hat{p}_c = 0.01 + 1.28$ s.e. Note that the standard error is different in each case. The first has $p = 0.015$; the second $p = 0.01$. So you get

$$0.015 - 0.25 \times 0.1216/\sqrt{n} = 0.01 + 1.28 \times 0.0995/\sqrt{n}$$

from which we see that you must take a sample of $n = 996$ parts. If the normal approximation for $\hat{p}$ is invalid, you must revert to the binomial model.

(b) The rejection region is $\{\hat{p}|\hat{p} > 0.0140\}$.

6.4.1 **(a)** 0.62% of all cups overflow on average.

(b) 90% chance that somewhere between 12.64% and 32.25% of all cups overflow (s.e. ≈ 0.0596).

(c) 90% chance that 1 to $2\frac{2}{3}$ cups overflow on average *per two days* ($s = 1.2906$, s.e. ≈ 0.2213). The 90% confidence interval: (0.6065, 1.3347).

(d) For the small sample case you would use Student's t-distribution, but that requires that you're sampling from a normal distribution. Here Y certainly would NOT be normally distributed. What's a reasonable distribution for Y? Think about it.[7]

6.4.2 Here you have complete information about the "population," all baggage for this airline. The question is about ΣX for a "sample." This reverses the usual pattern. The problem is just a calculation using the sampling distribution of the estimator $\overline{X}$, recalling that $\Sigma X = n\overline{X}$. To four places, the interval is (4212.8843, 4566.3157).

6.4.3 95% (18.6288, 41.4179), a prediction interval.

6.4.4 1.22% chance.

6.4.5 The p-value is

$$\begin{aligned} P(\Sigma X > 6425 | \mu_{\Sigma X} = 6000, n = 214) &= P(\overline{X} > 30.0234 | \mu_{\overline{X}} = 28.0374) \\ &= 0 \end{aligned}$$

with $Z = 5.01$. Zero is SMALL as a p-value and so the answer is NO.

[7] The data itself indicates that Y is skewed to the right. The reasonable model is Poisson! If λ is more than about nine, the Poisson distribution is approximately normal. But the data in part (c) would suggest λ is LESS THAN ONE.

6.4.6 The standard error for the difference in our two means is 0.3044, the square root of

$$4.269^2/429 + 4.761^2/452 = 0.0926.$$

So

$$p\text{-value} = \mathrm{P}(\overline{X}_1 - \overline{X}_2 > 3.523) = \mathrm{P}(Z > 11.58) = 0.$$

There is a criticism to be made here: Evidently, Barrow and Morrisey tested all students in the selected schools. Taking all students from randomly chosen schools does NOT provide a simple random sample of students, although they seem to have treated it as such. What kind of sample is that?

6.4.7 We're leaving this to you. Make sure you can do them!

6.4.8 Let X = "length of link."

(a) $\mathrm{P}(X > 0.71)$, where $X \sim N(0.7, 0.06)$. Note, you're assuming a normal distribution for X. Justify that assumption and be prepared to give an example of how it might fail.

(b) $\mathrm{P}(L > 57.51) = P(\overline{X} > 0.71)$, where $\overline{X} \sim N(0.7, 0.06/9)$. Justify using a normal distribution for $\overline{X}$.

(c) $\mathrm{P}(Y \geq 10)$, where $Y \sim B(81, p)$ with p calculated in part (a). This assumes a chain is a random sample of links, so our data becomes a sample from the distribution of "length" X.

(d) $\mathrm{P}(Y \geq 10)$, where $Y \sim B(100, p)$ with p = answer in (b). Assume the chains are randomly chosen.

(e) $\mathrm{P}(\hat{p} > 0.2)$ or $\mathrm{P}(X > 16 | n = 81, p = ?)$.

(f) Give a confidence interval based on one chain of 81 links as a sample, giving a sample mean of 0.71 cm. You must interpret the question as asking for the *mean* length of a link.

(g) Do a test of significance: Does the data (the chain) seem to challenge the hypothesis?

(h) Set up hypothesis tests for μ and σ^2. This is like Problem 6.3.5 parts (c) and (d). But here the problem is vaguely stated. You would need to clarify—the "boss" needs to clarify—exactly what is meant by "meets specifications" (at most?, at least? etc). You explore the possibilities.

6.4.9

(**a**) This is a problem about ΣX, convert it into a sample *mean* problem.

(**b**) CLT, n is large.

(**c**) H = "weight of haul" is ΣX, so $H = n\overline{X}$ from which $\mu_H = n\mu$ and $\sigma_H = n\sigma$. Or forget $\overline{X}$, the mean of ΣX is just $\Sigma \mu_X$; μ is added to itself n times, which is $n\mu$. And with independence the variance of ΣX is $n\sigma^2$. Here, you have complete information about all fish in these waters, so this is not the usual confidence interval. It's a question about a "sample" (see Problem 6.4.3). Conclusion: "We can be 95% sure today's haul will weigh between 12,998 and 13,413 pounds." To four places: (12,997.5841, 13,413.3583).

(**d**) $\Sigma X = H = n\overline{X}$, so $\sigma_H^2 = n^2\sigma_{\overline{X}}^2 = n^2(\sigma^2/n) = n\sigma^2$. Or forget the sample mean, the *variance* of ΣX is just $\Sigma\sigma_X^2$ which is just σ_X^2 added to itself n times. For this second approach, the various values of X in the sample must be independent—why is it true here that the values are independent?

(**e**) You're sampling from W = "weight of one fish." Random experiment: the DOING: "fish until you catch one"; clearly repeatable; an OUTCOME: a fish; clearly you cannot predict in advance what fish you'll get. W assigns to each fish (each outcome) its weight.

The weights of the fish in a haul can be written down in a list (an "ordered set of n values of W") and are generated by n repetitions of "fishing until you catch one," the random experiment for W. If these repetitions are independent, the definition of a "simple random sample from the distribution of W" is satisfied.

(**f**) If there are only, say, two kinds of fish and they travel in large schools, any single fish caught is likely to be the same type of fish as the previous one, the independence assumption would fail. But here you have a fleet of many boats presumably covering a large area with fish of many different types. This would seem to make the independence assumption more or less valid. For this situation, the "doing" would have to be more precisely defined as "a fisherman in *some boat somewhere* in the large area of your operations catches a fish." Note that for a fleet of commercial fishing boats where the fish are caught in large nets, this assumption of independence becomes more problematic. The same considerations will come up in considering the possibility that W is normally distributed—so far we've not needed that assumption.

(**g**) $\Sigma X = n\overline{X}$; it's of the form $Y = a + b\overline{X}$ with $a = 0$ and $b = n$. So the underlying experiment is the same as for the sample mean. What is that experiment? ΣX is approximately normally distributed, why? Its mean is n times the mean of the sample means—what's the formula? The variance? Note that this is all just redoing the analysis in part (d).

(**h**) The girl can be about 95% sure that the fish in these waters weigh somewhere between 4 lbs. 7 oz. and 7 lbs. 3 oz. on average. [Hint: $t = 3.1825, s^2 = 2.2707/3$].

(**i**) With a five percent risk of error, we can suppose her fish will weigh between 2 lbs. 11 oz. and 8 lbs. 14 oz. This is NOT "on average"! We've done a prediction interval. Our nonparametric prediction interval is not helpful here because it would give at most a confidence coefficient of 60%. Explain! Furthermore, to give that 60% interval would require more information (what?) than we have.

(**j**) First, her four fish must be assumed to be a *random sample* of all the fish in the waters of her area. Second, because n is small, the weight of ALL the fish (the population) must be *normally distributed*. Both of these assumptions might be questionable [see part (f)]—even if the assumptions are valid for the fleet's operation, the restricted area of the girl's fishing might make the assumptions fail in her case. Reread part (f) and see why this is true. Remember, for the weight of fish to be normally distributed, the difference in the weight of two fish must "look like random error." That is, the difference must be due to many independent random factors. This will fail if there is some systematic factor (one or more) *systematically* accounting for some of the differences.

But still, the assumption of normality might hold under appropriate conditions: Salmon, for example, live in the ocean waters for about two years and then return upstream to spawn. If the girl is fishing near the mouth of their river when they're there, preparing to enter the river and return to the spawning grounds, and if other fish are by comparison negligible in numbers, then the normality assumption would be quite reasonable: These salmon are virtually all about the same age and the differences in their weights would be due simply to many independent random factors.

You see here why we keep saying that statistics is NOT just a matter of numbers. The numbers taken out of context are meaningless! Informed judgment is unavoidable.

(**k**) The average weight of fish for the fleet is 7.5202 pounds. This number is out of the probable range for weight in the girl's area [see part (h)]. Maybe she fishes only in shallow water where the really big fish don't go, to name just one possibility.

(**l**) It's reasonable to assume that W = "weight of fish" to be normally distributed *in this case* (why?). So a confidence interval is NOT required because the population in question is completely described: normally distributed with $\mu = 7.5202$, $\sigma = 2.5311$. $P(W > 10) = P(Z > 0.98) = 0.1635$.

(**m**) Real world: "We can be about 95% sure that between 29% and 33% of the fish in the fleet's area of operation are flounder." Note that the real-world answer is *as complete in detail as the problem allows.*

(n) $1.96^2 pq/n = 0.0140^2$. For the worst case, $pq = 0.25$.

(o) At 4.5 pounds, $t = 2.9885$ giving a p-value between 2.5% and 5%. Because this is less than our given five percent criterion, the p-value IS SMALL. Real world: "Based on the girl's catch, it appears the old man's information is wrong. The fish in these waters ("around here") would seem to weigh more than five pounds, on average."

Note that for a test of significance you're required to do three things: calculate the p-value, say explicitly "small" or "not small," and then give the real-world interpretation. If no criterion of "small" is given, you decide. Of course, a p-value more than ten percent is never considered small and one less than one percent would almost certainly be considered small. Otherwise, either choice is possible.

(p) $H_o : p = 0.001$, and so on. The rejection region is $\{\hat{p}|\hat{p} > 0.0036\}$. The standard error here is 0.0016. For a significance level of FIVE percent, $Z = 1.645$. Make sure you can set up the hypotheses correctly with correct formal notation and can draw the pictures of the estimator and of Z.

(q) $P(Z > 2.96) = 0.0015$. Assumption? Reasonable? How could it fail?

(r) $n\mu = 2000 \times 7.5202 = 15,040.4$, so $15,040 \pm 1.96 \times 113.1942$ (95% sure).

(s) s.e. $= 0.0566$. $P(Z < -0.36) = 0.3594$. Draw the pictures!

(t) $1.96^2\sigma^2/n = 1^2$. Assume $\sigma = 2.5311$.

(u) The sample size is small. But then, σ is known.

(v) $H_A : \mu < 7$, the s.e. $= 0.0428$, and $z = 1.645$. Be sure you could set up the hypotheses completely and state the possible conclusions and errors both formally and in real-world terms. Also, you must be able to draw the pictures both of the estimator and of Z, indicating clearly in the picture the rejection region.

(w) "Today's haul does indeed suggest that there were fewer flounder in these waters a year ago." The p-value is small.

You have to reframe the question as: Does today's haul seem to challenge the contention that $p_1 - p_2 = 0$? Here, p_1 and p_2 are the proportion of flounder, respectively, last year and this year. Because you will assume the contention true to compute a p-value, that means the two p's will be assumed equal. Call that value just p . Then the squared standard error for the estimator $\hat{p}_1 - \hat{p}_2$ is just

$$pq/2144 + pq/1756 = 0.0010pq.$$

Thus, in the worst case, s.e. ≤ 0.0003 $(pq \leq 0.25)$.

In the less conservative approach, because you are assuming $p_1 = p_2$, you can estimate p by "pooling" the two samples (see Problems 6.2.9 and 6.2.12). Here, there are $601+547$ flounder out of $2144+1756$ fish, that is, $p \approx 1148/3900 = 0.2944$. This gives an estimate of 0.2077 for pq and so an estimate of the squared standard error of $0.2077 \times 0.0010 = 0.0002$.

Now, let's follow the less conservative approach. We'll see that the observed difference is "statistically significant." In other words, the p-value is small: $\hat{p}_1 - \hat{p}_2 = 0.2803 - 0.3115 = -0.0312$, so $Z = -2.13$. Here, we've done the s.e. calculations and stored it in memory without rounding—if you round the s.e. to four places, $z = -2.21$. Of course, -2.13 is more accurate (note the loss of accuracy from intermediate rounding!). This gives a p-value of 0.0136. By our criterion ($\alpha = 0.05$) this is small.

(x) $\mathrm{H_A} : \mu > 0.23$, s.e.$= 0.1266$, $z = 1.645$. Be sure . . . [see part (u)].

(y) We need the expected value of $X - 1$ where $X =$ number of fish you catch to obtain the first one which . . . . X is geometric with mean $1/p$, where $p = \mathrm{P}(W > 8) = \mathrm{P}(Z > 0.19) = 0.4247$. That's p, so $1/p = 2.3546$.

(z) In part (y), we saw that the proportion of fish over eight pounds in these waters is 0.4247. This is known (assuming normality of weight). So the question does not ask us to compare the data with an "hypothesis about p" because p is known. The question just says "compare the data with the *fact* that $p = 0.4247$". So we do a p-value calculation. This is another form of the test of significance where the question takes the form: "Does the true value of the parameter make our data look *nonrandom*?" Here $\hat{p} = 7/52$, s.e.$= 0.0685$, $z = -4.23$. This gives a p-value of zero.

Thus, we have far fewer fish weighing more than eight pounds than is reasonable for a random sample of fish from these waters. You should doubt that the 52 fish are a *random* sample and look for an explanation [see parts (f) and (j)]. Of course, the whole analysis is also based on an assumption of normality; maybe THAT's the assumption which is wrong! Or maybe we really did just get a very unusual sample! Note how the test of significance can be generalized to a number of different kinds of question.

6.4.10

(a) $\mathrm{P}(X > 2) = 1 - e^{-\lambda}[1 + \lambda + \lambda^2/2] = 0.9326$, with the expected number of bombs to fall in one region being $\lambda = 0.927083333$.

(b) There's no "unit of time" given in the problem. You're not observing a period of TIME, but rather of SPACE (a geographical region). The exponential model doesn't make sense here. To answer the question, you would need information about the number of bombs falling in a given period of time.

(c) For example, $P(X = 2) = e^{-\lambda}(\lambda^2/2) = 0.1701$, with $\lambda = 0.927083333$. Thus, the expected number of regions suffering two hits is seventeen percent of 576 (211 regions, exactly what was observed). So:

No. of hits	Observed frequency	Expected frequency	$\frac{(O-E)^2}{E}$
0	229	227	0.0176
1	211	211	0
2	93	98	0.2551
3	35	31	0.5161
≥ 4	8	9	0.1111
	576	576	0.8999

To measure the "goodness of fit" of the expected frequencies to the observed frequencies, you calculate $\chi^2 = 0.9000$ (this differs from the 0.8999 of the table because we preserved all the accuracy of the calculator). But with three degrees of freedom (see Problem 6.2.26), to get a p-value of 10% or less would require a χ^2 larger than 6.251. So the p-value is NOT SMALL and there's no evidence for lack of fit. Note, as always with goodness-of-fit tests, you do not "prove" fit, rather you simply show there's no evidence in the data for lack of fit.

This problem is treated by McPherson [Problem 16.5, page 474] using Analysis of Deviance [McPherson, p. 355], a test procedure that's more sensitive than χ^2.

6.4.11 (a)–(d) You complete these. Be sure you give the real-world conclusions.

(e) In part (a), you assumed a normal distribution for the weight of bags of frozen green peas from this distributor and you assumed a mean weight of 1.2 pounds (taking that as implicit from the label). Note that it's NOT correct to say "the peas are normally distributed." First, it's the BAGS we're talking about, not the peas. Second, physical objects (bags) cannot have "distributions." What assumptions did you make in the other parts of this problem? Be careful about assumptions of "randomness."

6.4.12 You complete this.

6.4.13

(a) $H_A : r > 0.95$. This controls the error of blocking an effective drug from the market. Surely in view of the seriousness of the present situation, this would be the error to control.

(b) $H_A : r < 0.05$. This controls the error of admitting an ineffective drug to market. Given the risk of side effects and the relative effectiveness of present treatment, this might well be the error to control.

6.4.14

(a) We leave this to you; the graph is very suggestive (see Phillips' article for a discussion).

(b) $\chi^2 = 1.9478 + 0.1733 + \cdots = 17.19$ (11 degrees of freedom). With a p-value greater than 10%, no one would say this data suggests any special "power" to postpone one's death.

Phillips has a different analysis of the data and draws a different conclusion! There's a discussion unfavorable to Phillips' analysis in the text *Statistical Reasoning* by Gary Smith (p. 442). But what about the strongly suggestive data of the sociologist Kunz in Problem 6.2.23? We need a much broader study in much greater depth than either of these!

(c) The p-value is about 3%, not dramatically conclusive!

(d) In Problem 6.2.23, we looked at the *three* months prior to the birthday. There are certainly other possibilities as well [see Phillips].

6.4.15

(a) 5; (b) 7; (c) 6; (d) 3; (e) 10; (f) 2.

Chapter 7

7.1.1

(a) Y has the right form because if $\epsilon = Y - \mu$, then $Y = \mu + \epsilon$, as required. Note, because μ is a constant, ϵ is a linear function of Y.

Now to say that ϵ is "like random error" just means it's normally distributed with mean zero. But the mean of $\epsilon = Y - \mu$ is just $\mu_Y - \mu$. And that's zero because μ IS μ_Y (we just used an abreviated notation). Once again, as so often before, we've used the fundamental equations for linearly related random variables from Section 1.3 of Chapter 1. We can take this further, the same argument says the variance of ϵ is equal to the variance of Y. So the variability in ϵ, the "error," is exactly the variability in Y. This makes perfect sense, see parts (b) and (c).

(**b**) The systematic part of Y is its mean. Note how this comes out in the examples: The systematic part of a measurement is the true dimension of the object being measured. Unless there's something wrong in the measurement process, the measurements should *on average* give that true dimension. For the drink machine, the systematic part of "fill" is the amount at which the fill mechanism is set. Again, if there's not something wrong with the fill mechanism, the on average fill (the mean) should be the amount at which the fill mechanism is set (the systematic part).

(**c**) If Y is normally distributed, the variability around its mean is just due to many independent random factors, many small independent influences which prevent the systematic effect from being completely exact. That inexactness is seen in the ϵ of part (a).

(**d**) Either show that the difference in two values of Y "looks like random error" or show that Y is a linear function of a normally distributed random variable. The second approach is more direct here: $Y = a + b\epsilon$, where $a = \mu$ and $b = 1$. Done!

Or use the first approach which is more basic: $Y_1 - Y_2 = (\mu + \epsilon_1) - (\mu + \epsilon_2) = \epsilon_1 - \epsilon_2$. But $\epsilon_1 - \epsilon_2$ is just a normally distributed random variable with mean zero! Why? So this difference "looks like random error."

(**e**) By looking at the *difference* in two values, our normality criterion wipes out the one fixed systematic effect—the mean—leaving only the purely random part. See part (d). That's why the difference in two values will "look like random error."

(**f**) First, the variable effect on Y, expressed by X, is required to affect only the mean of Y. It shouldn't affect Y in any other way. Second, that effect should be only through a linear function.

7.1.2 (**a**) The height of the plant is constant for the first few days after planting (it's zero!). Once the plant's fully grown, the height is again constant. A graph might look roughly something like:

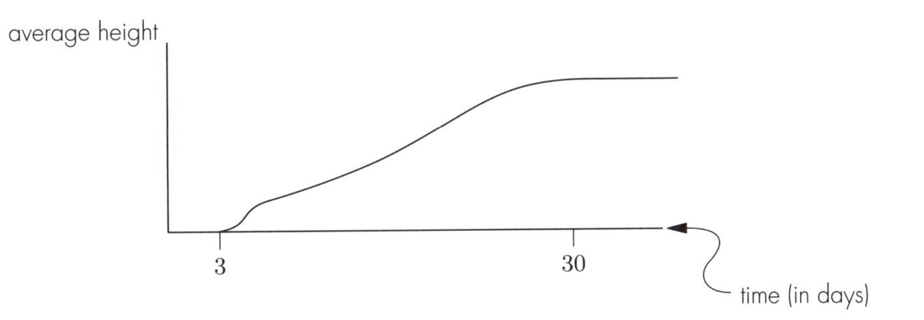

Similarly, a tiny amount of fertilizer should have no noticeable effect on crop yield. And there should be an optimal amount after which too much fertilizer might actually *decrease* the yield:

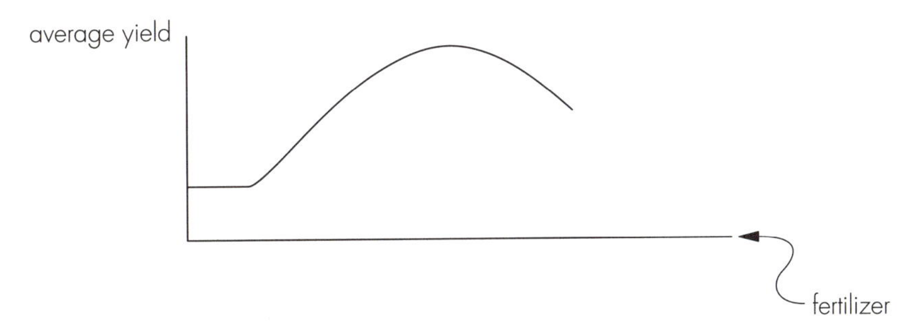

So, in both cases, to have a LINEAR relationship, you must restrict X to a smaller range of values over which these graphs seem to be approximately linear. We'll develop an hypothesis test later that serves as a check on the linearity of this relationship.

In the first example, you may have to restrict X to be somewhere between, let's say, five to 20 days. This would be appropriate if this particular type of bamboo attains a measurable height within five days and does not attain its full growth before 20 days. In the third example, you would restrict X to take values within the range of normal usage for this fertilizer.

(**b**) In the first example, the number of days since planting is certainly known and relevant to the height a given plant will attain. Also, it would seem that "height" would be the variable in question, not "number of days after planting." So it seems unlikely that we would want to reverse the roles of X and Y.

Similarly, you can possibly control the amount of chemical in the soil, but it's not likely you can control how much of that chemical gets into the plant. In all likelihood, it's the amount of chemical in the plant that would be in question, with the amount in the soil as relevant information. Again the roles of X and Y seem fairly clear-cut.

So also, for each of the other examples the role of X seems clear as "known" or in some sense "controlled," providing information about Y, whereas Y would be the variable of interest, in some way affected by X. Still, for none of the examples can we dogmatically rule out as impossible a situation in which the roles would be reversed.

(c)

3. The fertilizer by itself does not cause the crop yield—there will be some yield even with no fertilizer. But it certainly should be a cause for an increase in yield. So, X is not *the* cause of Y, but it docs cxcrcise a causal effect on the mean of Y.

4. Income does not cause expenditure! But it does make the expenditures possible. Moreover, there almost certainly would be a positive correlation of income with expenditure in the sense that, as income increases, we would expect to see household expenditures increase also ($\beta > 0$).

5. Obviously X, the production level (how many units you're producing), is one cause of production cost, Y.

6. Did an employee's outstanding test score *cause* her excellent performance on the job? Of course not! This is very typical: X and Y are correlated because they're both caused by some common third factor. In this case, both the test score and the job performance are caused by the employee's actual skill for that job.

7. Campaign expenditures make possible certain campaign strategies. Those strategies are themselves partial causes of the success of the candidate. Here you might say X is a secondary cause of Y—a cause of a cause.

8. This is like Example 7. Education is a cause of certain skills, capacities, knowledge, and so on, which are themselves partial causes of a later income level.

9. This is like Example 6. The height of the mother and the height of the daughter are both caused, in part, by common hereditary factors.

(d) In none of these examples can we dogmatically claim that X MUST be or CANNOT be random. It's always conceivable someone carries out an experiment in which X is the opposite of what we thought. Still, here's what one would think for our nine examples:

In Examples one, three, five, and eight, you would probably choose your sample data by determining in advance the values of X and then generating random values of Y. If so, X is not random. For example, you might go into the field every other day (X determined) and measure the height (Y random) of 15 randomly chosen bean plants. Or you might put a determined amount of fertilizer onto plots of ground and then observe the (random) crop yield. Note how this same principle would probably hold for Examples five and eight.

By contrast, in Example two the amount of toxic chemical in the plant and in the surrounding soil will probably be observed together, varying randomly from one observation to another. Similarly, the data available for Example nine may have resulted from a survey in which mother and daughter were chosen randomly and their heights recorded together. Example six is also probably like this. Examples four and seven might be of either type.

7.1.3

(a) Look at the difference in two values of $Y|X$. Because the effect of X is identical for both values (X is fixed), there's nothing systematic to account for this difference (X is the only systematic factor affecting $Y|X$). Heuristically, that says the difference is purely random, "due to many independent random factors." So by the normality criterion, $Y|X$ should be normally distributed.

(b) Because $\epsilon = Y|X - \mu_{Y|X}$, $a = -\mu_{Y|X}$, $b = 1$. As a linear function of a normally distributed random variable, ϵ is normallly distributed. Then,

$$\mu_\epsilon = a + b\mu_{Y|X} = 0$$

and

$$\sigma_\epsilon^2 = b^2\sigma_{Y|X}^2 = \sigma^2 \quad \text{remember } \sigma^2 \text{ means } \sigma_{Y|X}^2.$$

(c) X is the only systematic factor affecting the values of Y and it affects the mean Y only, through a linear function: $\mu_{Y|X} = \alpha + \beta X$. If $\beta = 0$, the only effect X has on Y has been completely wiped out:

$$\mu_{Y|X} = \alpha + \text{ZERO}, \quad \text{no } X \text{ in sight!}$$

7.1.4

(a) Y is the variable in question. X plays the role of a known "effect" on Y and so it plays the role of "input information" relevant to Y, the variable in question. Go through the nine examples given in the text to see the real-world meaning of this characterization of the roles of X and Y.

(b) The parameters are: α, β, and $\sigma_{Y|X}^2$. Of course, $\sigma_{Y|X}^2$ may be denoted simply by σ^2 if the context makes the reference clear.

(c)

First Assumption: There's one and only one X, which expresses a systematic effect on Y. Because there's only one X, the model is called "simple" linear regression.

Second Assumption: X is the only systematic effect on Y.

Third Assumption: The mean of Y is a linear function of X:

$$\mu_{Y|X} = \alpha + \beta X.$$

Consequences:

- For a fixed level of X, $Y|X$ is normally distributed. This is another way of saying X is the only systematic effect on Y.
- From one level of X to another, the $Y|X$'s are independent. This is implicit in saying X affects only the *mean* of Y.
- The variances of the different $Y|X$'s for different levels of X are the same. If we call that common variance σ^2, then $\sigma^2_{Y|X} = \sigma^2$ for all X. This too is implicit in the assumption that X affects only the mean of Y.
- The parameter β is not zero. Otherwise, X is wiped out of the model entirely! This is a consequence of the assumption that X is relevant at all. See Problem 7.1.3(c).

7.1.5

1. α is not meaningful in itself (why?). β is the daily increase in the average height of the plants. The word "daily" translates the phrase "unit increase in X." Briefly, β is "how fast the plants grow."

2. α should be zero. Presumably when there is no toxic chemical in the soil ($X = 0$), there would be none in the plants. β is the increase in the average amount of toxic chemical in the plants for a 1-kg increase of that chemical in the soil (assuming the chemical is measured in kilograms). You could express β as the average amount of chemical in the plant per kilogram of chemical in the soil.

3. α is the average crop yield when you use no fertilizer. β is the increase in average yield for a one-ton increase in fertilizer (assume fertilizer measured in tons). In other words, β is the average yield per ton of fertilizer.

4. α is the average household expenditure for those households which report no income (α might not be zero, depending on exactly what is meant by "expenditure" and "income"). If our data is in terms of "hundreds of dollars," β is the average increase in household expenditure for a \$100 increase in household income.

Note that β would not seem to be constant over a large range of incomes.

This suggests you would have to restrict this model to a fairly narrow range of household incomes to have the linearity condition hold.

5. α is the "fixed cost" of production, the costs incurred when you produce nothing. β is the increase in average cost of production when you increase production by one "unit." A unit of production might be "one machine part" if it's machine parts you're producing. Or a dozen machine parts if, for example, you market the parts in packages of 12.

6. α is the measure of average on-job performance of persons who score zero on the job-skills test. β is the increase in average on-job performance for each point increase on the job-skills test.

Note that here, as in all the previous examples, β would seem to be positive! But sometimes there are surprises! You might actually discover that your job-skills test is testing the wrong skills. This would certainly be the case if β were negative!

7. α is the percent of the vote captured on average by those candidates who spend nothing on their campaign. Of course, in most elections, this is a meaningless number—zero campaign expenditure is unheard of! β is the change (presumably an increase) in the percentage of the vote going to a candidate on average for a $1000 increase in campaign expenditure (assuming expenditures are measured in thousands of dollars).

8. α is the average income at age 45 for persons who have never attented school. β is the increase in the average income at age 45 for one more year of schooling.

9. α is meaningless! β is the increase in the average height of an adult daughter for a 1-in. increase in the height of her mother. That does NOT mean the mother grows an inch taller, by the way! It means if you look at two women whose mothers' heights differ by 1-in., then you would expect the daughters to differ in height by β inches.

Again, as in Example 4, for β to be constant you almost certainly would have to restrict the model to certain homogeneous groups of females. It seems likely that such a parameter might change its value depending on some cultural or ethnic factors, just to name two possibilities.

7.1.6

If β is variable, the mean of Y is affected by TWO variables, not one. So we're not talking about SIMPLE linear regression. And if you multiply β and X together, you would not even be talking about a linear relationship. Linear functions don't multiply the variables together.

7.1.7

(a) Suppose we're attempting to model "yield per acre" of a crop on the basis of two X's: "fertilizer level" and "rainfall." because there are two X's, the model would not be SIMPLE linear regression.

(**b**) Let Y be "daily household power consumption in January and Feburary" with X as "average daily temperature." Certainly, as the temperature rises, power consumption will decrease on average. As X increases, $\mu_{Y|X}$ should decrease, meaning: $\beta < 0$.

7.1.8

A quadratic regression model would have the means of the normal distributions lying along a parabola instead of along a line:

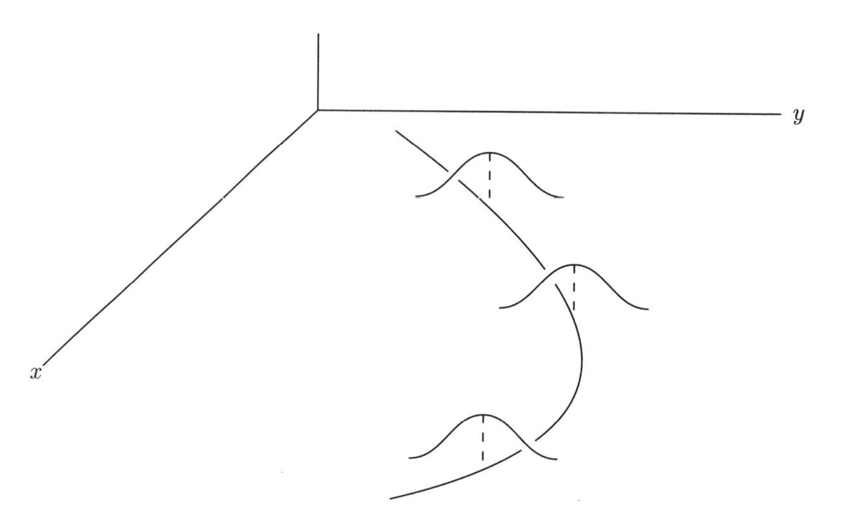

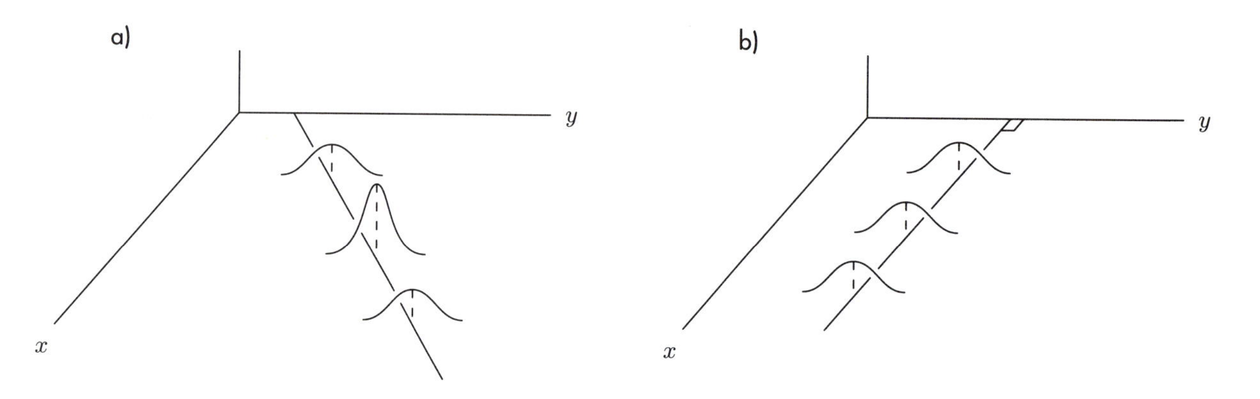

(**c**) This is just like Problem 7.1.8. There the relationship was not linear, it was quadratic.

7.1.10

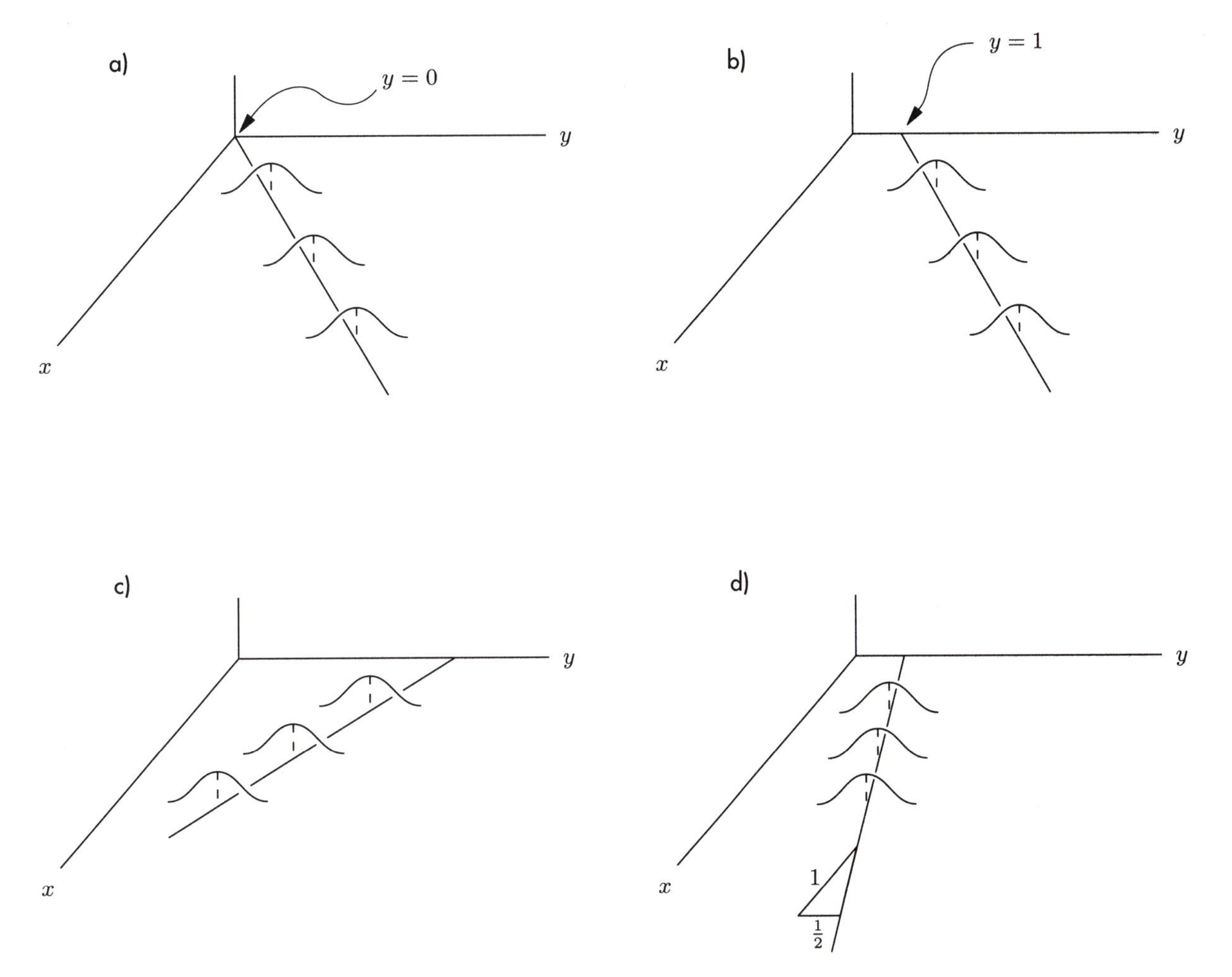

7.1.11

First, you have to determine if a simple linear regression model seems appropriate. Then you must fully specify the model. That means you must estimate the parameters of the model. After all, you can't actually do anything with

$$Y = \alpha + \beta X + \epsilon, \quad \epsilon \sim N(0, \sigma^2),$$

you need something with specific numbers in it, like

$$Y = 7 + 3X + \epsilon, \quad \epsilon \sim N(0, 1.2).$$

Then for a given X you can say something about Y.

7.2.1

(a) Given the roles of X and Y in the regression model, we should let the height of the shoots be Y and the number of days since planting be X. For (12, 21), go to the second column of the data where $X = 12$. You see 21 in the second position. That means (12, 21) refers to the second shoot observed on the twelveth day after planting. It had a height of 21 cm ($Y = 21$). And it's the fifth observation.

(b) The shoot was 53 cm tall.

(c) $n = 16$.

(d) (X_8, Y_8) is just (14, 26); the eighth observation was the second observation on the 14th day after planting. (X_3, Y_{16}) is nonsense! The notation requires that the two subscripts be the same. You could have (X_3, Y_3) or (X_{16}, Y_{16}). They are (10, 6) and (20, 110), respectively.

(e) Not all the values of X occur with the same frequency. The value X=10 occurs three times, $X = 12$ THREE times, $X = 14$ TWO times, and so on. There are 16 of these observations altogether. So $\overline{X} = (1/16)\Sigma X f$.

(f) $\overline{Y} = 51.3125$.

(g) The average daily height is $\overline{Y} = (1/16) \times 821 = 51.3125$. The daily *total* heights, respectively, are 20, 56, 60, 160, 178, and 347. To obtain the daily average heights, you must divide each daily total by THREE or TWO, depending on how many observations were obtained on that day. Then average these:

$$(1/6)\left[20/3 + 56/3 + 60/2 + 160/3 + 178/2 + 347/3\right]$$
$$= 52.2222, \quad \text{which is NOT } 51.3125.$$

In this calculation, some of the 16 observations are weighted by 1/18 and some are weighted by 1/12. In the average daily height, each observation is weighted by 1/16. That's why they are not the same.

7.2.2

(a) The scatter diagram is

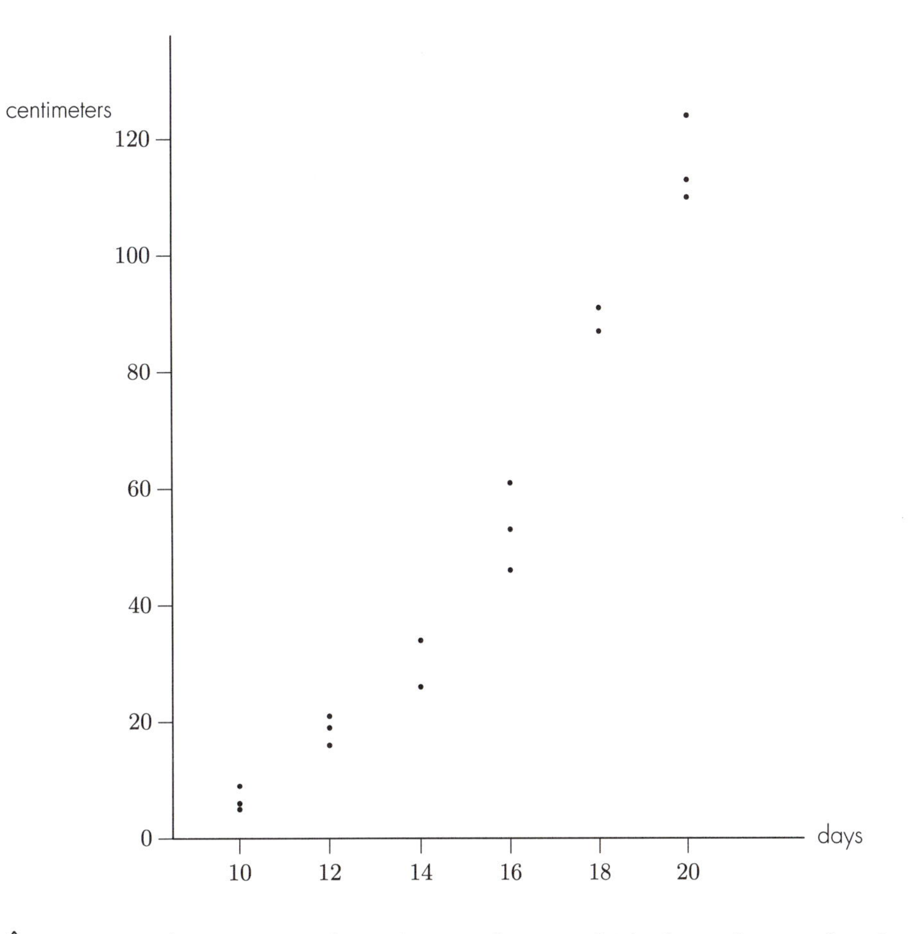

(b) $\hat{Y} = a + bX$, because a is the estimator for α and b is the estimator for β.

(c) $\hat{Y} = a + 13b = 30.6275$. But if $X = 6$, $\hat{Y} = a + 6b = -46.5958$. Six days after planting, we had a NEGATIVE average height! This is because $X = 6$ is out of the range of meaningful X's. The model is not valid for such X's. After all, why did you start making observations only on the tenth day? Precisely because only then had the shoots achieved a measurable height. Look again at Problem 7.1.2(a).

(d) The points (12, 19.5956) and (18, 85.7870) are on the line. When you plot these two points, use a different notation so they don't look like observed points. Note how we've used dots for the observed points and small crosses for these two theoretical points. Of course, a professionally drawn diagram or a diagram from a computer software package would not show them at all; they're eliminated once the line is obtained.

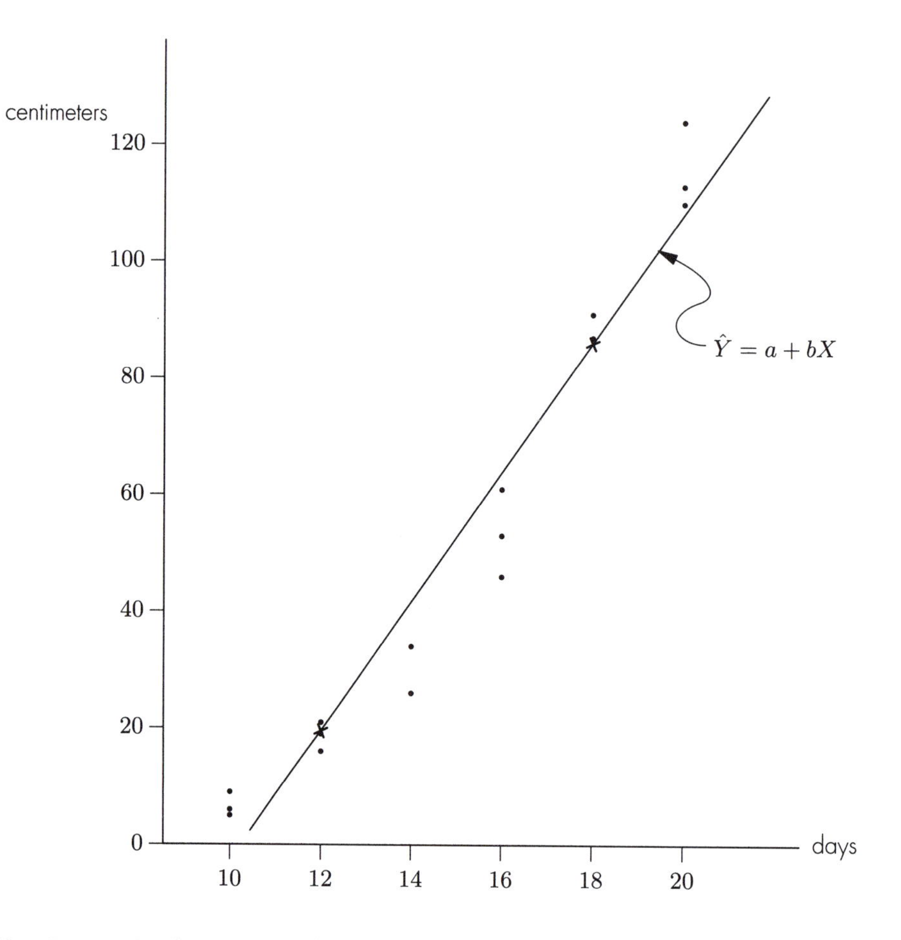

(e) For $X = 12$, the scatter diagram gives three actual heights 12 days after planting. By contrast, the model at $X = 12$ pictures a normal distribution for ALL POSSIBLE heights 12 days after planting. In general, the scatter diagram gives, for each X, several observations from the normal distribution of $Y|X$—several actual observations from *all theoretically possible* heights of this type of bamboo X days after planting. Note how theoretical the model is. Only in theory can you speak of "all possible" heights of such bamboo shoots.

(f) $\hat{Y}$ estimates $\mu_{Y|X}$. So the line in your scatter diagram is an *estimate* for the line in the simple linear regression model. Look at the picture of the model. The line in the model is the line containing the means of the normal distributions of the $Y|X$'s. It's a theoretical line estimated by a line which you've drawn based on actual observed data.

7.2.3 (a) Out of hundreds or even thousands of bamboo shoots on the hillside, our researcher has observed only two or three on any given day, only 16 altogether. Now the point estimate for β is $b = 11.0319$. If we took that as the approximate value of β, we would be saying that, on average, the shoots gain about 11 cm of height per day. But the 16 bamboo shoots this researcher observed could be very atypical of the hundreds or even thousands of such shoots. It could be that such bamboo shoots growing under similar conditions only gain on average about three centimeters per day and that the 16 shoots observed happened to be especially vigorous and fast growing. Of course, the researcher does not know β, and so she has no way to know if the observed 16 bamboo shoots were typical or not.

(b) Give a confidence interval estimate for β. The maximum error of the estimate (half the width of the interval) will measure "the accuracy of the estimate" and the confidence coefficient will measure "the certainty with which that accuracy is attained." We'll have an interval with (ideally) endpoints of the form

$$11.0319 \pm z\sigma_b,$$

where σ_b is the standard error of the estimator b. Here, $z\sigma_b$ is the maximum error of the estimate and z is determined by the confidence coefficient. We said "ideally" these are the endpoints because, in fact, the standard error will typically have to be estimated from the data and with fewer than 30 observations, we'll have to use t instead of Z.

(c) You might think that "properly generated" random data must be "typical" data! THAT'S NOT TRUE! Thinking that is precisely what could "go wrong."

(d) The total context of a point estimate is the sampling distribution of the estimator. One set of observed data by itself is meaningless; we require the entire THEORY of all possible such data sets. That theory is the probability distribution of the estimator, the sampling distribution. We'll see later that under the assumptions of our model the estimator b is normally distributed and unbiased ($\mu_b = \beta$) and we'll have a formula for estimating its standard error.

7.2.4

(a) Looking carefully at the scatter diagram, you'll see that TWO different lines seem to be suggested. The observations of the last three days seem to lie along a line with steeper slope than the first three days.

Compare the actual observations with the estimated regression line. Note how all three observations for day ten lie above that line and both observations for day 14 lie below the line. In other words, for the first three days observed, the average daily increase in height—the slope of the regression line for those three days—seems to be less than our b of 11.0319. Similarly, for the last three days observed, the daily rate of growth seems to be greater than our b. Look at the last three observed days, the observations for day 16 are all BELOW the line, and for day 20 are all ABOVE the line. So it appears there may be TWO DIFFERENT β's for the model.

(**b**) In fact—or so we can imagine for our fictitious data—when our researcher investigated this problem, she discovered that about two weeks after planting, the gardener had received some fertilizer which had been on order. With great relief that finally it had arrived, he immmediately spread the hillside with that fertilizer. Of course, this increased the rate of growth for the bamboo. And so after day 14, we have a different β.

(**c**) You could restrict the model so that there's a *constant* β. Restrict either to days before the fertilizer was spread on the hillside or to the days after that.

(**d**) Our research worker wanted to study growth of this type of bamboo during the first three weeks of its growth cycle, yet the restricted model would only be valid for the first two weeks.

Remember why the model must be restricted. You can only use the model after you have collected data to estimate the unknown parameters of the model, α, β, $\sigma^2_{Y|X}$. But these estimates are not valid outside the range of X's you've actually observed. So you cannot use the model to draw conclusions beyond those observed X's. The researcher had better start over!

7.2.5

(**a**) The bottom point is $(x, \hat{y})$ and the vertical distance is the difference in the y-values. That distance is just $y - \hat{y}$.

(**b**) Here's the picture:

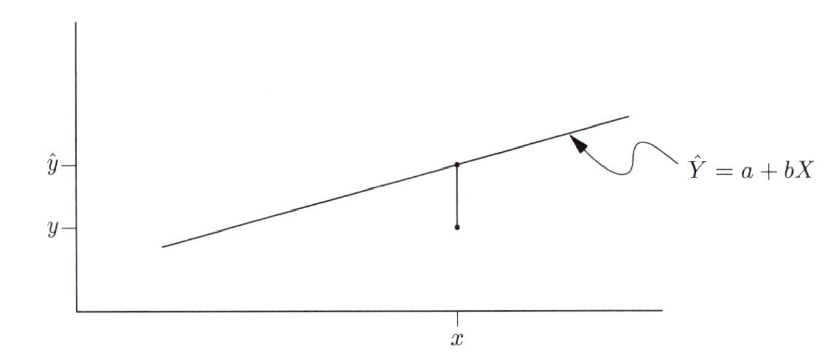

Note that $\hat{y}$ is *larger than* y here, so $y - \hat{y}$ is negative. But distances are never negative.

(c) We squared it! So here we should look at $(y-\hat{y})^2$. This will change its value, of course. This is no longer the *distance* between y and $\hat{y}$; it's the *squared distance*. But it still serves as a measure of how far apart they are without introducing those awful absolute values!

(d) "Total" means "add." Here's the total: $\Sigma(y - \hat{y})^2$.

7.2.6 $\overline{X} = 14.875$, $\overline{Y} = 51.3125$. We only need to verify that

$$51.3125 \overset{?}{=} -112.7872 + 11.0319 \times 14.875.$$

The right-hand side is not exactly 51.3125; the difference is due to rounding error. But you can't know that for sure at this point. You should be very suspicious!

7.2.7 The variable X is a known quantity which is supposed to help us say something about Y. It plays the role of "input information" relevant to Y. Y is the variable in question. So it's the variability of Y that's important. By looking only in the *vertical* direction, we're ignoring changes in X and concentrate on changes in Y only.

7.2.8 **(a)** $\Sigma(y - \hat{y})^2$, the total squared deviations away from the line (see Problem 7.2.5).

(b) $\Sigma(y - \hat{y})^2$ measures the variability of Y about the line. So the principle of least squares finds the line which "best fits" the data in this sense: For the observed values, the variability away from the line in the vertical direction (the direction of Y) is as small as possible.

7.2.9

X	Y	XY	X^2	Y^2
10	7	70	100	49
10	11	110	100	121
10	6	60	100	36
12	22	264	144	484
12	21	⋮	⋮	⋮
12	25			
14	43			
14	38			
14	33			
16	62	⋮	⋮	⋮
16	48			
16	50			
18	70			
18	74			
20	84			
20	82	⋮	⋮	⋮
20	68	1360	400	4624
252	744	12484	3936	43566

$$b = \frac{17 \times 12484 - 252 \times 744}{17 \times 3936 - 252^2}$$
$$= 7.259389671$$

and

$$a = 43.76470588 - 14.82352941\, b$$
$$= -63.84507042$$

7.2.10

(a) Y is the "variable in question," X is input information relevant for answering that question. So X must be "water content of snow on April 1" and Y must be "April to July water yield in the Snake River watershed in Wyoming."

(b) $\hat{Y} = 0.7254 + 0.4981X$.

(c) $\hat{Y} = 0.52X$.

(d) We leave this to you.

7.2.11 (a) Because $a = \overline{Y} - b\overline{X}$, we get

$$a + b\overline{X} = (\overline{Y} - b\overline{X}) + b\overline{X} = \overline{Y} \quad \text{since the } b\overline{X}\text{'s cancel!}$$

So the equation is satisfied.

(b)
$$\begin{aligned}\Sigma(y - \hat{y}) = \Sigma\big[y - (a + bx)\big] &= \Sigma y - na - b\Sigma x \\ &= \Sigma y - b\Sigma x - na \\ &= n\overline{Y} - bn\overline{X} - na \\ &= n(\overline{Y} - b\overline{X}) - na \\ &= na - na \\ &= \text{ZERO.}\end{aligned}$$

(c) It's useless to try measuring the variability of the Y's around the estimated mean $\hat{Y}$ (the regression line) by looking at the average deviations, $(1/n)\Sigma(y - \hat{y})$. That's always zero! We get around this problem by looking at the *squared* deviations. That brings in SSE. The regression line is the line for which SSE is as small as possible. You'll soon see how SSE comes into consideration as a measure of spread for the Y's.

This is parallel to the characterization of the arithmetic mean as that number for which the squared deviations are minimum. Using techniques of calculus, you can prove that $\Sigma(X - c)^2$ takes on its smallest possible value when c is just $\overline{X}$.

(d) The first of the normal equations gives

$$na = \Sigma\, y - b\,\Sigma\, x,$$

so when you divide by n you get the equation $a = \overline{Y} - b\overline{X}$.

To obtain the equation for b multiply the first equation by Σx and the second equation by n and then subtract

$$\begin{aligned}\Sigma y \Sigma x &= na\Sigma x + b(\Sigma x)^2 \\ n\Sigma xy &= na\Sigma x + nb\Sigma x^2.\end{aligned}$$

When you subtract the second from the first, you get

$$\Sigma\, y\, \Sigma\, y - n\,\Sigma\, xy = b\Big[(\Sigma\, x)^2 - n\,\Sigma\, x^2\Big]$$

so

$$b = \frac{\Sigma y \Sigma y - n\Sigma xy}{(\Sigma x)^2 - n\Sigma x^2}.$$

Now, if you just multiply the numerator and denominator by -1, you'll get the formula in the text.

7.2.12 **(a)** So $\alpha = \ln(Np)$ and $\beta = \ln(q)$.

(b) $\hat{Y} = 4.0886 - 0.5841X$.

(c) $Np \approx e^a = 59.6563$.

(d) $q \approx e^b = 0.5576$ and $p \approx 0.4424$. This is the estimate for p we gave you to use in Problem 3.6.16.

(e) $N \approx 59.6563/0.4424 = 134.8470$. That's pretty close to the true 135!

(f) $N = Np + Nq = Np + 76 \approx e^a + 76 = 135.6551$. Or $Nq = \Sigma_{x \neq 0} Y = 76$ and so $N = 76/q \approx 76/e^b = 136.2910$. Still pretty close to the true 135.

(g) $N \approx e^a/\hat{p} = 111.5$, far from 135. Why? Because we've mixed two unrelated ways of estimating the parameters. We've used the MLE for p by itself, assuming a geometric model, and we've used (a, b), the least squares estimate (and so, it's the MLE) for (α, β), assuming a regression model. No wonder the estimate is off! Note one interesting fact which is beyond us to prove: The least squares estimate for the regression model is the MLE.

(h) We're doing regression for $\ln(Y)$, associating the random error ϵ with $\ln(Y)$. But it's Y not $\ln(Y)$ that would be subject to random fluctuation about the projected model. Furthermore, there's no reason to think the $Y|X$'s are independent. Also, because we have only one value of ln(Y) for each X, it's impossible to get any sense from the data for whether the variance is constant from one level of X to another. There are other problems as well. For example, (a, b) is the maximum likelihood estimator for (α, β), so we're using a standard estimation procedure known to be "good" in a number of senses. But we're not actually using a and b; we're using e^a and e^b. Do those estimators have any decent properties? Are they reasonable?

The study by Edwards and Eberhardt was just exploratory, looking at a number of different models. Of all the models they considered for their specific data, this analysis gave the best estimate for N. We can say that only because we know the value of N. But with new data, one of the other models might give a closer approximation. Justifying one of Edwards and Eberhardt's models as "good" (let alone "best') would require extensive further study. There has been an enormous amount of work since then on the question of estimating population size for animal populations, in particular on capture–recapture techniques. See the many books in your library on "population biology." We've shown you Edwards and Eberhardt's analysis just by way of an introduction.

7.3.1

(**a**) The observed value of the estimator b standardizes to $t = 15.9643$, *much* larger than the critical t_c of 2.1315. So you should reject H_0.

With a five percent risk of error, we conclude that there is a linear relationship between height of bamboo shoots and the number of days after planting, at least within the range of ten to twenty days after planting.

(**b**) This is not a monitoring situation. You've got a set of data and a hypothesis ($\beta \neq 0$) and you want to see if the data "seems to challenge the hypothesis." That's a test of significance.

(**c**) Because the standard error will change with each new set of data, it's easier to state the decision rule in terms of t. The decision rule is: Calculate $t = (b-8)/s_b$ from the data for that three-day period. If the result is less than -1.753, alter the watering schedule.

(**d**) Alter the watering schedule, $t = -1.935 < -1.753$.

(**e**) Testing the "linearity assumption" of the model is theoretical. We'll use the model provided $\beta \neq 0$. If it is positive or negative, still we'll use the model. The test in part (c) is real world. Usually, a real-world test will involve a different action when you think the parameter is small from the action you would take if the parameter is large. So a real-world test will usually be one-tailed. In part (c), you're not worried about the bamboo growing too fast; if β is large, that's fine. No action required.

7.3.2

(**a**) Y is the variable in question; X is information relevant to answering that question. So for Forbes, X should be "boiling point" and Y "atmospheric pressure."

(**b**) The model is not using *altitude* to predict pressure. As altitude increases, boiling point decreases and so does pressure. So, boiling point and pressure increase and decrease together. Therefore, b should be positive.

(**c**) $\hat{Y} = 0.5222X - 80.9220$.

(**d**) p-value $= \mathrm{P}(b > 0.5222|\beta = 0) = \mathrm{P}(t > 51.7073|\text{d.f.} = 15) = 0$.

(e) We give a confidence interval for $-\beta$, the average decrease in pressure for a one degree decrease in boiling point. Note that we're interested in increasing altitude and, therefore we're interested in *decreases* in boiling point and pressure. So get the endpoints of a confidence interval for β and, then, converting to $-\beta$, answer the question. The endpoints for a 99% confidence interval for β are $0.5222 \pm 2.9467 \times 0.0101$. This gives the on average increase in pressure for one degree increase in boiling point, so for $-\beta$ you get the interval $(-0.5520, -0.4925)$. This translates into a *decrease* of between 0.4925 and 0.552 inches. in the reading of mercury on a barometer for a one degree decrease in boiling point.

7.3.3

$r^2 = b^2\Sigma(X - \overline{X})^2/\Sigma(Y - \overline{Y})^2 = b^2\Sigma(X - \overline{X})^2/\left[b^2\Sigma(X - \overline{X})^2\right] = 1.$ In fact, more can be said: If the observed values of Y all lie along ANY line, it will be the line $a + bX$ and so still $r^2 = 1$. And conversely, if $r^2 = 1$, then all the points lie along a line (not so easy to prove). This explains why r^2 or r are properly described as measures of LINEAR correlation.

7.3.4

$\hat{Y} - \overline{Y} = a + bX - \overline{Y} = \overline{Y} - b\overline{X} + bX - \overline{Y} = b(X - \overline{X})$. Recall that the averages on the individual tests are $\overline{X}$ and $\overline{Y}$, respectively, for the first and second tests. Because $b < 1$, you expect the second test score to be "less far above average" than the first score was. For low scorers, the situation is just the reverse. You would expect their scores on the second test to be higher.

7.3.5

We're leaving this to you. Be sure you do it, it's very instructive! By way of a hint, here's the appropriate table for the computations for picture (f)

X	Y	XY	X^2	Y^2
2	2	4	4	4
3	2	6	9	4
3	4	12	9	16
4	4	16	16	16
12	12	38	38	40

7.3.6

(a) $r^2 = 0.5229^2 \times 530.7824/145.9378 = 0.9945$. For the observed data, almost all the variability in barometric reading is accounted for by the boiling point of water. One might suspect a deterministic relationship, in which case the observed variability is only due to measurement error. Yet neither variable is determined exactly by altitude; they're determined largely by atmospheric pressure which itself varies with weather conditions. Still, an essentially deterministic relation may exist between our two variables through atmospheric pressure, as opposed to altitude. Further study is required.

(b) Edwards and Eberhardt were using the regression model in a very abstract way. It was just a tool to estimate the parameters of another model, the geometric random variable. They were not interested in "explaining" the number of rabbits caught five times by the number five, or two times by the number two, and so on. That doesn't even make sense!

(c) For Wilm's situation, $\alpha = 0$ and so special formulas are required.

7.3.7

(a) If there are only two observations in the scatter diagram, the observations will "fit" a line perfectly! When you go for the "best-fitting" line, you'll just get the line determined by your two observations. Because all observations lie on a line, $r^2 = 1$.

(b) With $n = 2$, you would think ANY X and Y were perfectly correlated even though they might be totally unrelated. To generalize, with a small sample, r^2 might be close to one even though X and Y were not meaningfully related.

(c) Two variables might be strongly correlated because of some third variable. For example, a high score on a job-skills test does not CAUSE a worker to do well on the job. Rather both effects are the result of the worker's skill. Or think of "height of mother" and "height of daughter." They should be strongly correlated but the mother's height does not CAUSE the daughter's height. Both are the result of certain common genes. See Problem 7.1.2(c). Again, for a large number of married couples, no doubt there would be a strong correlation between the ages of spouses. But the wife's age does not cause the husband's age or vice versa!

(d) This is an example of the "fallacy of ecological correlation." The grouping together that occurs with the averages introduces a completely spurious correlation. Clearly, the original data has very little correlation. Y is pretty much the same, no matter what value of X you're looking at. But the scatter diagram of averages lies very tightly along a line, just because of the grouping effect. The moral to this story is: The coefficient of determination (or the correlation coefficient) is misleading when applied to averages or percentages.

(e) Again, $r^2 = 0.9$, suggesting a strong correlation. A SPURIOUS correlation. This is the ecological fallacy again, now with proportions instead of averages. You can see some real-life examples of this kind of fallacy in the text of Freedman, Pisani, and Purves.

(f) Wrong! $r^2 = 0$ suggests there's no LINEAR relationship between X and Y. Some NONlinear relationship might exist. With the given data, evidently there's a relationship; it just isn't given by one single line. A quadratic relationship might provide a reasonable approximation. The point is: The coefficient of determination (or the correlation coefficient) is a measure of LINEAR correlation. Higher-order correlations—quadratic, cubic, and so on—will not be detected by r^2. They'll give an r^2 close to zero because the relationship isn't LINEAR.

7.3.8

(a) The standard error here is 5.5767. We can be about 95% sure that 15 days after planting, this little shoot will be between 33.1591 cm and 56.9325 cm tall.

(b) We did this in the text!

(c) Now the standard error is 2.4346. So we can be about 95% sure that in another eight days (15 days after planting), these seven little shoots will average somewhere between 39.8564 cm and 50.2352 cm tall.

7.3.9

(a) For predicting the mean of all Y's, the squared standard error is given by (1) below. For a particular Y, Y_p, it's given by (2). The difference is exactly $s^2_{Y|X}$. And for the average of a sample of m it's given by (3):

$$1/n(s^2_{Y|X}) + \text{etc.}, \tag{1}$$

$$s^2_{Y|X} + 1/n(s^2_{Y|X}) + \text{etc.}, \tag{2}$$

$$1/m(s^2_{Y|X}) + 1/n(s^2_{Y|X}) + \text{etc.} \tag{3}$$

This all makes perfect sense: How much harder is it to predict a particular value of Y than the average of all Y's? That's measured by $s^2_{Y|X}$, the estimate for the spread of the normal distribution of $Y|X$. If you're predicting the average of a sample of m, the problem will be correspondingly easier, so you have to add in only a fraction of $s^2_{Y|X}$: add in $(1/m)(s^2_{Y|X})$.

(b) More information, more accurate solution, smaller standard error! The contribution to the standard error from n is $1/n$. When n is large, $1/n$ will be small and the standard error correspondingly smaller.

(c) X is input information relevant to Y. If the particular X you're asking about, X_p, is far from what's typical, $\overline{X}$, of the information available, your answer should be correspondingly less accurate. So if $(X_p - \overline{X})^2$ is large, the standard error should be correspondingly large, as it will be.

(d) No, you don't always want variability to be small! A gambler who wants to beat the odds wants lots of varibility in the game so she has a good chance to get far ahead of the odds. That's when she'll quit (or so she says!). A testing service wants a test with good variability in the scores, otherwise, the test is useless. Suppose everybody makes the same score. That tells you nothing about the ability of the testees!

Here, X is input information relevant to Y. Clearly, the more ground your information covers—the more variable X is—the better.

7.3.10 We will leave this to you.

7.3.11 **(a)** Only about 15% of spinning quality as measured by skein strength is accounted for by fiber length, indicating a weak relationship ($r^2 = 0.1478$). Thinking of a hypothesis test with $H_o : \beta = 0$, we get $b/s_b = 1.0202$, giving a p-value greater than 10%. Again a weak relationship, it's entirely possible $\beta = 0$. A helpful equation: $b/s_b = \sqrt{(n-2)}\, r/\sqrt{(1-r^2)}$.

In fact, the government's full data set of 183 measurements still gives a fairly weak relationship: $r^2 \approx 36\%$. It would appear that fiber length by itself is not a very adequate measure of skein strength. A much stronger model ($r^2 \approx 80\%$) is obtained if, in addition to length, you introduce fiber tensile strength. See Duncan for a study of the multiple regression model with three regressors for fiber: length, tensile strength, and fineness.

(b) Calculating r^2 amounts to giving a point estimate for ρ, whereas the hypothesis test is a more complete analysis which takes into account sampling error in the data (see Problem 7.2.3).

(c) You're asked for a confidence interval for β. The endpoints are $0.4524 \pm 2.4469 \times 0.4434$. Real-world answer?

(d) $68.3317 \pm 2.4469 \times 30.6144$. The interval is (0, 143). We round the answer to an integer because the data was given in integers. The interval must begin at zero because S is evidently not negative. Note that the answer in part (b) could be negative. Why? What would that mean?

7.4.1 "We can be about 90% sure that quarterly maintenance supply costs for one of Residential Management, Inc.'s large apartment complexes would be between \$1480 and \$3000 ." [s.e. = 4.1425 for a prediction interval]

7.4.2 The data standardizes to $t = -2.84$, giving a p-value between a half and one percent (s.e. = 11.4564 with $s = 36.2284$). This p-value is SMALL (less than 10%) and, so, "It would seem based on this data there are fewer than 200 residents per complex on average in Residential Management's large apartment complexes."

7.4.3 p-value 0.2639 from the BINOMIAL model. Because $nq = 1$ (NOT ≥ 5), the normal approximation is not valid. This is a NOT SMALL p-value, so, "This data provides no evidence that fewer than 90% of the large complexes have more than 130 residents." Here's a hint: $P(X \leq 8) = 1 - P(X \geq 9) = 1 - [10qp^9 + p^{10}] = 1 - p^9[10q + p] = 1 - 1.9p^9$.

7.4.4 $\hat{B} = 5.6677 + 0.0999C$. Or you can say $\hat{Y} = 5.6677 + 0.0999X$. This model seems good because 84% of quarterly maintenance supply costs for our ten observations is explained by the number of residents on the first day of the quarter.

7.4.5 $H_o : \beta = 0$, $t_c = 1.8595$ (10% two-tailed test puts 5% in right tail); $H_A : \beta \neq 0$, b standardizes to 6.4870 ($s_b = 0.0154$) so REJECT H_o.

Here $\Sigma(X - \overline{X})^2 = 11,812.5$, $s^2_{Y|X} = 2.8163$. There's no reason to doubt the linearity assumption of the model. However, there's a ten percent chance to obtain data like ours even though "number of residents" does not determine "average maintenance supply costs" through a linear function.

7.4.6 Here 18.6547 ± 1.8595 s.e. $s^2_{Y|X}\left[1.1 + (130 - 167.5)^2/11812.5\right] = 3.4332$. "Based on this quarter's data, maintenance supply costs for this complex in the coming quarter should, with a ten percent risk of error, be between \$1521 and \$2210."

7.4.7 "We can be about 90% sure that the maintenance supply costs incurred with one new resident for one of the complexes run by Residential Management, Inc. should increase by at least \$7.13 and not more than \$12.85."

7.4.8 "We can be about 90% sure there are, on average, between 55 and 88 units in any one of the large apartment complexes run by Residential Management, Inc." (55.5928, 87.2072)

7.4.9 (a) $n\hat{p} = 3$; so we can't say $np > 5$.

(b) Estimating p with $\hat{p} = 0.3$, $n = 17$ is required.

(c) $n = 140$. Round UP from 139.2 ($Z = 2.575$).

(d) The normal approximation is good for p as small as 0.0357.

7.4.10 Here $\Sigma(X-\overline{X})^2 = 6692.4$, $s^2_{Y|X} = 506.8575$, $\overline{Y} = 90.6312+50\times 1.0766 = 144.4609$, $s^2_{Y|X}[0.1 + (50 - 71.4)^2/6692.4] = 85.3699$. So the answer: (127.2799, 161.6419). You give the real-world answer. The model seems moderately good with an r^2 of only 66%.

7.4.11 $H_o : \mu = 25$ ($\mu < 25$ irrelevant) [here, $t_c = 1.383$]. $H_A : \mu > 25$. "If the data standardizes to a value larger than 1.383, then a careful examination of the records of all complexes should be made. There's a ten percent risk of doing this when, in fact, it's unnecessary."

7.4.12 Same variables as in Problem 7.4.10, but $X_p = 93$. So $\overline{Y} = 190.7550$ and the standard error squared is 592.8788. Answer: (145.4778, 236.0322). Real-world??

7.4.13 "We can be about 90% sure that the increase in average quarterly maintenance supply costs for one more unit is at least \$7 and not more than \$17.34." $b \pm ts_b$; $s^2_{Y|X} = 5.1619$, $s_b = 0.0278$. The model looks good because about 71% of the observed quarterly maintenance supply costs is explained by the number of units. Note (you were not asked for this) that the p-value for $\beta = 0$ is less than half a percent, confirming that the linearity assumption seems valid (this test takes sampling error in the data into consideration, r_2 does not).

7.4.14 $(22.4 - 25)/\text{s.e.} = -2.0817 = t$ with $df = 9$ and s.e. $= s/\sqrt{10} = 1.2490$. The p-value is SMALL, between two and a half and five percent (less than 10%). "Yes, on the basis of this data, it looks as if this quarter's average maintenance supply costs are below \$2500 per complex on average."

7.4.15 $X = B$ and $Y = C$; $X_p = 20$. $s^2_{Y|X}[0.1 + (20 - 22.4)^2/140.4]$ with $s^2_{Y|X} = 236.8901$. "We can be about 90% sure that as of this quarter there are 136–158 residents on average in those of Residential Management's large apartment complexes which incurred maintenance supply costs of \$2000."

The model seems fine because $r^2 = 84\%$. Note that this is the same value of r^2 as computed in Problem 7.4.4. Interchanging the roles of the two variables in simple linear regression gives the same r^2. Look carefully at the formula; you can see that it's symmetric in the two variables.

7.4.16 $H_o : \beta = 0.13$, $t = (0.0999 - 0.13)/s_b = -1.95$, and so on; $H_A : \beta > 0.13$ this is not needed because $b = 0.0999$ is *less* than 0.13.

Note that 0.0999 in real-world terms means \$10. Now, observed costs of \$10 per resident in the sample could hardly suggest that the true cost per resident has risen MORE than \$13! Report: "No further investigation is required. Our data provides no evidence that the increase in average maintenance supply costs this quarter for one more resident has risen above $13."

7.4.17 $71.4 \pm 1.8331s/\sqrt{n}$ "She can be about 90% sure that there are on average between 55 and 88 units per complex among the large apartment complexes managed by Residential Management, Inc." This is the same as Problem 7.4.8. You don't know the total number of complexes managed by Residential Management and so you can only answer on a "per complex" basis.

7.4.18 p-value $= \text{P}(\hat{p} \leq 0.2|p = 0.5)$. The normal approximation is valid here (just barely, $np = 5$), but there's no need to use it because you can easily calculate

$$\text{P}(X \leq 2|n=10, p=0.5) = q^{10} + 10q^9p + 45q^8p^2 = 0.5^{10}(1+10+45) = 0.0547.$$

SMALL p-value (less than 10%): "Our data does seem to suggest that less than half of the complexes this quarter incurred maintenance supply costs in excess of $2500."

7.4.19 $\hat{p} = 0.2$ is certainly consistent with $p = 0.2$. But it's also consistent with lots of other possible values of p. "The data is inconclusive as to whether no more than twenty percent of the complexes this quarter incurred maintenance supply costs in excess of $2500." This is equivalent, of course, to saying, "This data provides no evidence for the manager's claim that no more than 20% . . . "

7.4.20 p-value $= \text{P}(X \leq 2|n = 10, p = 0.3) = 0.3828$. The normal approximation is not valid (why?). The p-value is NOT SMALL. "The data provides no evidence that no more than 30% of the complexes this quarter incurred maintenance supply costs in excess of $2500."

7.4.21 We require a confidence interval for p with $\hat{p} = 0.2$, $n = 10$. Because this suggests as a rough approximation that np is around TWO, we cannot justify the normal approximation. But we have no technique for generating a confidence interval using the binomial tables. We have NO TECHNIQUE.

7.4.22 Most of the calculations are in Problem 7.4.15. $X_p = 25$ and $s^2_{Y|X}[0.1 + (25 - 22.4)^2/140.4] = 35.0948$. The endpoints are 189.3529 ± 1.8595s.e. "We can be about 90% sure that there are between 178 and 200 residents on average this quarter in those of Residential Management, Inc.'s complexes which reported $2500 in quarterly maintenance supply costs."

7.4.23 $\text{P}(\overline{X} \leq 22.4|\mu = 23.87, \sigma = 3.18) = \text{P}(Z < -1.46) = 0.0721$. No real-world answer is required because the question just asked to calculate a certain probabilty. Because $n < 30$, you must assume that the quarterly maintenance supply costs per complex for Residential Management Inc.'s large apartment complexes are normally distributed (population normally distributed so that the sample mean will be normally distributed). This could fail if—just to give one possible example—some of these complexes have garden maintenance counted into these costs and some don't.

7.4.24 Here, the normal approximation is valid because $nq > np = 9.25 > 5$. $\mathrm{P}(\hat{p} < 0.5 | n = 25, p = 0.37) = \mathrm{P}(Z < 1.35) = 0.9115$. "There's about a 91% chance that less than half of a sample of 25 complexes would report maintenance supply costs in excess of $2500 if, in fact, the true proportion for all complexes is 37%."

7.4.25 Six.

Appendix

The Tables

Table A. Cumulative probability distribution for Z, giving $P(Z < z)$.

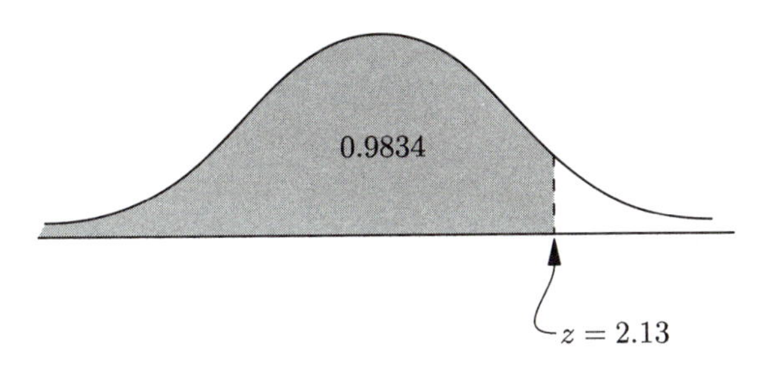

z	0.00	0.01	0.02	0.03	0.04	0.05	0.06	0.07	0.08	0.09	z
0.00	.5000	.5040	.5080	.5120	.5160	.5199	.5239	.5279	.5319	.5359	0.00
0.10	.5398	.5438	.5478	.5517	.5557	.5596	.5636	.5675	.5714	.5753	0.10
0.20	.5793	.5832	.5871	.5910	.5948	.5987	.6026	.6064	.6103	.6141	0.20
0.30	.6179	.6217	.6255	.6293	.6331	.6368	.6406	.6443	.6480	.6517	0.30
0.40	.6554	.6591	.6628	.6664	.6700	.6736	.6772	.6808	.6844	.6879	0.40
0.50	.6915	.6950	.6985	.7019	.7054	.7088	.7123	.7157	.7190	.7224	0.50
0.60	.7257	.7291	.7324	.7357	.7389	.7422	.7454	.7486	.7517	.7549	0.60
0.70	.7580	.7611	.7642	.7673	.7704	.7734	.7764	.7764	.7823	.7852	0.70
0.80	.7881	.7910	.7939	.7967	.7995	.8023	.8051	.8078	.8106	.8133	0.80
0.90	.8159	.8186	.8212	.8238	.8264	.8289	.8315	.8340	.8365	.8389	0.90
1.00	.8413	.8438	.8461	.8485	.8508	.8531	.8554	.8577	.8599	.8621	1.00
1.10	.8643	.8665	.8686	.8708	.8729	.8749	.8770	.8790	.8810	.8830	1.10
1.20	.8849	.8869	.8888	.8907	.8925	.8944	.8962	.8980	.8997	.9015	1.20
1.30	.9032	.9049	.9066	.9082	.9099	.9115	.9131	.9147	.9162	.9177	1.30
1.40	.9192	.9207	.9222	.9236	.9251	.9265	.9279	.9292	.9306	.9319	1.40
1.50	.9332	.9345	.9357	.9370	.9382	.9394	.9406	.9418	.9429	.9441	1.50
1.60	.9452	.9463	.9474	.9484	.9495	.9505	.9515	.9525	.9535	.9545	1.60
1.70	.9554	.9564	.9573	.9582	.9591	.9599	.9608	.9616	.9625	.9633	1.70
1.80	.9641	.9649	.9656	.9664	.9671	.9678	.9686	.9693	.9699	.9706	1.80
1.90	.9713	.9719	.9726	.9732	.9738	.9744	.9750	.9756	.9761	.9767	1.90
2.00	.9772	.9778	.9783	.9788	.9793	.9798	.9803	.9808	.9812	.9817	2.00
2.10	.9821	.9826	.9830	.9834	.9838	.9842	.9846	.9850	.9854	.9857	2.10
2.20	.9861	.9864	.9868	.9871	.9875	.9878	.9881	.9884	.9887	.9890	2.20
2.30	.9893	.9896	.9898	.9901	.9904	.9906	.9909	.9911	.9913	.9916	2.30
2.40	.9918	.9920	.9922	.9925	.9927	.9929	.9931	.9932	.9934	.9936	2.40
2.50	.9938	.9940	.9941	.9943	.9945	.9946	.9948	.9949	.9951	.9952	2.50
2.60	.9953	.9955	.9956	.9957	.9959	.9960	.9961	.9962	.9963	.9964	2.60
2.70	.9965	.9966	.9967	.9968	.9969	.9970	.9971	.9972	.9973	.9974	2.70
2.80	.9974	.9975	.9976	.9977	.9977	.9978	.9979	.9979	.9980	.9981	2.80
2.90	.9981	.9982	.9982	.9983	.9984	.9984	.9985	.9985	.9986	.9986	2.90
3.00	.9987	.9987	.9987	.9988	.9988	.9989	.9989	.9989	.9990	.9990	3.00
3.10	.9990	.9991	.9991	.9991	.9992	.9992	.9992	.9992	.9993	.9993	3.10
3.20	.9993	.9993	.9994	.9994	.9994	.9994	.9994	.9995	.9995	.9995	3.20
3.30	.9995	.9995	.9995	.9996	.9996	.9996	.9996	.9996	.9996	.9997	3.30
3.40	.9997	.9997	.9997	.9997	.9997	.9997	.9997	.9997	.9997	.9998	3.40
3.50	.9998	.9998	.9998	.9998	.9998	.9998	.9998	.9998	.9998	.9998	3.50
3.60	.9998	.9998	.9999	.9999	.9999	.9999	.9999	.9999	.9999	.9999	3.60
3.70	.9999	.9999	.9999	.9999	.9999	.9999	.9999	.9999	.9999	.9999	3.70
3.80	.9999	.9999	.9999	.9999	.9999	.9999	.9999	.9999	.9999	.9999	3.80

Table B. Cumulative probability distribution for χ^2. Degrees of freedom are given in the left column, and cumulative probabilities across the top row.

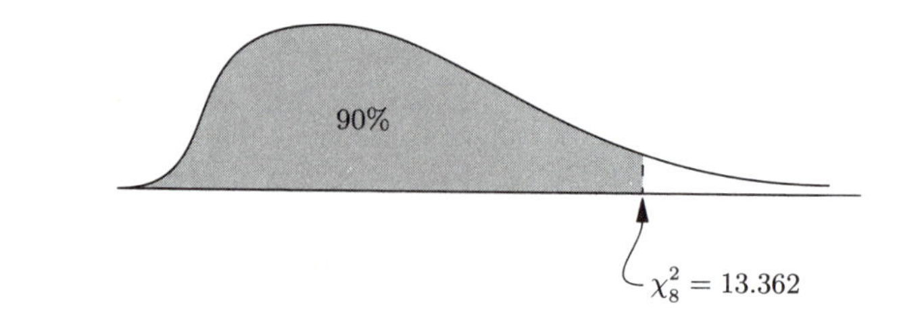

d.f.	0.005	0.025	0.05	0.90	0.95	0.975	0.99	0.995
1	0.0000393	0.000982	0.00393	2.706	3.841	5.024	6.635	7.879
2	0.0100	0.0506	0.103	4.605	5.991	7.378	9.210	10.597
3	0.0717	0.216	0.352	6.251	7.815	9.348	11.345	12.838
4	0.207	0.484	0.711	7.779	9.488	11.143	13.277	14.860
5	0.412	0.831	1.145	9.236	11.070	12.832	15.086	16.750
6	0.676	1.237	1.635	10.645	12.592	14.449	16.812	18.548
7	0.989	1.690	2.167	12.017	14.067	16.013	18.475	20.278
8	1.344	2.180	2.733	13.362	15.507	17.535	20.090	21.955
9	1.735	2.700	3.325	14.684	16.919	19.023	21.666	23.589
10	2.156	3.247	3.940	15.987	18.307	20.483	23.209	25.188
11	2.603	3.816	4.575	17.275	19.675	21.920	24.725	26.757
12	3.074	4.404	5.226	18.549	21.026	23.336	26.217	28.300
13	3.565	5.009	5.892	19.812	22.362	24.736	27.688	29.819
14	4.075	5.629	6.571	21.064	23.685	26.119	29.141	31.319
15	4.601	6.262	7.261	22.307	24.996	27.488	30.578	32.801
16	5.142	6.908	7.962	23.542	26.296	28.845	32.000	34.267
17	5.697	7.564	8.672	24.769	27.587	30.191	33.409	35.718
18	6.265	8.231	9.390	25.989	28.869	31.526	34.805	37.156
19	6.844	8.907	10.117	27.204	30.144	32.852	36.191	38.582
20	7.434	9.591	10.851	28.412	31.410	34.170	37.566	39.997
21	8.034	10.283	11.591	29.615	32.671	35.479	38.932	41.401
22	8.643	10.982	12.338	30.813	33.924	36.781	40.289	42.796
23	9.260	11.688	13.091	32.007	35.172	38.076	41.638	44.181
24	9.886	12.401	13.848	33.196	36.415	39.364	42.980	45.558
25	10.520	13.120	14.611	34.382	37.652	40.646	44.314	46.928
26	11.160	13.844	15.379	35.563	38.885	41.923	45.642	48.290
27	11.808	14.573	16.151	36.741	40.113	43.194	46.963	49.645
28	12.461	15.308	16.928	37.916	41.337	44.461	48.278	50.993
29	13.121	16.047	17.708	39.087	42.557	45.722	49.588	52.336
30	13.787	16.791	18.493	40.256	43.773	46.979	50.892	53.672
35	17.192	20.569	22.465	46.059	49.802	53.203	57.342	60.275
40	20.707	24.433	26.509	51.805	55.758	59.342	63.691	66.766
45	24.311	28.366	30.612	57.505	61.656	65.410	69.957	73.166
50	27.991	32.357	34.764	63.167	67.505	71.420	76.154	79.490
60	35.535	40.482	43.188	74.397	79.082	83.298	88.379	91.952
70	43.275	48.758	51.739	85.527	90.531	95.023	100.425	104.215
80	51.172	57.153	60.391	96.578	101.879	106.629	112.329	116.321
90	59.196	65.647	69.126	107.565	113.145	118.136	124.116	128.299
100	67.328	74.222	77.929	118.498	124.342	129.561	135.807	140.169

Table C. Cumulative probability distribution for Student's t. Degrees of freedom are given in the left column, and cumulative probabilities across the top row.

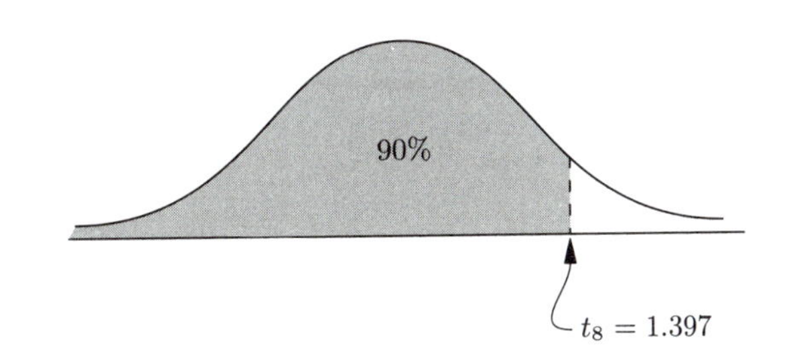

d.f.	0.90	0.95	0.975	0.99	0.995
1	3.078	6.3138	12.706	31.821	63.657
2	1.886	2.9200	4.3027	6.965	9.9248
3	1.638	2.3534	3.1825	4.541	5.8409
4	1.533	2.1318	2.7764	3.747	4.6041
5	1.476	2.0150	2.5706	3.365	4.0321
6	1.440	1.9432	2.4469	3.143	3.7074
7	1.415	1.8946	2.3646	2.998	3.4995
8	1.397	1.8595	2.3060	2.896	3.3554
9	1.383	1.8331	2.2622	2.821	3.2498
10	1.372	1.8125	2.2281	2.764	3.1693
11	1.363	1.7959	2.2010	2.718	3.1058
12	1.356	1.7823	2.1788	2.681	3.0545
13	1.350	1.7709	2.1604	2.650	3.0123
14	1.345	1.7613	2.1448	2.624	2.9768
15	1.341	1.7530	2.1315	2.602	2.9467
16	1.337	1.7459	2.1199	2.583	2.9208
17	1.333	1.7396	2.1098	2.567	2.8982
18	1.330	1.7341	2.1009	2.552	2.8784
19	1.328	1.7291	2.0930	2.539	2.8609
20	1.325	1.7247	2.0860	2.528	2.8453
21	1.323	1.7207	2.0796	2.518	2.8314
22	1.321	1.7171	2.0739	2.508	2.8188
23	1.319	1.7139	2.0687	2.500	2.8073
24	1.318	1.7109	2.0639	2.492	2.7969
25	1.316	1.7081	2.0595	2.485	2.7874
26	1.315	1.7056	2.0555	2.479	2.7787
27	1.314	1.7033	2.0518	2.473	2.7707
28	1.313	1.7011	2.0484	2.467	2.7633
29	1.311	1.6991	2.0452	2.462	2.7564
30	1 310	1.6973	2.0423	2.457	2.7500
35	1.3062	1.6896	2.0301	2.438	2.7239
40	1.3031	1.6839	2.0211	2.423	2.7045
45	1.3007	1.6794	2.0141	2.412	2.6896
50	1.2987	1.6759	2.0086	2.403	2.6778
100	1.2901	1.6602	1.9840	2.364	2.6260
160	1.2869	1.6545	1.9749	2.350	2.6070
200	1.2858	1.6525	1.9719	2.345	2.6006
z	1.282	1.645	1.96	2.326	2.576

Bibliography

Amemiya, Takeshi [1985] *Advanced Econometrics*. Harvard University Press, Cambridge, MA.

Barrow, Lloyd H., and Morrisey, J. Thomas [1989] "Energy Literacy of Ninth-Grade Students: A Comparison Between Maine and New Brunswick." *J. Environment. Educ.* 20(2).

Bickel, Peter J. and Doksum, Kjell A. [1977] *Mathematical Statistics*. Holden-Day, Oakland, CA.

Clarke, R. D. [1946] "An Application of the Poisson Distribution." *J. Instit. Actuaries* 72.

Ehrenberg, A. S. C. [1972] *Repeat Buying*. North-Holland, Amsterdam/London.

Finkelstein, Michael O. and Levin, Bruce [1990] *Statistics for Lawyers*. Springer-Verlag, New York.

Fleiss, Joseph L. *Statistical Methods for Rates and Proportions*. 2nd ed. Wiley, New York.

Freedman, David, Pisani, Robert, and Purves, Roger [1980] *Statistics*. 1st ed. W. W. Norton & Co., New York.

Ham, Richard E. [1990] "What is Stuttering: Variations and Stereotypes." *J. Fluency Disorders* 15(5/6), 259.

Hoaglin, C. David and Moore, David S. [1992] *Perspectives on Contemporary Statistics*. Mathematical Association of America, Washington, DC.

Larsen, Richard J. and Marx, Morris L. [1986] *An Introduction to Mathematical Statistics and Its Applications*. 2nd ed. Prentice-Hall, Englewood Cliffs, NJ.

Madow, William G., Nisselson, Harold, and Olkin, Ingram, eds. [1983] *Incomplete Data in Sample Surveys*. Academic Press, New York.

Mendel, Gregor [1965] *Experiments in Plant Hybridisation*, J. H. Bennett, ed. With commentary by Sir Ronald A. Fisher. Oliver and Boyd, Edinburg/London.

Mendenhall, William, Scheaffer, Richard L., and Wackerly, Dennis D. [1981] *Mathematical Statistics with Applications*. 2nd ed. Duxbury Press, Boston, MA.

Monahan, John and Walker, Laurens [1985] *Social Science in Law*. The Foundation Press, Inc. Mineola, NY.

McPherson, Glen [1990] *Statistics in Scientific Investigation*. Springer-Verlag, New York/Heidelberg/Berlin.

Ryan, Barbara F., Joiner, Brian L., and Ryan, Thomas A., Jr. [1985] *MINITAB Handbook*, 2nd ed. PWS-Kent Publishing Co., Boston, MA.

Sachs, Lothar [1982] *Applied Statistics*. Springer-Verlag. New York/ Heidelberg/Berlin.

Sieh, K E. [1984] "Lateral Offsets and Revised Dates of Large Prehistoric Earthquakes at Pallett Creek, Southern California." *J. Geophys. Res.* 89, 7641–7670.

Smith, Gary [1988] *Statistical Reasoning*, 2nd ed. Allyn and Bacon, Boston, MA.

Stigler, Stephen M. [1986] *The History of Statistics*. The Belknap Press of Harvard University Press, Cambridge, MA.

Stopher, Peter R. and Meyburg, Arnim H. [1979] *Survey Sampling and Multivariate Analysis for Social Scientists and Engineers*. D. C. Heath and Co., Lexington, MA.

Tanur, J. M., et al. [1972] *Statistics, a Guide to the Unknown*, 1st ed. Holden-Day, San Francisco, CA.

Tanur, J. M., et al. [1989] *Statistics, a Guide to the Unknown*, 3rd ed. Wadsworth and Brooks/Cole, Pacific Grove, CA.

Tufte, Edward R. [1983] *The Visual Display of Quantitative Information*. Graphics Press, Cheshire, CT.

Vardeman, Stephen B. [1992] "What About the Other Intervals?" *American Statistician* 46(3).

Index of Notation

Author Index

Subject Index

Entries for a term are for the first occurrence only within a discussion of that term. The discussion may continue on the next page or next several pages. A term appearing in a problem may also appear in the solution without a separate listing here.

I

J

K

L

M

N

O

P

Springer Texts in Statistics *(continued from page ii)*

Madansky	Prescriptions for Working Statisticians
McPherson	Statistics in Scientific Investigation: Its Basis, Application, and Interpretation
Nguyen and Rogers	Fundamentals of Mathematical Statistics: Volume I: Probability for Statistics
Nguyen and Rogers	Fundamentals of Mathematical Statistics: Volume II: Statistical Inference
Noether	Introduction to Statistics: The Nonparametric Way
Peters	Counting for Something: Statistical Principles and Personalities
Pfeiffer	Probability for Applications
Pitman	Probability
Santner and Duffy	The Statistical Analysis of Discrete Data
Saville and Wood	Statistical Methods: The Geometric Approach
Sen and Srivastava	Regression Analysis: Theory, Methods, and Applications
Whittle	Probability via Expectation, Third Edition
Zacks	Introduction to Reliability Analysis: Probability Models and Statistical Methods

8/26/94